Springer Series in Advanced Microelectronics

Volume 65

The *Springer Series in Advanced Microelectronics* provides systematic information on all the topics relevant for the design, processing, and manufacturing of microelectronic devices. The books, each prepared by leading researchers or engineers in their fields, cover the basic and advanced aspects of topics such as wafer processing, materials, device design, device technologies, circuit design, VLSI implementation, and sub-system technology. The series forms a bridge between physics and engineering, therefore the volumes will appeal to practicing engineers as well as research scientists.

Alexander Trubin

Introduction to the Theory of Dielectric Resonators

Springer

Alexander Trubin
Telecommunication System Institute
National Technical University of Ukraine
Kyiv, Ukraine

ISSN 1437-0387 ISSN 2197-6643 (electronic)
Springer Series in Advanced Microelectronics
ISBN 978-3-031-65395-7 ISBN 978-3-031-65396-4 (eBook)
https://doi.org/10.1007/978-3-031-65396-4

This Springer imprint is published by the registered company Springer Nature Switzerland AG
The registered company address is: Gewerbestrasse 11, 6330 Cham, Switzerland

Preface

For the first time, the problem of scattering electromagnetic waves on a dielectric body of spherical shape was posed and solved at the beginning of the 20th century [1–3]. Based on the type of solutions obtained, it was found that the most significant contribution to the expansion of the scattered field is made by terms close in frequency to a discrete series determined by the shape, size and material of the scattering body, consisting of a dielectric with a low loss value. Later [4], such bodies were called dielectric resonators (DR). Interest in dielectric resonators has grown significantly after the development of new thermally stable materials with low losses [5]. Later, at the end of the 60s, the attention of researchers was attracted by resonators with whispering gallery oscillations [6, 7]. Whispering gallery oscillations have extremely low radiation losses and are suitable for use in the millimeter and shorter wavelength ranges [8–11]. Since then, the scope of DR applications has been constantly expanding. In the 70–80s of the last century, dielectric resonators began to be more widely used in a variety of devices in the microwave wavelength range–filters, antennas, generators, etc. [12, 13]. At this stage, the lower types of their own oscillations were most often used. Today, DRs are one of the basic elements for constructing a wide class of devices in the terahertz, infrared and visible wavelength ranges [10]. More complex structures with dielectric resonators are being studied. First of all, these are various DR lattices with the properties of metamaterials, shapers of twisted waves and Airy rays, different encoding devices, lattices with "topological" properties, add-drop filters, multi-element antennas, etc.

The accumulated experience in the development of various devices made using DR has shown the high sensitivity of their electromagnetic scattering parameters both from the properties of materials, relative position, shape and types of oscillations of the resonators themselves and from the shape and spatial dimensions of the external electrodynamics' structure containing them. Therefore, the calculation and optimization of their characteristics play an important role in the development of devices with DR.

A popular direction in calculating the parameters of complex resonant dielectric structures, which have become widespread recently, is the use of various numerical methods for solving boundary value problems for the system of Maxwell's equations. The use of standard packages for numerical modeling makes it possible to establish most physically significant dependencies, which visibly facilitates the practical design of complex devices. A significant drawback of this direction is that the obtained dependences are essentially only a way of visualizing the laws of change in various parameters of electromagnetic processes occurring in the structures under study, and tell us little about the essence of physical phenomena.

A complementary alternative to numerical solution methods is the development of more rigorous theoretical models for the analysis of dielectric structures, based on the use of methods of the physical theory of diffraction, constructed on the basis of analytical solutions of the corresponding boundary value problems for Maxwell's equations. Quite often, such boundary value problems contain complex multiply connected interfaces as boundary conditions, so "accurate" calculation of electromagnetic fields in such structures is not possible, sometimes even with the use of modern computing tools. In this regard, the development of electrodynamic models that reflect the main features of electromagnetic processes is of significant interest. In turn, the development of electromagnetic scattering models allows us to better understand physical laws and use the acquired knowledge for further practical design and optimization.

The consistent development of physically adequate models for describing the scattering of electromagnetic waves on dielectric structures also makes it possible to consider from a unified standpoint the entire variety of structures, while simultaneously taking into account the properties of both transmission lines and the types of DRs used, when both basic and higher types of natural oscillations are excited in them.

It is known that dielectric resonators near their natural frequencies are characterized by a higher concentration of the stored field. This allows us to formulate a general "locality condition" for them—the assumption that the stray field in the volume occupied by the dielectric differs slightly from the field of natural oscillations. This condition makes it possible to apply perturbation theory [14, 15] and, in the most general form, derive equations that determine the field amplitudes in each of the partial DR structures, both for natural oscillations of the entire system as a whole, and in the case of solving the scattering problem. The proposed approach made it possible to obtain a clear and fairly complete physically visual representation of the processes occurring in devices with an arbitrary number of resonators. The limitations of this approach are also analyzed in this monograph.

Another important physical assumption commonly used in deriving equations for the field amplitudes of a DR system is the assumption that the scattered field is determined mainly by their coupled oscillations. This assumption, in combination with the Galerkin method, in which the system of basis functions is a superposition of the natural oscillations of all resonators of the structure, in most practically interesting cases also makes it possible to present the solution to the scattering problem in a final closed analytical form.

When constructing the general scattering theory, we neglected the contribution of non-resonant components expressed in terms of functions with a continuous spectrum distribution [16]. This monograph analyzes the scattering properties due only to the contribution of the eigen functions of resonator systems in frequency bands near their coupled oscillations. Specific results obtained on the basis of the assumptions made, in most cases, are in good agreement with experimental data and numerical calculations.

The main style of presentation used in this monograph involves a transition from the general to the specific. The DR theory is one of the particular applications of electrodynamics and optics, applied to solving problems of resonant scattering of electromagnetic waves on dielectric bodies in different microwave and optical transmission lines, and therefore it, can be considered from the most general point of view—the physical theory of diffraction. At the same time, it seems most natural to us to further develop the classical method of natural frequencies; therefore, first we consider the natural oscillations of single DRs of various shapes—spherical, cylindrical, rectangular, and only then, the problem of coupled oscillations; after which we move on to solving problems of scattering of waves and field pulses on systems consisting of a large number of resonators. The data from experiments and numerical calculations presented from time to time serve only as a clear illustration of the reliability of the theoretical calculations performed.

When describing harmonic oscillations of the electromagnetic field (**E**, **H**), solutions of Maxwell's equations were used in complex form with time dependence t in the form: $\mathbf{E} = \mathbf{E}_0 e^{i\omega t}$; $\mathbf{H} = \mathbf{H}_0 e^{i\omega t}$, where ω—is the circular frequency.

To obtain the necessary initial relations, we used surface and volume integrals throughout. Presentation of various electrodynamic parameters—such as, for example, the expansion coefficients of the DR field over the waves of the transmission line, or the field of the wave zone of the DR, in quadratures allows us to obtain the final results in the most convenient form for analysis. Since in most cases the integration was carried out without additional approximations, this made it possible to minimize the error in the introduced physical assumptions, while the numerical calculations are characterized by noticeably less time. The proposed approach provides a significant gain in speed, which is most important for subsequent optimization of the parameters of structures containing a big number of DRs.

This book complements and significantly expands previously published results [14, 15]. New material is presented on coupled oscillations of rotating DRs, DRs with whispering gallery oscillations, resonators detuned in frequency and resonators with loss and gain, as well as on the scattering of electromagnetic waves on various DR lattices in open space. New designs of antenna-filters with better frequency characteristics, filters with coupled DR subsystems, as well as lattices and filters on rotating resonators are considered. The results obtained earlier on the scattering of pulses on systems of coupled DRs are generalized. The application of the constructed theory of scattering on systems of coupled microcavities with whispering gallery oscillations is considered from a unified perspective. The use of a new approach to calculating scattering characteristics made it possible to construct new electromagnetic models to describe a wide class of notch and various types of Add-drop filters [17] used in the infrared and optical wavelength ranges.

This monograph is intended both for specialists who have been working with dielectric resonators for a long time, and for graduate students who want to deepen their knowledge in this field. It will also be useful to university lecturers teaching courses in electrodynamics and optics.

The author expresses deep gratitude to his colleagues, whose practical assistance at different times was invaluable in conducting research and confirming the results obtained.

Kyiv, Ukraine Alexander Trubin

References

1. G. Mie, Beltrage sur Optik truber Medien. Ann. Phys. (Leipzig). **25**, 377–445 (1908)
2. P. Debye, Der Zichtruch auf Kugeln von belibigem Material. Ibid. **30**, 57–136 (1909)
3. M. Born, E. Wolf, *Principles of Optics* (Pergamon Press, 1964), 855 p
4. R.D. Richtmyer, Dielectric resonators. J. Appl. Phys. **10**(6), 391–398 (1939)
5. A. Okaya, The rutil microwave resonator. Proc. IRE **48**(11), 1921 (1960)
6. J. Wait, Electromagnetic whispering gallery waves in dielectric rod. Radio Sci. **2**(9), 1005–1017 (1967)
7. S. Vlasov, Oscillations of the "whispering gallery" in open resonators with a dielectric rod. Radio Eng. Electron. **12**(3), 572–573 (1967)
8. H.A. Haus, M.A. Popovic, M.R. Watts, C. Manolatou, B.E. Little, S.T. Chu, Optical resonators and filters, in *Optical Microcavities*, chap. 00, ed. by K. Vahala (California Institute of Technology, USA, 2004), 516 p
9. A.B. Matsko, V.S. Ilchenko, Optical resonators with whispering gallery modes—Part 1: Basics. IEEE J. Sel. Top. Quantum Electr. **12**(1), 3–14 (2006)
10. A.B. Matsko, V.S. Ilchenko, Optical resonators with whispering gallery modes—Part 2: Applications. IEEE J. Sel. Top. Quantum Electr. **12**(1), 15–32 (2006)
11. M.L. Gorodetsky, *Optical Microcavities with a Giant Q-Factor* (Fizmatlit, Moscow, 2011), 416 p (in Russian)
12. D. Kajfez, P. Guillon, *Dielectric Resonators* (Artech House, 1986), 201 p
13. M.E. Ilchenko (ed.), *Dielectric Resonators* (Radio and Communication, Moscow, 1989), 327 p (in Russian)
14. M.E. Ilchenko, A.A. Trubin, *Electrodynamics of Dielectric Resonators* (Naukova Dumka, Kiev, 2004), 265 p (in Russian)
15. A.A. Trubin, *Lattices of Dielectric Resonators* (Springer International Publishing, Switzerland, 2016), 171 p
16. L.A. Veinshtein, *Open Resonators and Open Waveguides*. Golem Series in Electromagnetics, vol. 2 (Golem Press, 1969), 439 p
17. I. Chremmos, O. Schwelb, N. Uzunoglu (ed.), *Photonic Microresonator Research and Applications*. Series in Optical Sciences, vol. 156 (Springer International Publishing, Switzerland, 2010), 517 p

Contents

Chapter 1
Natural Oscillations of Isolated Dielectric Resonators

Important parameters of dielectric resonators (DR) are the characteristics of their natural oscillations: resonant frequencies, partial quality factors, field distributions. This chapter discusses the natural oscillations of isolated dielectric resonators in open space. The frequency spectrum and fields of spherical, cylindrical and rectangular resonators are examined in detail. The main attention is paid to the lower types of natural oscillations, as well as high-quality oscillations of higher types—whispering gallery modes (WGM). Using the example of a spherical DR, the influence of dielectric and metallic inhomogeneities on frequencies and quality factors was studied.

Calculation of the parameters of an isolated DR in open space is carried out by solving the system of Maxwell's equations (C.2). In the simplest cases, problems of natural oscillations are solved using the method of partial regions. It's assumed that we know electric and magnetic fields both in the dielectric region $\left(\mathbf{e}^1, \mathbf{h}^1\right)$ with dielectric constant $\varepsilon_1 = \varepsilon_1' - i\varepsilon_1''$ and outside the resonator $\left(\mathbf{e}^0, \mathbf{h}^0\right)$. In this chapter we will assume that there are no losses in the dielectric: $\varepsilon_1'' = 0$.

In the general case, the field in the dielectric and near the surface of the resonator is subject to the condition of finiteness of the stored energy in any arbitrarily small region [1].

In a spherical coordinate system (r, θ, φ), the resonator field requires the fulfillment of an additional condition of finite power flow in the far zone for $r \to \infty$, known as the radiation condition [1]:

$$\mathbf{e}^0 \sim \mathbf{e}^\infty = \mathbf{e}_0 \cdot e^{-ik_0 r}/r; \quad \mathbf{h}^0 \sim \mathbf{h}^\infty = \mathbf{h}_0 \cdot e^{-ik_0 r}/r. \tag{1.1}$$

where functions $\mathbf{e}_0 = \mathbf{e}_0(\theta, \varphi)$, $\mathbf{h}_0 = \mathbf{h}_0(\theta, \varphi)$ depend only on angular variables (θ, φ); where $k_0 = \omega(\varepsilon_0\mu_0)^{1/2}$—is the wave number; $\omega = 2\pi f$; f—frequency; ε_0—dielectric constant, μ_0—magnetic permeability of vacuum.

The requirement that the tangent field components on the resonator surface be equal $\mathbf{e}_\tau^1 = \mathbf{e}_\tau^0$; $\mathbf{h}_\tau^1 = \mathbf{h}_\tau^0$ leads to a system of linear equations for the desired

A. Trubin, *Introduction to the Theory of Dielectric Resonators*, Springer Series in Advanced Microelectronics 65, https://doi.org/10.1007/978-3-031-65396-4_1

amplitudes, which, in turn, determines the frequencies and fields of homogeneous DR.

If there is a metal ideally conducting surface near the resonator, or in its volume, the field of its natural oscillations must satisfy the additional condition on it: $\mathbf{e}_{\tau}^{0,1} = 0$.

The material presented in this chapter is the basis for understanding the further content of the book.

1.1 Spherical Dielectric Resonator

The possibility of obtaining accurate "single-wave" analytical solutions of Maxwell's equations in a spherical coordinate system (r, θ, φ) makes it possible to construct a relatively simple representation of the field of natural oscillations of a DR in the form of a dielectric ball disposed in open space. The obtained solutions allow us to study in detail the main physical features inherent in open resonators [2].

Let us first consider the natural oscillations of a homogeneous spherical DR in open space (Fig. 1.1a). In the general case, solutions of Maxwell's equations in a spherical coordinate system are divided into two types.

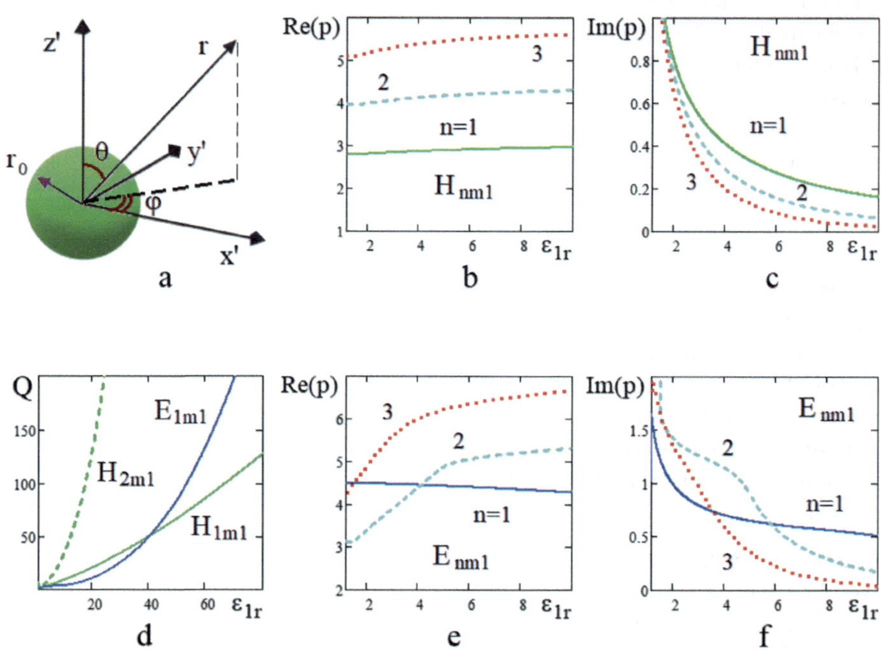

Fig. 1.1 Spherical dielectric resonator (**a**). Dependences of the real (**b**, **e**) and imaginary parts (**c**, **f**) of the characteristic parameter $p = k_1 r_0$ on the relative dielectric constant of the resonator material for lower types of oscillations. Dependence of quality factor on dielectric constant (**d**)

The field of the first type is characterized by the radial electric component being equal to zero. Oscillations of this type are called magnetic:

$$
\begin{aligned}
e_r &= 0;\\
e_\theta &= -i\omega\mu_0 f_\nu^{(s)}(k_s r)\mu\frac{P_\nu^\mu(\cos\theta)}{\sin\theta}\begin{bmatrix}\cos\\ -\sin\end{bmatrix}\mu\varphi;\\
e_\varphi &= i\omega\mu_0 f_\nu^{(s)}(k_s r)\frac{dP_\nu^\mu(\cos\theta)}{d\theta}\begin{bmatrix}\sin\\ \cos\end{bmatrix}\mu\varphi;\\
h_r &= \frac{\nu(\nu+1)}{r}f_\nu^{(s)}(k_s r)P_\nu^\mu(\cos\theta)\begin{bmatrix}\sin\\ \cos\end{bmatrix}\mu\varphi;\\
h_\theta &= \frac{1}{r}\frac{d}{dr}\left\{rf_\nu^{(s)}(k_s r)\right\}\frac{dP_\nu^\mu(\cos\theta)}{d\theta}\begin{bmatrix}\sin\\ \cos\end{bmatrix}\mu\varphi;\\
h_\varphi &= \frac{1}{r}\frac{d}{dr}\left\{rf_\nu^{(s)}(k_s r)\right\}\mu\frac{P_\nu^\mu(\cos\theta)}{\sin\theta}\begin{bmatrix}\cos\\ -\sin\end{bmatrix}\mu\varphi.
\end{aligned}
\tag{1.2}
$$

The field of the second type is characterized by the equality to zero of the radial magnetic field $h_r = 0$. Such oscillations are called electrical:

$$
\begin{aligned}
e_r &= \frac{\nu(\nu+1)}{r}f_\nu^{(s)}(k_s r)P_\nu^\mu(\cos\theta)\begin{bmatrix}\sin\\ \cos\end{bmatrix}\mu\varphi;\\
e_\theta &= \frac{1}{r}\frac{d}{dr}\left\{rf_\nu^{(s)}(k_s r)\right\}\frac{dP_\nu^\mu(\cos\theta)}{d\theta}\begin{bmatrix}\sin\\ \cos\end{bmatrix}\mu\varphi;\\
e_\varphi &= \frac{1}{r}\frac{d}{dr}\left\{rf_\nu^{(s)}(k_s r)\right\}\mu\frac{P_\nu^\mu(\cos\theta)}{\sin\theta}\begin{bmatrix}\cos\\ -\sin\end{bmatrix}\mu\varphi;\\
h_r &= 0;\\
h_\theta &= i\omega\varepsilon_s f_\nu^{(s)}(k_s r)\mu\frac{P_\nu^\mu(\cos\theta)}{\sin\theta}\begin{bmatrix}\cos\\ -\sin\end{bmatrix}\mu\varphi;\\
h_\varphi &= -i\omega\varepsilon_s f_\nu^{(s)}(k_s r)\frac{dP_\nu^\mu(\cos\theta)}{d\theta}\begin{bmatrix}\sin\\ \cos\end{bmatrix}\mu\varphi.
\end{aligned}
\tag{1.3}
$$

Here $s = 0, 1$; $P_\nu^\mu(\cos\theta)$—associated Legendre function [3]; $f_\nu^{(s)}(z)$- one of the Bessel functions in Sommerfeld notations:

$$
\begin{aligned}
j_\nu(z) &= (\pi/2z)^{1/2}J_{\nu+1/2}(z);\ y_\nu(z) = (\pi/2z)^{1/2}Y_{\nu+1/2}(z);\\
h_\nu^{(1)}(z) &= (\pi/2z)^{1/2}H_{\nu+1/2}^{(1)}(z);\ h_\nu^{(2)}(z) = (\pi/2z)^{1/2}H_{\nu+1/2}^{(2)}(z).
\end{aligned}
\tag{1.4}
$$

It is easy to verify that functions (1.4) and their derivatives for any ν satisfy the same recurrence relations as the spherical Bessel functions [3].

In the case of a homogeneous dielectric ball in open space, the indices μ and ν can take only integer values $\nu = n$, $\mu = m$, then functions (1.4) become spherical. At that: $n \geq 1$, and $|m| \leq n$ (if $|m| > n$ $P_n^m(x) = 0$). The values of n, m determine the number of field variations, respectively, in the meridional plane: $\varphi = const$ and in the azimuthal plane: $\theta = \pi/2$.

Inside the dielectric ball ($r \leq r_0$) (Fig. 1.1a), the solution (1.2), (1.3) is selected taking into account the requirement that the field is finite at the origin of coordinates $(r = 0)$:$f_n^{(1)}(k_1 r) = a_1 j_n(k_1 r)$, and in external space ($r > r_0$)—based on the requirement that the radiation conditions (1.1): $r \to \infty f_n^{(0)}(k_0 r) \sim e^{-ik_0 r}/r$ are valid, i.e. at should be a diverging spherical wave:$f_n^{(0)}(k_0 r) = a_0 h_n^{(2)}(k_0 r)$. Here $k_0 = \omega(\mu_0 \varepsilon_0)^{1/2}$, $k_1 = \omega(\mu_0 \varepsilon_1)^{1/2}$—wave numbers; ε_s—dielectric constant of vacuum for $s = 0$; dielectric constant of the resonator material for $s = 1$; μ_0—magnetic permeability (we will only consider non-magnetic materials); a_0, a_1—unknown amplitudes.

The condition of continuity of the tangent field components on the surface of the dielectric: $r = r_0$ leads to relationships between the amplitudes:

For natural oscillations of magnetic types H_{nml}:

$$a_1 = h_1; \quad a_0 = \frac{j_n(p)}{h_n^{(2)}(q)} h_1; \tag{1.5}$$

for natural oscillations of electrical types E_{nml}:

$$a_1 = e_1; \quad a_0 = \varepsilon_{1r} \frac{j_n(p)}{h_n^{(2)}(q)} e_1. \tag{1.6}$$

where $\varepsilon_{1r} = \varepsilon_1/\varepsilon_0$—relative dielectric constant of the resonator material.

Dimensionless numbers $p = k_1 r_0$, $q = k_0 r_0$, henceforth we will call the characteristic parameters of a spherical resonator.

Nontrivial solutions of the form (1.5), (1.6) exist if the characteristic parameters p, q, satisfy the equations:

$$\varepsilon_{1r}^{-1/2} j_n(p) h_{n-1}^{(2)}(q) = j_{n-1}(p) h_n^{(2)}(q) \tag{1.7}$$

for natural oscillations of magnetic types H_{nml} and

$$\varepsilon_{1r}^{-1} \frac{d}{dp}\{p j_n(p)\} h_n^{(2)}(q) = j_n(p) \frac{d}{dq}\left\{q h_n^{(2)}(q)\right\} \tag{1.8}$$

for natural oscillations of electrical types E_{nml}.

Characteristic Eqs. (1.7), (1.8) do not depend on m, this means that the natural oscillations of the dielectric ball are degenerate, their complex frequencies at given n coincide for any values $|m| \leq n$. Equations (1.7), (1.8) are usually solved numerically, taking into account the relationship between the parameters: $q = \varepsilon_{1r}^{-1/2} p$ and have complex roots. For every fixed value n there is a countable set of roots. We will number the roots of Eqs. (1.7), (1.8) by letter l in ascending order of their real part. The number l determines the number of field variations in the radial direction: $\theta = const$, $\varphi = const$.

The real part $p_R = \mathrm{Re}(p)$ of the characteristic parameter determines the natural frequency of oscillation of the dielectric ball in free space, and the imaginary part $p_J = Im(p)$ determines the losses, generally due to losses in the dielectric, as well as radiation. The radiation quality factor of a given natural oscillation for a resonator without losses in the dielectric is determined from the relation $Q = Q^{\Sigma} = p_R/2p_J$.

Let us consider in more detail the general patterns of changes in characteristic parameters with increasing relative dielectric constant of the ball material. In the case of $\varepsilon_{1r} \gg 1$ for $\varepsilon_1'' = 0$, solutions of Eqs. (1.7), (1.8) can be found explicitly by expanding the characteristic parameters p, q into an asymptotic series in powers of the small parameter $\varepsilon_{1r}^{-1/2}$:

$$p = p_0 + \varepsilon_{1r}^{-1/2} p_1 + \varepsilon_{1r}^{-1} p_2 + \cdots \tag{1.9}$$

Substituting (1.9) into (1.7), (1.8) and using Neumann's addition theorem for Bessel functions [3], we also expand each of the Eqs. (1.7), (1.8) in powers of the small parameter $\varepsilon_{1r}^{-1/2}$. Comparing the terms with each other, with the same powers $\varepsilon_{1r}^{-1/2}$, we get:

$$p_R = j_{n-1,l}\left\{1 - \frac{1}{\varepsilon_{1r}(2n-1)} - \frac{1}{\varepsilon_{1r}^2(2n-1)^2}\left[n - 1 + \frac{j_{n-1,l}}{(2n-3)}\right] + \cdots\right\};$$

$$p_J = \frac{j_{n-1,l}^2}{\varepsilon_{1r}^{3/2}}\left\{\delta_{n1} + \frac{j_{n-1,l}^2}{\varepsilon_{1r}(2n-1)^2(2n-3)}(\delta_{n1} + \delta_{n2}) + \cdots\right\} \tag{1.10}$$

for natural oscillations of magnetic types H_{nml};

$$p_R = j_{n,l}\left\{1 - \frac{1}{\varepsilon_{1r} n} - \frac{1}{\varepsilon_{1r}^2 n^2}\left[\frac{j_{n,l}^2}{(2n-1)} - 1\right] + \cdots\right\};$$

$$p_J = \frac{j_{n,l}^4}{\varepsilon_{1r}^{5/2}}\delta_{n1} + \cdots \tag{1.11}$$

for natural oscillations of electrical types E_{nml} of the dielectric ball in open space.

The relations found show that the real part of the characteristic parameters takes a value slightly less than the l—th root of the spherical Bessel function of the first kind [3]: $j_{n-1,l}$—for magnetic-type oscillations and $j_{n,l}$—for electric-type oscillations. These roots are the limiting values of the parameter at $\varepsilon_{1r} = \infty$. The imaginary parts of the characteristic parameters decrease with increasing dielectric constant both $\varepsilon_{1r}^{-3/2}$ for the lowest frequency magnetic oscillations and $\varepsilon_{1r}^{-5/2}$ for the lowest electrical oscillations. Accordingly, the radiation quality factor Q^{Σ} increases proportionally $\varepsilon_{1r}^{3/2}$ for magnetic type oscillations H_{1m1}, and proportionally $\varepsilon_{1r}^{5/2}$ for electrical type oscillations E_{1m1}.

The dependences of the characteristic parameters of the ball on the dielectric constant are shown in Fig. 1.1; distribution of the electric field modulus for three oscillations of lower types for $\varepsilon_{1r} = 10$ in Fig. 1.2. It can be seen that the field of these oscillations in the near zone is concentrated mainly in the volume of the dielectric.

The distribution of the field of oscillations of lower types corresponds to the distribution of the magnetic dipole field for H_{1m1} (a–d) and also the distribution of the electric dipole field for E_{1m1}, therefore these oscillations are sometimes called dipole. The field distributions of dipole types of oscillations differ from each other only in spatial rotation by an angle $\pi/2$. The frequencies of these oscillations coincide due to the spherical symmetry of the structure. Approximately the same statements remain valid for part of the oscillations with $n > 1$ (In the case of $n = 2$, see Fig. 1.2e–j), however, the structure of the field of degenerate types may already be different.

Since the imaginary parts of the characteristic parameters are positive (Fig. 1.1c), the field of the resonators in the far zone increases exponentially at $r \to \infty$. As noted

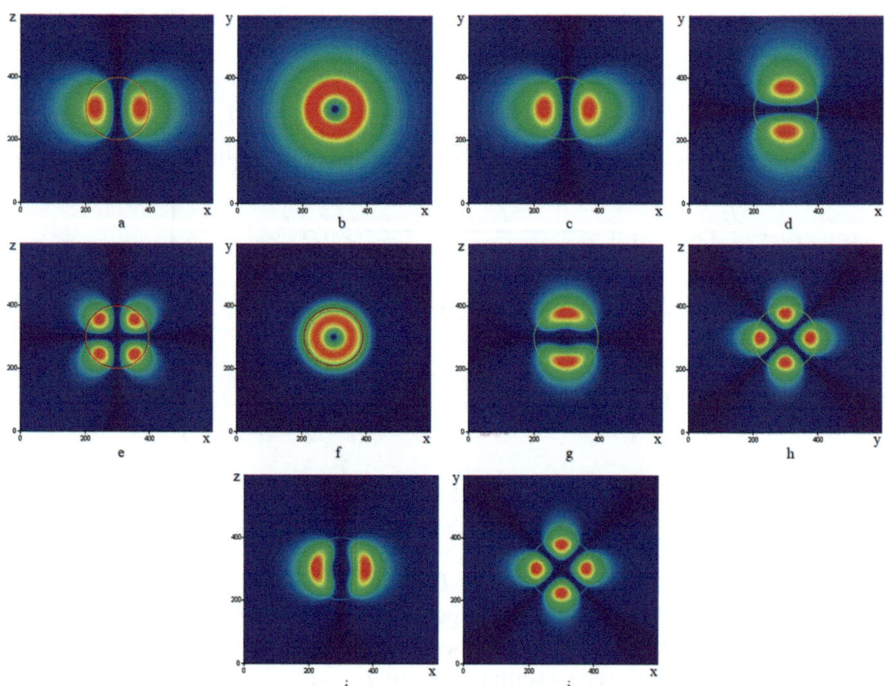

Fig. 1.2 Distribution of the electric field modulus |**e**| of the main oscillations of the magnetic type: dipole H_{101} (**a**, **b**), H_{111} (**c**, **d**); quadrupole H_{201} (**e**, **f**); H_{211} (**g**, **h**); H_{221} (**i**, **j**) spherical DR with $\varepsilon_{1r} = 10$

in [2], this is due to the fact that the finite amplitude of harmonic field oscillations in the near zone of any open resonator with losses is possible only if it was larger in the past. Earlier times for the emitted waves correspond to a greater distance from the resonator, so it is necessary that the field increases with distance from its center.

At the same time, the exponential increase in the field in the far zone of the DR clearly contradicts the radiation condition (1.1). This contradiction is removed if we take into account the fact that the radiation condition is usually formulated for the problem of scattering by a body of finite dimensions. In the case of natural oscillations of an open resonator, the radiation condition must be formulated in a different form.

Let us consider the natural oscillation field of an open resonator made of a lossless dielectric at a distance significantly greater than its maximum size: i.e. when the condition is met: $r/r_0 \gg 1$. In this case, a more correct statement is the requirement for an exponential increase in its field:

$$\mathbf{e}^0 \sim \mathbf{e}^\infty = \mathbf{e}_0 \cdot e^{-iqr/r_0}/(r/r_0);$$
$$\mathbf{h}^0 \sim \mathbf{h}^\infty = \mathbf{h}_0 \cdot e^{-iqr/r_0}/(r/r_0),\ \text{if}\, r/r_0 \to \infty$$

due to radiation losses [2]. Here $k_0 r => qr/r_0$. From where, taking into account the complexity of the characteristic parameters: $q = q' + iq''$, we come to the requirement: $q'' > 0$.

The characteristic parameters become real if only the radiation losses of oscillations in an open resonator are compensated by an influx of energy from some external source, for example an incident wave, or external field.

To verify what has been said, let us imagine, for example, the external field of a dielectric ball in the form: $f_n^{(0)}(k_0 r) = a_0 y_n(k_0 r)$. As follows from the relation $y_n(x) = [h_n^{(1)}(x) - h_n^{(2)}(x)]/2i$: and the asymptotic representation of the Hankel functions [3] for $x \to \infty$, such a replacement corresponds to the transformation of the problem of natural oscillations into the problem of scattering of a spherical electromagnetic wave on a dielectric ball. In this case, the "characteristic equations" take the form:

$$\varepsilon_{1r}^{-1/2} j_n(p) y_{n-1}(q) = j_{n-1}(p) y_n(q) \tag{1.12}$$

for scattered waves of magnetic types H_{nml};

$$\varepsilon_{1r}^{-1} \frac{d}{dp}\{p j_n(p)\} y_n(q) = j_n(p) \frac{d}{dq}\{q y_n(q)\}; \tag{1.13}$$

for scattered waves of electrical types E_{nml}.

Equations (1.12), (1.13) have purely real solutions: $q'' = 0$, since the losses associated with radiation, in this case, are compensated by the incoming spherical wave, determined by the dependence proportional to the function $h_n^{(1)}(k_0 r)$.

Equations (1.12), (1.13) can also be interpreted as an approximate representation of solutions to the problem of natural oscillations corresponding to (1.7), (1.8) for

$\varepsilon_{1r} \gg 1$, since in this case, for oscillations with $|q| \ll 1$, Hankel spherical functions in the near zone of the resonator can be approximately represented by Neumann functions: $h_n^{(2)}(q) \approx y_n(q)$.

As was established in [4], the characteristic Eqs. (1.7), (1.8) also have an additional countable set of solutions, the lowest types of which are shown in (Fig. 1.3a–b). The field of these oscillations is stored predominantly in the outer region of the resonator (Fig. 1.3d–g). The quality factor of such external oscillations remains low, even with an increase in the dielectric constant of the material (Fig. 1.3c).

In the case of higher-type oscillations at $n \gg l$, the field is concentrated mainly in the dielectric near the interface between the media. Such oscillations, called "whispering gallery" oscillations, have a very high quality factor [5].

On Fig. 1.4 shows several examples of solving Eqs. (1.7), (1.8) for oscillations of a whispering gallery ($n = 20 \ldots 22$, $l = 1$). Magnetic type oscillations H_{nml} are shown in green, and electrical type oscillations E_{nml} are shown in blue lines (a, b). It can also be seen that the characteristic parameters of the ball for oscillations of the whispering gallery change noticeably only in the region of small dielectric constants. At the same time, the "density" of the frequency distribution remains approximately constant and quite high: $\delta\omega_n/\overline{\omega}_n \cdot 100\% < 4,1\%$ for $n > 20$ here $\delta\omega_n = \omega_n^E - \omega_n^H$ is the distance between the frequencies of neighboring high-Q oscillations of the resonator (Fig. 1.4a); $\overline{\omega}_n = 1/2(\omega_n^E + \omega_n^H)$.

The distribution of the field of natural oscillations depends on the values of the azimuthal indices m. On Fig. 1.4c–e, examples of the spatial distribution of the electric field modulus are given for $n = 25$; and $m = 0, 1, 25$.

If $l \gg n$, the field of natural oscillations is formed by standing waves directed predominantly along the radius of the ball. The wave vector is directed orthogonally to the surface of the dielectric, so the field penetrates into open space. Such oscillations, called "bouncing ball" oscillations [2], have a low quality factor (Fig. 1.5b).

Using asymptotic approximations for cylindrical functions [3] obtained for $|z| \gg n$: $j_n(z) \approx 1/z \sin(z - 1/2n\pi)$; $h_n^{(2)}(z) \approx 1/z e^{-i[z-1/2(n+1)\pi]}$, we obtain from (1.7), (1.8):

for magnetic oscillations H_{nml}:

$$p \approx \frac{1}{2}(n+2s)\pi + arctg(i\varepsilon_{1r}^{1/2}); \tag{1.14}$$

for electrical oscillations E_{nml}

$$tg(p - \frac{1}{2}n\pi) \approx \frac{i\varepsilon_{1r}^{-1/2} p}{p - in(\varepsilon_{1r}^{1/2} - \varepsilon_{1r}^{-1/2})}. \tag{1.15}$$

If the condition $p \gg n\varepsilon_{1r}^{1/2}$ is met, the last equation is also solved explicitly:

$$p \approx \frac{1}{2}(n+2s)\pi + arctg(i\varepsilon_{1r}^{-1/2}) \tag{1.16}$$

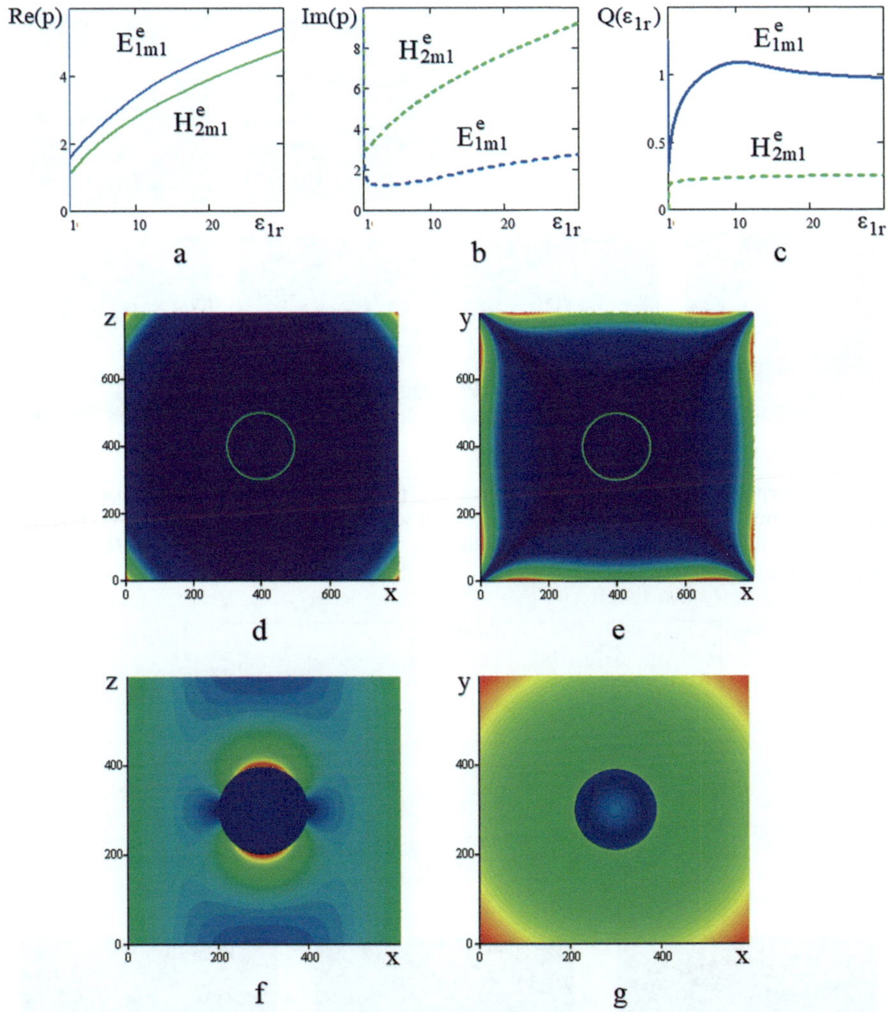

Fig. 1.3 Dependences of the real (**a**) and imaginary part (**b**) of the characteristic parameter p on the relative dielectric constant of the resonator for external oscillations. Dependence of quality factor on dielectric constant for external oscillations (**c**). Distribution of the electric field modulus for the lowest external oscillations of the ball ($\varepsilon_{1r} = 9$).H^e_{221} (**d**, **e**); E^e_{101} (**f**, **g**)

From (1.14), (1.15) it follows that the resonance of a bouncing ball oscillations occurs when: $(n + 2s)\lambda_1/2 \approx 2r_0$, i.e., when an integer number of half-waves fit along the diameter of the ball. Here λ_1 is the dielectric wavelength. On Fig. 1.5 shows the distribution of the electric field modulus for oscillations of a spherical DR: H_{105} (c); E_{105} (d), calculated from Eqs. (1.7), (1.8). Expressions (1.14), (1.16) were used as an initial approximation when solving characteristic equations.

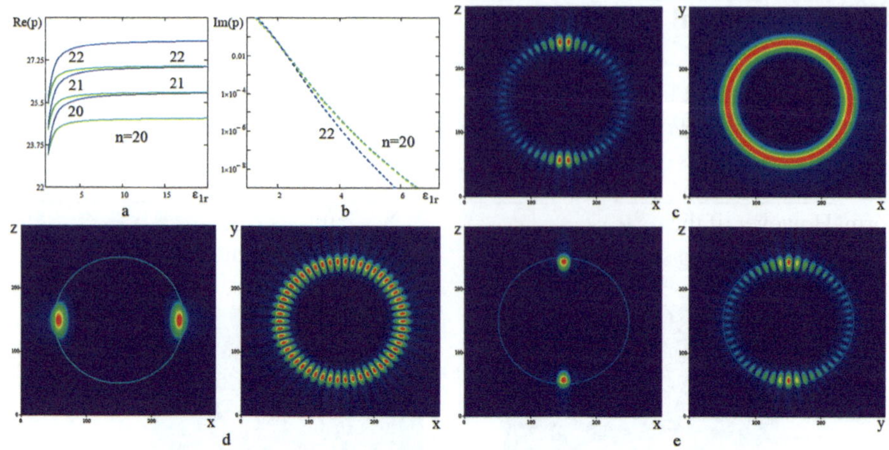

Fig. 1.4 Dependences of the real (**a**) and imaginary parts (**b**) of the characteristic parameters p on the dielectric constant of the resonator for several whispering gallery oscillations. Distribution of the electric field modulus of magnetic oscillations of a spherical DR $H_{25,0,1}$ (**c**); $H_{25,25,1}$ (**d**); $H_{25,1,1}$ (**e**) ($\varepsilon_{1r} = 2, 2$)

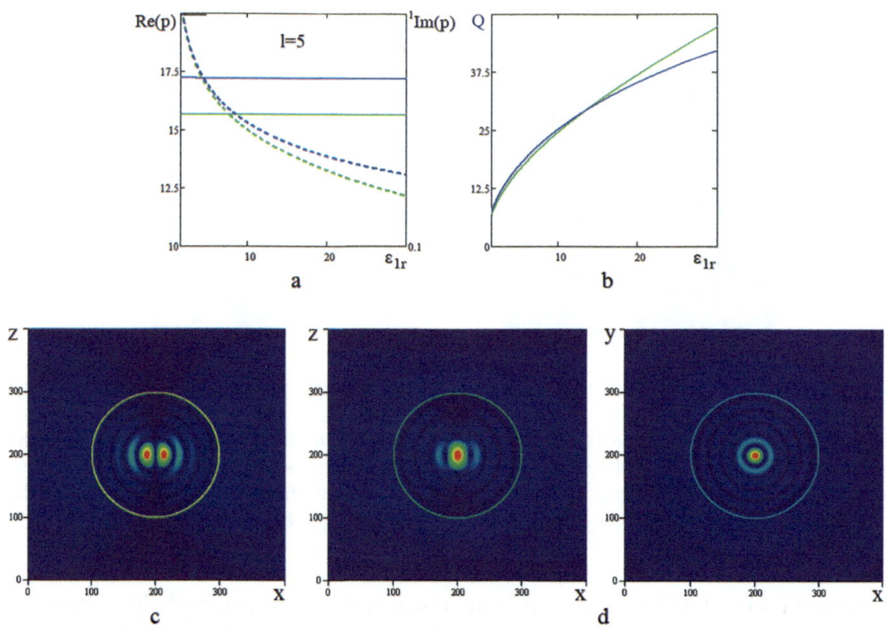

Fig. 1.5 Dependences of the real (**a**—solid curves) and imaginary parts (**a**—dotted curves) of characteristic parameters p on the dielectric constant of the resonator for oscillations of the "bouncing ball" type: $l \gg n$. Dependence of radiation quality factor on dielectric constant (**b**). Distribution of the electric field modulus of oscillations of a spherical DR: H_{105} (**c**); E_{105} (**d**); ($\varepsilon_{1r} = 36$)

1.2 Cylindrical Dielectric Resonator

For a cylindrical DR located in open space, constructing an exact "single-wave" solution of the problem of natural oscillations that satisfies all the necessary conditions in the external region is impossible. The surface of a cylindrical DR is piecewise smooth and the solution to Maxwell's equations cannot be represented in a simple analytical form. However, if the relative dielectric permittivity of the resonator material is large ($\varepsilon_{1r} \gg 1$) and the losses in the dielectric are small $\left(\varepsilon_1'' \ll \varepsilon_1\right)$, then the field of natural high-Q oscillations inside the dielectric is quite accurately approximated by solutions obtained in coordinate systems, the coordinate surfaces of which correspond to the interfaces between the media. In this case, as a rule, it is not possible to satisfy one or more boundary conditions in the external space, however, due to the high quality factor of such oscillations, the field of the external region is small and the inaccuracy of frequency calculations usually does not exceed 5–8%. Using the indicated approximate representation of the field in the case of a dielectric cylinder-shaped resonator allows one to calculate its characteristics quite accurately. This representation can also serve as a first approximation for more rigorous methods, for example, the Weinstein method of intermediate problems [6] or the Galerkin method [7].

Let us denote by L the height, and by r_0 the radius of the dielectric cylinder (Fig. 1.6a). In the general case, the field of natural oscillations inside the DR: $\rho \leq r_0$; $|z| \leq L/2$ can be approximately represented as a standing wave of a circular dielectric waveguide in a cylindrical coordinate system (ρ, φ, z):

$$\begin{aligned}
e_\rho &= \left[e_1 \frac{\beta_z}{\beta} J_m'(\beta\,\rho) + h_1 m \frac{i\omega\mu_0}{\beta} \cdot \frac{J_m(\beta\,\rho)}{\beta\,\rho}\right] \begin{Bmatrix} \sin m\varphi \\ \cos m\varphi \end{Bmatrix} \begin{bmatrix} \cos\beta_z z \\ -\sin\beta_z z \end{bmatrix}; \\
e_\varphi &= \left[e_1 m \frac{\beta_z}{\beta} \cdot \frac{J_m(\beta\,\rho)}{\beta\,\rho} + h_1 \frac{i\omega\mu_0}{\beta} \cdot J_m'(\beta\,\rho)\right] \begin{Bmatrix} \cos m\varphi \\ -\sin m\varphi \end{Bmatrix} \begin{bmatrix} \cos\beta_z z \\ -\sin\beta_z z \end{bmatrix}; \\
e_z &= e_1 J_m(\beta\,\rho) \begin{Bmatrix} \sin m\varphi \\ \cos m\varphi \end{Bmatrix} \begin{bmatrix} \sin\beta_z z \\ \cos\beta_z z \end{bmatrix}; \\
h_\rho &= \left[e_1 \frac{i\omega\varepsilon_1}{\beta} \cdot \frac{J_m(\beta\,\rho)}{\beta\,\rho} - h_1 \frac{\beta_z}{\beta} \cdot J_m'(\beta\,\rho)\right] \begin{Bmatrix} \cos m\varphi \\ -\sin m\varphi \end{Bmatrix} \begin{bmatrix} \sin\beta_z z \\ \cos\beta_z z \end{bmatrix}; \\
h_\varphi &= \left[-e_1 \frac{i\omega\varepsilon_1}{\beta} \cdot J_m'(\beta\,\rho) + h_1 m \frac{\beta_z}{\beta} \cdot \frac{J_m(\beta\,\rho)}{\beta\,\rho}\right] \begin{Bmatrix} \sin m\varphi \\ \cos m\varphi \end{Bmatrix} \begin{bmatrix} \sin\beta_z z \\ \cos\beta_z z \end{bmatrix}; \\
h_z &= h_1 J_m(\beta\,\rho) \begin{Bmatrix} \cos m\varphi \\ -\sin m\varphi \end{Bmatrix} \begin{bmatrix} \cos\beta_z z \\ -\sin\beta_z z \end{bmatrix};
\end{aligned} \qquad (1.17)$$

where $J_m'(x)$—derivative of the Bessel function [3]; e_1, h_1—amplitudes of the electric and magnetic fields, respectively; β, β_z—"radial" and "longitudinal" wave numbers. Constants e_1, h_1, β, β_z are determined in various ways depending on the representation of the external field and the method of solving the problem of natural oscillations.

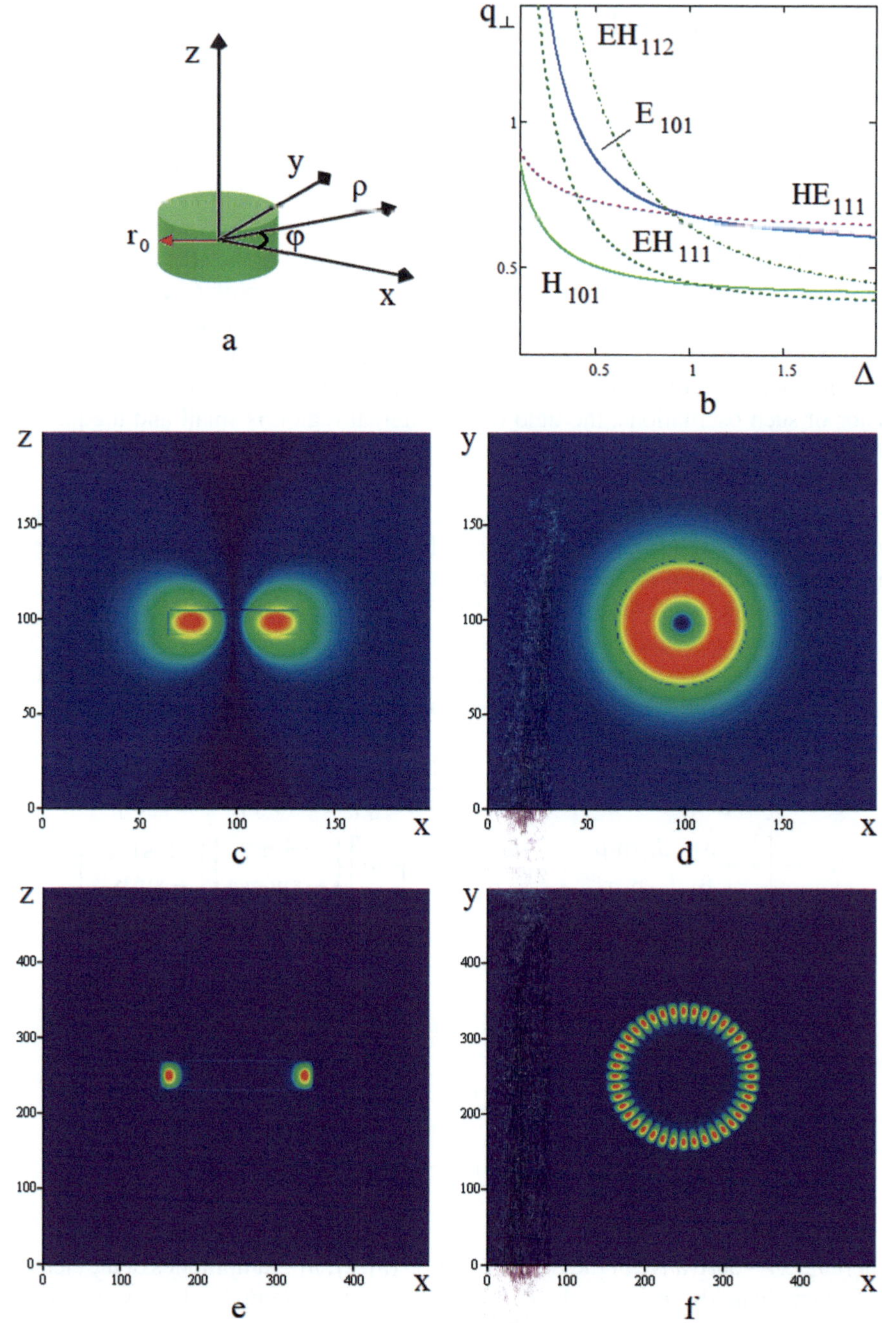

Fig. 1.6 Cylindrical DR (**a**). Dependence of the characteristic parameter $q_\perp = k_0 r_0$ on the relative dimensions of the resonator $\Delta = L/2r_0$ for $\varepsilon_{1r} = 36$. Distribution of the electric field modulus of the main azimuthally uniform oscillations H_{101} (**c**, **d**), of the whispering gallery mode $EH_{1,20,1}$ (**e**, **f**) ($\varepsilon_{1r} = 4$)

Let us construct an approximate solution to the problem for high-quality oscillations of a resonator by approximating its external field in the form of damped waves of a segment of a circular dielectric waveguide. In this case, the solutions of Maxwell's equations contain all six field components. In the area: $\rho > r_0$; $|z| \le L/2$ let us choose radial dependencies proportional to the modified Bessel function $K_m(\beta_0\rho)$ [3], and in the region: $\rho \le r_0$: $|z| > L/2$ by replacing harmonic functions in (1.17) with exponentially decreasing. $\exp(-\beta_{0z}|z|)$. At that, the constants $\beta, \beta_0, \beta_{0z}$ must satisfy the additional equations:

$$\beta^2 + \beta_z^2 = k_1^2; \quad -\beta_0^2 + \beta_z^2 = k_0^2; \quad \beta^2 - \beta_{0z}^2 = k_0^2; \tag{1.18}$$

The condition of continuity of the tangent components of the internal and external fields (1.17) on the side surface of the resonator leads to a linear system of equations for unknown amplitudes. The existence of a nontrivial solution to this system of equations determines another relationship between its characteristic parameters:

$$\left[\frac{\varepsilon_{1r}}{\beta}\frac{J_m'(\beta\, r_0)}{J_m(\beta\, r_0)} + \frac{1}{\beta_0}\frac{K_m'(\beta_0\, r_0)}{K_m(\beta_0\, r_0)}\right]\cdot\left[\frac{1}{\beta}\frac{J_m'(\beta\, r_0)}{J_m(\beta\, r_0)} + \frac{1}{\beta_0}\frac{K_m'(\beta_0\, r_0)}{K_m(\beta_0\, r_0)}\right]$$
$$= \left(\frac{m\beta_z}{k_0 r_0}\right)^2\left(\frac{1}{\beta^2} + \frac{1}{\beta_0^2}\right)^2. \tag{1.19}$$

Also equating the dominant tangential components of the internal and external fields on the end surfaces of the dielectric cylinder, we obtain equations that determine the approximate continuity:

If the longitudinal components of the magnetic field are predominant in the resonator material:

$$\beta_{0z} = \beta_z \cdot \begin{bmatrix} tg\beta_z L/2 \\ -ctg\beta_z L/2 \end{bmatrix}. \tag{1.20}$$

We will denote such oscillations as $HE_{nml}^{\pm}$.

If the longitudinal components of the electric field are predominant in the resonator material:

$$\varepsilon_{1r}\beta_{0z} = \beta_z \cdot \begin{bmatrix} -ctg\beta_z L/2 \\ tg\beta_z L/2 \end{bmatrix}. \tag{1.21}$$

We will denote such oscillations, as $EH_{nml}^{\mp}$.

Here and below, the index n determines the root number in the radial direction of the dielectric cylinder, and the root number l for solutions along the axis z. The indices ($\pm$) determine the parity of the prevailing z-components of the field in the direction of the z-axis (1.17).

The solution for field amplitudes in a dielectric (1.17) has the form:

$$e_1 = a_0 \frac{m\beta_z}{r_0}\left(\frac{1}{\beta^2} + \frac{1}{\beta_0^2}\right); \; h_1 = a_0\left[\frac{i\omega\varepsilon_1}{\beta} \cdot \frac{J'_m(\beta\, r_0)}{J_m(\beta\, r_0)} + \frac{i\omega\varepsilon_0}{\beta_0} \cdot \frac{K'_m(\beta_0\, r_0)}{K_m(\beta_0\, r_0)}\right]. \quad (1.22)$$

In the case of $m \geq 1$ Eq. (1.19) it is convenient to divide into two branches:

$$\frac{J'_m(\beta r_0)}{m\beta r_0 J_m(\beta r_0)} = -\frac{1}{2}\left(\frac{1}{\varepsilon_{1r}} + 1\right)\frac{K'_m(\beta_0 r_0)}{m\beta_0 r_0 K_m(\beta_0 r_0)}$$

$$\pm\left\{\frac{1}{4}\left(\frac{1}{\varepsilon_{1r}} - 1\right)^2\left[\frac{K'_m(\beta_0 r_0)}{m\beta_0 r_0 K_m(\beta_0 r_0)}\right]^2 + \frac{1}{\varepsilon_{1r}}\left(\frac{\beta_z}{k_0 r_0^2}\right)^2\left(\frac{1}{\beta^2} + \frac{1}{\beta_z^2}\right)^2\right\}^{1/2}, \quad (1.23)$$

corresponding to EH_{nml} or HE_{nml} oscillations.

For azimuthally homogeneous oscillations ($m = 0$), Eq. (1.19) breaks down into two independent:

$$\frac{1}{\beta} \cdot \frac{J_1(\beta\, r_0)}{J_0(\beta\, r_0)} + \frac{1}{\beta_0} \cdot \frac{K_1(\beta_0\, r_0)}{K_0(\beta_0\, r_0)} = 0, \quad (1.24)$$

corresponding to oscillations of magnetic types H_{n0l};

$$\frac{\varepsilon_{1r}}{\beta} \cdot \frac{J_1(\beta\, r_0)}{J_0(\beta\, r_0)} + \frac{1}{\beta_0} \cdot \frac{K_1(\beta_0\, r_0)}{K_0(\beta_0\, r_0)} = 0, \quad (1.25)$$

corresponding to oscillations of electric types E_{n0l}.

In this case, we find the required field components from (1.17) by formally setting $e_1 = 0$ for H_{nol} oscillations and $h_1 = 0$ for oscillations E_{n0l}.

Calculation of characteristic parameters $p_\perp = \beta r_0$; $p_z = \beta_z L/2$; $q_\perp = k_0 L/2$; $q_z = k_0 L/2$ is carried out by numerically solving the system of Eqs. (1.23) or (1.24), (1.25) together with (1.20) or (1.21) and (1.18).

The constructed approximate field representations are not defined in corner regions $\rho \geq r_0$; $|z| \geq L/2$; and, in addition, they do not satisfy the radiation conditions (1.1). This leads to the fact that the error in calculating the frequencies of the main natural oscillations reaches 8% in some cases.

The error in calculating the characteristic parameters can be reduced by using other approximate conditions for the field on the end surfaces of the dielectric cylinder. For example, the accuracy of calculating the frequency of asymmetric oscillations of a dielectric cylinder increases if, instead of relations (1.20), (1.21), we use the condition of a magnetic wall at the ends of the DR:

$$\beta_z L/2 = \pi/2$$

for EH_{111} oscillating,

$$\Delta\beta_{0z} = \beta_z tg\beta_z L/2. \quad (1.26)$$

for HE_{111} oscillating:

The accuracy of frequency calculation using these conditions is already 2.8% for oscillating HE_{111} and 3% for oscillations EH_{111}.

The result of the calculation of the characteristic parameters of the DR, made from the dielectric $\varepsilon_{1r} = 36$, is shown on Fig. 1.6b. As can be seen with $\Delta = L/(2r_0) \approx 0,44$ oscillations EH_{111} and HE_{111} are degenerates (their frequencies coincide). At the point $\Delta \approx 1$ there is a degeneracy of oscillations of the main types: H_{101} and EH_{111} The structure of the field EH_{111} is similar to the oscillation field H_{101}, but differs from it by rotating in space by an angle $\pi/2$ relative to the axis z, therefore, when $\Delta \approx 1$ the frequencies of this natural oscillations coincide. The fields of natural oscillations E_{101} and HE_{111}, are also close to each other when $\Delta \approx 1$; their frequencies also coincide.

In a homogeneous cylindrical DR, along with other modes of higher types, high-Q oscillations of a whispering gallery are possible too. As follows from (1.17), whispering gallery oscillations are possible for fields of both magnetic and electric types, if the conditions: $m \gg n$ and $m \gg l$ are met. In this case, the quality factor and frequency of whispering gallery oscillations are determined mainly by the values of azimuthally numbers m, dielectric constant ε_1 and, to a lesser extent, the relative dimensions of the resonator $\Delta = L/2r_0$. The oscillation field of the whispering gallery mode is concentrated mainly in the region of the generatrix of the dielectric cylinder (see for example Fig. 1.6e, f).

In hybrid-integrated devices, the DR is often located between parallel metal and dielectric plates; in this case, it is necessary to take into account the influence of the substrate, screen and stand (Fig. 1.7) [8]. These inhomogeneities are located in close proximity to the surface of the resonator and significantly change its properties. In this case, an approximate solution to the problem of natural oscillations of a cylindrical DR can also be obtained using the partial domain method. In particular, for azimuthally homogeneous oscillations H_{101}, longitudinal components of the magnetic field can be represented in the form:

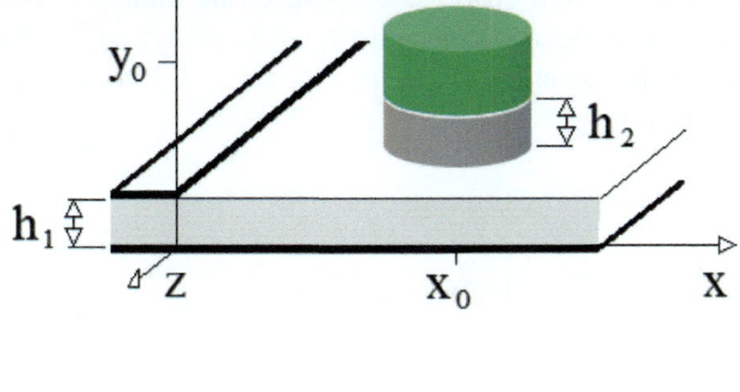

Fig. 1.7 Cylindrical DR on the substrate of a shielded asymmetrical microstrip line

inside a dielectric cylinder:

$$h_z = J_0(\beta\ \rho) \cdot [a \cos \beta_z z - b \sin \beta_z z]; \tag{1.27}$$

and in the region: $\rho \le r_0$ (Fig. 1.7):

$$h_z^{(s)} = J_0(\beta\ \rho) \cdot \left[a_s e^{\beta_z^{(s)} z} + b_s e^{-\beta_z^{(s)} z}\right], s = (1, 2, 3). \tag{1.28}$$

Equating the tangent field components at the interfaces between media: $y = 0; h_1; h_1 + h_2; h_1 + h_2 + L; H$; we get the equation:

$$\begin{aligned}&\left[\beta_z^{(3)} - \beta_z tg\beta_z(h_1 + h_2 + L) th\beta_z^{(3)}(H - h_1 - h_2 - L)\right]\\ &\quad \cdot \left[\beta_z\delta^+ - \beta_z^{(2)} tg\beta_z(h_1 + h_2)\delta^-\right]\\ &= \left[\beta_z^{(3)} tg\beta_z(h_1 + h_2 + L) + \beta_z th\beta_z^{(3)}(H - h_1 - h_2 - L)\right]\\ &\quad \cdot \left[\beta_z\delta^+ tg\beta_z(h_1 + h_2) + \beta_z^{(2)} \cdot \delta^-\right];\end{aligned} \tag{1.29}$$

as well as expressions for amplitudes (1.27):

$$\begin{aligned}a &= \beta_z^{(3)} D^+ \sin \beta_z L/2 - \beta_z D^- \cos \beta_z L/2;\\ b &= \beta_z^{(3)} D^+ \cos \beta_z L/2 + \beta_z D^- \sin \beta_z L/2,\end{aligned}$$

where

$$\begin{aligned}\delta^\pm &= e^{\beta_z^{(2)}(h_1+h_2)} \pm e^{\beta_z^{(2)}(h_1-h_2)} \frac{\beta_z^{(2)} th\beta_z^{(1)} h_1 - \beta_z^{(1)}}{\beta_z^{(2)} th\beta_z^{(1)} h_1 + \beta_z^{(1)}};\\ D^\pm &= e^{\beta_z^{(3)L/2}} \pm e^{\beta_z^{(3)}(2H-3L/2-2h_1-2h_2)}\end{aligned} \tag{1.30}$$

In this case, Eqs. (1.29) must be solved together with (1.23) and

$$\beta^2 + \beta_z^2 = k_1^2; \quad -\beta_0^2 + \beta_z^2 = k_0^2; \quad \beta^2 - \left(\beta_z^{(s)}\right)^2 = k_s^2. \tag{1.31}$$

Here $k_s = \omega(\varepsilon^{(s)}\mu_0)^{1/2}$; $\varepsilon^{(1)}$—substrate dielectric constant ($0 \le y \le h_1$); $\varepsilon^{(2)}$—pedestal dielectric constant ($h_1 \le y \le h_2$); $\varepsilon^{(3)}$—dielectric constant of the layer above the resonator ($h_1 + h_2 + L \le y \le H$); ω—circular frequency; μ_0—magnetic permeability.

1.3 Rectangular Dielectric Resonators

Let us now consider the natural oscillations of a rectangular dielectric resonator, with the L height and $a_0 \times b_0$ cross-section of the base (Fig. 1.8a).

A simple approximation of solutions of Maxwell's equations is to represent the field in the dielectric region in the form of a standing wave in a rectangular coordinate system located at the center of the resonator:

For magnetic type oscillations H_{nml}:

$$e_x = -h_1 \frac{i\omega\mu_0\beta_y}{\left(\beta_x^2+\beta_y^2\right)} \cdot \begin{pmatrix} \sin\beta_x x \\ \cos\beta_x x \end{pmatrix} \begin{bmatrix} \cos\beta_y y \\ -\sin\beta_y y \end{bmatrix} \begin{Bmatrix} \sin\beta_z z \\ \cos\beta_z z \end{Bmatrix};$$

$$e_y = h_1 \frac{i\omega\mu_0\beta_x}{\left(\beta_x^2+\beta_y^2\right)} \cdot \begin{pmatrix} \cos\beta_x x \\ -\sin\beta_x x \end{pmatrix} \begin{bmatrix} \sin\beta_y y \\ \cos\beta_y y \end{bmatrix} \begin{Bmatrix} \sin\beta_z z \\ \cos\beta_z z \end{Bmatrix};$$

$$e_z = 0;$$

$$h_x = h_1 \frac{\beta_x\beta_z}{\left(\beta_x^2+\beta_y^2\right)} \cdot \begin{pmatrix} \cos\beta_x x \\ -\sin\beta_x x \end{pmatrix} \begin{bmatrix} \sin\beta_y y \\ \cos\beta_y y \end{bmatrix} \begin{Bmatrix} \cos\beta_z z \\ -\sin\beta_z z \end{Bmatrix};$$

$$h_y = h_1 \frac{\beta_y\beta_z}{\left(\beta_x^2+\beta_y^2\right)} \cdot \begin{pmatrix} \sin\beta_x x \\ \cos\beta_x x \end{pmatrix} \begin{bmatrix} \cos\beta_y y \\ -\sin\beta_y y \end{bmatrix} \begin{Bmatrix} \cos\beta_z z \\ -\sin\beta_z z \end{Bmatrix};$$

$$h_z = h_1 \cdot \begin{pmatrix} \sin\beta_x x \\ \cos\beta_x x \end{pmatrix} \begin{bmatrix} \sin\beta_y y \\ \cos\beta_y y \end{bmatrix} \begin{Bmatrix} \sin\beta_z z \\ \cos\beta_z z \end{Bmatrix}. \tag{1.32}$$

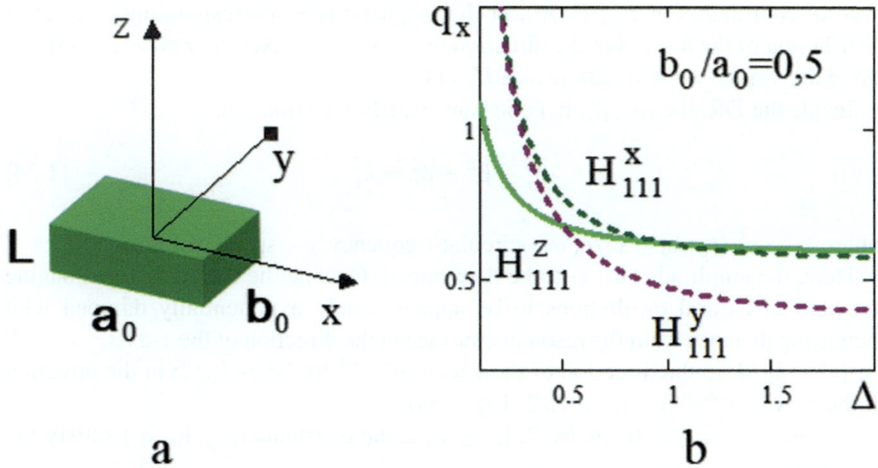

Fig. 1.8 Rectangular DR (**a**). Dependence of the characteristic parameter $q_x = k_0 a_0/2$ on the relative dimensions $\Delta = a_0/L$ of the resonator for $b_0/a_0 = 0, 5$ and $\varepsilon_{1r} = 36$

For electric type oscillations E_{nml}:

$$e_x = e_1 \frac{\beta_x \beta_z}{\left(\beta_x^2 + \beta_y^2\right)} \cdot \begin{pmatrix} \cos \beta_x x \\ -\sin \beta_x x \end{pmatrix} \begin{bmatrix} \sin \beta_y y \\ \cos \beta_y y \end{bmatrix} \begin{Bmatrix} \cos \beta_z z \\ -\sin \beta_z z \end{Bmatrix};$$

$$e_y = e_1 \frac{\beta_y \beta_z}{\left(\beta_x^2 + \beta_y^2\right)} \cdot \begin{pmatrix} \sin \beta_x x \\ \cos \beta_x x \end{pmatrix} \begin{bmatrix} \cos \beta_y y \\ -\sin \beta_y y \end{bmatrix} \begin{Bmatrix} \cos \beta_z z \\ -\sin \beta_z z \end{Bmatrix};$$

$$e_z = e_1 \cdot \begin{pmatrix} \sin \beta_x x \\ \cos \beta_x x \end{pmatrix} \begin{bmatrix} \sin \beta_y y \\ \cos \beta_y y \end{bmatrix} \begin{Bmatrix} \sin \beta_z z \\ \cos \beta_z z \end{Bmatrix};$$

$$h_x = e_1 \frac{i\omega\varepsilon_1 \beta_y}{\left(\beta_x^2 + \beta_y^2\right)} \cdot \begin{pmatrix} \sin \beta_x x \\ \cos \beta_x x \end{pmatrix} \begin{bmatrix} \cos \beta_y y \\ -\sin \beta_y y \end{bmatrix} \begin{Bmatrix} \sin \beta_z z \\ \cos \beta_z z \end{Bmatrix};$$

$$h_y = -e_1 \frac{i\omega\varepsilon_1 \beta_x}{\left(\beta_x^2 + \beta_y^2\right)} \cdot \begin{pmatrix} \cos \beta_x x \\ -\sin \beta_x x \end{pmatrix} \begin{bmatrix} \sin \beta_y y \\ \cos \beta_y y \end{bmatrix} \begin{Bmatrix} \sin \beta_z z \\ \cos \beta_z z \end{Bmatrix};$$

$$h_z = 0. \tag{1.33}$$

The n, m, l indices indicate the number of the root of the characteristic equations derived below, which determine the distribution of the field of natural oscillations, respectively, in the directions x, y, z (Fig. 1.8a). In some cases, for convenience, for oscillations of different types of symmetry, we will denote them by indices s, or c. The index s corresponds to the distribution of odd z—field components proportional to the sin function, and the c index corresponds to the distribution of the corresponding even z—field components proportional to the cos function. For example, we will denote the electric type of oscillations corresponding to an odd e_z field distribution over all coordinates as E_{sss} (1.33), or the magnetic type, corresponding to an even distribution of the h_z field in the direction of the x- and z- axes, as well as an odd one along the y-axis, we will denote as a H_{csc} (1.32).

Inside the DR, the $(\beta_x, \beta_y, \beta_z)$ constants satisfy the equation:

$$\beta_x^2 + \beta_y^2 + \beta_z^2 = k_1^2. \tag{1.34}$$

where $k_1 = \varepsilon_{1r}^{1/2} k_0$; $k_0 = \omega/c$; ω—circular frequency; c—speed of light.

Here, the amplitudes h_1, e_1 to be determined. Outside the dielectric, we imagine the field of natural oscillations to be approximately exponentially damped with increasing distance from the resonator surface in the direction of the x-axis $\sim e^{-\beta_{0x}|x|}$ for $|x| > a_0/2$; in the direction of the y-axis: $e^{-\beta_{0y}|y|}$ for $|y| > b_0/2$; in the direction of the z-axis: $e^{-\beta_{0z}|z|}$ for $|z| > L/2$ (Fig. 1.8a).

In area $|x| > a_0/2$; $|y| \le b_0/2$; $|z| \le L/2$, the constants $(\beta_{0x}, \beta_y, \beta_z)$ satisfy the equation:

$$-\beta_{0x}^2 + \beta_y^2 + \beta_z^2 = k_0^2; \tag{1.35}$$

In area: $|x| < a_0/2$; $|y| \geq b_0/2$; $|z| \leq L/2$

$$\beta_x^2 - \beta_{0y}^2 + \beta_z^2 = k_0^2 \tag{1.36}$$

And in area: $|x| < a_0/2$; $|y| \leq b_0/2$; $|z| \geq L/2$

$$\beta_x^2 + \beta_y^2 - \beta_{0z}^2 = k_0^2 \tag{1.37}$$

Matching the tangent field components at the dielectric-air interfaces approximately determines the field amplitudes and wave numbers inside and outside the resonator.

The simplest approximate method for calculating the frequencies and characteristic parameters of a rectangular DR involves excluding the field in the region $|x| > a_0/2$; $|y| \geq b_0/2$, from consideration, imposing on the characteristic parameters the condition:

$$p_x = p_y = \pi/2 \tag{1.38}$$

($p_x = \beta_x a_0/2$; $p_y = \beta_y b_0/2$), for the main type of oscillation H_{ccc}, known as the "magnetic wall" condition on the lateral surface,

and

$$p_x = p_y = \pi \tag{1.39}$$

for the main type of electric oscillation E_{ccc}. For an approximate calculation in this case of the magnetic and electric components of the field, it is sufficient to require that boundary conditions be met only on the end surfaces: $|z| = L/2$. Equating the tangent field components at the ends, we arrive at the expressions:

$$\beta_{0z} = \beta_z tg(p_z) \tag{1.40}$$

for magnetic type oscillations H_{ccc}^z;

for H_{ccs}^z:

$$\beta_{0z} = -\beta_z ctg(p_z). \tag{1.41}$$

For electric type oscillations E_{ccc}^z

$$\varepsilon_{1r}\beta_{0z} = \beta_z tg(p_z); \tag{1.42}$$

and for E_{ccs}^z:

$$\varepsilon_{1r}\beta_{0z} = -\beta_z ctg(p_z). \tag{1.43}$$

To calculate the characteristic parameters of a rectangular DR with (1.38), or (1.39) condition, Eqs. (1.40), or (1.41) should be solved together with (1.34), (1.37) for oscillations of magnetic types and (1.42), or (1.43) together with (1.34), (1.37) for electrical types.

Calculation of the field of natural oscillations rotated relative to the z- axis by an angle $\pi/2$ is simpler in the case of a rectangular resonator, due to the symmetry of the structure, to do this, just swap the sizes $a_0 \rightleftarrows L$, or $b_0 \rightleftarrows L$, or, for example, for the magnetic type, change the conditions on the end and side surfaces (1.40), or (1.41), (1.38) to.

$$p_z = p_y = \pi/2 \tag{1.44}$$

and solve Eqs. (1.34), (1.35) together with

$$\beta_{0x} = \beta_x tg(p_x) \tag{1.45}$$

for H^x_{ccc}. In such cases, for more precise designation of oscillations, we will indicate the polarization of the magnetic (electric) field at the center of the resonator with a superscript, for example H^x_{ccc} (E^x_{ccc}).

In Fig. 1.8b shows the dependence of the characteristic parameter $q_x = k_0 a_0/2$ on the relative dimensions of the resonator for the main magnetic oscillations H^x_{ccc}, H^y_{ccc}, H^z_{ccc}, calculated from the conditions of the magnetic wall.

In the general case, these oscillations are not degenerate. Degeneration of oscillations occurs in two special cases: in the first case for H^x_{ccc} and H^y_{ccc}, if the "cross section" of the prism has the same dimensions: $a_0 = b_0$ and in the second case for H^x_{ccc}, H^y_{ccc}, H^z_{ccc}, if the resonator has the shape of a cube: $a_0 = b_0 = L$.

1.4 Spherical Dielectric Resonator on a Metal Wedge

Dielectric resonators are open structures; the appearance of various types of inhomogeneities in the dielectric, or near the resonator, can significantly change the spectrum of its natural oscillations.

We investigate the influence of a metal ideally conducting wedge (in the limiting case of a metal half-plane) on the spectrum of natural oscillations of a dielectric ball in open space (Fig. 1.9a). Let us write the solution of Maxwell's equations in a spherical coordinate system in the general case in the form (1.2), (1.3), while assuming that the μ, v indices can now take continuous values.

Let us present μ, v in the form [11]: $\mu = -\eta$, $v = \eta + n$, where η is the real parameter to be determined, and n is a positive integer. The choice of this type of indices for Legendre functions is due to the requirement that the condition of uniqueness at the tip of the wedge be fulfilled [1]:

$$P^{-\eta}_{\eta+n}(\pm 1) = 0 \tag{1.46}$$

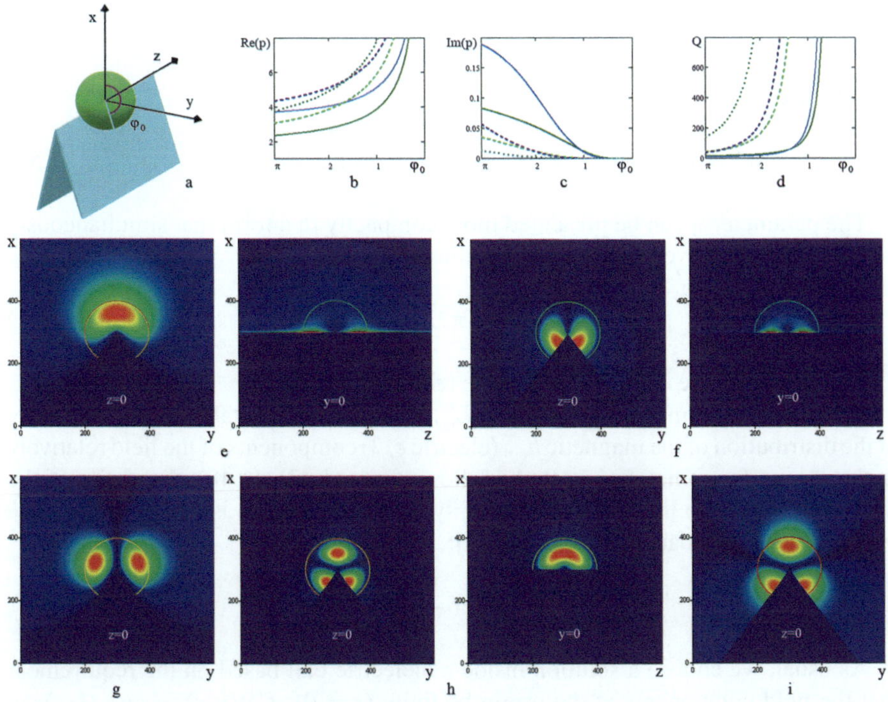

Fig. 1.9 Spherical DR on a metal wedge (**a**). Dependence of the real and imaginary part of the characteristic parameter $p = k_1 r_0$ (**b, c**), quality factor Q (**d**) on the wedge opening angle for $H^+_{\eta,-\eta,1}$, $E^+_{\eta,-\eta,1}$ oscillations ($s = 1$—solid curves); ($s = 2$—dotted curves); $s = 3$—points). Distribution of the modulus of the electric field of oscillations $H^+_{\eta,-\eta,1}$ (**e, i**); $H^-_{\eta,-\eta,1}$ (**g**); magnetic field module for oscillations $E^+_{\eta,-\eta,1}$ (**f**); $E^-_{\eta,-\eta,1}$ (**h**) ($\varepsilon_{1r} = 36$)

Indeed, from [3]

$$P^{-\eta}_{\eta+n}(\cos\theta) = \frac{n!}{(2\eta+1)(2\eta+2)\cdots(2\eta+n)} C_n^{(\eta+1/2)}(\cos\theta) P^{-\eta}_{\eta}(\cos\theta) \quad (1.47)$$

and

$$P^{-\eta}_{\eta}(\cos\theta) = \frac{2^{-\eta}}{\Gamma(\eta+1)}(\sin\theta)^{\eta} \quad (1.48)$$

it follows that justice (1.46). Here $C_n^{(\eta+1/2)}(x)$—is the Gegenbauer polynomial [3].

The requirement that the tangential components of the electric field on the metal wedge surface vanish determines the dependence η on the opening angles: $\pm\varphi_0$ (Fig. 1.9a):

For magnetic oscillations $H^{(\pm)}_{nsl}$

$$\eta^+ = (\pi/\varphi_0)s; \quad \eta^- = (\pi/\varphi_0)(s + 1/2); \quad (s = 0, 1, 2, \ldots); \tag{1.49}$$

For electrical oscillations $E_{nsl}^{(\pm)}$

$$\eta^+ = (\pi/\varphi_0)(s + 1/2); \quad \eta^- = (\pi/\varphi_0)s; \quad (s = 1, 2, \ldots) \tag{1.50}$$

The parameter η can be presented more compactly in a form that simultaneously takes into account even and odd oscillations:

$$\eta = (\pi/2\varphi_0)s \tag{1.51}$$

Conditions on the edge [1] lead to the requirement: $\eta > 0$.

The parity of magnetic $H_{nsl}^{(\pm)}$ (electric $E_{nsl}^{(\pm)}$) oscillations is established by the parity of the distribution of the magnetic $h_{r,\theta}$ (electric $e_{r,\theta}$) components of the field relatively to the plane of symmetry $\varphi = 0$. As follows from (1.47), (1.48), the parity of the field with respect to the second plane of symmetry: $\theta = \pi/2$ is determined by the parity of the Gegenbauer polynomials [3]:

$$C_n^{(\eta+1/2)}(-x) = (-1)^n C_n^{(\eta+1/2)}(x).$$

As usual, we choose a solution inside a dielectric ball based on the requirement that the field components at the origin be finite ($r = 0$): $f_\nu^{(1)}(k_1 r) = a_1 j_\nu(k_1 r)$; in external space—from the requirements of meeting radiation conditions: $f_\nu^{(0)}(k_0 r) = a_0 h_\nu^{(2)}(k_0 r)$, where a_0, a_1—amplitudes to be determined; $k_0 = \omega(\mu_0\varepsilon_0)^{1/2}$, $k_1 = \omega(\mu_0\varepsilon_1)^{1/2}$.

The continuity condition of the tangent field components (1.2), (1.3) at the dielectric boundary $r = r_0$ leads to the following relations between the amplitudes:

for magnetic oscillations:

$$a_1 = h_1; \quad a_0 = j_\nu(p)/h_\nu^{(2)}(q)h_1; \tag{1.52}$$

for electrical oscillations

$$a_1 = e_1; \quad a_0 = \varepsilon_{1r} j_\nu(p)/h_\nu^{(2)}(q)e_1. \tag{1.53}$$

Here, as in (1.1), $p = k_1 r_0$, $q = k_0 r_0$—characteristic parameters of the resonator ($p = \varepsilon_{1r}^{1/2} q$).

Characteristic parameters p, q satisfy the equations:

$$p j_{\eta+n-1}(p) h_{\eta+n}^{(2)}(q) = j_{\eta+n}(p) q h_{\eta+n-1}^{(2)}(q); \tag{1.54}$$

for magnetic oscillations $H_{nsl}^{(\pm)}$ and

$$\frac{d}{dp}\{pj_{\eta+n}(p)\}q^2h^{(2)}_{\eta+n}(q) = \frac{d}{dq}\left\{qh^{(2)}_{\eta+n}(q)\right\}p^2j_{\eta+n}(p) \tag{1.55}$$

for electrical oscillations $E^{(\pm)}_{nsl}$.

From (1.49) to (1.51) it follows that, in contrast to a homogeneous dielectric ball, the natural oscillations of a ball on a metal wedge are in the general case not degenerate. At the same time, from (1.2) we note that for particular values $\mu = m = 0$; $\nu = n$—the field of magnetic azimuthally uniform oscillations H_{nol} of the ball is not disturbed by the metal wedge. In this case, the boundary conditions of the wedge surface for these types of oscillations are satisfied automatically for any values φ_0.

The numerical solution of Eqs. (1.54), (1.55) also shows that the formation of a "thin" metal wedge in the DR material leads to a decrease in the frequencies of oscillations of lower types and, at the same time, a decrease in their quality factor (Fig. 1.9b–d). This phenomenon is caused by the excitation of so-called "edge waves" concentrated near the peak of the metal wedge (Fig. 1.9e, f) and characterized by high radiation losses. The distribution of the field of fundamental oscillations, calculated using formulas (1.2), (1.3), taking into account (1.51)–(1.55) are shown in Fig. 1.9e–i.

1.5 Dielectric Resonators with Metal Inhomogeneities of a Conical Shape

Let us consider the natural oscillations of a dielectric ball in the presence of metal ideally conducting cones (Figs. 1.10 and 1.13a) [12, 13] and show that the formation of a thin metal cone in a spherical DR material also leads to a decrease in the frequency of electrical azimuthally uniform oscillations.

Let a dielectric ball of radius r_0 with a relative dielectric constant ε_{1r} be located on a metal ideally conducting cone, the surface of which in a spherical coordinate

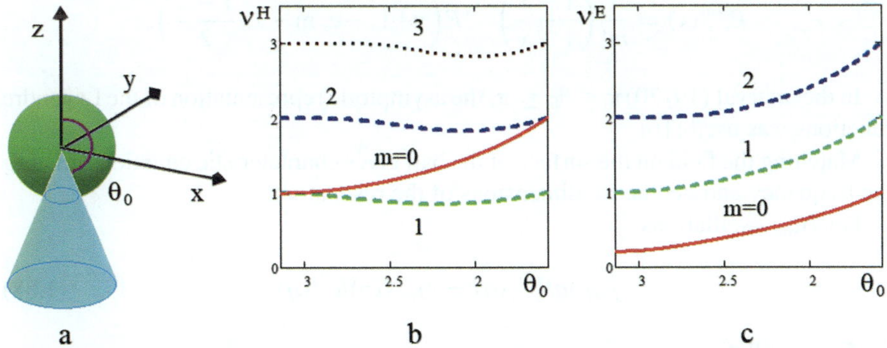

Fig. 1.10 Spherical DR on a metal cone (**a**). Dependences of meridional parameters ν for the main magnetic (**b**), electrical (**c**) oscillations

system is determined by the equation: $\theta = \theta_0$ (Fig. 1.10a). All boundary surfaces of this structure have symmetry of rotation about the z axis, therefore, to fulfill the conditions of unambiguity, the azimuthal indices of the solution (1.2–1.3) must take only integer values. The values of the meridional index ν will be determined based on the requirement that boundary conditions be met on the surface of the cone:

For magnetic oscillations $H_{\nu ml}$

$$\frac{d}{d\theta}P_\nu^{-m}(\cos\theta_0) = 0; \tag{1.56}$$

For electrical oscillations $E_{\nu ml}$

$$P_\nu^{-m}(\cos\theta_0) = 0; \tag{1.57}$$

A particular solution to Eqs. (1.56) and (1.57) for $m = 0$ is given in [14]; overview of numerical solutions in the general case in [15]. Considering equality [14]

$$\frac{d}{d\theta}P_\nu(\cos\theta) = -\nu(\nu+1)P_\nu^{-1}(\cos\theta),$$

as well as (1.2), (1.3), we come to the conclusion that in a particular case the functions $\nu_{0,s}^H(\theta_0)$ of magnetic azimuthally uniform oscillations coincide with the functions $\nu_{1,s}^E(\theta_0)$ of electrical oscillations.

The solution to Eqs. (1.56) is shown in Fig. 1.10b, and (1.57)—in Fig. 1.10c. $\nu^H(\theta_0)$ and $\nu^E(\theta_0)$ functions for $m > 0$ take integer values at $\theta_0 = \pi$ (thin metal rod) and at $\theta_0 = \pi/2$ (metal plane). These limiting values correspond to those oscillations of a homogeneous dielectric ball that satisfy the boundary conditions on the metal rod and plane.

In the interval $\theta_0 < (19/20)\pi$, Legendre functions were calculated based on their representation in the form of a hypergeometric series [17]

$$P_\nu^{-m}(x) = \frac{1}{m!}\left(\frac{1-x}{1+x}\right)^{m/2} F\left(\nu+1, -\nu, \mathrm{m}+1, \frac{1-\mathrm{x}}{2}\right).$$

In the interval $(19/20)\pi \le \theta_0 \le \pi$, the asymptotic representation of the Legendre functions was used [16].

Matching the field on the surface of the ball gives characteristic equations relating the frequency and geometric dimensions of the resonators:

For $H_{\nu ml}$ oscillations:

$$j_\nu(p)qh_{\nu-1}^{(2)}(q) = pj_{\nu-1}(p)h_\nu^{(2)}(q) \tag{1.58}$$

for $E_{\nu ml}$ oscillations:

$$\frac{d}{dp}\{pj_\nu(p)\}q^2h_\nu^{(2)}(q) = \frac{d}{dq}\left\{qh_\nu^{(2)}(q)\right\}p^2j_\nu(p), \tag{1.59}$$

where also $p = k_1r_0$, $q = k_0r_0$.

Equation (1.58) is solved together with (1.56), and (1.59) with (1.57). In this case, the condition at the tip of the cone [1] limits the range of order changes $\nu > -0, 5$, which, as can be seen from the calculations (Fig. 1.10b, c), is satisfied.

As can be seen from the data shown in Fig. 1.11a–c, a slight decrease in the frequency of the fundamental oscillation $H_{\nu,1,1}$ is also accompanied by a decrease in their quality factor. The greatest field disturbance occurs for oscillations $H_{\nu,1,1}$ and $E_{\nu,0,1}$ (Fig. 1.11e, f).

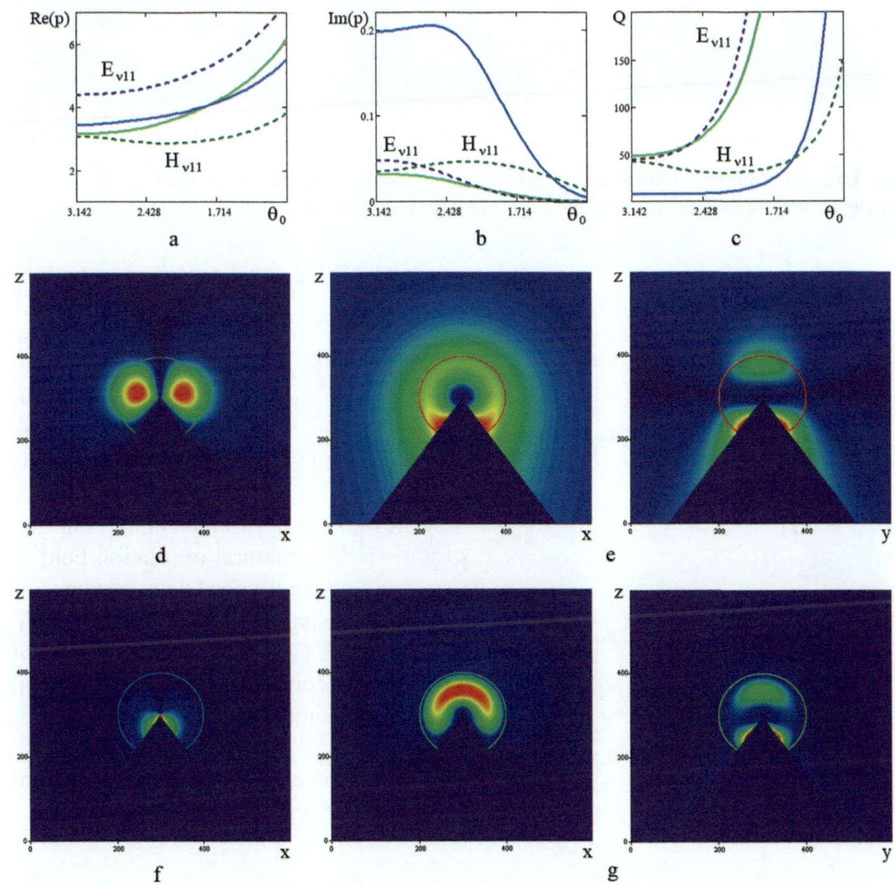

Fig. 1.11 Dependence of the real (**a**) and imaginary (**b**) parts of the characteristic parameter p, as well as the quality factor (**c**) on the cone opening angle for $H_{\nu,0,1}$, $E_{\nu,0,1}$ oscillations - solid curves; $H_{\nu,1,1}$, $E_{\nu,1,1}$ oscillations—dotted curves. Distribution of the modulus of the electric field for oscillations $H_{\nu,0,1}$ (**d**); $H_{\nu,1,1}$ (**e**); magnetic field module for oscillations $E_{\nu,0,1}$ (**f**); $E_{\nu,1,1}$ (**g**) ($\varepsilon_{1r} = 36$; $\theta_0 = 0, 8\pi$)

In the case of resonances of higher types, whispering gallery oscillations are of greatest interest. For magnetic oscillations, the numerical solution of Eq. (1.56) gives the values $\nu^H < m$, which leads to a field distribution whose maximum values lie near the surface of the cone (Fig. 1.12a). For electrical oscillations, $m \sim \nu$, the field distribution (Fig. 1.12b, c) is more like the field of a homogeneous dielectric ball with $m = n$ (Fig. 1.4d).

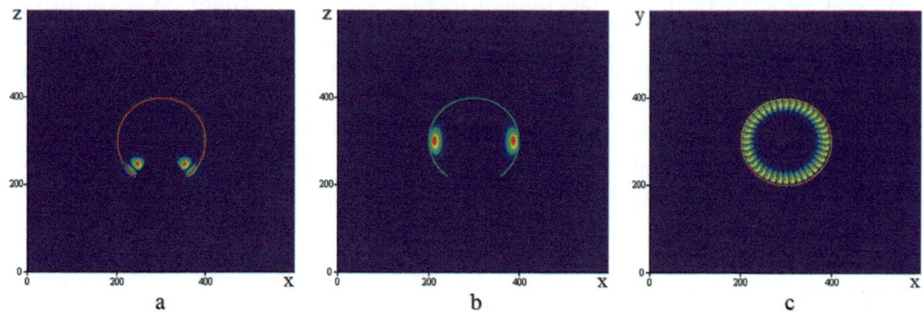

Fig. 1.12 Distribution of the electric field modulus of whispering gallery oscillations: $H_{\nu,20,1}$ (**a**); magnetic field module for oscillations $E_{\nu,20,1}$ (**b**); (**c**) ($\varepsilon_{1r} = 9$; $\theta_0 = 0, 8\pi$)

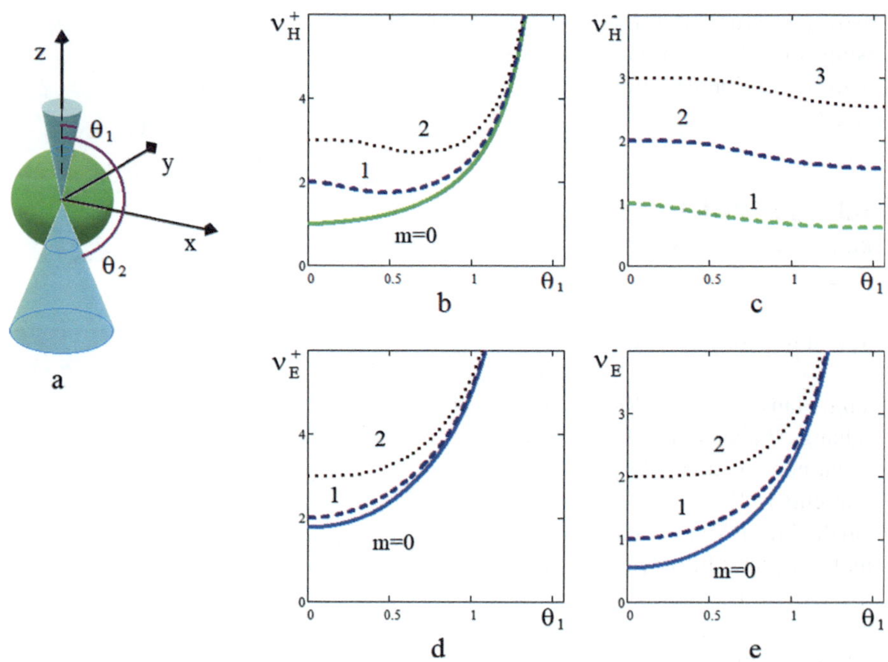

Fig. 1.13 Spherical DR on a metal bicone (**a**). Dependences of meridional parameters ν on opening angles of a symmetrical bicone for magnetic oscillations $H^{\pm}_{\nu ml}$ (**b, c**); electrical oscillations $E^{\pm}_{\nu ml}$ (**d, e**)

To solve the problem of natural oscillations of a spherical DR on a metal bicone (Fig. 1.13a) in relations (1.2), (1.3), one should move from the Legendre function to their linear combination, consisting of two independent solutions: $\alpha P_\nu^m(\cos\theta) + \beta P_\nu^m(-\cos\theta)$ [16]. In this case, the boundary conditions on the bicone surface ($\theta = \theta_1, \theta_2$) determine the values of the coefficients:

$$\alpha = \frac{dP_\nu^m(-\cos\theta_1)}{d\theta}; \quad \beta = -\frac{dP_\nu^m(\cos\theta_1)}{d\theta} \tag{1.60}$$

for magnetic oscillations $H_{\nu ml}$;

$$\alpha = P_\nu^m(-\cos\theta_1); \beta = -P_\nu^m(\cos\theta_1) \tag{1.61}$$

for electrical oscillations $E_{\nu ml}$.

The requirement for the existence of a nontrivial solution to the system of equations with respect to the unknowns coefficients α, β, leads to a dependence between the parameters ν, θ_1, θ_2, which takes the form:

for $H_{\nu ml}$ oscillations:

$$\frac{dP_\nu^m(\cos\theta_1)}{d\theta} \cdot \frac{dP_\nu^m(-\cos\theta_2)}{d\theta} = \frac{dP_\nu^m(-\cos\theta_1)}{d\theta} \cdot \frac{dP_\nu^m(\cos\theta_2)}{d\theta}; \tag{1.62}$$

for $E_{\nu ml}$ oscillations:

$$P_\nu^m(\cos\theta_1)P_\nu^m(-\cos\theta_2) = P_\nu^m(-\cos\theta_1)P_\nu^m(\cos\theta_2). \tag{1.63}$$

The characteristic equations obtained as a result of field matching at the dielectric-air interface do not change—(1.58), (1.59).

Equations (1.62) and (1.63) are simplified for a symmetric bicone: $\theta_2 = \pi - \theta_1$. In this case, symmetry of the structure relative to the plane $\theta = \pi/2$ arises, so the field of natural oscillations can be characterized by a certain parity. Relations (1.62), (1.63) break down into equations of two types, corresponding, respectively, to even and odd oscillations:

$$\frac{dP_\nu^m(\cos\theta_1)}{d\theta} + \frac{dP_\nu^m(-\cos\theta_1)}{d\theta} = 0 \tag{1.64}$$

and $\alpha = \beta$ for $H_{\nu ml}^+$;

$$\frac{dP_\nu^m(\cos\theta_1)}{d\theta} - \frac{dP_\nu^m(-\cos\theta_1)}{d\theta} = 0 \tag{1.65}$$

and $\alpha = -\beta$ for $H_{\nu ml}^-$;

$$P_\nu^m(\cos\theta_1) + P_\nu^m(-\cos\theta_1) = 0 \tag{1.66}$$

and $\alpha = \beta$ for $E^{+}_{\nu ml}$;

$$P^{m}_{\nu}(\cos\theta_1) - P^{m}_{\nu}(-\cos\theta_1) = 0 \tag{1.67}$$

and $\alpha = -\beta$ for $E^{-}_{\nu ml}$.

In this case, it is convenient to distinguish even and odd terms in the Legendre functions. We obtain the necessary representations using relations connecting Legendre functions with hypergeometric functions $F(a, b; c; z)$ [16]:

$$\begin{aligned} &2^{-m}\pi^{-1/2}\left(1-x^2\right)^{m/2}\left[P^{m}_{\nu}(x) - P^{m}_{\nu}(-x)\right] \\ &\quad = -4xF\left(\frac{1-\nu-m}{2}, \frac{2+\nu-m}{2}; \frac{3}{2}; x^2\right)\Big/\Gamma\left(\frac{1-\nu-m}{2}\right)\Gamma\left(-\frac{\nu+m}{2}\right); \end{aligned} \tag{1.68}$$

$$\begin{aligned} &2^{-m}\pi^{-1/2}\left(1-x^2\right)^{m/2}\left[P^{m}_{\nu}(x) + P^{m}_{\nu}(-x)\right] \\ &\quad = 2F\left(-\frac{m+\nu}{2}, \frac{1+\nu-m}{2}; \frac{1}{2}; x^2\right)\Big/\Gamma\left(\frac{1-\nu-m}{2}\right)\Gamma\left(\frac{1+\nu-m}{2}\right). \end{aligned} \tag{1.69}$$

Substituting (1.68), (1.69) into (1.64), (1.67) and using expressions for derivatives through adjacent Legendre functions [16], we obtain:

For even oscillations $H^{+}_{\nu ml}$

$$\begin{aligned} &\sin^2\theta_1 F\left(\frac{1}{2} - \frac{\nu+m}{2}, 1 + \frac{\nu-m}{2}; \frac{1}{2}; \cos^2\theta_1\right) \\ &\quad = -F\left(-\frac{1}{2} - \frac{\nu+m}{2}, \frac{\nu-m}{2}; \frac{1}{2}; \cos^2\theta_1\right); \end{aligned} \tag{1.70}$$

for odd oscillations $H^{-}_{\nu ml}$

$$\begin{aligned} &(\nu-m+1)(\nu+m)\sin^2\theta_1 F\left(1 - \frac{\nu+m}{2}, \frac{3}{2} + \frac{\nu-m}{2}; \frac{3}{2}; \cos^2\theta_1\right) \\ &\quad = -(\nu+m+1)(\nu-m)F\left(-\frac{\nu+m}{2}, \frac{1}{2} + \frac{\nu-m}{2}; \frac{3}{2}; \cos^2\theta_1\right). \end{aligned} \tag{1.71}$$

As well as for even oscillations $E^{+}_{\nu ml}$

$$F\left(-\frac{\nu+m}{2}, \frac{1}{2} + \frac{\nu-m}{2}; \frac{1}{2}; \cos^2\theta_1\right) = 0; \tag{1.72}$$

and for odd oscillations $E^{-}_{\nu ml}$

$$F\left(\frac{1}{2} - \frac{\nu+m}{2}, 1 + \frac{\nu-m}{2}; \frac{3}{2}; \cos^2\theta_1\right) = 0. \tag{1.73}$$

The dependence of the parameters $\nu_{ms}^{\pm}$ on the opening angles of a symmetrical bicone is shown in Fig. 1.13b–e. As follows from the results obtained, some of the parameters $\nu_{ms}^{\pm}$ (Fig. 1.13b, d, e) as well as p (Fig. 1.14a) tend to infinity at $\theta_1, \theta_2 \to \pi/2$. An increase in the meridional functions at a fixed value of the radial root leads to the fact that the field is concentrated near the boundary of the dielectric, and in the center of the dielectric ball it decreases: the region between the cones becomes, as it were, below cutoff. At the same time, as follows from the calculations performed, Fig. 1.13c, there are oscillations whose parameters weakly depends on the opening angles of the bicone. The meridional functions of such oscillations decrease at $\theta_1 \to \pi/2$ and tend to the limiting values:

$$\nu_{ml}^{-}(\pi/2) = \left[-1 + \left(1 + 4m^2\right)^{1/2}\right]/2; \; (m \geq 1).$$

In the region $\theta_1 > \pi/4$ (Fig. 1.14a), the spectrum of the lowest natural oscillations of the dielectric ball becomes rarefied.

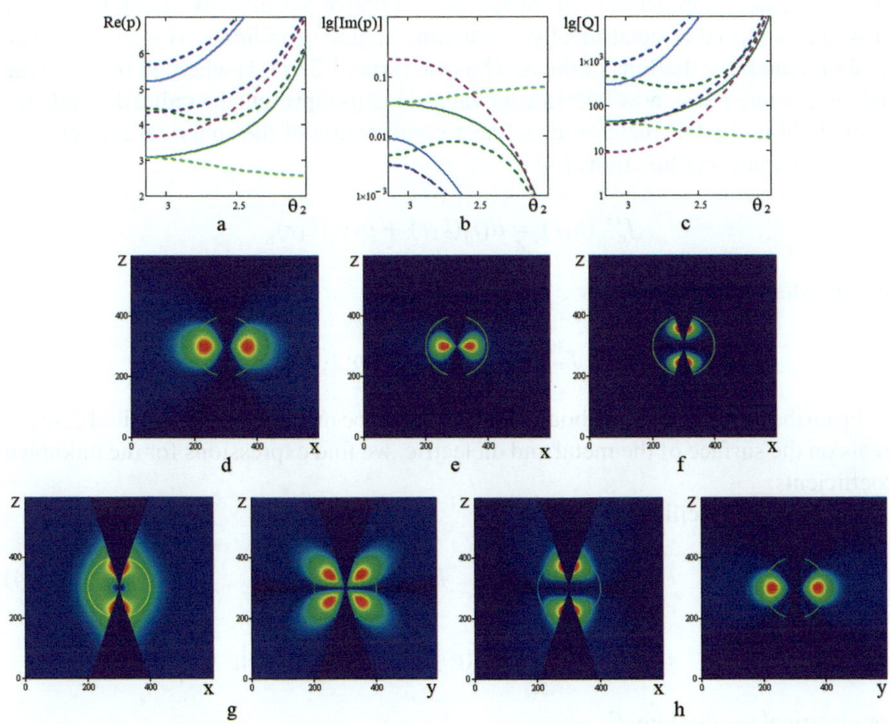

Fig. 1.14 Dependence of the real (**a**) and imaginary (**b**) parts of the characteristic parameter p, logarithm of the quality factor (**c**) on the opening angles of the symmetrical ($\theta_2 = \pi - \theta_1$) bicone for oscillations $H_{\nu,0,1}^{\pm}$, $E_{\nu,0,1}^{\pm}$—solid curves; $H_{\nu,1,1}^{\pm}$, $E_{\nu,1,1}^{\pm}$ oscillations—dotted curves. Distribution of the electric field modulus for oscillations $H_{\nu,0,1}^{+}$ (**d**); $H_{\nu,1,1}^{-}$ (**g**) and $H_{\nu,1,1}^{+}$ (**h**); magnetic field module for oscillations $E_{\nu,0,1}^{+}$ (**e**); $E_{\nu,0,1}^{-}$ (**f**); ($\varepsilon_{1r} = 36$; $\theta_1 = 0, 1\pi$)

1.6 Layered Spherical Dielectric Resonators

A known disadvantage of DR is the increased frequency density of oscillations of higher types. The frequency spectrum of a three-dimensional resonator can be rarefied in various ways: by exciting oscillations of the desired type; by reducing the quality factor of oscillations of higher types, as well as by detuning or combining frequencies. The last two methods are most simply implemented on quasi-one-dimensional spherically symmetric structures. Note that this type of resonator is also of interest due to the fact that the partial quality factors of the main oscillations of spherical DR are also close to the maximum [9, 10]. The presented data determine a possible direction for further improvement of spectral parameters by studying the influence of various types of spherically symmetric metallic and dielectric inhomogeneities on the natural oscillations of spherical DRs.

Let us consider in more detail the possibilities of improving the frequency spectrum of natural oscillations of a dielectric ball by creating metallic or dielectric inhomogeneities [18, 19] not breaking the spherical symmetry of the structure. Let us first consider a dielectric ball of radius r_2, relative permittivity ε_{1r}, in the center of which is placed a metal ideally conducting sphere of radius r_1 (Fig. 1.15a). The field of natural oscillations of the ball has the form (1.2), (1.3) where in this case the indices $\mu = m$, $\nu = n$ now take integer values. Let us represent the radial dependence of the field in the DR material as a linear combination of the spherical Bessel $j_n(x)$ and Neumann $y_n(x)$ functions [3]:

$$f_n^{(1)}(k_1 r) = a_1 j_n(k_1 r) + b_1 y_n(k_1 r),$$

and in external space:

$$f_n^{(0)}(k_0 r) = a_0 h_n^{(2)}(k_0 r).$$

From the requirement that boundary conditions be met for the tangent field components on the surface of the metal and dielectric, we find expressions for the unknown coefficients;

for magnetic oscillations H_{nml}

$$a_1 = y_n(s)h_0; \quad b_1 = -j_n(s)h_0; \tag{1.74}$$

$$a_0 = h_0/h_n^{(2)}(q)\left[j_n(p)y_n(s) - j_n(s)y_n(p)\right];$$

for electrical oscillations E_{nml}

$$a_1 = d/ds\{sy_n(s)\}e_0; \quad b_1 = -d/ds\{sj_n(s)\}e_0; \tag{1.75}$$

$$a_0 = \varepsilon_{1r}/h_n^{(2)}(q)\left[j_n(p)d/ds\{sy_n(s)\} - y_n(p)d/ds\{sj_n(s)\}\right]e_0,$$

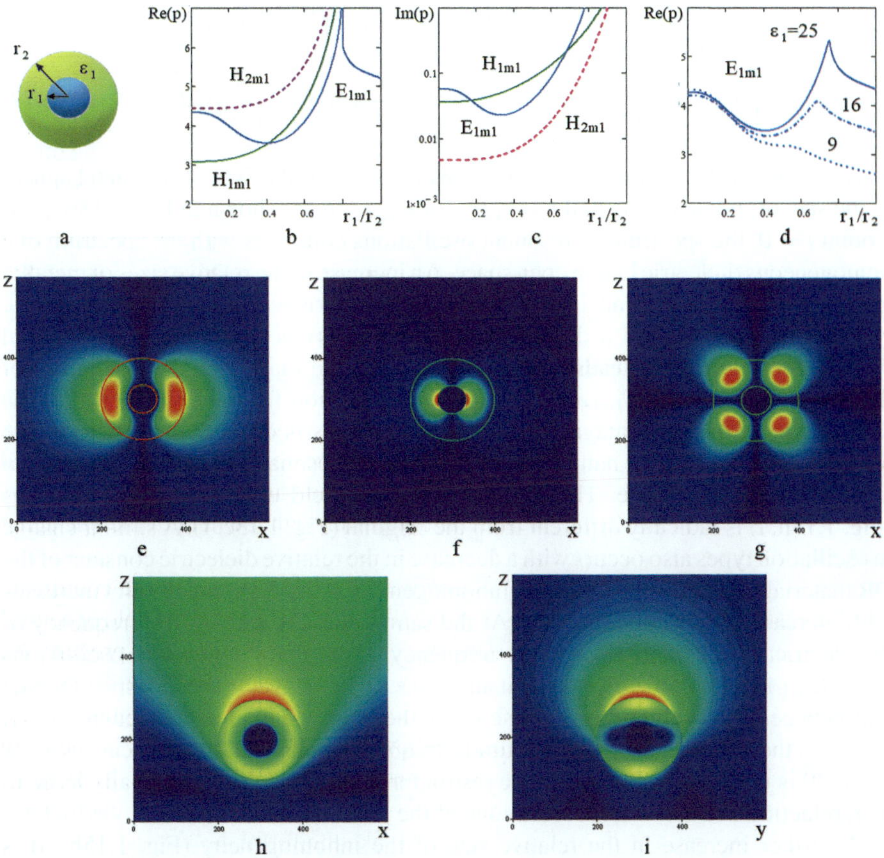

Fig. 1.15 Spherical DR with a metal ball (**a**). Dependence of the real (**b, d**) and imaginary (**c**) parts of the characteristic parameter p on the relative sizes of the dielectric and metal ball for $\varepsilon_{1r} = 36$—(**b, c**), for different dielectric constant (**d**). Distribution of the electric field modulus for oscillations H_{101} (**e**); H_{201} (**g**); magnetic field modulus for E_{101} (**f**), and also for degenerate oscillations $H_{111} + E_{111}$ (**h, i**) ($\varepsilon_{1r} = 10$; $r_1/r_2 = 0, 338$)

and also find equations connecting the characteristic parameters of the resonator: $s = k_1 r_1; p = k_1 r_2; q = k_0 r_2$:

in case of magnetic oscillations

$$\begin{aligned}&\left[p j_{n-1}(p) h_n^{(2)}(q) - j_n(p) q h_{n-1}^{(2)}(q)\right] y_n(s)\\&= \left[p y_{n-1}(p) h_n^{(2)}(q) - q h_{n-1}^{(2)}(q) y_n(p)\right] j_n(s);\end{aligned} \tag{1.76}$$

in case of electrical oscillations:

$$\left[p^2 j_n(p)\frac{d}{dq}\{qh_n^{(2)}(q)\} - q^2\frac{d}{dp}\{pj_n(p)\}h_n^{(2)}(q)\right]\frac{d}{ds}\{sy_n(s)\}$$
$$= \left[p^2 y_n(p)\frac{d}{dq}\{qh_n^{(2)}(q)\} - q^2 h_n^{(2)}(q)\frac{d}{dp}\{py_n(p)\}\right]\frac{d}{ds}\{sj_n(s)\}. \quad (1.77)$$

Solutions of the characteristic equations for varying the size of the metal sphere to the size of the dielectric ball $t = r_1/r_2$ for $\varepsilon_{1r} = 36$ are shown in Fig. 1.15b, c. At a point $t = 0$, the spectrum of resonant oscillations coincides with the spectrum of a homogeneous dielectric ball in open space. An increase in the relative sizes of metallic inhomogeneity leads to an increase in the characteristic parameters of magnetic types of oscillations and a decrease in the characteristic parameters of electrical types, which ultimately leads to a replacement of the main type of oscillations of the resonator $H_{1m1} \Leftrightarrow E_{1m1}$. In this case, the condition for the coincidence of the frequencies of the main magnetic and electrical types is of particular interest. The lowest frequency type of natural oscillations of the resonator at this point turns out to be six fold degenerate. The distribution of the field of degenerate oscillations (Fig. 1.15h, i) is radically different from the original (Fig. 1.15e, f). A similar change in oscillation types also occurs with a decrease in the relative dielectric constant of the DR material. The size of the metallic inhomogeneity at the degeneracy point increases with increasing dielectric constant. At the same time, the ratio of the frequency of the electrical oscillation E_{1m2} to the frequency of the main degenerate oscillations H_{1m1}, E_{1m1} is approximately constant and exceeds an octave. Indeed, despite the fact that between the oscillations of these types there are oscillation frequencies H_{2m1}, E_{2m1}, ..., their influence on the amplitude-frequency characteristics of scattering for $\varepsilon_{1r} > 30$ is usually insignificant. The restructuring of the spectrum actually leads to its rarefaction near the main oscillations of the DR.

A further increase in the relative size of the inhomogeneity (Fig. 1.15b, d) is accompanied by a restructuring of the field of electrical oscillations associated with the transition to natural oscillations of a metal ball covered with a thin film of dielectric.

Thus, the calculations show that the formation of metallic inhomogeneities in the resonator material allows, under certain conditions, to effectively rarefy the spectrum near its fundamental oscillations, however, due to finite conductivity, it usually leads to a noticeable decrease in the quality factor. Rarefaction of the frequency spectrum of the resonator without deteriorating its quality factor can be achieved by introducing into the resonator material a dielectric inhomogeneity that does not violate the spherical symmetry of the shape and, as a result, does not remove the degeneracy of the frequencies of natural oscillations. Let us explore this possibility for a dielectric ball containing a spherically symmetric inhomogeneity of the dielectric (Fig. 1.16a).

Let us assume that the radius of the central spherical inhomogeneity is equal to r_1 and of the outer layer r_2; the relative dielectric constant of the inhomogeneity is ε_{1r}, and of the ball—ε_{2r} and represent the radial dependence of the field in three regions in the form:

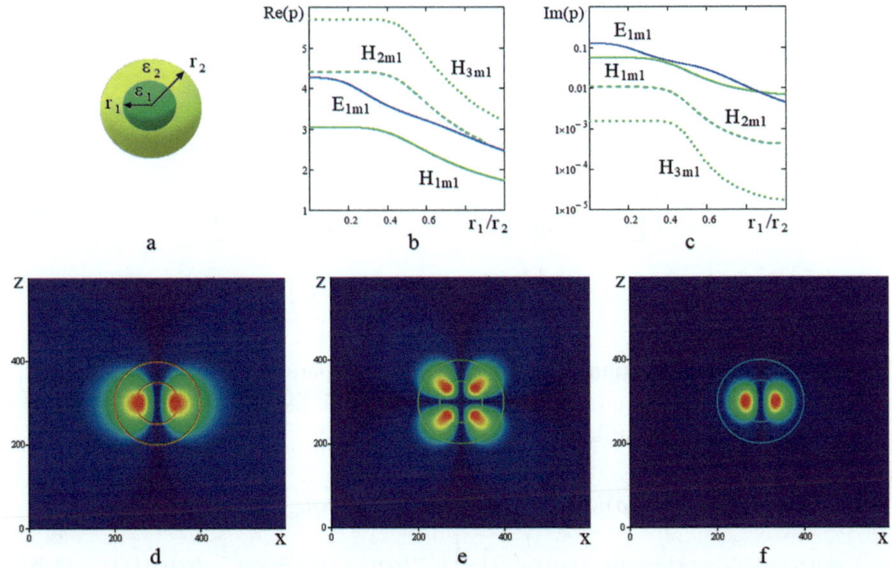

Fig. 1.16 Double layer spherical DR (**a**). Dependence of the real (**b**) and imaginary (**c**) parts of the characteristic parameter p on the relative dimensions of the dielectric ball for $\varepsilon_{1r} > \varepsilon_{2r}$ ($\varepsilon_{1r} = 81$; $\varepsilon_{2r} = 25$—(**b**, **c**). Distribution of the electric field modulus for oscillations H_{101} (**d**); H_{201} (**e**), magnetic field modulus for oscillations E_{101} (**f**) ($r_1/r_2 = 0, 5$)

$$f_n^{(1)}(k_1 r) = a_1 j_n(k_1 r);$$

$$f_n^{(2)}(k_2 r) = a_2 j_n(k_2 r) + b_2 y_n(k_2 r); \tag{1.78}$$

$$f_n^{(0)}(k_0 r) = a_3 h_n^{(2)}(k_0 r);$$

where a_1, a_2, b_2, a_3—constants determined from the conditions of equality of the tangential components of the field at the interfaces between media: $r = r_1$; $r = r_2$.

For magnetic oscillations H_{nml}:

$$a_1 = h_0$$

$$a_2 = s\left[j_n(p) s y_{n-1}(s) - p j_{n-1}(p) y_n(s)\right] h_0; \tag{1.79}$$

$$b_2 = s\left[p j_{n-1}(p) j_n(s) - j_n(p) s j_{n-1}(s)\right] h_0;$$
$$a_3 = 1/h_n^{(2)}(q)\left[a_2 j_n(t) + b_2 y_n(t)\right];$$

For electrical oscillations E_{nml}:

$$a_1 = e_0;$$

$$a_2 = \frac{1}{s}\left[p^2 j_n(p)\frac{d}{ds}\{sy_n(s)\} - s^2 y_n(s)\frac{d}{dp}\{pj_n(p)\}\right]e_0;$$

$$b_2 = \frac{1}{s}\left[s^2 j_n(s)\frac{d}{dp}\{pj_n(p)\} - p^2 j_n(p)\frac{d}{ds}\{sj_n(s)\}\right]e_0; \tag{1.80}$$

$$a_3 = \frac{\varepsilon_{2r}}{h_n^{(2)}(q)}\left[a_2 j_n(t) + b_2 y_n(t)\right].$$

At that, the equations connecting the characteristic parameters $p = k_1 r_1; s = k_2 r_1; t = k_2 r_2; q = k_0 r_2; (k_s = \omega(\varepsilon_s \mu_0)^{1/2}; s = 0, 1, 2)$ look like:

in case of magnetic oscillations

$$\begin{aligned}&\left[j_n(p)sj_{n-1}(s) - pj_{n-1}(p)j_n(s)\right]\cdot\left[h_n^{(2)}(q)ty_{n-1}(t) - qh_{n-1}^{(2)}(q)j_n(t)\right]\\&= \left[j_n(p)sy_{n-1}(s) - pj_{n-1}(p)y_n(s)\right]\cdot\left[h_n^{(2)}(q)tj_{n-1}(t) - qh_{n-1}^{(2)}(q)j_n(t)\right];\end{aligned} \tag{1.81}$$

in case of electrical oscillations:

$$\begin{aligned}&\left[p^2 j_n(p)\frac{d}{ds}\{sj_n(s)\} - s^2 j_n(s)\frac{d}{dp}\{pj_n(p)\}\right]\\&\cdot\left[q^2 h_n^{(2)}(q)\frac{d}{dt}\{ty_n(t)\} - t^2 y_n(t)\frac{d}{dq}\{qh_n^{(2)}(q)\}\right]\\&= \left[p^2 j_n(p)\frac{d}{ds}\{sy_n(s)\} - s^2 y_n(s)\frac{d}{dp}\{pj_n(p)\}\right]\\&\cdot\left[q^2 h_n^{(2)}(q)\frac{d}{dt}\{tj_n(t)\} - t^2 j_n(t)\frac{d}{dq}\{qh_n^{(2)}(q)\}\right].\end{aligned} \tag{1.82}$$

At $r_1 = 0$, or $r_1 = r_2$, Eqs. (1.81), (1.82) transform into characteristic equations of resonant oscillations of a homogeneous dielectric ball (1.7), (1.8), therefore, in the limiting cases $\iota = r_1/r_2 = 0, 1$, the frequency spectrum of a two-layer ball should coincide with the spectrum of a homogeneous sphere. At the same time, if $\iota \to 0$, then $t \to p_2$ (p_2—characteristic parameter of a homogeneous dielectric ball of radius r_2: $p_2 = k_2 r_2$); if $\iota \to 1$, then $p \to p_1 = k_1 r_1$. From here we find: $t_{\iota\to 1} \approx p/\iota\, n_{12}$, where $n_{12} = (\varepsilon_1/\varepsilon_2)^{1/2}$. Thus, depending on the ratio $\varepsilon_1/\varepsilon_2$, the characteristic curves $t(\iota)$ will decrease if $\varepsilon_1 > \varepsilon_2$ (Fig. 1.16b, c), or increase (Fig. 1.17b, c), if $\varepsilon_1 < \varepsilon_2$ with increasing ι.

An example of calculating the characteristic parameters when varying the relative sizes of the inhomogeneity to the size of the ball is shown in Fig. 1.16 for case $\varepsilon_{1r} > \varepsilon_{2r}$.

The presence of dielectric inhomogeneity in the resonator material disturbs mainly the electric component of the field. Thus, for electrical types of oscillations E_{1m1}, the

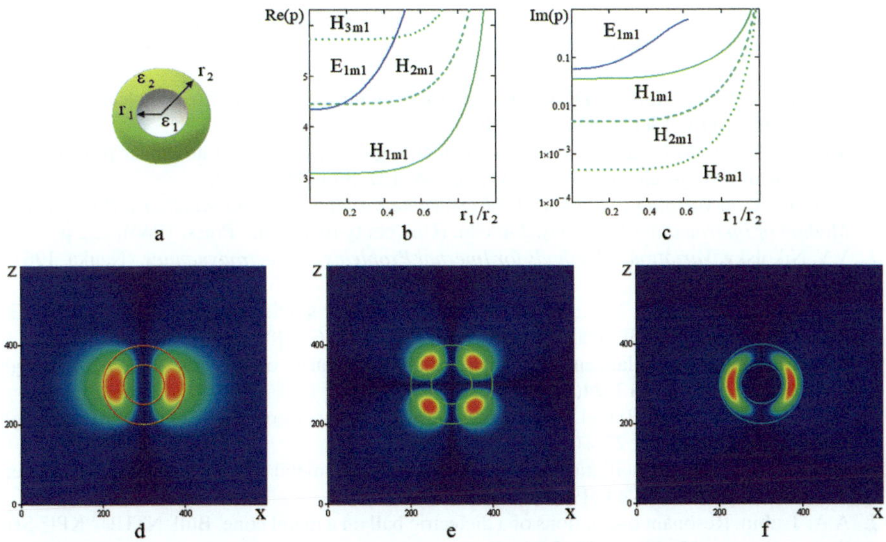

Fig. 1.17 Hollow spherical DR (**a**). Dependence of the real (**b**) and imaginary (**c**) parts of the characteristic parameter p on the relative dimensions of the cavity; $\varepsilon_{1r} = 1$; $\varepsilon_{2r} = 36$—(**b**, **c**). Distribution of the electric field modulus for oscillations H_{101} (**d**); H_{201} (**e**), and magnetic field modulus for oscillations E_{101} (**f**) ($r_1/r_2 = 0, 5$)

maximum of the electric field is located in the center of the ball, therefore the appearance of inhomogeneity of the dielectric in the center leads to a greater disturbance in the frequency of natural oscillations. For H_{1m1} oscillations, the maximum magnetic field is located in the center of the ball; the appearance of dielectric inhomogeneity practically does not disturb its field. It should be expected that such a perturbation will increase with increasing dielectric constant ratio, for example, when a cavity appears in the center of the DR (Fig. 1.17).

Solutions of characteristic Eqs. (1.81), (1.82) for hollow spherical DR: $\varepsilon_{1r} = 1$; $\varepsilon_{2r} = 81$ are shown in Fig. 1.17b, c.

The difference between this case and solutions with a finite permittivity of dielectric inhomogeneity is that the characteristic curves do not have limit points at $\iota = 1$. The ratio of the E_{1m1} oscillation frequency to the frequency of the fundamental oscillation H_{1m1} when $\iota > 0, 5$, exceeds an octave. Spatial distributions of the field of natural oscillations of the main types are shown in Fig. 1.16, Fig. 1.17d–f.

References

1. R. Mittra, S.W. in *Lee Analytical Techniques in the Theory of Guided Waves* (The Macmillan Company, 1971), 302 p
2. L.A. Veinshtein. Open resonators and open waveguides, in *Golem series in electromagnetics,* vol. 2 (Golem Press, 1969), 439 p

3. M. Abramowitz, I.A. Stegun (eds.), *Handbook of Mathematical Functions with Formulas, Graphs and Mathematical Tables* (National Bureau of Standards. Applied Mathematics Series 55, 1964), 830 p
4. M. Gastine, L. Courtois, L. Dormann, Electromagnetic resonances of free dielectric spheres. IEEE Trans. MTT. **15**(12), 694–700 (1967)
5. V.B. Braginsky, M.L. Gorodetsky, V.S. Ilchenko, Quality-factor and nonlinear properties of optical whispering-gallery modes. Phys. Lett. A, **137**, 393–397 (1989)
6. S.H. Gould, in *Variational Methods for Eigenvalue Problems: An Introduction to the Weinstein Method of Intermediate Problems*, 2nd edn. (University of Toronto Press. 1966), 292 p
7. V.V. Nikolsky, *Variational Methods for Internal Problems of Electrodynamics*. (Nauka, 1967), 460 p (in Russian)
8. Y. Kobayashi, T. Aoki, Y. Kabe, Influence of conductor shields on the Q-factors of a TE0 dielectric resonator. IEEE Trans. MTT. **33**(12), 1361–1366 (1985)
9. J. Van Bladel, M. Verplanken, The electric-dipol resonances of ring resonator of very high permittivity. Ibid. V. MTT **24**(2), 108–112 (1976)
10. M. Verplanken, J. Van Bladel, The magnetic-dipol resonances of ring resonators of very high permittivity. Ibid. MTT. **27**(4), 328–333 (1979)
11. A.A. Trubin, Natural oscillations of a dielectric ball on a metal wedge. Bull. NTUU KPI Ser. Radiotechnique. **23**, 41–44 (1986). (in Russian)
12. A.A. Trubin, Resonant oscillations of a dielectric ball on a metal cone. Bull. NTUU "KPI" Ser. Radiotechnique **22**, 40–43 (1985). (in Russian)
13. A.A. Trubin, Natural oscillations of a dielectric ball on a metal bicone. Bull. NTUU "KPI" Ser. Radiotechnique. **26**, 33–35 (1989). (in Russian)
14. E. Jahnke, F. Emde, *Tables of Functions with Formulas and Curves* (Dover Publications. New York, 1945), 382 p
15. Y.L. Luke, *Mathematical Functions and their Approximations* (Academic Press Inc. New York, San Francisco, London, 1975), 586 p
16. H. Bateman, A. Erdelyi, *Higher Transcendental Functions,* vol. 1 (Mc Graw-Hill Book Company Inc, 1953), 302 p
17. I.S. Gradshteyn, I.M. Ryzhik, *Table of Integrals, Series, and Products,* 7th edn (Academic Press, 2007), 1171 p
18. A.A. Trubin, On the influence of metallic inhomogeneity on the resonant oscillations of a dielectric ball. Bull. NTUU "KPI" Ser. Radiotechnique. **21**, 81–82 (1984). (in Russian)
19. A.A. Trubin, Resonant oscillations of open two-layer spherical structures, in *Abstract of the Report. Scientific and Technical Conference: Design and Application of Radio-Electronic Devices Based on Dielectric Waveguides and Resonators* (1983), pp. 32–33. (in Russian)

Chapter 2
Dielectric Resonators in Regular Metal Waveguides

It's convenient to express the amplitude-frequency characteristics of the reflection and transmission parameters of various waveguide structures with DR in terms of coupling coefficients. We define the coupling coefficient of the resonator with the microwave transmission line as the reciprocal of its radiation quality factor: $\tilde{k}_s = 1/Q_s^{\Sigma}$. Calculation of coupling coefficients determined by the s-th type of oscillations and the field of propagating waves for specific forms of DR and transmission line is an independent task.

This chapter calculates and analyzes the coupling coefficients of various shapes DRs—spherical, cylindrical, and rectangular with hollow metal waveguides of different cross sections. General analytical expressions for the expansion of the DR field into line waves, suitable for calculating the coupling coefficients of resonators of both basic and higher types of oscillations are obtained. The conditions for increasing coupling when varying the parameters of resonators and waveguides are studied.

2.1 Field of Natural Oscillations of a Dielectric Resonator in a Waveguide

Let us consider in general the natural oscillations of an isolated DR in a regular waveguide (Fig. 2.1). Let us denote by $(\mathbf{e}^1, \mathbf{h}^1)$ the field in the dielectric, and by $(\mathbf{e}^0, \mathbf{h}^0)$ the field outside the resonator. For simplicity, we'll assume here that only one type of the wave $(\mathbf{E}_l^{\pm}, \mathbf{H}_l^{\pm})$ (l—multi-index) can propagate in the waveguide in the positive (+ sign) or negative (−) direction relative to the z-axis. All components of the external field of the resonator must satisfy the radiation condition in the waveguide:

$$\mathbf{e}^0_{z\to\pm\infty} \approx c_l^{\pm}\mathbf{E}_l^{\pm}; \quad \mathbf{h}^0_{z\to\pm\infty} \approx c_l^{\pm}\mathbf{H}_l^{\pm}, \tag{2.1}$$

as well as continuity conditions

A. Trubin, *Introduction to the Theory of Dielectric Resonators*, Springer Series in Advanced Microelectronics 65, https://doi.org/10.1007/978-3-031-65396-4_2

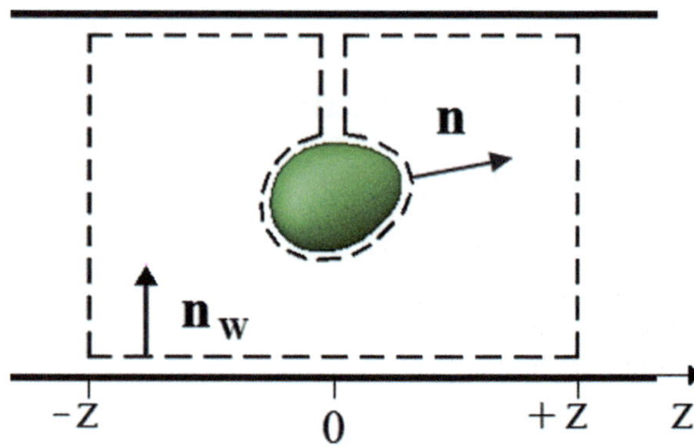

Fig. 2.1 Dielectric resonator in a transmission line

on the dielectric surface:

$$\left[\mathbf{e}^1 - \mathbf{e}^0, \mathbf{n}\right] = 0; \quad \left[\mathbf{h}^1 - \mathbf{h}^0, \mathbf{n}\right] = 0$$

and on waveguide surface Σ:

$$\left[\mathbf{e}^0, \mathbf{n}\right]_\Sigma = 0. \tag{2.2}$$

Here **n**—normal to the surface of the dielectric and waveguide, respectively.

If we know the field inside the dielectric, then in the general case, the field of the s-th oscillation of the DR in the region $|z| > d_z/2$ (d_z—is the maximum longitudinal size of the resonator along the z-axis) can be expanded into a series in terms of the natural magnetic and electric waves $\left(\mathbf{E}_t^{\pm}, \mathbf{H}_t^{\pm}\right)$ of the waveguide:

$$\mathbf{e}_s^0 = \sum_{t=1}^{\infty} c_s^{t\pm}\mathbf{E}_t^{\pm} + \sum_{t=1}^{\infty} d_s^{t\pm}\mathbf{E}_t^{\pm}; \quad \mathbf{h}_s^0 = \sum_{t=1}^{\infty} c_s^{t\pm}\mathbf{H}_t^{\pm} + \sum_{t=1}^{\infty} d_s^{t\pm}\mathbf{H}_t^{\pm} \tag{2.3}$$

Here s, t are multi-indices characterizing the type of DR oscillations and the type of waveguide wave.

The expansion coefficients $c_s^{t\pm}$, $d_s^{t\pm}$ in magnetic and electric waves, respectively, are determined by using the Lorentz lemma for the resonator field and the waveguide field, integrating over the volume limited by the dielectric surface s_R, the waveguide surface s_w and sections: $\pm z$ (Fig. 2.1). Using the orthogonality conditions over the cross section for natural waveguide waves (A.1), (2.1), and the associative property of the mixed product of vectors, we find [1]:

$$c_s^{t\pm} = -\frac{1}{2}\oint_{s_R} \left\{[\mathbf{e}_s^1, \mathbf{n}]\left(\mathbf{H}_t^{\pm}\right)^* + [\mathbf{n}, \mathbf{h}_s^1]\left(\mathbf{E}_t^{\pm}\right)^*\right\} ds \tag{2.4}$$

for $\omega > \omega_{cw}^t$ and

$$\begin{pmatrix} c_s^{t\pm} \\ d_s^{t\pm} \end{pmatrix} = (\mp)\frac{i}{2} \oint_{S_R} \left\{ [\mathbf{e}_s^1, \mathbf{n}]\left(\mathbf{H}_t^{\mp}\right)^* + [\mathbf{n}, \mathbf{h}_s^1]\left(\mathbf{E}_t^{\mp}\right)^* \right\} ds$$

for $\omega \le \omega_{cw}^t$.

Here ω_{cw}^t—critical frequency of the wave $\left(\mathbf{E}_t^{\pm}, \mathbf{H}_t^{\pm}\right)$; $c_s^{t\pm}$—expansion coefficients of the DR field into series of magnetic (− sign in parentheses); $d_s^{t\pm}$—electric (+ sign) waveguide waves; *—is the complex conjugation symbol.

For a rectangular DR, it's more convenient to calculate the expansion coefficients using volumetric integrals. In this case, we should move from (2.4) to the relations following from the Ostrogradsky-Gauss theorem:

$$c_s^{t\pm} = \frac{i}{2}\omega(\varepsilon_1 - \varepsilon_0) \int_{v_R} (\mathbf{e}_s^1, (\mathbf{E}_t^{\pm})^*) dv \tag{2.5}$$

for $\omega > \omega_{cw}^t$;

$$c_s^{t\pm} = (\mp)\frac{1}{2}\omega(\varepsilon_1 - \varepsilon_0) \int_{v_R} (\mathbf{e}_s^1, (\mathbf{E}_t^{\mp})^*) dv \tag{2.6}$$

for $\omega \le \omega_{cw}^t$.

Relations (2.4)–(2.6) are suitable for expansions (2.3) in the region $|z| > d_z/2$, where d_z—maximum size of DR in the direction of z axis. If this condition is violated, in the region $|z| \le d_z/2$, the expansion coefficients become functions of the coordinates.

If the center of the resonator is shifted (Fig. 2.1) by a distance z_0, and its transverse coordinates do not change, then, as follows from (2.4) to (2.5), for $\omega > \omega_{cw}^t$, the expansion coefficients take on a new form in the region $|z - z_0| > d_z/2$:

$$c_s^{t\pm} = (c_s^{t\pm})_0 e^{\pm i\Gamma z_0} \tag{2.7}$$

Here $(c_s^{t\pm})_0$ value of $c_s^{t\pm}$ at $z_0 = 0$.

The power emitted by the DR into the transmission line is calculated taking into account the accepted normalization (A.1):

$$P^{\Sigma} = \left|c^{\pm}\right|^2 \tag{2.8}$$

Energy stored in the resonator material:

$$w = 1/4 \int_{v_R} (\varepsilon_1 \left|\mathbf{e}^1\right|^2 + \mu_0 \left|\mathbf{h}^1\right|^2) dv; \tag{2.9}$$

here v_R,—is the DR volume. From where, according to the above definitions, we calculate the coupling coefficient of the DR with the waveguide on $\omega = \omega_0$:

$$\tilde{k} = \left|c^{\pm}\right|^2/(\omega_0 w). \tag{2.10}$$

here ω_0—is the resonant frequency.

2.2 Coupling Coefficients of Dielectric Resonators with a Rectangular Waveguide

Let us apply the relations obtained above to calculate the coupling coefficients of various types of DR with metal waveguides.

Let us first consider a dielectric ball of radius r_0, consisting of a material of relative dielectric constant $\varepsilon_{1r} = \varepsilon_1/\varepsilon_0$, placed in a rectangular metal waveguide with a cross-section $a \times b$ (Fig. 2.2a) [7]. We assume that the axes of the local rectangular coordinate system, at the center of the ball $\left(x', y', z'\right)$ (Fig. 1.1), are parallel to the coordinate axes of the waveguide (x, y, z).

To calculate the power emitted by the ball, we substitute the expressions for the field on its surface (1.2) or (1.3) as well as the waveguide field (A.2; A.3) into (2.4) and integrate the resulting expressions over the surface of the dielectric taking into account (D.1)–(D.5), as well as (D.11). The analytical expressions for the expansion coefficients (2.3) obtained as a result of algebraic transformations have the form:

in the case of magnetic oscillations of a dielectric ball H_{nml}:

$$c_{su}^{nml\pm} = 2\pi i^{n-m} w_0 h_{su}^0 \left[\frac{j_n(p)}{y_n(q)}\right] \cdot \frac{1-\gamma^2}{|\gamma|} \cdot \frac{dP_n^m(\gamma)}{d\gamma}$$
$$\cdot \begin{bmatrix} \sin\chi_{uy}y_0\left\{\sin\chi_{sx}x_0 \sin m\psi_0 \cos^2 m\frac{\pi}{2} - \cos\chi_{sx}x_0 \cos m\psi_0 \sin^2 m\frac{\pi}{2}\right\} \\ \cos\chi_{uy}y_0\left\{\cos\chi_{sx}x_0 \cos m\psi_0 \cos^2 m\frac{\pi}{2} - \sin\chi_{sx}x_0 \sin m\psi_0 \sin^2 m\frac{\pi}{2}\right\} \end{bmatrix}; \tag{2.11}$$

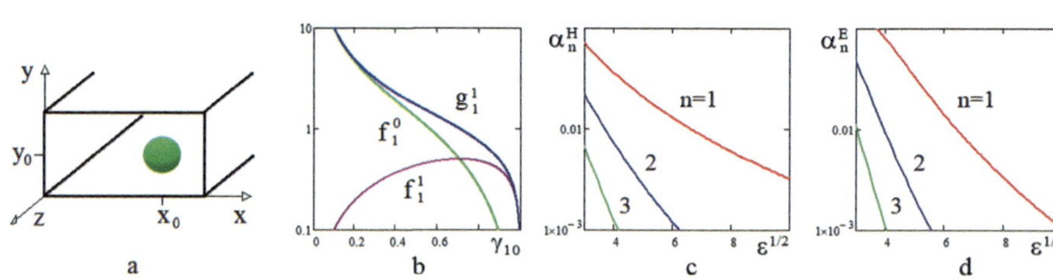

Fig. 2.2 Spherical DR in a regular rectangular waveguide (**a**). Dependences of the spherical DR coupling coefficient factors on the longitudinal wave number $\gamma_{10} = \Gamma_{10}/k_0$ (**b**); relative dielectric constant (**c, d**)

$$d_{su}^{nml\pm} = \pm 2\pi i^{n-m} e_{su}^{0} \cdot \begin{bmatrix} j_n(p) \\ y_n(q) \end{bmatrix} \cdot m \frac{P_n^m(\gamma)}{|\gamma|} \cdot \begin{bmatrix} \sin\chi_{nu}y_0\{\sin\chi_{sx}x_0\cos m\psi_0\cos^2 m\frac{\pi}{2} + \cos\chi_{sx}x_0\sin m\psi_0\sin^2 m\frac{\pi}{2}\} \\ -\cos\chi_{nu}y_0\{\sin\chi_{sx}x_0\cos m\psi_0\sin^2 m\frac{\pi}{2} + \cos\chi_{sx}x_0\sin m\psi_0\cos^2 m\frac{\pi}{2}\} \end{bmatrix}; \tag{2.12}$$

in case of electrical oscillations E_{nml}:

$$c_{su}^{nml\pm} = \pm 2\pi i^{n-m} h_{su}^{0} \varepsilon_{1r} \begin{bmatrix} j_n(p) \\ y_n(q) \end{bmatrix} \cdot m \frac{P_n^m(\gamma)}{|\gamma|} \cdot \begin{bmatrix} \cos\chi_{uy}y_0\{\cos\chi_{sx}x_0\cos m\psi_0\cos^2 m\frac{\pi}{2} - \sin\chi_{sx}x_0\sin m\psi_0\sin^2 m\frac{\pi}{2}\} \\ -\sin\chi_{uy}y_0\{\sin\chi_{sx}x_0\sin m\psi_0\cos^2 m\frac{\pi}{2} - \cos\chi_{sx}x_0\cos m\psi_0\sin^2 m\frac{\pi}{2}\} \end{bmatrix}; \tag{2.13}$$

$$d_{su}^{nml\pm} = 2\pi i^{n-m} \frac{e_{su}^{0}}{w_0} \varepsilon_{1r} \begin{bmatrix} j_n(p) \\ y_n(q) \end{bmatrix} \cdot \frac{1-\gamma^2}{|\gamma|} \cdot \frac{dP_n^m(\gamma)}{d\gamma} \cdot \begin{bmatrix} \cos\chi_{uy}y_0\{\sin\chi_{sx}x_0\cos m\psi_0\sin^2 m\frac{\pi}{2} + \cos\chi_{sx}x_0\sin m\psi_0\cos^2 m\frac{\pi}{2}\} \\ \sin\chi_{uy}y_0\{\sin\chi_{sx}x_0\cos m\psi_0\cos^2 m\frac{\pi}{2} + \cos\chi_{sx}x_0\sin m\psi_0\sin^2 m\frac{\pi}{2}\} \end{bmatrix}. \tag{2.14}$$

Here also c_{su}^{nml}—expansion coefficients of the DR field in magnetic and d_{su}^{nml}—in electric waves of a rectangular waveguide (2.4); h_{su}^{0}, e_{su}^{0}—normalizing factors (A.1); $\chi_{sx} = s\pi/a$; $\chi_{uy} = u\pi/b$; $\Gamma = \sqrt{k_0^2 - \chi_{sx}^2 - \chi_{uy}^2}$; $\gamma = \pm\Gamma/k_0$; $\sin\psi_0 = \chi_{sx}/\chi$; $\chi = (\chi_{sx}^2 + \chi_{uy}^2)^{1/2}$; $w_0 = (\mu_0/\varepsilon_0)^{1/2}$; $k_0 = \omega(\varepsilon_0\mu_0)^{1/2}$.

In the particular case of coupling the resonator with the main wave H_{10} of a rectangular waveguide $\psi_0 = \pi/2$; relations (2.11)–(2.14) are transformed to a simpler form:

For magnetic oscillations H_{nml}:

$$c_{10}^{nml\pm} = 2\pi i^{n-m} w_0 h_{10}^{0} \begin{bmatrix} j_n(p) \\ y_n(q) \end{bmatrix} \frac{1-\gamma^2}{|\gamma|} \cdot \frac{dP_n^m(\gamma)}{d\gamma} \begin{bmatrix} 0 \\ \cos(\chi_{1x}x_0 + m\frac{\pi}{2}) \end{bmatrix}; \tag{2.15}$$

For electrical oscillations E_{nml}:

$$c_{10}^{nml\pm} = \pm 2\pi i^{n-m} h_{10}^{0} \varepsilon_{1r} \begin{bmatrix} j_n(p) \\ y_n(q) \end{bmatrix} m \frac{P_n^m(\gamma)}{|\gamma|} \begin{bmatrix} \cos(\chi_{1x}x_0 + m\frac{\pi}{2}) \\ 0 \end{bmatrix}. \tag{2.16}$$

The expressions in square brackets correspond to the azimuthal distribution of the natural oscillation field of the spherical resonator (1.2) and (1.3).

We calculate the stored energy in the ball material, neglecting the influence of the walls of the waveguide:

$$w = \frac{\pi r_0}{2}\begin{pmatrix}\mu_0|h_1|^2\\ \varepsilon_1|e_1|^2\end{pmatrix}(1+\delta_{m0})\frac{n(n+1)}{2n+1}\cdot\frac{(n+m)!}{(n-m)!}$$
$$\cdot\left\{[p^2 - n(n+1)]j_n^2(p) + [nj_n(p) - pj_{n-1}(p)]^2\right\}. \tag{2.17}$$

Here, respectively, for H_{nml} (upper expression in parentheses); for E_{nml} oscillations (lower).

Using (2.10), (2.17) we find for any non-degenerate oscillations:
of magnetic types H_{nml}

$$\tilde{k} = \frac{16}{(1+\delta_{m0})}\cdot\frac{2n+1}{n(n+1)}\cdot\frac{(n-m)!}{(n+m)!}\cdot\frac{1}{k_0 b}\cdot\frac{|\gamma|}{\sqrt{1-\gamma^2}}$$
$$\cdot\left[\frac{(1-\gamma^2)}{|\gamma|}\cdot\frac{dP_n^m(\gamma)}{d\gamma}\right]^2\cdot\alpha_n^H(p,q)\cdot\begin{bmatrix}0\\ \cos^2(\chi_{1x}x_0 + m\pi/2)\end{bmatrix}; \tag{2.18}$$

electrical types E_{nml}

$$\tilde{k} = \frac{16}{(1+\delta_{m0})}\cdot\frac{2n+1}{n(n+1)}\cdot m^2\cdot\frac{(n-m)!}{(n+m)!}\cdot\frac{1}{k_0 b}\cdot\frac{|\gamma|}{\sqrt{1-\gamma^2}}$$
$$\cdot\left[\frac{P_n^m(\gamma)}{|\gamma|}\right]^2\cdot\alpha_n^E(p,q)\cdot\begin{bmatrix}\cos^2(\chi_{1x}x_0 + m\pi/2)\\ 0\end{bmatrix}. \tag{2.19}$$

The dependencies that determine the relationship with the parameters of the dielectric ball have the form:

$$\alpha_n^H(p,q) = \frac{1}{q}\left|\frac{j_n(p)}{y_n(q)}\right|^2 / \left\{[p^2 - n(n+1)]j_n^2(p) + \left[nj_n(p) - pj_{n-1}(p)\right]^2\right\};$$
$$\alpha_n^E(p,q) = \frac{\varepsilon_{1r}}{q}\left|\frac{j_n(p)}{y_n(q)}\right|^2 / \left\{[p^2 - n(n+1)]j_n^2(p) + \left[nj_n(p) - pj_{n-1}(p)\right]^2\right\}. \tag{2.20}$$

To simplify calculations, we replaced the spherical Hankel functions of the second kind $h_n^{(2)}(q)$ with Neumann functions $y_n(q)$, as well as the characteristic parameters with their real parts: $\mathrm{Re}(p)$ and $\mathrm{Re}(q)$. As follows from (1.5), (1.6), functions (2.20) are proportional to the square of the ratio of the field amplitudes outside and inside the DR. The dependences of the coupling functions on the dielectric constant are shown in Fig. 2.2c, d.

The calculations show that an increase in the radial index l leads to an increase coupling coefficients, and an increase in the meridional index n reduces the coupling coefficients of the resonator (Fig. 2.2c, d). This is simply explained by the fact that at $l \gg n$, the distribution of the field of natural oscillations inside the dielectric in the radial direction has the character of a standing wave (see Fig. 1.5). In this case, the ray trajectories of partial waves are directed relative to the interface between the media at angles close to the normal (oscillations of the "bouncing ball" type). The reflection coefficients of such waves are small, so the field stored by the DR decreases, and the coupling coefficients increase. In case of $n \gg l$, partial waves in the resonator material are incident on the surface of the dielectric at angles close to $\pi/2$ with respect to the normal (oscillations of the "whispering gallery" type (see Fig. 1.4)). Their reflection coefficients are high, and their coupling coefficients decrease.

In general, from the data presented, a practically important conclusion follows that an increase in the relative dielectric constant of the resonator material leads to a rapid decrease in its coupling coefficients and a rarefaction of the spectrum of higher-type oscillations, mainly due to a decrease in the coupling of "whispering gallery" oscillations.

The presence of a waveguide violates the spherical symmetry of the structure, which leads to the appearance of a dependence of the coupling on the azimuthal indices m. As follows from (2.18), (2.19), the coupling coefficients of oscillations characterized by even values $m = 0, 2, 4, \ldots$ take maximum values at the walls of the waveguide; oscillations with odd azimuthal indices $m = 1, 3, \ldots$ are characterized by a coupling maximum in its plane of symmetry $x_0 = a/2$.

The dependence of the coupling coefficients on the parameters of the waveguide is determined by the factors:

$$f_n^m(\gamma) = \frac{|\gamma|}{\sqrt{1-\gamma^2}}\left[\frac{1-\gamma^2}{\gamma}\cdot\frac{dP_n^m(\gamma)}{d\gamma}\right]^2; \tag{2.21}$$

for magnetic and

$$g_n^m(\gamma) = \frac{|\gamma|}{\sqrt{1-\gamma^2}}\left[\frac{P_n^m(\gamma)}{\gamma}\right]^2 \tag{2.22}$$

for electrical oscillations of the boll. As can be seen from the data in Fig. 2.2b, coupling coefficients increase with $\omega \to \omega_{cw}^{10}$ for H_{101}, E_{111} oscillation, where ω_{cw}^{10}— critical frequency of the H_{10} wave. For H_{111} oscillations, the maximum coupling is achieved at $\gamma = 1/\sqrt{2}$, that is, when $\omega = \sqrt{2}\cdot\omega_{cw}^{10}$.

The coupling coefficients of the cylindrical DR with the H_{10} wave were calculated for three fixed positions of the dielectric cylinder relative to the waveguide axis (Fig. 2.3a–c), called below a, b, c, respectively. Neglecting the influence of the metal walls of the waveguide on the distribution of the natural oscillation field in the DR material, we again use relations (2.4), as well as (1.17), (A.2), (A.3). Let's integrate them over the surface of the cylinder, applying (D.1)–(D.8) for each of the standard positions of the DR in a rectangular waveguide.

We present the expansion coefficients in the field of the main wave H_{10} obtained as a result of transformations in the form:

for a—position of the DR (Fig. 2.3a);

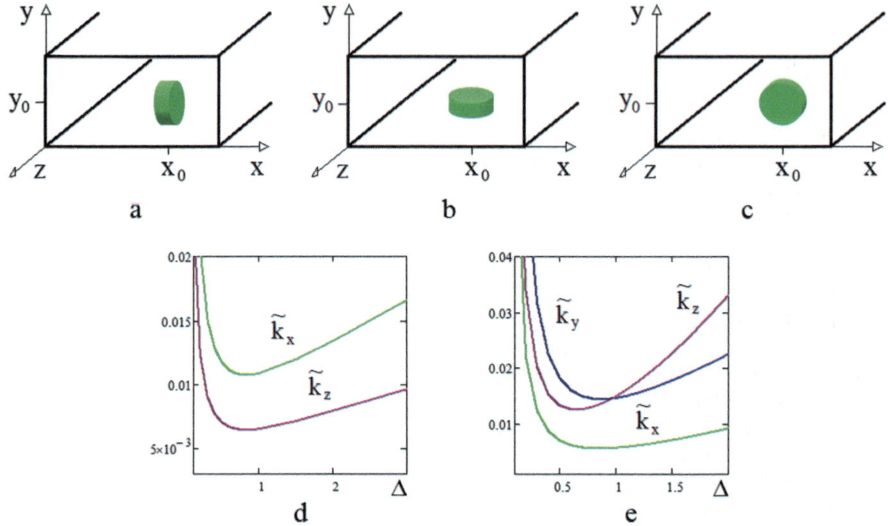

Fig. 2.3 Standard positions of a cylindrical DR in a regular rectangular waveguide (**a**, **b**, **c**). Dependences of coupling coefficients on the relative sizes of a cylindrical DR $\Delta = L/2r_0$ for main magnetic oscillation H_{101}—(**d**); EH_{111}^{+}—(**e**) and $f_0 = 7$ GHz; $a \times b = 35 \times 15$ mm; $\varepsilon_{1r} = 81$; for (**a**, **b** positions) $x_0 = a/2$; for (**c** position) $x_0 = 0$; $y_0 = b/2$ ($\tilde{k}_x$—corresponds to the position of the DR (**a**); $\tilde{k}_y$—(**b**); $\tilde{k}_z$—(**c**))

$$c_{10}^{nml\pm} = 2\pi r_0 \frac{\chi_{1x}}{\beta|\Gamma|} h_{10}^0 e^{\pm i(m-1)\frac{\pi}{2}} \frac{(k_1^2 - k_0^2)}{(\beta_z^2 - \chi_{1x}^2)} \cdot \begin{Bmatrix} 1 \\ 0 \end{Bmatrix}$$
$$\cdot \left[h_1 \frac{i\omega\mu_0}{\beta^2 - \Gamma^2} \{\beta J_m(p_\perp) J_m'(\Gamma r_0) - \Gamma J_m'(p_\perp) J_m(\Gamma r_0)\} + e_1 m \frac{J_m(p_\perp) J_m(\Gamma r_0)}{p_\perp \Gamma} \beta_z \right]$$
$$\cdot \begin{bmatrix} \sin \chi_{1x} x_0 \{\beta_z \sin p_z \cos \chi_{1x} L/2 - \chi_{1x} \cos p_z \sin \chi_{1x} L/2\} \\ \cos \chi_{1x} x_0 \{\beta_z \cos p_z \sin \chi_{1x} L/2 - \chi_{1x} \sin p_z \cos \chi_{1x} L/2\} \end{bmatrix}; \tag{2.23}$$

for b—position (Fig. 2.3b):

$$c_{10}^{nml\pm} = 2\pi r_0 \frac{\chi_{1y}}{\beta_z|\Gamma|} h_{10}^0 \frac{(k_1^2 - k_0^2)}{(\beta^2 - k_0^2)}$$
$$\cdot \begin{bmatrix} \cos \chi_{1x} x_0 \{\cos m\psi_0 \sin^2 m\frac{\pi}{2} \pm i \sin m\psi_0 \cos^2 m\frac{\pi}{2}\} \\ \sin \chi_{1x} x_0 \{\cos m\psi_0 \cos^2 m\frac{\pi}{2} \pm i \cos m\psi_0 \sin^2 m\frac{\pi}{2}\} \end{bmatrix}$$
$$\cdot e_1 \left[\beta J_m'(p_\perp) J_m(k_0 r_0) - k_0 J_m(p_\perp) J_m'(k_0 r_0)\right] \begin{bmatrix} 0 \\ \sin p_z \end{bmatrix}, \tag{2.24}$$

where $\sin \psi_0 = \Gamma/k_0$;

for c—position (Fig. 2.3c):

$$c_{10}^{nml\pm} = 2\pi r_0 \frac{\chi_{1x}}{\beta|\Gamma|} h_{10}^0 \frac{\left(k_1^2 - k_0^2\right)}{\left(\beta_z^2 - \Gamma^2\right)} \left\{ \begin{matrix} \cos\left(\chi_{1x} x_0 + m\frac{\pi}{2}\right) \\ 0 \end{matrix} \right\}$$
$$\cdot \left[\frac{i\omega\mu_0}{\beta^2 - \chi_{1x}^2} h_1 \{\beta J_m(p_\perp) J_m'(\chi_{1x} r_0) - \chi_{1x} J_m'(p_\perp) J_m(\chi_{1x} r_0)\} + e_1 m \frac{J_m(p_\perp) J_m(\chi_{1x} r_0)}{\chi_{1x} p_\perp} \beta_z \right]$$
$$\cdot \left[\begin{matrix} \Gamma \cos p_z \sin \Gamma L/2 - \beta_z \sin p_z \cos \Gamma L/2 \\ \pm i(\Gamma \sin p_z \cos \Gamma L/2 - \beta_z \cos p_z \sin \Gamma L/2) \end{matrix} \right]. \qquad (2.25)$$

Here $J_m'(x) = d/dx J_m(x)$—derivatives of Bessel functions [2].

Let us calculate the expressions for the electromagnetic energy stored in the dielectric using the approximation $w_{nml} \approx 2w_{nml}^E$ where w_{nml}^E—energy of the electric component of the field stored in the resonator:

$$w_{nml} = \frac{\pi r_0^2 \varepsilon_1}{8\beta_z}(1 + \delta_{m0}) \cdot \left\{ \frac{1}{2}\{(|e_1|\frac{\beta_z}{\beta} - \frac{\omega\mu_0}{\beta}|h_1|)^2 [J_{m-1}^2(p_\perp) - J_m(p_\perp) J_{m-2}(p_\perp)] \right.$$
$$+ (|e_1|\frac{\beta_z}{\beta} + \frac{\omega\mu_0}{\beta}|h_1|)^2 [J_{m+1}^2(p_\perp) - J_m(p_\perp) J_{m+2}(p_\perp)]\} \cdot \left[\begin{matrix} 2p_z + \sin(2p_z) \\ 2p_z - \sin(2p_z) \end{matrix} \right]$$
$$\left. + |e_1|^2 [J_m^2(p_\perp) - J_{m-1}(p_\perp) J_{m+1}(p_\perp)] \cdot \left[\begin{matrix} 2p_z - \sin(2p_z) \\ 2p_z + \sin(2p_z) \end{matrix} \right] \right\}. \qquad (2.26)$$

As a result, the coupling coefficients of a cylindrical DR with a rectangular waveguide obtained on the basis of (2.23)–(2.26) and (2.10) take the form:

In a position:

$$\tilde{k} = \frac{64\pi}{(1 + \delta_{m0})} \cdot \frac{(\varepsilon_{1r} - 1)^2}{\varepsilon_{1r}} \cdot \frac{p_z q_z^4}{(k_0^2 ab) p_\perp^2 s_\perp \Delta^3} \cdot \frac{1}{(p_z^2 - s_z^2)^2} \left\{ \begin{matrix} 1 \\ 0 \end{matrix} \right\}$$
$$\cdot \left\{ e_0 m \frac{J_m(p_\perp) J_m(s_\perp)}{s_\perp} - h_0 \frac{p_\perp}{p_\perp^2 - s_\perp^2} [p_\perp J_m(p_\perp) J_m'(s_\perp) - s_\perp J_m'(p_\perp) J_m(s_\perp)] \right\}^2$$
$$\cdot \left[\begin{matrix} \sin \chi_{1x} x_0 (p_z \sin p_z \cos s_z - s_z \cos p_z \sin s_z) \\ \cos \chi_{1x} x_0 (p_z \cos p_z \sin s_z - s_z \sin p_z \cos s_z) \end{matrix} \right]^2 / v_m(p_\perp, p_z); \qquad (2.27)$$

in b position:

$$\tilde{k} = \frac{64\pi}{(1 + \delta_{m0})} \cdot \frac{(\varepsilon_{1r} - 1)^2}{\varepsilon_{1r}} \cdot \frac{q_\perp^4 \Delta}{(k_0 ab) p_z s_\perp (p_\perp^2 - q_\perp^2)^2} \left(\Delta \frac{p_\perp}{p_z} e_0 \right)^2$$
$$\cdot \left[p_{1\perp} J_m'(p_\perp) J_m(q_\perp) - q_\perp J_m(p_\perp) J_m'(q_\perp) \right]^2$$
$$\cdot \left[\begin{matrix} 0 \\ \sin^2 p_z \end{matrix} \right] \cdot \left\{ \begin{matrix} \cos^2 \chi_{1x} x_0 \sin^2 m\left(\psi_0 + \frac{\pi}{2}\right) \\ \sin^2 \chi_{1x} x_0 \cos^2 m\left(\psi_0 + \frac{\pi}{2}\right) \end{matrix} \right\} / v_m(p_\perp, p_z); \qquad (2.28)$$

in c position:

$$\tilde{k} = \frac{64\pi}{(1+\delta_{m0})} \cdot \frac{(\varepsilon_{1r}-1)^2}{\varepsilon_{1r}} \cdot \frac{p_z q_z^4}{(k_0^2 ab)\Delta^2 p_\perp^2 t_z} \cdot \frac{1}{(p_z^2 - t_z^2)^2}$$
$$\cdot \left\{ e_0 m \frac{J_m(p_\perp) J_m(t_\perp)}{t_\perp} - h_0 \frac{p_\perp}{p_\perp^2 - t_\perp^2} \left[p_\perp J_m(p_\perp) J_m'(t_\perp) - t_\perp J_m'(p_\perp) J_m(t_\perp) \right] \right\}^2$$
$$\cdot \begin{bmatrix} p_z \sin p_z \cos t_z - t_z \cos p_z \sin t_z \\ p_z \cos p_z \sin t_z - t_z \sin p_z \cos t_z \end{bmatrix}^2 \cdot \begin{Bmatrix} \cos^2\left(\chi_{1x} x_0 + m\frac{\pi}{2}\right) \\ 0 \end{Bmatrix} / \nu_m(p_\perp, p_z). \tag{2.29}$$

Here $\Delta = L/2r_0$; $q_\perp = k_0 r_0$; $q_z = k_0 \mathrm{L}/2$; $s_\perp = \Gamma r_0$; $sin\, \psi_0 = s_\perp / q_\perp$; $t_\perp = \chi_{1x} r_0$; $t_z = \Gamma L/2$;

$$\nu_m(p_\perp, p_z) = \left\{ (e_0 - h_0)^2 \left[J_{m-1}^2(p_\perp) - J_m(p_\perp) J_{m-2}(p_\perp) \right] \right.$$
$$\left. + (e_0 + h_0)^2 \left[J_{m+1}^2(p_\perp) - J_m(p_\perp) J_{m+2}(p_\perp) \right] \right\} \begin{bmatrix} p_z + \sin p_z \cos p_z \\ p_z - \sin p_z \cos p_z \end{bmatrix}$$
$$+ 2\left(\Delta \frac{p_\perp}{p_z} e_0 \right)^2 \left[J_m^2(p_\perp) - J_{m-1}(p_\perp) J_{m+1}(p_\perp) \right] \begin{bmatrix} p_z - \sin p_z \cos p_z \\ p_z + \sin p_z \cos p_z \end{bmatrix}. \tag{2.30}$$

It is convenient to normalize the amplitudes of the resonator's natural oscillations in such a way that:

$$e_0 - h_0 = \frac{J_{m+1}(p_\perp)}{p_\perp J_m(p_\perp)} + \frac{K_{m+1}(p_{0\perp})}{p_{0\perp} K_m(p_{0\perp})};$$
$$e_0 + h_0 = \frac{J_{m-1}(p_\perp)}{p_\perp J_m(p_\perp)} - \frac{K_{m-1}(p_{0\perp})}{p_{0\perp} K_m(p_{0\perp})}. \tag{2.31}$$

The factors of the coupling coefficients in curly brackets (2.27)–(2.29) exactly correspond to the parity of the distribution of the field of natural oscillations of the DR adopted in (1.17).

From the obtained expressions it follows that H_{10} wave is coupled with natural oscillations characterized by the maximum of z-component of the magnetic field directed along the wide wall of the waveguide in the DR center, in positions a, c. In position b, the maximum coupling is achieved for oscillations with a maximum z-component of the electric field at the center of the resonator.

In special cases of azimuthally homogeneous oscillations ($m = 0$), formulas (2.27)–(2.29) are simplified and take the form:

for a DR position:

and for $H_{nol}^{\pm}$

$$\tilde{k} = 16\pi\frac{(\varepsilon_{1r}-1)^2}{\varepsilon_{1r}}\cdot\frac{p_z q_z^4}{(k_0^2 ab)s_\perp\Delta^3}\cdot\frac{1}{(p_\perp^2-s_\perp^2)^2(p_z^2-s_z^2)^2}$$
$$\cdot\begin{Bmatrix}1\\0\end{Bmatrix}\cdot\frac{\left[p_\perp J_0(p_\perp)J_1(s_\perp)-s_\perp J_1(p_\perp)J_0(s_\perp)\right]^2}{J_1^2(p_\perp)-J_0(p_\perp)J_2(p_\perp)}$$
$$\cdot\begin{bmatrix}\sin^2\chi_{1x}x_0(p_z\sin p_z\cos s_z - s_z\cos p_z\sin s_z)^2/(p_z+\sin p_z\cos p_z)\\ \cos^2\chi_{1x}x_0(p_z\cos p_z\sin s_z - s_z\sin p_z\cos s_z)^2/(p_z-\sin p_z\cos p_z)\end{bmatrix}; \quad (2.32)$$

for $E_{nol}^{\mp}$ $\tilde{k}=0$;for b DR position:
and for $H_{nol}^{\pm}$ $\tilde{k}=0$; for $E_{nol}^{\mp}$:

$$\tilde{k} = 16\pi\frac{(\varepsilon_{1r}-1)^2}{\varepsilon_{1r}}\cdot\frac{p_\perp^2 q_\perp^4\Delta^3}{(k_0^2 ab)p_z^3 s_\perp}\cdot\frac{1}{(p_\perp^2-q_\perp^2)^2}$$
$$\cdot\left[p_\perp J_1(p_\perp)J_0(q_\perp)-q_\perp J_0(p_\perp)J_1(q_\perp)\right]^2\begin{bmatrix}0\\ sin^2 p_z\end{bmatrix}\begin{Bmatrix}0\\ sin^2\chi_{1x}x_0\end{Bmatrix}$$
$$/\left\{\left[J_1^2(p_\perp)-J_0(p_\perp)J_2(p_\perp)\right]\begin{bmatrix}p_z+\sin p_z\cos p_z\\ p_z-\sin p_z\cos p_z\end{bmatrix}\right.$$
$$\left.+\left(\Delta\frac{p_\perp}{p_z}\right)^2\left[J_0^2(p_\perp)+J_1^2(p_\perp)\right]\begin{bmatrix}p_z-\sin p_z\cos p_z\\ p_z+\sin p_z\cos p_z\end{bmatrix}\right\}; \quad (2.33)$$

For c DR position
and for $H_{nol}^{\pm}$:

$$\tilde{k} = 16\pi\frac{(\varepsilon_{1r}-1)^2}{\varepsilon_{1r}}\cdot\frac{p_z q_z^4}{(k_0^2 ab)\Delta^2 t_z}\cdot\frac{1}{(p_\perp^2-t_\perp^2)^2(p_z^2-t_z^2)^2}$$
$$\cdot\begin{Bmatrix}\cos^2\chi_{1x}x_0\\ 0\end{Bmatrix}\cdot\frac{\left[p_\perp J_0(p_\perp)J_1(t_\perp)-t_\perp J_0(t_\perp)J_1(p_\perp)\right]^2}{J_1^2(p_\perp)-J_0(p_\perp)J_2(p_\perp)}$$
$$\cdot\begin{bmatrix}(p_z\sin p_z\cos t_z - t_z\cos p_z\sin t_z)^2/(p_z+\sin p_z\cos p_z)\\ (p_z\cos p_z\sin t_z - t_z\sin p_z\cos t_z)^2/(p_z-\sin p_z\cos p_z)\end{bmatrix}; \quad (2.34)$$

for $E_{nol}^{\mp}$: $\tilde{k}=0$.

From the obtained relations it follows that magnetic azimuthally homogeneous oscillations are coupled with the H_{10} wave of a rectangular waveguide only in the DR positions a, c. In this case, even-numbered in z-axes H_{nol}^{+} oscillations are characterized by maximum coupling in the center of the rectangular waveguide, and odd-numbered H_{nol}^{-} oscillations take on maximum values near the plane of the narrow walls. In position c, even magnetic oscillations take on maximum values near the plane of the side walls (Fig. 2.3d). The coupling coefficients of even-numbered in z-axes E_{101}^{+} electrical oscillations take maximum values at the center of the waveguide only in

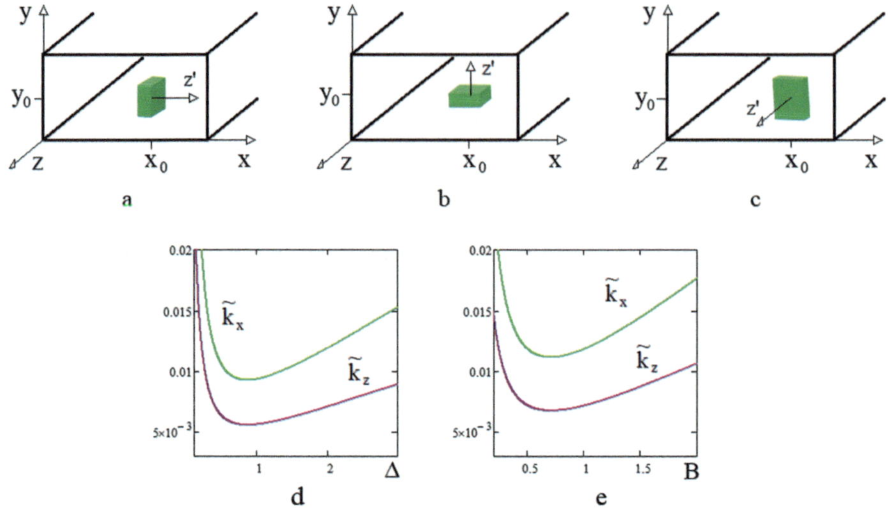

Fig. 2.4 Standard positions of a rectangular DR in a waveguide (**a**, **b**, **c**). Dependences of the coupling coefficients of a rectangular DR on the relative sizes for the main magnetic oscillations H_{111} on $\Delta = L/a_0$, for $b_0/a_0 = 1$—(**d**); on $B = b_0/a_0$ for $\Delta = 0,4$—(**e**) and $f_0 = 7\,\text{GHz}$; $a \times b = 35 \times 15\,\text{mm}$; $\varepsilon_{1r} = 81$; for (**a**, **b**) $x_0 = a/2$; for (**c**) $x_0 = 0$; $y_0 = b/2$ ($\tilde{k}_x$—corresponds to the orientation of the DR z' axis (a); $\tilde{k}_z$—(**c**))

position b. Odd in z-axes E^-_{101} electrical oscillations of a cylindrical DR, as follows from (2.33)–(2.34), are not coupled with the H_{10} wave of a rectangular waveguide in any position. Magnetic azimuthally inhomogeneous oscillations H^+_{n1l} are coupled with the H_{10} waveguide wave for all three positions of the DR axis (Fig. 2.3e). In all considered cases, the coupling coefficients take minimum values for oblate resonator shapes: in the region: $0,5 < \Delta < 1$ (Fig. 2.3d, e).

We will also calculate the coupling coefficients of the rectangular DR for three fixed positions of the resonator relative to the waveguide axis (Fig. 2.4a–c). Let us assume that the field of natural oscillations of the magnetic type in the local coordinate system (x', y', z') associated with the center of the DR has the form (1.32). Then the expansion coefficients obtained using (2.5) can be expressed in general form:

for the position shown in Fig. 2.4a:

$$c^{\pm}_{su} = 4c_0\left\{E^*_{y0}\beta_y \cdot \omega_x(\eta_{uy}) \cdot \overline{\omega}_y(\pm\gamma) + iE^*_{z0}\beta_x \cdot \overline{\omega}_x(\eta_{uy}) \cdot \omega_y(\pm\gamma)\right\} \cdot \omega_z(\xi_{sx}) \cdot \begin{pmatrix} \cos\chi_{sx}x_0 \\ i\sin\chi_{sx}x_0 \end{pmatrix} \cdot \begin{bmatrix} i\sin\chi_{uy}y_0 \\ \cos\chi_{uy}y_0 \end{bmatrix} \cdot e^{\pm i\Gamma z_0}; \qquad (2.35)$$

for the position shown in Fig. 2.4b:

$$c_{su}^{\pm} = 4c_0\left\{E_{x0}^{*}\beta_y \cdot \omega_x(\xi_{sx}) \cdot \overline{\omega}_y(\pm\gamma) - iE_{z0}^{*}\beta_x \cdot \overline{\omega}_x(\xi_{sx}) \cdot \omega_y(\pm\gamma)\right\} \cdot \omega_z(\eta_{uy}) \cdot \begin{pmatrix} i\sin\chi_{sx}x_0 \\ \cos\chi_{sx}x_0 \end{pmatrix} \cdot \begin{bmatrix} \cos\chi_{uy}y_0 \\ i\sin\chi_{uy}y_0 \end{bmatrix} \cdot e^{\pm i\Gamma z_0}; \quad (2.36)$$

for the position shown in Fig. 2.4c:

$$c_{su}^{\pm} = 4c_0\left\{E_{x0}^{*}\beta_y \cdot \omega_x(\xi_{sx}) \cdot \overline{\omega}_y(\eta_{uy}) - iE_{z0}^{*}\beta_x \cdot \overline{\omega}_x(\xi_{sx}) \cdot \omega_y(\eta_{uy})\right\} \cdot \omega_z(\pm\gamma) \cdot \begin{pmatrix} i\sin\chi_{sx}x_0 \\ \cos\chi_{sx}x_0 \end{pmatrix} \cdot \begin{bmatrix} i\sin\chi_{uy}y_0 \\ \cos\chi_{uy}y_0 \end{bmatrix} \cdot e^{\pm i\Gamma z_0}. \quad (2.37)$$

Here and below we denote

for H-waves of the rectangular waveguide:

$$E_{0x} = ih_{su}^{0}\left|w_{su}^{H}\right|\chi_{uy}; \quad E_{0y} = -ih_{su}^{0}\left|w_{su}^{H}\right|\chi_{sx}; \quad E_{0z} = 0;$$

for E-waves of the rectangular waveguide:

$$E_{0x} = -ie_{su}^{0}\chi_{sx}\Gamma/|\Gamma|; \quad E_{0y} = -ie_{su}^{0}\chi_{uy}\Gamma/|\Gamma|; \quad E_{0z} = e_{su}^{0}\chi^{2}/|\Gamma|;$$

$$\omega_\nu(\xi) = \frac{1}{p_\nu^2 - (q_\nu\xi)^2} \cdot \begin{pmatrix} -i[p_\nu\cos p_\nu\sin(q_\nu\xi) - q_\nu\xi\sin p_\nu\cos(q_\nu\xi)] \\ p_\nu\sin p_\nu\cos(q_\nu\xi) - q_\nu\xi\cos p_\nu\sin(q_\nu\xi) \end{pmatrix};$$

$$\varpi_\nu(\xi) = \frac{1}{p_\nu^2 - (q_\nu\xi)^2} \cdot \begin{pmatrix} p_\nu\sin p_\nu\cos(q_\nu\xi) - q_\nu\xi\cos p_\nu\sin(q_\nu\xi) \\ i[p_\nu\cos p_\nu\sin(q_\nu\xi) - q_\nu\xi\sin p_\nu\cos(q_\nu\xi)] \end{pmatrix}; \quad (2.38)$$

$\nu = (x, y, z)$; where $q_x = k_0a_0/2$; $q_y = k_0b_0/2$; $q_z = k_0L/2$; $\chi_{sx} = s\pi/a$; $\chi_{uy} = u\pi/b$; $\Gamma = \sqrt{k_0^2 - \chi_{sx}^2 - \chi_{uy}^2}$.

Energy (2.9) stored in the material of a rectangular DR for magnetic oscillations (1.32):

$$w = \frac{\mu_0}{4} \cdot \frac{|h_1|^2}{\beta_x\beta_y\beta_z} \cdot \frac{1}{(k_1^2 - \beta_z^2)} \cdot \upsilon H, \quad (2.39)$$

where

$$\upsilon H = \varepsilon_{1r}k_0^2[\beta_y^2\pi_x^{\mp}\pi_y^{\pm}\pi_z^{\mp} + \beta_x^2\pi_x^{\pm}\pi_y^{\mp}\pi_z^{\mp}] + \beta_x^2\beta_z^2 \cdot \pi_x^{\pm}\pi_y^{\mp}\pi_z^{\pm} + \beta_y^2\beta_z^2 \cdot \pi_x^{\mp}\pi_y^{\pm}\pi_z^{\pm} + (\beta_x^2 + \beta_y^2)^2 \cdot \pi_x^{\mp}\pi_y^{\mp}\pi_z^{\mp};$$

here $\pi_\nu^{\pm} = p_\nu \pm \sin p_\nu \cos p_\nu$.

Substituting (2.35)–(2.39) into (2.9), we find for the H_{10} waveguide wave and magnetic oscillations H_{nml} of a rectangular DR:

In the position shown in Fig. 2.4a:

$$\tilde{k}_x = \tilde{k}_0 \beta_x^2 \left| \overline{\omega}_x(\gamma) \omega_y(0) \omega_z(\xi_{1x}) \right|^2 \begin{bmatrix} 0 \\ 1 \end{bmatrix} \begin{Bmatrix} \cos^2 \chi_{1x} x_0 \\ \sin^2 \chi_{1x} x_0 \end{Bmatrix} \tag{2.40}$$

In the position shown in Fig. 2.4b:

$$\tilde{k}_y = 0 \tag{2.41}$$

In the position shown in Fig. 2.4c:

$$\tilde{k}_z = \tilde{k}_0 \beta_x^2 \left| \overline{\omega}_x(\xi_{1x}) \omega_y(0) \omega_z(\gamma) \right|^2 \begin{Bmatrix} \sin^2 \chi_{1x} x_0 \\ \cos^2 \chi_{1x} x_0 \end{Bmatrix}. \tag{2.42}$$

Here $\xi_{1x} = \chi_{1x}/k_0$; $\gamma = \Gamma/k_0$;

$$\tilde{k}_0 = 2(\varepsilon_{1r} - 1)^2 \cdot k_0^4 \cdot (\beta_x \beta_y \beta_z) \cdot \frac{1}{|\Gamma| ab} \cdot \frac{(a_0 b_0 L)^2}{\upsilon H}. \tag{2.43}$$

On Fig. 2.4 shows the dependences of the coupling coefficients of a rectangular DR on the relative sizes for the main H_{111} magnetic oscillations, calculated using formulas (2.40), (2.42). It can be seen that compression or stretching of the DR in any direction is accompanied by field redistribution to the external region, which leads to an increase in coupling coefficients (Fig. 2.4d, e).

2.3 Coupling Coefficients Between Dielectric Resonators and Circular, Coaxial and Sector Waveguides

When a plane of symmetry appears, the field of natural oscillations in it can take only certain values, characterized by the condition of a magnetic or electric wall. In the case of a non-coaxial location of the spherical DR in a circular waveguide (Fig. 2.5a) with the coordinates of the center (ρ_0, α_0, z_0), such symmetry planes will be the planes: $z = z_0$ and $\alpha = \alpha_0$, $\alpha = \alpha_0 + \pi$. Therefore, in the original expressions (2.4), we represent the field of natural oscillations of the DR in the form (1.2), or (1.3) as well as (A.5), (A.6), (A.12), taking into account (A.13), where plane: $\varphi = 0$ in the local coordinate system DR now coincides, taking into account the above, with the plane $\alpha = \alpha_0$ and satisfies the conditions of a magnetic or electric wall. In this case, z' axis of the local spherical coordinate system associated with the center of the ball is parallel to the axis of the waveguide. To carry out integration over the surface of the ball, we express the field of the waveguide in the coordinate system associated with the center of the resonator and expand it into a series, applying the Graph addition theorem [2]. We write the general expressions for coupling coefficients obtained as a result of transformations in the form:

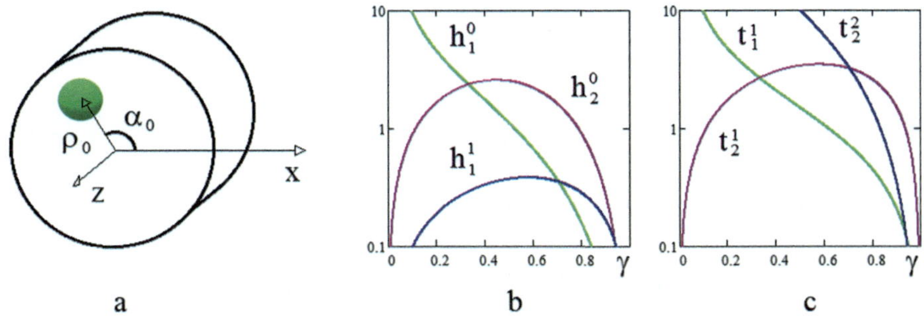

Fig. 2.5 Spherical DR in a regular circular waveguide (**a**). Dependences of coupling coefficient multipliers on $\gamma = \Gamma/k_0$ for magnetic h_n^m (**b**); electrical t_n^m oscillations (**c**)

when decomposed by waves H_{su} with $H_z \sim \cos u\alpha$ (A.5)—lower and $H_z \sim \sin u\alpha$—top,

for H_{nml}:

$$\tilde{k} = \begin{bmatrix} \{J_{m+u}(\chi\rho_0) - (-1)^m J_{-m+u}(\chi\rho_0)\}^2 \sin^2 u\alpha_0 \\ \{J_{m+u}(\chi\rho_0) + (-1)^m J_{-m+u}(\chi\rho_0)\}^2 \cos^2 u\alpha_0 \end{bmatrix} \cdot \frac{\tilde{k}(0)}{(1+\delta_{u0})(1+\delta_{m0})}; \tag{2.44}$$

for E_{nml} oscillations:

$$\tilde{k} = \begin{bmatrix} \{J_{m+u}(\chi\rho_0) + (-1)^m J_{-m+u}(\chi\rho_0)\}^2 \cos^2 u\alpha_0 \\ \{J_{m+u}(\chi\rho_0) - (-1)^m J_{-m+u}(\chi\rho_0)\}^2 \sin^2 u\alpha_0 \end{bmatrix} \cdot \frac{\tilde{k}(0)}{(1+\delta_{u0})}; \tag{2.45}$$

by waves E_{su} with $E_z \sim \sin u\alpha$ (A.6)—lower and $E_z \sim \cos u\alpha$—top;

for H_{nml}

$$\tilde{k} = \begin{bmatrix} \{J_{m+u}(\chi\rho_0) + (-1)^m J_{-m+u}(\chi\rho_0)\}^2 \sin^2 u\alpha_0 \\ \{J_{m+u}(\chi\rho_0) - (-1)^m J_{-m+u}(\chi\rho_0)\}^2 \cos^2 u\alpha_0 \end{bmatrix} \cdot \frac{\tilde{k}(0)}{(1+\delta_{u0})}; \tag{2.46}$$

for E_{nml}

$$\tilde{k} = \begin{bmatrix} \{J_{m+u}(\chi\rho_0) - (-1)^m J_{-m+u}(\chi\rho_0)\}^2 \cos^2 u\alpha_0 \\ \{J_{m+u}(\chi\rho_0) + (-1)^m J_{-m+u}(\chi\rho_0)\}^2 \sin^2 u\alpha_0 \end{bmatrix} \cdot \frac{\tilde{k}(0)}{(1+\delta_{u0})(1+\delta_{m0})}; \tag{2.47}$$

Here $\tilde{k}(0)$ the values of the coupling coefficients of the DR located on the axis of the circular waveguide [3]:

For magnetic oscillations H_{nml} and magnetic waves H_{su}:

$$\tilde{k}(0) = \frac{4}{(j_{m,s}^{\prime 2} - m^2) J_m^2(j_{m,s}^{\prime})} \cdot \frac{2n+1}{n(n+1)} \cdot \frac{(n-m)!}{(n+m)!} \cdot \frac{1}{|\gamma|} \left[(1-\gamma^2) \frac{dP_n^m(\gamma)}{d\gamma} \right]^2 \alpha_n^H(p,q); \tag{2.48}$$

For magnetic oscillations H_{nml} and electritical waves E_{su}

$$\tilde{k}(0) = \frac{4m^2}{j_{m,s}^2 J_m^{\prime 2}(j_{m,s})} \cdot \frac{2n+1}{n(n+1)} \cdot \frac{(n-m)!}{(n+m)!} \cdot \frac{1}{|\gamma|} \left[P_n^m(\gamma) \right]^2 \alpha_n^H(p,q); \tag{2.49}$$

For electritical oscillations E_{nml} and magnetic waves H_{su}:

$$\tilde{k}(0) = \frac{4m^2}{(j_{m,s}^{\prime 2} - m^2) J_m^2(j_{m,s}^{\prime})} \cdot \frac{2n+1}{n(n+1)} \cdot \frac{(n-m)!}{(n+m)!} \cdot \frac{1}{|\gamma|} \left[P_n^m(\gamma) \right]^2 \alpha_n^E(p,q); \tag{2.50}$$

For electritical oscillations E_{nml} and electritical waves E_{su}:

$$\tilde{k}(0) = \frac{4}{j_{m,s}^2 J_m^{\prime 2}(j_{m,s})} \cdot \frac{2n+1}{n(n+1)} \cdot \frac{(n-m)!}{(n+m)!} \cdot \frac{1}{|\gamma|} \cdot \left[(1-\gamma^2) \frac{dP_n^m(\gamma)}{d\gamma} \right]^2 \alpha_n^E(p,q). \tag{2.51}$$

From the obtained expressions it follows that if the coordinates of the center of the DR lie on the axis of the waveguide (2.44–2.47), then only the coupling coefficients of natural oscillations with $m = u$ remain nonzero, since $J_\nu(0) \neq 0$ only for $\nu = 0$. A dielectric ball located on the axis of a circular waveguide emits waves only of the same azimuthal indices m. In this case, as follows from (2.49), (2.50), magnetic azimuthally homogeneous oscillations of the DR excite only magnetic azimuthally homogeneous waves of a circular waveguide and electrical azimuthally homogeneous oscillations excite only electrical azimuthally homogeneous waves. The dependence of the coupling coefficients on the parameters of the material of the dielectric ball is also determined by the functions $\alpha_n^{H,E}(p,q)$ (2.20). This coincidence for waveguides of different cross sections is explained by the fact that in both cases the influence of metal walls on the field structure of the dielectric resonator is not taken into account.

The dependence of the coupling coefficients on the parameters of the circular waveguide is determined by the functions:

$$h_n^m(\gamma) = \frac{1}{|\gamma|} \left[(1-\gamma^2) \frac{dP_n^m(\gamma)}{d\gamma} \right]^2; \quad t_n^m(\gamma) = \frac{1}{|\gamma|} \left[P_n^m(\gamma) \right]^2. \tag{2.52}$$

Multipliers (2.52) for the main oscillations of a spherical DR are shown in Fig. 2.5b, c. Just as for a rectangular waveguide, the coupling coefficients increase at $\omega \to \omega_{cw}^{11}$ in the case of oscillations H_{10l}, E_{11l} and are characterized by a maximum for oscillations H_{11l}, which is achieved at the frequency $\omega = (3/2)^{1/2}\omega_{cw}^{11}$. Here ω_{cw}^{11}—is the critical frequency of the fundamental wave H_{11} of a circular waveguide.

If a spherical DR is located on the axis of a circular waveguide, then the coupling coefficients of its main magnetic and electrical oscillations with the H_{11} wave take a simpler form:

for H_{111} oscillations:

$$\tilde{k} = \frac{3}{(j_{11}^{\prime 2} - 1)J_1^2(j_{1,1}^{\prime})} \cdot |\gamma|(1-\gamma^2) \cdot \alpha_1^H(p, q); \tag{2.53}$$

for E_{111}

$$\tilde{k} = \frac{3}{(j_{1,1}^{\prime 2} - 1)J_1^2(j_{1,1}^{\prime})} \cdot \frac{1-\gamma^2}{|\gamma|} \cdot \alpha_1^E(p, q). \tag{2.54}$$

To calculate the coupling of a spherical DR with the TEM wave of a coaxial waveguide (A.10), (A.11); (Fig. 2.6a) it is necessary to use the integral (D.12), as well as the limiting value of the Graph expansion of the first-order Neumann function

$$\frac{1}{\rho}\begin{bmatrix} \sin(\alpha-\alpha_0) \\ \cos(\alpha-\alpha_0) \end{bmatrix} = \frac{1}{\rho_0}\sum_{t=0}^{\infty}\left(\frac{r}{\rho_0}\right)^t \begin{bmatrix} \sin t\varphi \\ \cos t\varphi \end{bmatrix}. \quad (r < \rho_0)$$

Here ρ, r, ρ_0—are sides of the triangle; φ—angle between sides r, ρ_0, and $(\alpha - \alpha_0)$—angle between sides ρ, ρ_0. The resulting expressions have the form [4]:

for H_{nml}:

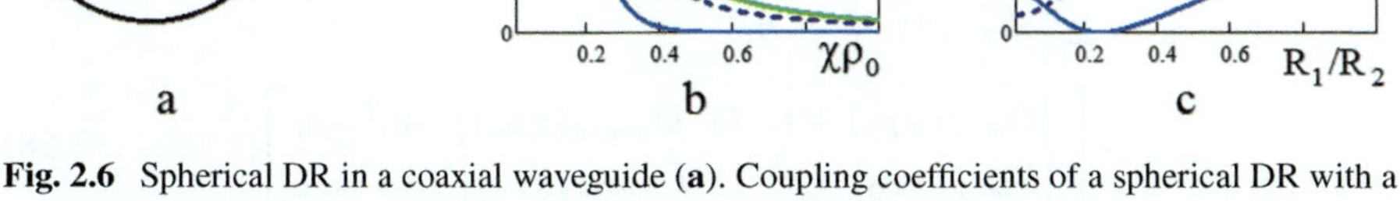

Fig. 2.6 Spherical DR in a coaxial waveguide (**a**). Coupling coefficients of a spherical DR with a coaxial waveguide for magnetic H_{111} oscillations and *TEM* waves (solid green curves); H_{11} waves dotted curves; for electrical E_{111} oscillations and H_{11} waves, solid blue curves. $\varepsilon_{1r} = 81$; $k_0R_2 = 2$; $R_1/R_2 = 0,5$ (**b**); $\chi\rho_0 = 0,4$ (**c**)

$$\tilde{k} = (1 - \delta_{m0})\frac{2n+1}{n(n+1)} \cdot \frac{(n+m)!}{(n-m)!} \cdot \frac{1}{(\ln R_2/R_1)(k_0\rho_0)^{2m}}\begin{bmatrix}1\\0\end{bmatrix} \cdot \alpha_n^H(p,q); \quad (2.55)$$

for E_{nml}:

$$\tilde{k} = (1 + \delta_{m0})\frac{2n+1}{n(n+1)} \cdot \frac{(n+m)!}{(n-m)!} \cdot \frac{1}{(\ln R_2/R_1)(k_0\rho_0)^{2m}}\begin{bmatrix}0\\1\end{bmatrix} \cdot \alpha_n^E(p,q); \quad (2.56)$$

From (2.55) and (2.56) it follows that the TEM wave of a coaxial waveguide is coupled only with H_{nml} oscillations of a dielectric ball, characterized by a maximum of the magnetic field (1.2) directed along α, as well as E_{nml} oscillations (1.3) with a maximum radial component of the electric field directed along ρ. In this case, the coupling coefficients do not depend on the angular coordinate and decrease with increasing radial coordinate of the center of the DR, the faster the larger the azimuthal number m.

Similarly, we find the coupling coefficients of the spherical DR with higher types of waves of a regular coaxial waveguide (A.5)–(A.9):

for H_{nml} oscillations and H_{su} waves:

$$\begin{aligned}\tilde{k} = {} & \frac{4}{(1+\delta_{m0})(1+\delta_{u0})} \cdot \frac{2n+1}{n(n+1)} \cdot \frac{(n-m)!}{(n+m)!} \cdot \\ & \frac{1}{C_u^2(\chi R_2)\left[(\chi R_2)^2 - u^2\right] - C_u^2(\chi R_1)\left[(\chi R_1)^2 - u^2\right]} \cdot \\ & \frac{1}{|\gamma|}\left[(1-\gamma^2)\frac{dP_n^m(\gamma)}{d\gamma}\right]^2 \alpha_n^H(p,q) \\ & \cdot \begin{bmatrix}\left\{C_{m+u}(\chi\rho_0) - (-1)^m C_{-m+u}(\chi\rho_0)\right\}^2 \sin^2 u\alpha_0 \\ \left\{C_{m+u}(\chi\rho_0) + (-1)^m C_{-m+u}(\chi\rho_0)\right\}^2 \cos^2 u\alpha_0\end{bmatrix}; \end{aligned} \quad (2.57)$$

for H_{nml} oscillations and E_{su} waves:

$$\begin{aligned}\tilde{k} = {} & \frac{4}{(1+\delta_{u0})} \cdot \frac{2n+1}{n(n+1)} \cdot \frac{(n-m)!}{(n+m)!} \\ & \cdot \frac{m^2}{[\chi R_2 D_{u-1}(\chi R_2)]^2 - [\chi R_1 D_{u-1}(\chi R_1)]^2} \\ & \cdot \frac{1}{|\gamma|}\left[P_n^m(\gamma)\right]^2 \alpha_n^H(p,q) \\ & \cdot \begin{bmatrix}\left\{D_{m+u}(\chi\rho_0) + (-1)^m D_{-m+u}(\chi\rho_0)\right\}^2 \sin^2 u\alpha_0 \\ \left\{D_{m+u}(\chi\rho_0) - (-1)^m D_{-m+u}(\chi\rho_0)\right\}^2 \cos^2 u\alpha_0\end{bmatrix}; \end{aligned} \quad (2.58)$$

for E_{nml} oscillations and H_{su} waves:

$$\tilde{k} = \frac{4}{(1+\delta_{u0})} \cdot \frac{2n+1}{n(n+1)} \cdot \frac{(n-m)!}{(n+m)!} \cdot \frac{m^2}{C_u^2(\chi R_2)\left[(\chi R_2)^2 - u^2\right] - C_u^2(\chi R_1)\left[(\chi R_1)^2 - u^2\right]} \cdot \frac{1}{|\gamma|}\left[P_n^m(\gamma)\right]^2 \alpha_n^E(p,q) \cdot \begin{bmatrix} \left\{C_{m+u}(\chi\rho_0) + (-1)^m C_{-m+u}(\chi\rho_0)\right\}^2 \cos^2 u\alpha_0 \\ \left\{C_{m+u}(\chi\rho_0) - (-1)^m C_{-m+u}(\chi\rho_0)\right\}^2 \sin^2 u\alpha_0 \end{bmatrix}; \tag{2.59}$$

for E_{nml} oscillations and E_{su} waves:

$$\tilde{k} = \frac{4}{(1+\delta_{m0})(1+\delta_{u0})} \cdot \frac{2n+1}{n(n+1)} \cdot \frac{(n-m)!}{(n+m)!} \cdot \frac{1}{[\chi R_2 D_{u-1}(\chi R_2)]^2 - [\chi R_1 D_{u-1}(\chi R_1)]^2} \cdot \frac{1}{|\gamma|} \cdot \left[(1-\gamma^2)\frac{dP_n^m(\gamma)}{d\gamma}\right]^2 \alpha_n^E(p,q) \cdot \begin{bmatrix} \left\{D_{m+u}(\chi\rho_0) - (-1)^m D_{-m+u}(\chi\rho_0)\right\}^2 \cos^2 u\alpha_0 \\ \left\{D_{m+u}(\chi\rho_0) + (-1)^m D_{-m+u}(\chi\rho_0)\right\}^2 \sin^2 u\alpha_0 \end{bmatrix}. \tag{2.60}$$

Here $\gamma = \Gamma/k_0 = \left[1 - (\chi/k_0)^2\right]^{1/2}$; $\chi = \kappa_{su}/R_2$—transverse propagation constant determined from equations (A.8) or (A.9).

On Fig. 2.6 shows the dependences of the coupling coefficients of the spherical DR with the lower waves of the coaxial waveguide. Let us pay attention to the noticeable increase in coupling with increasing thickness of the central conductor (c).

In the particular case of $R_1 = 0$, the resulting relations (2.57)–(2.60) transform into (2.44)–(2.47) for a circular waveguide.

When calculating the coupling coefficients with a sector waveguide (Fig. 2.7a), we assume that the surface of the dielectric ball does not touch the tip of the wedge: $r_0 < \rho_0$. In this case, the condition for applying the addition theorem for Bessel functions is also satisfied, and the final result has the form (2.44) to (2.47), where the factors $k(0)$ differ from (2.48) to (2.51) by normalization (A.13), and in addition, the order of cylindrical functions (A.12). The appearance of a metal wedge, in the general case, eliminates the plane of symmetry $\alpha = \alpha_0$, $\alpha = \alpha_0 + \pi$, so the question of the "natural polarization" of dielectric ball natural oscillations in this structure remains open (see. 4.1).

Generalization of relations (2.44)–(2.47) to the case of cupling of a spherical DR with a sector waveguide is possible under the condition $\alpha_0 = 0$; for azimuthally homogeneous oscillations ($m = 0$) or oscillations of a given type.

On Fig. 2.7 shows the calculated dependences of the coupling coefficients on the parameters and coordinates of the ball. The appearance of a metal wedge near the

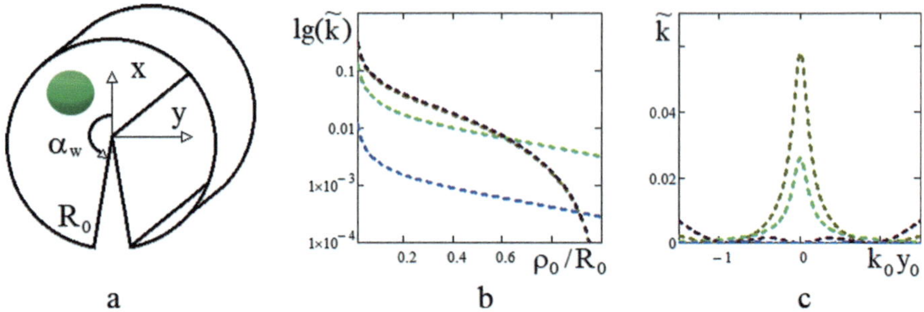

Fig. 2.7 Spherical DR in a sector waveguide (**a**). Coupling coefficients of a spherical DR with a sector waveguide (**b, c**) for $m = n = 1$. $\alpha_w = 3/4\pi$; $k_0R_0 = 1, 5$; for magnetic H_{111} oscillations and magnetic waves—dotted green; electric waves—black; for electrical E_{111} oscillations and electric waves—dotted blue and magnetic waves—purple

surface of the DR leads to an increase in the coupling of transverse magnetic-type oscillations (c).

2.4 Dielectric Resonators in a Planar Waveguide

The analytical expressions found in the previous sections can serve as the basis for deriving new useful relationships [5, 6] for the coupling coefficients of DR in more open structures obtained by removing one or more waveguide walls and passing to a continuous spectrum. Indeed, if the expansion of any external parameter of the DR (field components, coupling coefficient, etc.) is known into a series in terms of the system of natural waves of a metal waveguide, for example a rectangular one, with a cross section $a \times b$:

$$f = \frac{\pi^2}{k_0^2 ab} \sum_{s,u=(0)}^{\infty} g(\chi_{sx}, \chi_{uy}, \ldots), \tag{2.61}$$

where $g(\ldots)$—known functions that determine the dependence f on the parameters of the resonator and waveguide, then series (2.61) can be summed up if one or both dimensions of the waveguide tend to infinity. So, for example, in the limit $a = \infty$, expansion (2.61) takes the form:

$$\varphi = \frac{\pi}{k_0 b} \sum_{u=0}^{\infty} \alpha(\chi_{uy}, \ldots), \tag{2.62}$$

where the sum over s is converted into an integral:

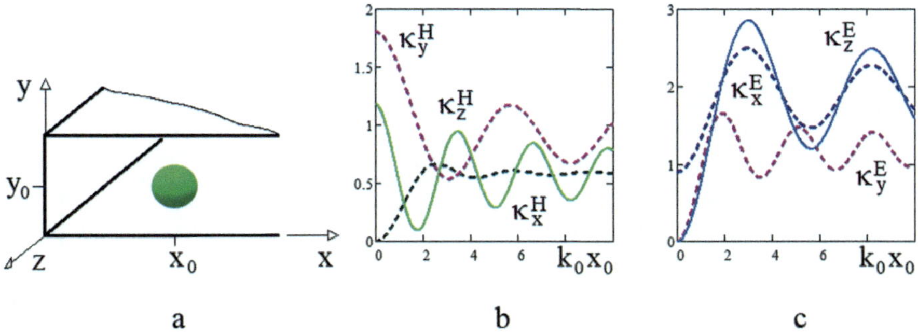

Fig. 2.8 Coupling coefficients of a spherical DR with a flat waveguide and a metal wall (**a**). $\kappa_s^H = \tilde{k}_s^H/2\alpha_1^H$; $\kappa_s^E = \tilde{k}_s^E/2\alpha_1^E$ $(s = x, y, z)$; for oscillations H_{101} (**b**), E_{101} (**c**)—solid curves; for oscillations H_{111} (**b**), E_{111} (**c**)—dotted curves ($k_0 b = 4$)

$$\alpha = \int_0^\infty g(\xi, \chi_{uy}, \ldots) d\xi. \tag{2.63}$$

Here $\lim_{s,k_0a\to\infty} s\pi/k_0a \to \xi$; $\lim_{s,k_0a\to\infty}[(s+1)\pi/k_0a - s\pi/k_0a] = \lim_{k_0a\to\infty} \pi/k_0a = d\xi$. Integral (2.63) can often be calculated in quadratures, taking into account the relationship between the wave numbers: $\gamma = \lim_{s,k_0a\to\infty} \Gamma/\mathrm{k}_0 = \left[1 - (u\pi/k_0b)^2 - \xi^2\right]^{1/2}$.

Let us consider, as an example, the field of a spherical DR located in a semi-infinite trench (Fig. 2.8a). We will proceed from the expansion of the DR field (2.11) and (2.12) in a rectangular waveguide (2.3). For magnetic azimuthally uniform oscillations H_{10l}:

$$\mathbf{e}^0 = \begin{pmatrix} e_x^0 \\ e_y^0 \\ e_z^0 \end{pmatrix} = \frac{8\pi}{ab} w_0 \left[\frac{j_1(p)}{y_1(q)}\right] \sum_{s,u=(0)}^{\infty} \frac{1}{1+\delta_{so}+\delta_{uo}} \frac{e^{-i\Gamma(z-z_0)}}{k_0|\Gamma|}$$
$$\cdot \begin{pmatrix} \chi_{uy}\cos\chi_{sx}x_0\cos\chi_{sx}x\cos\chi_{uy}y_0\sin\chi_{uy}y \\ \chi_{sx}\sin\chi_{sx}x\cos\chi_{sx}x_0\cos\chi_{uy}y_0\cos\chi_{uy}y \\ 0 \end{pmatrix}.$$

Passing to the limit $a = \infty$ using (D.20) leads to:

$$\mathbf{e}^0 = \frac{2\pi}{b} w_0 \left[\frac{j_1(p)}{y_1(q)}\right] \cdot \sum_{u=0}^{\infty} \begin{pmatrix} \frac{\chi_{uy}}{k_0}\left[H_0^{(2)}\left(\Gamma_{II}\rho^-\right) + H_0^{(2)}\left(\Gamma_{II}\rho^+\right)\right] \sin\chi_{uy}y\cos\chi_{uy}y_0 \\ -\frac{\Gamma_{II}}{k_0}\left[\frac{x+x_0}{\rho^-}H_1^{(2)}\left(\Gamma_{II}\rho^-\right) + \frac{x-x_0}{\rho^+}H_1^{(2)}\left(\Gamma_{II}\rho^+\right)\right]\cos\chi_{uy}y\cos\chi_{uy}y_0 \\ 0 \end{pmatrix}, \tag{2.64}$$

where now the external field of the resonator is expressed through cylindrical waves of a plane waveguide with longitudinal wave numbers $\Gamma_{II} = \left[k_0^2 - (u\pi/b)^2\right]^{1/2}$. Here $H_m^{(2)}(x)$ is the Hankel function of the second kind [2]. The presence of a metal wall in the plane $x = 0$ leads to the appearance of an "image" of the DR, located at a distance $\rho^- = \left[(z - z_0)^2 + (x + x_0)^2\right]^{1/2}$ from the observation point $(x,\ y,\ z)$ symmetrically to the center of the resonator, located at a distance $\rho^+ = \left[(z - z_0)^2 + (x - x_0)^2\right]^{1/2}$.

Passing to the limit $x_0 = \infty$ under the condition of $x - x_0$ finiteness, leads to relations for the field of this DR in a plane waveguide. In this case $\rho^- \to \infty$, and due to the behavior of the Hankel functions, the terms containing ρ^-, vanish.

It is easy to verify that the tangential components of the electric field (2.64) satisfy the boundary conditions on the surface of the metal trench: $y = 0$, b; x = 0. When $\rho^- = (x^2 + y^2)^{1/2} \to \infty$ the cavity field is in the form of diverging cylindrical waves, that is, it satisfies the radiation conditions in this structure. On the DR axis, due to the singularity of the Hankel functions [2], series (2.64) goes to infinity. This phenomenon is due to the incorrectness of the expansion of the form (2.3) in the region of the waveguide, including the volume of the dielectric. On the $(x,$ z) plane, as a result of transformation, such an area takes the shape of a cylindrical.

For a similar transformation of the coupling coefficients of spherical DR with a rectangular waveguide in the sum (2.61), it is necessary to take into account only those types of waves that become propagating when $a \to \infty$. In this case, the integration of the sum of the coupling coefficients of the resonator, found on the basis of (2.11)–(2.14), is carried out over the real γ ones using (D.17) and leads to the expressions:

for H_{10l} oscillations

$$\tilde{k} = \frac{6\pi}{k_0 b} \cdot \alpha_1^H(p, q) \cdot \sum_{u=0}^{(\chi_{uy}/k_0)<1} \frac{\cos^2\chi_{uy}y_0}{(1 + \delta_{u0})} \{J_0(2\Gamma_{11}x_0) - \frac{\Gamma_{II}}{2k_0^2 x_0} J_1(2\Gamma_{II}x_0) + \frac{1}{2}\left[1 + \left(\chi_{uy}/k_0\right)^2\right]\}; \tag{2.65}$$

For H_{11l} oscillations:

$$\tilde{k} = \frac{6\pi}{k_0 b} \cdot \alpha_1^H(p, q) \cdot \sum_{u=0}^{(\chi_{uy}/k_0)<1} \frac{1}{(1+\delta_{u0})}$$
$$\cdot \begin{bmatrix} \sin^2 \chi_{uy} y_0 \left[1 - (\chi_{uy}/k_0)^2\right][1 + J_0(2\Gamma_{II} x_0)]; \\ \cos^2 \chi_{uy} y_0 \left\{ \frac{1}{2}\left[1 - \left(\frac{\chi_{uy}}{k_0}\right)^2\right]\left[1 - \frac{1}{\Gamma_{II} x_0} J_1(2\Gamma_{II} x_0)\right] + \left(\frac{\chi_{uy}}{k_0}\right)^2 [1 - J_0(2\Gamma_{II} x_0)] \right\} \end{bmatrix}; \quad (2.66)$$

for E_{10l} oscillations:

$$\tilde{k} = \frac{6\pi}{k_0 b} \cdot \alpha_1^E(p, q) \cdot \sum_{u=0}^{(\chi_{uy}/k_0)<1} \sin^2 \chi_{uy} y_0 \left\{ \frac{1}{2}\left[1 + (\chi_{uy}/k_0)^2\right] \right.$$
$$\left. - J_0(2\Gamma_{II} x_0) + \frac{\Gamma_{II}}{2k_0^2 x_0} J_1(2\Gamma_{II} x_0) \right\}; \quad (2.67)$$

for E_{11l} oscillations:

$$\tilde{k} = \frac{6\pi}{k_0 b} \alpha_1^E(p, q) \cdot \sum_{u=0}^{(\chi_{uy}/k_0)<1} \frac{1}{(1+\delta_{u0})}.$$
$$\begin{bmatrix} \cos^2 \chi_{uy} y_0 \left[1 - (\chi_{uy}/k_0)^2\right] \cdot [1 - J_0(2\Gamma_{II} x_0)]; \\ \sin^2 \chi_{uy} y_0 \left\{ \frac{1}{2}\left[1 - \left(\frac{\chi_{uy}}{k_0}\right)^2\right] \cdot \left[1 + \frac{1}{\Gamma_{II} x_0} J_1(2\Gamma_{II} x_0)\right] + \left(\frac{\chi_{uy}}{k_0}\right)^2 [1 - J_0(2\Gamma_{II} x_0)] \right\} \end{bmatrix}. \quad (2.68)$$

The results of the calculation using formulas (2.65)–(2.68) of the normalized dependences of the coupling coefficients for main magnetic and electrical oscillations at $y_0 = b/2$ on coordinates are shown in Fig. 2.8b, c, respectively, where $\kappa_s^{E,H} = \tilde{k}_s^{E,H}/2\alpha_1^{E,H}$. The radiation of a dielectric ball in the space of a semi-infinite trough, as follows from (2.64), is equivalent to the radiation of a system of two spherical DRs with coordinates $(\pm x_0, y_0, z_0)$ in a flat waveguide, therefore the wave nature of the change in coupling can be explained by field interference when the relative distance between their centers is varied. In the limit $x_0 = \infty$, relations (2.65)–(2.68) transform into coupling coefficients of a spherical DR with a flat waveguide.

Conversion to the limit $a = \infty$ for a cylindrical DR in a rectangular waveguide is much more difficult to implement due to the complexity of the integrands. Nontrivial integration in the case of a short-circuited plane waveguide can be carried out for azimuthally uniform oscillations of a dielectric cylinder in position b (Fig. 2.9b), for example, for H_{nol} oscillations:

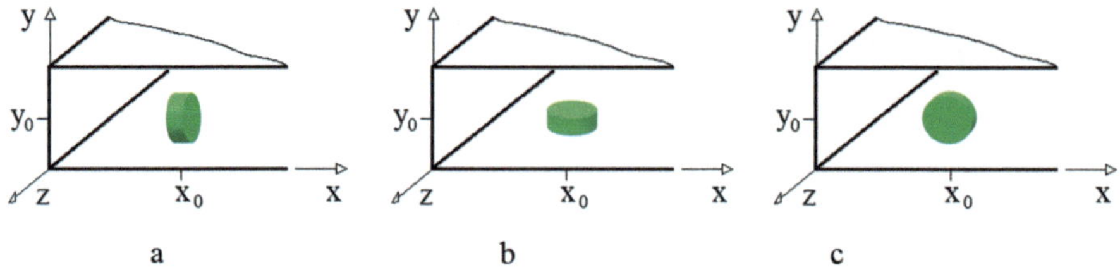

Fig. 2.9 Standard positions of a cylindrical DR in a flat waveguide (**a**, **b**, **c**) with a metal wall

$$\tilde{k} = \frac{16\pi\beta_z}{k_1^2 b} \cdot \frac{(\varepsilon_{1r} - 1)^2}{[J_1^2(p_\perp) - J_0(p_\perp)J_2(p_\perp)]} \Big/ \left[\frac{2p_z + \sin 2p_z}{2p_z - \sin 2p_z}\right]$$
$$\cdot \sum_{u=0}^{(\chi_{uy}/k_0)<1} \frac{1 + J_0(2\Gamma_{II}x_0)}{\left[\left(\frac{\beta}{k_0}\right)^2 - \left(\frac{\Gamma_{II}}{k_0}\right)^2\right]^2 \left[\left(\frac{\beta_z}{k_0}\right)^2 - \left(\frac{\chi_{uy}}{k_0}\right)^2\right]^2} \cdot \frac{1}{(1+\delta_{u0})}$$
$$\cdot \left\{\frac{\beta}{k_0} J_0(p_\perp)J_1(\Gamma_{II}r_0) - \frac{\Gamma_{II}}{k_0} J_1(p_\perp)J_0(\Gamma_{II}r_0)\right\}^2$$
$$\cdot \left[\begin{array}{c} \sin^2 \chi_{uy}y_0 \left(\frac{\beta_z}{k_0} \sin p_z \cos \chi_{uy}\frac{L}{2} - \frac{\chi_{uy}}{k_0} \cos p_z \sin \chi_{uy}\frac{L}{2}\right)^2 \\ \cos^2 \chi_{uy}y_0 \left(\frac{\beta_z}{k_0} \cos p_z \sin \chi_{uy}\frac{L}{2} - \frac{\chi_{uy}}{k_0} \sin p_z \cos \chi_{uy}\frac{L}{2}\right)^2 \end{array}\right]. \tag{2.69}$$

The transition from the obtained expressions to the coupling coefficients of a cylindrical DR with a flat waveguide is quite simple: in (2.69) we should also put $J_0(2\Gamma_{II}x_0) = 0$.

2.5 Rotations of Dielectric Resonators in a Regular Rectangular Waveguide

Rotations of the DR in different structures give us an additional degree of freedom when designing and configuring various devices. For rotating dielectric resonators, a calculation of the coupling coefficients usually leads to extremely cumbersome analytical expressions. However, in some cases, obtaining compact analytical expressions is possible. To calculate the expansion coefficients of the DR field over the waveguide field, we will use representation (2.5), which we will consider in the coordinate system associated with the center of the cylindrical resonator (1.17). After integration (2.5), we return to the waveguide coordinate system. For simplicity, we will consider only the main magnetic type of oscillations H_{101} and rotation of the DR only relative to one of the axes of the waveguide coordinate system (Fig. 2.10a–c).

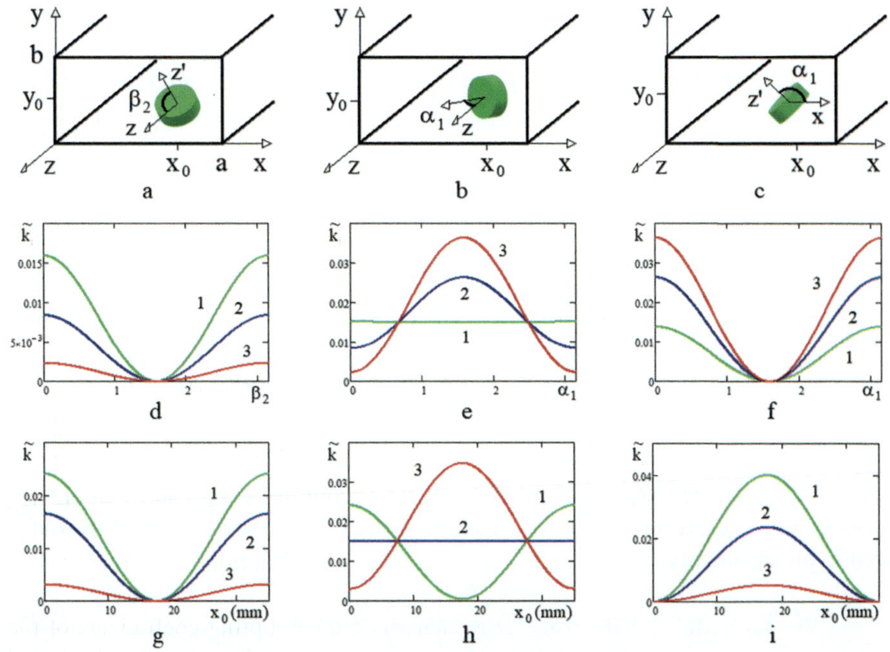

Fig. 2.10 Rotation of the cylindrical dielectric resonator relatively the x-axis (**a**); y-axis (**b**); z-axis (**c**). Dependence of the DR coupling coefficient on the rotation angle β_2 (**d**): ($y_0 = 0, 5b$; 1—$x_0 = 0, 2a$; 2—$x_0 = 0, 3a$; 3—$x_0 = 0, 4a$). Dependence of the DR coupling coefficient on the rotation angle α_1 relatively the y-axis (**e**): ($y_0 = 0, 5b$; 1—$x_0 = 7, 341$ mm; 2—$x_0 = 0, 3a$; 3—$x_0 = 0, 4a$). The dependence of the coupling coefficient of the DR on the angle of rotation α_1 for the initial position $z' \| x$ (**f**) ($y_0 = 0, 5b$; 1—$x_0 = 0, 2a$; 2—$x_0 = 0, 3a$; 3—$x_0 = 0, 4a$). Dependence of the coupling coefficient on the DR coordinate $x_0(y_0 = 0, 5b)$ (**g**): (1—$\beta_2 = 0, 1$; 2—$\beta_2 = 0, 6$; 3—$\beta_2 = 1, 2$); (**h**): (1—$\alpha_1 = 0, 1$; 2—$\alpha_1 = 0, 659$; 3—$\alpha_1 = 1, 2$); (**i**): (1—$\alpha_1 = 0, 1$; 2—$\alpha_1 = 0, 7$; 3—$\alpha_1 = 1, 2$)

The results of calculating the coupling coefficients are most conveniently presented in the form:

(a) In the case of rotation of the DR axis about the x axis of the waveguide (Fig. 2.10a)

$$\tilde{k} = 4\tilde{k}_0 \chi_{1x}^2 \cos^2\beta_2 |\upsilon(\chi_{1x}, \Gamma\sin\beta_2, \Gamma\cos\beta_2)|^2 \cdot \cos^2(\chi_{1x}x_0); \quad (2.70)$$

(b) Rotation of the resonator relative to the y axis of the waveguide (Fig. 2.10b)

$$\begin{aligned}\tilde{k} = \tilde{k}_0 \big|&(\chi_{1x}\cos\alpha_1 \pm \Gamma\sin\alpha_1)\upsilon((\chi_{1x}\cos\alpha_1 \pm \Gamma\sin\alpha_1), 0, (\chi_{1x}\sin\alpha_1 \mp \Gamma\cos\alpha_1))e^{-i\chi_{1x}x_0} \\ &+ (\chi_{1x}\cos\alpha_1 \mp \Gamma\sin\alpha_1) \quad \upsilon((\chi_{1x}\cos\alpha_1 \mp \Gamma\sin\alpha_1), 0, (\chi_{1x}\sin\alpha_1 \pm \Gamma\cos\alpha_1))e^{i\chi_{1x}x_0}\big|^2;\end{aligned} \quad (2.71)$$

(c) Rotation of the resonator relative to the z axis of the waveguide (Fig. 2.10c)

$$\tilde{k} = 4\tilde{k}_0|\Gamma|^2\cos^2\alpha_1|\upsilon(\Gamma,\ \chi_{1x}\sin\alpha_1,\ \chi_{1x}\cos\alpha_1)|^2\cdot\sin^2(\chi_{1x}x_0). \quad (2.72)$$

where

$$\upsilon(s_x, s_y, s_z) = \frac{[p_\perp J_1(\sqrt{s_x^2+s_y^2}r_0)J_0(p_\perp) - \sqrt{s_x^2+s_y^2}r_0 J_0(\sqrt{s_x^2+s_y^2}r_0)J_1(p_\perp)]}{(s_x^2+s_y^2)r_0^2 - p_\perp^2}\frac{\omega_z^*(s_z)}{\sqrt{s_x^2+s_y^2}}; \quad (2.73)$$

$$\omega_z(s_z) = \frac{1}{p_z^2-(s_zL/2)^2}\begin{Bmatrix} -i[p_z\cos p_z\sin(s_zL/2) - s_zL/2\sin p_z\cos(s_zL/2)] \\ [p_z\sin p_z\cos(s_zL/2) - s_zL/2\cos p_z\sin(s_zL/2)] \end{Bmatrix};$$

$$\tilde{k}_0 = 4\pi q_\perp^2 p_z\frac{(\varepsilon_{1r}-1)^2}{\varepsilon_{1r}}\frac{L}{|\Gamma|abv_0}; \quad (2.74)$$

$v_0 = [J_1^2(p_\perp) - J_0(p_\perp)J_2(p_\perp)](2p_z + \sin 2p_z)$; $*$—complex conjugate symbol.

As can be seen from (2.73), for basic magnetic oscillations H_{101} the function is even in all arguments:

$$\upsilon(-s_x, s_y, s_z) = \upsilon(s_x, -s_y, s_z) = \upsilon(s_x, s_y, -s_z) = \upsilon(s_x, s_y, s_z). \quad (2.75)$$

The results of calculating the dependences of the coupling coefficients on the coordinates and the angle of rotation of the resonator are shown in Fig. 2.10d–i for H_{101} oscillations and for $\varepsilon_{1r} = 36$; $\Delta = L/2r_0 = 0,4$; $f_0 = 7$ GHz; $a \times b = 35\times 15mm^2$. In Fig. 2.10d, f shows the possibility of changing the coupling coefficient by rotating the resonator about the x and y axes, as well as by varying the coordinate of the resonator center along the x-axis.

But the most nontrivial results are obtained by the rotation of the resonator about the y-axis (Fig. 2.10e, h). In the first case (Fig. 2.10h, e, line 1), we see that the coupling coefficient does not depend on the orientation of the DR axis at the points of "circular polarization" of the magnetic field of the waveguide:

$$x_{01} = \frac{1}{\chi_{1x}}arcctg\frac{|\Gamma|}{\chi_{1x}}, \quad (2.76)$$

and for $x_{02} = a - x_{01}$.

In the second case (Fig. 2.10e), the coupling coefficient does not depend on the transverse coordinates (Fig. 2.10h line 2) when the axis of the DR is rotated by an angle:

$$\alpha_C = \pm arcctg\frac{\Gamma}{\chi_{1x}}. \quad (2.77)$$

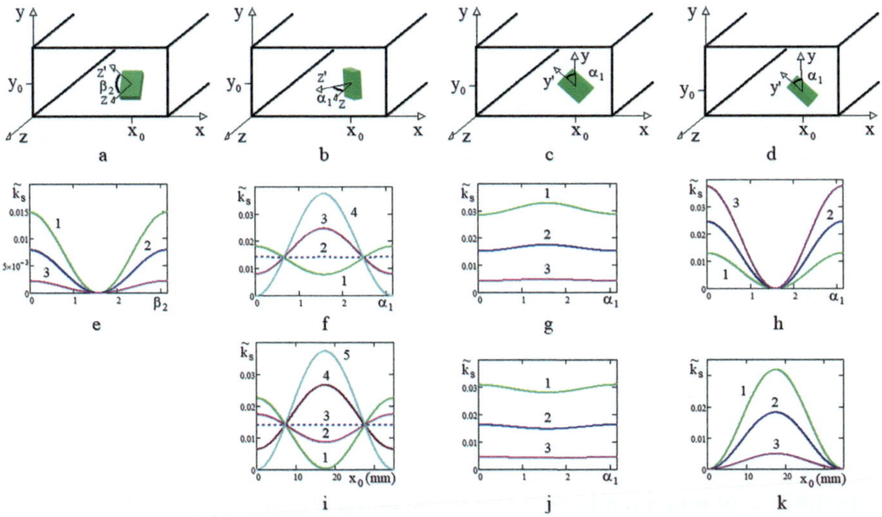

Fig. 2.11 Rotation of the rectangulare dielectric resonator relatively the x-axis (**a**); y-axis (**b**); z-axis (**c, d**). Dependence of the DR coupling coefficient on the x-rotation angle β_2 (**e**): ($y_0 = 0,5b$; 1—$x_0 = 0,2a$; 2—$x_0 = 0,3a$; 3—$x_0 = 0,4a$). Dependence of the DR coupling coefficient on the y-rotation angle (**f**): (1—$x_0 = 0,15a$; 2—$x_0 = 7,341$ mm; 3—$x_0 = 0,3a$; 4—$x_0 = 0,5a$). Dependence of the coupling coefficient on the DR coordinate (**i**): (1—$\alpha_1 = 0,1$; 2—$\alpha_1 = 0,5$; 3—$\alpha_1 = 0,659$; 4—$\alpha_1 = 1$; 5—$\alpha_1 = 1,5$). Dependence of the DR coupling coefficient on the z-rotation angle (**c**): (1- $x_0 = 0,2a$; 2- $x_0 = 0,3a$; 3- $x_0 = 0,4a$) for $b_0/a_0 = 0,2$ (**g**); $b_0/a_0 = 3$ (**j**). Dependence of the coupling coefficient on the z-rotation (**d**) angle (**h**): (1—$x_0 = 0,2a$; 2—$x_0 = 0,3a$; 3—$x_0 = 0,5a$). Dependence of the coupling coefficient on the DR coordinate (**k**): (1—$\alpha_1 = 0,4$; 2—$\alpha_1 = 0,8$; 3—$\alpha_1 = 1,2$)

In limiting cases $\alpha_1, \beta_2 = 0, \pm\pi/2$, expressions (2.70–2.72) coincide with the coupling coefficients for standard positions of the resonator in the waveguide (2.32), (2.34).

Let's now consider the rotation of a rectangular resonator relatively the x-axis of the waveguide (Fig. 2.11a). In this case, the coupling coefficient of the DR with the H_{10} wave of a rectangular waveguide has the form:

$$\tilde{k} = \tilde{k}_0 \cdot \beta_x^2 \cdot \cos^2\beta_2 \left|\overline{\omega}_x(\chi_{1x}/k_0)\omega_y(\Gamma/k_0 \sin\beta_2)\omega_z(\Gamma/k_0 \cos\beta_2)\right|^2 \cdot \begin{pmatrix} \sin^2 \chi_{1x}x_0 \\ \cos^2 \chi_{1x}x_0 \end{pmatrix}, \tag{2.78}$$

The coupling coefficient of a DR rotated relatively the y-axis with the H_{10} wave of a rectangular waveguide at the initial position shown in Fig. 2.11b, has the form:

$$\tilde{k} = \tilde{k}_0 \cdot \frac{\beta_x^2}{4} \cdot \left|\omega_y(0)\right|^2 \cdot \left|e^{-i\chi_{1x}x_0}\overline{\omega}_x(-(\chi_{1x}/k_0 \cos\alpha_1 + \Gamma/k_0 \sin\alpha_1))\right.$$
$$\cdot\, \omega_z(-(\chi_{1x}/k_0 \sin\alpha_1 - \Gamma/k_0 \cos\alpha_1)) - e^{i\chi_{1x}x_0}\overline{\omega}_x(\chi_{1x}/k_0 \cos\alpha_1 - \Gamma/k_0 \sin\alpha_1)$$
$$\left. \cdot\, \omega_z(\chi_{1x}/k_0 \sin\alpha_1 + \Gamma/k_0 \cos\alpha_1)\right|^2; \tag{2.79}$$

In the case of rotation of the dielectric resonator relatively z-axis at the initial position shown in Fig. 2.11c:

$$\tilde{k} = \tilde{k}_0 \cdot \left|\beta_y \sin\alpha_1 \omega_x(\chi_{1x}/k_0 \cos\alpha_1)\overline{\omega}_y(\chi_{1x}/k_0 \sin\alpha_1)\right.$$
$$\left. + \beta_x \cos\alpha_1\, \overline{\omega}_x(\chi_{1x}/k_0 \cos\alpha_1)\omega_y(\chi_{1x}/k_0 \sin\alpha_1)\right|^2 \cdot |\omega_z(\Gamma/k_0)|^2 \cos^2\chi_{1x}x_0. \tag{2.80}$$

In the case of rotation around z-axis at the initial position shown in Fig. 2.11, d:

$$\tilde{k} = \tilde{k}_0 \cdot \left|\beta_x \cos\alpha_1 \cdot \overline{\omega}_x(\Gamma/k_0)\omega_y(\chi_{1x}/k_0 \sin\alpha_1)\omega_z(\chi_{1x}/k_0 \cos\alpha_1)\right|^2 \cdot \sin^2\chi_{1x}x_0. \tag{2.81}$$

Where (x_0, y_0, z_0) is the coordinates of the center of the resonator in the waveguide. All other parameters defined in (2.38), (2.39) and (2.43).

In Fig. 2.11 depending on the coupling coefficients on the angles of rotation and transverse coordinates are plotted for the rectangular resonator with $\varepsilon_{1r} = 36$; $a_0 = b_0$; $L/a_0 = 0, 4$. The cross section of the waveguide was $a \times b = 35 \times 15$ mm; frequency of the fundamental magnetic oscillation of the resonator H_{111} (1.32) $f_0 =$ 7 GHz.

At the rotation of the DR relatively the y-axis, as in the case of cylindrical DR (2.76), (Fig. 2.11b, curve 2), we see that the coupling coefficient also does not depend on the orientation of the DR axis at the point of "circular polarization" of the magnetic field of the waveguide.

In the second case (Fig. 2.11i), the coupling coefficient (2.79) does not depend on the transverse coordinates when the axis of the DR is rotated by an angle defined by (2.77).

In other cases, the dependences of the coupling coefficients on the angles of rotation change according to non-harmonic laws. In limiting cases $\alpha_1, \beta_2 = 0, \pm\pi/2$, expressions (2.78–2.81) coincide with the coupling coefficients (2.40)–(2.42) of the rectangular DR for it standard positions (Fig. 2.4) in the waveguide.

References

1. M.E. Ilchenko, A.A. Trubin, Scattering of electromagnetic waves by a dielectric resonator in a waveguide. Izv. vuzov. Radioelectronics **24**(11), 91–93 (1981). (In Russian)
2. M. Abramowitz, I.A. Stegun, (eds.) Handbook of mathematical functions with formulas, graphs and mathematical tables. National Bureau of Standarts. Appl. Math. Ser. **55**, 830 (1964)
3. A.A. Trubin, Dielectric ball in a circular waveguide. Bull. NTUU "KPI" Ser. Radiotechnique. **27**, 31–34 (1990). (in Russian)
4. A.A. Trubin, Dielectric ball in a coaxial waveguide. Bull. NTUU "KPI" Ser. Radiotechnique. **30**, 43–50 (1993). (in Russian)
5. A.A. Trubin, Coupling parameters of spherical dielectric resonators in semi-infinite rectangular metal structures. Bull. NTUU "KPI" Ser. Radiotechnique. **29**, 18–28 (1992). (in Russian)
6. A.A. Trubin, Coupling coefficients of cylindrical dielectric resonators in free space in the presence of metal planes. Bull. NTUU "KPI" Ser. Radiotechnique. **29**, 28–36 (1992). (in Russian)
7. M.E. Ilchenko, A.A. Trubin, Resonance characteristics of a dielectric ball in a waveguide. Izv. vuzov. Radioelectronics **27**(1), 72–74 (1984). (In Russian)

Chapter 3
Mutual Coupling Coefficients of Dielectric Resonators in Cut-Off Waveguides

One of the parameters that plays an important role in calculating the amplitude-frequency characteristics of wave scattering on systems of coupled DRs in a line are the mutual coupling coefficients. The mutual coupling coefficients determine the degree of influence of the field of one resonator on another.

The coefficient of mutual coupling between DR 1 and 2 in waveguide is most conveniently expressed through the field expansion coefficients (2.4), (2.6) for non-propagating waves:

$$k_{12} = \frac{-1}{\omega_0 w_2} \cdot \left[\sum_{t=1}^{\infty} (c_t^{1+})_0 (c_t^{2-})_0^* e^{-\Gamma|\Delta z|} - \sum_{t=1}^{\infty} (d_t^{1+})_0 (d_t^{2-})_0^* e^{-\Gamma|\Delta z|} \right]. \tag{3.1}$$

where t—multi-index characterizing the type of wave in the line; w_2—energy stored in the dielectric of the 2nd resonator; Γ—longitudinal wave number of the line. In (3.1), the dependence on the longitudinal coordinates of the DR from the expansion coefficients of the DR fields in magnetic ($c_t^{1,2\pm}$) and electric ($d_t^{1,2\pm}$) waves (see (2. 4), (2.6)) is expressed in the form of exponential factors. The longitudinal distance between the centers of the DR is indicated by $\Delta z = z_2 - z_1$.

As follows from (3.1), for unequal resonators the mutual coupling coefficients are not symmetrical: $k_{12} \neq k_{21}$. If the resonators have different shapes, or are made of different dielectrics, it is necessary to calculate both mutual coupling coefficients k_{12} and k_{21}.

The mutual coupling coefficient is simply measured through the relative splitting of the frequencies of coupled oscillations $\omega^{\pm}$ of a pair of identical DRs in a evanescent waveguide: $k_{12} = (\omega^+ - \omega^-)/\omega_0$.

In this chapter, the coefficients of mutual coupling of various shapes DRs located in hollow metal waveguides are calculated. The found analytical expressions are suitable for calculating the mutual coupling relationships of DR with both basic and higher types of magnetic and electrical oscillations. The dependences of the coupling on the structure parameters, as well as the relative coordinates of the DR in the line

A. Trubin, *Introduction to the Theory of Dielectric Resonators*, Springer Series in Advanced Microelectronics 65, https://doi.org/10.1007/978-3-031-65396-4_3

are studied. The coupling of cylindrical DRs of various orientations, as well as the mutual coupling coefficients of the DRs of various shapes—cylindrical and spherical, are considered. The existence of spatial surfaces of zero coupling, corresponding to the transition from negative to positive coupling coefficients, has been established.

3.1 Coupled Oscillations of Spherical Dielectric Resonators in a Rectangular Waveguide

The general analytical expressions obtained in Sect. 2.2 for the expansion coefficients of the field of a spherical DR over the waves of a rectangular waveguide make it possible, using (3.1), to simply calculate the mutual coupling coefficients for any non-degenerate oscillations of magnetic or electric type. As above, for $\varepsilon_{1r} \gg 1$, we use expressions for the internal field of the DR (1.2) and (1.3), without taking into account the influence of the walls of the waveguide.

Using also (2.11)–(2.14), as well as (3.1), after the necessary transformations we obtain for two spherical DR in a rectangular waveguide [1], (Fig. 3.1a):

when the same magnetic types of oscillations are excited in both resonators H_{nml}

$$k = \alpha_n^H(p, q)T_n^m(k_0x_1, k_0x_2; k_0y_1, k_0y_2; k_0\Delta z); \tag{3.2}$$

when the same electrical oscillations are excited in both resonators E_{nml}:

$$k = \alpha_n^E(p, q)U_n^m(k_0x_1, k_0x_2; k_0y_1, k_0y_2; k_0\Delta z); \tag{3.3}$$

when magnetic H_{nml} oscillations are excited in one resonator and electrical E_{nml} oscillations in the other:

$$k = \alpha_n^{HE,EH}(p_1, q_1; p_2, q_2)R_n^m(k_0x_1, k_0x_2; k_0y_1, k_0y_2; k_0\Delta z). \tag{3.4}$$

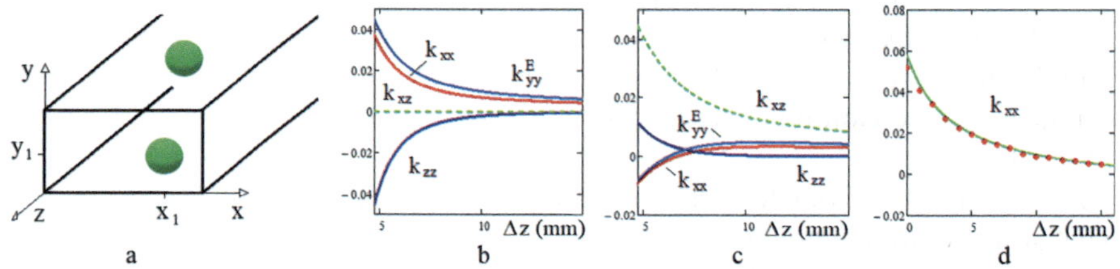

Fig. 3.1 Spherical DRs in a rectangular waveguide (**a**). Dependence of the mutual coupling coefficients for the main magnetic H_{1m1} oscillations of spherical DR on the distance along the waveguide axis for $x_1 = x_2 = a/2$ (**b**); $x_1 = a/2$; $x_2 = 3a/4$ (**c**); ($y_1 = y_2 = b/2$); red dots—measurement result $a = 20$ mm; $b = 15$ mm; $f_0 = 7$ GHz; $\varepsilon_{1r} = 81$ (**d**)

where $\alpha_n^H(p,q)$; $\alpha_n^E(p,q)$ are the functions determined from (2.20). The functions $\alpha_n^{EH}(p_1,q_1;p_2,q_2)$; $\alpha_n^{HE}(p_1,q_1;p_2,q_2)$ depend on the characteristic parameters of the DR and the types of excited oscillations:

$$\begin{aligned}
\alpha_n^{EH}(p_1,q_1;p_2,q_2) &= \frac{\varepsilon_{1r} j_n(p_1)}{y_n(q_1)} \cdot \frac{j_n(p_2)}{q_2 y_n(q_2)} \\
&\cdot \frac{1}{\left\{\left[p_2^2 - n(n+1)\right] j_n^2(p_2) + \left[n j_n(p_2) - p_2 j_{n-1}(p_2)\right]^2\right\}}; \\
\alpha_n^{HE}(p_1,q_1;p_2,q_2) &= \frac{j_n(p_1)}{y_n(q_1)} \cdot \frac{j_n(p_2)}{\varepsilon_{2r} q_2 y_n(q_2)} \\
&\cdot \frac{1}{\left\{\left[p_2^2 - n(n+1)\right] j_n^2(p_2) + \left[n j_n(p_2) - p_2 j_{n-1}(p_2)\right]^2\right\}}.
\end{aligned} \tag{3.5}$$

The functions

$$\begin{aligned}
T_n^m(k_0x_1, k_0x_2; k_0y_1, k_0y_2; k_0\Delta z) &= -\frac{32\pi}{1+\delta_{m0}} \frac{(2n+1)}{n(n+1)} \cdot \frac{(n-m)!}{(n+m)!} \cdot \frac{1}{k_0^2 ab} \\
&\cdot \left\{ \sum_{s,u=(0)}^{\infty} \frac{(1-\gamma^2) e^{-|\gamma k_0 \Delta z|}}{(1+\delta_{s0}+\delta_{u0})|\gamma|} \left[\frac{dP_n^m(\gamma)}{d\gamma}\right]^2 \right. \\
&\cdot \prod_{v=1}^{2} \begin{bmatrix} \sin\chi_{uy} y_v \left\{\sin\chi_{sx} x_v \sin m\Psi_0 \cos m\frac{\pi}{2} - \cos\chi_{sx} x_v \cos m\Psi_0 \sin m\frac{\pi}{2}\right\} \\ \cos\chi_{uy} y_v \left\{\cos\chi_{sx} x_v \cos m\Psi_0 \cos m\frac{\pi}{2} - \sin\chi_{sx} x_v \sin m\Psi_0 \sin m\frac{\pi}{2}\right\} \end{bmatrix} \\
&+ m^2 \sum_{s,u=1}^{\infty} \frac{e^{-|\gamma k_0 \Delta z|}}{|\gamma|} \left[\frac{P_n^m(\gamma)}{\sqrt{1-\gamma^2}}\right]^2 \\
&\cdot \left. \prod_{v=1}^{2} \begin{bmatrix} \sin\chi_{uy} y_v \left\{\sin\chi_{sx} x_v \cos m\Psi_0 \cos m\frac{\pi}{2} + \cos\chi_{sx} x_v \sin m\Psi_0 \sin m\frac{\pi}{2}\right\} \\ \cos\chi_{uy} y_v \left\{\sin\chi_{sx} x_v \cos m\Psi_0 \sin m\frac{\pi}{2} + \cos\chi_{sx} x_v \sin m\Psi_0 \cos m\frac{\pi}{2}\right\} \end{bmatrix} \right\};
\end{aligned} \tag{3.6}$$

$$U_n^m(k_0x_1, k_0x_2; k_0y_1, k_0y_2; k_0\Delta z) = -\frac{32\pi}{1+\delta_{m0}} \frac{(2n+1)}{n(n+1)} \cdot \frac{(n-m)!}{(n+m)!} \cdot \frac{1}{k_0^2 ab}$$

$$\cdot \left\{ \sum_{s,u=1}^{\infty} \frac{(1-\gamma^2)\mathrm{e}^{-|\gamma k_0 \Delta z|}}{|\gamma|} \left[\frac{\mathrm{d}P_n^m(\gamma)}{\mathrm{d}\gamma} \right]^2 \right.$$

$$\cdot \prod_{v=1}^{2} \begin{bmatrix} \cos \chi_{uy} y_v \left\{ \sin \chi_{sx} x_v \cos m\Psi_0 \sin m\frac{\pi}{2} + \cos \chi_{sx} x_v \sin m\Psi_0 \cos m\frac{\pi}{2} \right\} \\ \sin \chi_{uy} y_v \left\{ \sin \chi_{sx} x_v \cos m\Psi_0 \cos m\frac{\pi}{2} + \cos \chi_{sx} x_v \sin m\Psi_0 \sin m\frac{\pi}{2} \right\} \end{bmatrix}$$

$$+ m^2 \sum_{s,u=(0)}^{\infty} \frac{e^{-|\gamma k_0 \Delta z|}}{(1+\delta_{so}+\delta_{uo})|\gamma|(1-\gamma^2)} \left[P_n^m(\gamma) \right]^2$$

$$\left. \cdot \prod_{v=1}^{2} \begin{bmatrix} \cos \chi_{uy} y_v \left\{ \sin \chi_{sx} x_v \sin m\Psi_0 \sin m\frac{\pi}{2} + \cos \chi_{sx} x_v \cos m\Psi_0 \cos m\frac{\pi}{2} \right\} \\ \sin \chi_{uy} y_v \left\{ \cos \chi_{sx} x_v \cos m\Psi_0 \sin m\frac{\pi}{2} + \sin \chi_{sx} x_v \sin m\Psi_0 \cos m\frac{\pi}{2} \right\} \end{bmatrix} \right\}; \tag{3.7}$$

$$R_n^m(k_0x_1, k_0x_2; k_0y_1, k_0y_2; k_0\Delta z) = -\frac{32\pi}{k_0^2 ab} m \cdot \frac{(2n+1)}{n(n+1)} \cdot \frac{(n-m)!}{(n+m)!} \cdot$$

$$\left\{ \sum_{s,u=(0)}^{\infty} \frac{e^{-|\gamma k_0 \Delta z|}}{(1+\delta_{so}+\delta_{u0})} \frac{P_n^m(\gamma)}{|\gamma|} \cdot \frac{dP_n^m(\gamma)}{d\gamma} \cdot \right.$$

$$\cdot \begin{bmatrix} \sin \chi_{uy} y_1 \left\{ \sin \chi_{sx} x_1 \sin m\Psi_0 \cos^2 m\frac{\pi}{2} - \cos \chi_{sx} x_1 \cos m\Psi_0 \sin^2 m\frac{\pi}{2} \right\} \\ \cos \chi_{uy} y_1 \left\{ \cos \chi_{sx} x_1 \cos m\Psi_0 \cos^2 m\frac{\pi}{2} - \sin \chi_{sx} x_1 \sin m\Psi_0 \sin^2 m\frac{\pi}{2} \right\} \end{bmatrix}$$

$$\cdot \begin{bmatrix} \cos \chi_{uy} y_2 \left\{ \cos \chi_{sx} x_2 \cos m\Psi_0 \cos^2 m\frac{\pi}{2} - \sin \chi_{sx} x_2 \sin m\Psi_0 \sin^2 m\frac{\pi}{2} \right\} \\ \sin \chi_{uy} y_2 \left\{ \cos \chi_{sx} x_2 \cos m\Psi_0 \sin^2 m\frac{\pi}{2} - \sin \chi_{sx} x_2 \sin m\Psi_0 \cos^2 m\frac{\pi}{2} \right\} \end{bmatrix}$$

$$+ \sum_{s,u=1}^{\infty} e^{-|\gamma k_0 \Delta z|} \frac{P_n^m(\gamma)}{|\gamma|} \cdot \frac{dP_n^m(\gamma)}{d\gamma}$$

$$\cdot \begin{bmatrix} \sin \chi_{uy} y_1 \left\{ \sin \chi_{sx} x_1 \cos m\Psi_0 \cos^2 m\frac{\pi}{2} + \cos \chi_{sx} x_1 \sin m\Psi_0 \sin^2 m\frac{\pi}{2} \right\} \\ - \cos \chi_{uy} y_1 \left\{ \sin \chi_{sx} x_1 \cos m\Psi_0 \sin^2 m\frac{\pi}{2} + \cos \chi_{sx} x_1 \sin m\Psi_0 \cos^2 m\frac{\pi}{2} \right\} \end{bmatrix}$$

$$\left. \cdot \begin{bmatrix} \cos \chi_{uy} y_2 \left\{ \sin \chi_{sx} x_2 \cos m\Psi_0 \sin^2 m\frac{\pi}{2} + \cos \chi_{sx} x_2 \sin m\Psi_0 \cos^2 m\frac{\pi}{2} \right\} \\ \sin \chi_{uy} y_2 \left\{ \sin \chi_{sx} x_2 \cos m\Psi_0 \cos^2 m\frac{\pi}{2} + \cos \chi_{sx} x_2 \sin m\Psi_0 \sin^2 m\frac{\pi}{2} \right\} \end{bmatrix} \right\}. \tag{3.8}$$

are determined by the geometric parameters of the structure: the transverse coordinates of the DR in the waveguide; longitudinal distance $\Delta z = z_2 - z_1$ and waveguide dimensions $a \times b$. Here $\gamma = i|\Gamma/k_0|$; $\chi_{sx} = s\pi/a$; $\chi_{uy} = u\pi/b$; $\chi = \left[\chi_{sx}^2 + \chi_{uy}^2\right]^{1/2}$; $\psi_0 = \arcsin(\chi_{sx}/\chi)$. For brevity, relations (3.2)–(3.3) are written for the case of two identical DR.

The expressions found show that non-degenerate oscillations of spherical DRs exist only if both resonators are located on the axis of a rectangular waveguide. From (3.8) it follows that the coupling between magnetic and electric types of oscillations in a waveguide are possible only for azimuthally inhomogeneous modes: $m \neq 0$.

In Fig. 3.1b, c, the dependences of the coupling functions on the relative longitudinal distance between the centers of the DR are given for the main magnetic and electrical non-degenerate oscillations. The coupling coefficient indices determine the types of oscillations in which the magnetic (electric) field vector for H_{1m1} (for E_{1m1}) at the center of the resonators is directed along one of the axes of the waveguide coordinate system. Let us note that these functions can vanish for certain values of the relative distance between the resonators (c). The calculations also indicate the possibility of effective excitation of coupled oscillations of magnetic and electrical types H_{111} and E_{111}. The measurements performed showed good accuracy of calculations (Fig. 3.1d), performed on the basis of the assumption of the locality of the field of natural oscillations in the DR material without taking into account the influence of the walls of the waveguide.

3.2 Mutual Coupling Coefficients of Cylindrical Dielectric Resonators in a Rectangular Waveguide

Let us consider the mutual coupling coefficients of cylindrical DR located in a waveguide of rectangular cross-section. We will assume that each of the resonators can occupy one of the three previously specified fixed positions (Fig. 2.3). Then the independent positions of a pair of cylindrical DRs can be denoted by: aa, ab, ac and cc (Fig. 3.2).

Let us assume that there are two identical cylindrical DRs of radius r_0, height L, consisting of a material with dielectric constant ε_{1r}. Using relations (2.4), we find the expansion coefficients of the external field of azimuthally homogeneous magnetic

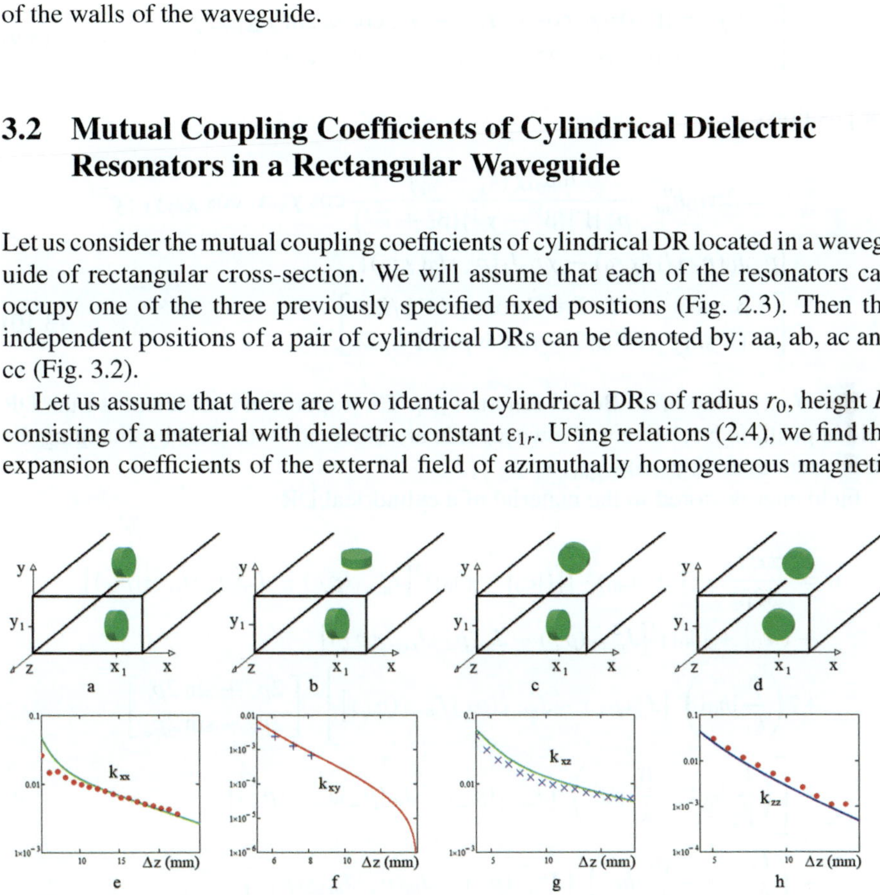

Fig. 3.2 Positions of cylindrical DRs in a rectangular waveguide (a–d). Comparison of theoretical and experimental data on the mutual coupling coefficients of identical cylindrical DRs with H_{101}^{+} oscillations and $\varepsilon_{1r} = 81$; $\Delta = 0,4$; $f_0 = 7,081$ GHz in a waveguide with the cross-section $a = 18$ mm, $b = 11$ mm. $x_1 = x_2 = a/2$; $y_1 = y_2 = b/2$ (a–e; d–h); $x_1 = 5$ mm; $x_2 = 7$ mm; $y_1 = 4$ mm; $y_2 = 5$ mm (b–f); $x_1 = 9$ mm; $x_2 = 12,5$ mm; $y_1 = y_2 = b/2$ (c–g)

oscillations $H_{nol}^{\pm}$ of the DR over the waveguide field (A.2), (A.3) taking into account (D.1)–(D.8): for position a:

$$\begin{pmatrix} c_{su}^{nol\pm} \\ d_{su}^{nol\pm} \end{pmatrix} = \pm 2\pi i r_0 \begin{pmatrix} i h_{su}^0 \chi_{sx} \\ \frac{e_{su}^0}{|w_{su}^E|} \chi_{xy} \end{pmatrix} e^{\mp\Gamma z_1}$$
$$\cdot \frac{\omega\mu_0 (k_1^2 - k_0^2)}{\beta\sqrt{k_0^2 - \chi_{sx}^2}(\beta^2 + \chi_{sx}^2 - k_0^2)(\beta_z^2 - \chi_{sx}^2)} \cos\chi_{uy} y_1$$
$$\cdot \left\{ \beta J_0(p_\perp) I_1\left(r_0\sqrt{\chi_{sx}^2 - k_0^2}\right) - \sqrt{\chi_{sx}^2 - k_0^2} J_1(p_\perp) I_0\left(r_0\sqrt{\chi_{sx}^2 - k_0^2}\right) \right\}$$
$$\cdot \begin{bmatrix} \sin\chi_{sx} x_1 (\beta_z \sin p_z \cos\chi_{sx} L/2 - \chi_{sx}\cos p_z \sin\chi_{sx} L/2) \\ \cos\chi_{sx} x_1 (\beta_z \cos p_z \sin\chi_{sx} L/2 - \chi_{sx}\sin p_z \cos\chi_{sx} L/2) \end{bmatrix}. \tag{3.9}$$

for position c:

$$c_{su}^{nol\pm} = 2\pi r_0 h_{su}^0 \cdot \frac{\omega\mu_0 \chi (k_1^2 - k_0^2)}{p_\perp |\Gamma| (\beta^2 - \chi^2)(\beta_z^2 + \Gamma^2)} \cos\chi_{sx} x_1 \cos\chi_{uy} y_1 \cdot e^{\mp\Gamma z_1}$$
$$\cdot \{ p_\perp J_0(p_\perp) J_1(\chi r_0) - \chi r_0 J_1(p_\perp) J_0(\chi r_0) \}$$
$$\cdot \begin{bmatrix} \beta_z \sin p_z ch\Gamma L/2 + \Gamma\cos p_z sh\Gamma L/2 \\ \pm(\Gamma\sin p_z ch\Gamma L/2 - \beta_z \cos p_z sh\Gamma L/2) \end{bmatrix}. \tag{3.10}$$

Here $I_n(x)$—modified Bessel function [2]. The expansion coefficients of the DR field at position b relative to the waveguide are obtained from (3.9) by changing the parameters $a \Leftrightarrow b$; $\chi_{sx} \Leftrightarrow \chi_{uy}$; $x_1 \Leftrightarrow y_1$.

Field energy stored in the material of a cylindrical DR:

$$w = \frac{\pi r_0^2 \varepsilon_1}{32\beta_z} \cdot (1 + \delta_{m0}) \cdot \Big\{ \Big[(|e_0| - |h_0|)^2 \big[J_{m-1}^2(p_\perp) - J_m(p_\perp) J_{m-2}(p_\perp) \big]$$
$$+ (|e_0| + |h_0|)^2 \big[J_{m+1}^2(p_\perp) - J_m(p_\perp) J_{m+2}(p_\perp) \big]$$
$$+ 2\left(\frac{\beta}{k_1}|h_0|\right)^2 \big[J_m^2(p_\perp) - J_{m-1}(p_\perp) J_{m+1}(p_\perp) \big] \Big] \cdot \begin{bmatrix} 2p_z + \sin 2p_z \\ 2p_z - \sin 2p_z \end{bmatrix}$$
$$+ \Big[\left(\frac{k_1}{\beta_z}|e_0| - \frac{\beta_z}{k_1}|h_0|\right)^2 \big[J_{m-1}^2(p_\perp) - J_m(p_\perp) J_{m-2}(p_\perp) \big]$$
$$+ \left(\frac{k_1}{\beta_z}|e_0| + \frac{\beta_z}{k_1}|h_0|\right)^2 \big[J_{m+1}^2(p_\perp) - J_m(p_\perp) J_{m+2}(p_\perp) \big]$$
$$+ 2\left(\frac{\beta}{\beta_z}|e_0|\right)^2 \big[J_m^2(p_\perp) - J_{m-1}(p_\perp) J_{m+1}(p_\perp) \big] \Big] \begin{bmatrix} 2p_z - \sin 2p_z \\ 2p_z + \sin 2p_z \end{bmatrix} \Big\}. \tag{3.11}$$

In the special case of magnetic azimuthally uniform oscillations $H_{nol}^{\pm}$

$$w = \frac{\pi r_0^2 \mu_0}{8\beta_z} \cdot s_{nol}(p_\perp, p_z), \tag{3.12}$$

where

$$s_{nol}(p_\perp, p_z) = \left\{ \left(\frac{k_1}{\beta}\right)^2 \left[J_1^2(p_\perp) - J_0(p_\perp)J_2(p_\perp)\right] + J_0^2(p_\perp) + J_1^2(p_\perp) \right\} \cdot \begin{bmatrix} 2p_z + \sin 2p_z \\ 2p_z - \sin 2p_z \end{bmatrix} + \left(\frac{\beta_z}{\beta}\right)^2 \left[J_1^2(p_\perp) - J_0(p_\perp)J_2(p_\perp)\right] \cdot \begin{bmatrix} 2p_z - \sin 2p_z \\ 2p_z + \sin 2p_z \end{bmatrix} \tag{3.13}$$

Here we used amplitude normalization (2.31).

Now substituting (3.9), (3.10), as well as (3.12) into (3.1), we obtain the mutual coupling coefficients of cylindrical DR with azimuthally homogeneous magnetic oscillations $H_{nol}^{\pm}$:

for position aa:

$$k_{xx} = \frac{128\pi k_0 \beta_z \left(k_1^2 - k_0^2\right)^2}{\beta^2 abs_{nol}(p_\perp, p_z)} \sum_{s,u=(0)}^{\infty} \left\{ \left[\frac{|\gamma| \chi_{sx}^2}{1 + \delta_{so} + \delta_{uo}} - \frac{\chi_{uy}^2}{|\gamma|}(1 - \delta_{so} - \delta_{uo}) \right] \cdot \frac{e^{-|\gamma k_0 \Delta z|}}{\left[\chi u_{sx}\left(\beta^2 + u_{sx}^2\right)\left(\beta_z^2 - \chi_{sx}^2\right)\right]^2} \begin{bmatrix} \sin \chi_{sx} x_1 \sin \chi_{sx} x_2 \\ \cos \chi_{sx} x_1 \cos \chi_{sx} x_2 \end{bmatrix} \cos \chi_{uy} y_1 \cos \chi_{uy} y_2 \cdot \left\{\beta J_0(p_\perp) I_1(u_{sx} r_0) - u_{sx} J_1(p_\perp) I_0(u_{sx} r_0)\right\}^2 \cdot \begin{bmatrix} \beta_z \sin p_z \cos \chi_{sx} L/2 - \chi_{sx} \cos p_z \sin \chi_{sx} L/2 \\ \beta_z \cos p_z \sin \chi_{sx} L/2 - \chi_{sx} \sin p_z \cos \chi_{sx} L/2 \end{bmatrix}^2 \right\}; \tag{3.14}$$

for position ab:

$$k_{xy} = \frac{128\pi k_0 \beta_z \left(k_1^2 - k_0^2\right)^2}{\beta^2 abs_{nol}(p_\perp, p_z)} \cdot \sum_{s,u=1}^{\infty} \left\{ \frac{\chi_{sx}\chi_{uy}}{\chi^2} \cdot \frac{|\gamma|\left(1 - 1/|\gamma|^2\right) e^{-|\gamma k_0 \Delta z|}}{u_{sx} u_{uy}\left(\beta^2 + u_{sx}^2\right)\left(\beta^2 + u_{uy}^2\right)\left(\beta_z^2 - \chi_{sx}^2\right)\left(\beta_z^2 - \chi_{uy}^2\right)} \cdot \left\{\beta J_0(p_\perp) I_1(u_{sx} r_0) - u_{sx} J_1(p_\perp) I_0(u_{sx} r_0)\right\} \cdot \left\{\beta J_0(p_\perp) I_1\left(u_{uy} r_0\right) - u_{uy} J_1(p_\perp) I_0\left(u_{uy} r_0\right)\right\} \cdot \begin{bmatrix} \sin \chi_{sx} x_1 (\beta_z \sin p_z \cos \chi_{sx} L/2 - \chi_{sx} \cos p_z \sin \chi_{sx} L/2) \\ \cos \chi_{sx} x_1 (\beta_z \cos p_z \sin \chi_{sx} L/2 - \chi_{sx} \sin p_z \cos \chi_{sx} L/2) \end{bmatrix} \cdot \begin{bmatrix} \cos \chi_{sx} x_2 \\ 0 \end{bmatrix} \cdot \begin{bmatrix} \cos \chi_{uy} y_1 \\ 0 \end{bmatrix} \cdot \begin{bmatrix} \sin \chi_{uy} y_2 \left(\beta_z \sin p_z \cos \chi_{uy} L/2 - \chi_{uy} \cos p_z \sin \chi_{uy} L/2\right) \\ \cos \chi_{uy} y_2 \left(\beta_z \cos p_z \sin \chi_{uy} L/2 - \chi_{uy} \sin p_z \cos \chi_{uy} L/2\right) \end{bmatrix} \right\}; \tag{3.15}$$

for position ac:

$$k_{xz} = \frac{128\pi k_0 \beta_z \left(k_1^2 - k_0^2\right)^2}{\beta^2 p_\perp abs_{nol}(p_\perp, p_z)} \sum_{s,u=(0)}^{\infty} \frac{1}{(1 + \delta_{so} + \delta_{u0})}$$
$$\cdot \frac{\chi_{sx} e^{-|\gamma k_0 \Delta z|}}{\chi u_{sx}\left(\beta^2 + u_{sx}^2\right)\left(\beta^2 - \chi^2\right)\left(\beta_z^2 - \chi_{sx}^2\right)\left(\beta_z^2 + \Gamma^2\right)}$$
$$\cdot \begin{bmatrix} \beta_z \sin p_z ch\Gamma L/2 + \Gamma \cos p_z sh\Gamma L/2 \\ \Gamma \sin p_z ch\Gamma L/2 - \beta_z \cos p_z sh\Gamma L/2 \end{bmatrix}$$
$$\cdot \{\beta J_0(p_\perp) I_1(u_{sx} r_0) - u_{sx} J_1(p_\perp) I_0(u_{sx} r_0)\}$$
$$\cdot \{p_\perp J_0(p_\perp) J_1(\chi r_0) - \chi r_0 J_1(p_\perp) J_0(\chi r_0)\}$$
$$\cdot \begin{bmatrix} \sin \chi_{sx} x_1 (\beta_z \sin p_z \cos \chi_{sx} L/2 - \chi_{sx} \cos p_z \sin \chi_{sx} L/2) \\ \cos \chi_{sx} x_1 (\beta_z \cos p_z \sin \chi_{sx} L/2 - \chi_{sx} \sin p_z \cos \chi_{sx} L/2) \end{bmatrix} \cdot \cos \chi_{sx} x_2$$
$$\cdot \begin{bmatrix} \cos \chi_{uy} y_1 \\ 0 \end{bmatrix} \cdot \cos \chi_{uy} y_2; \tag{3.16}$$

for position cc:

$$k_{zz} = -\frac{128\pi \beta_z \left(k_1^2 - k_0^2\right)^2}{k_0 p_\perp^2 abs_{nol}(p_\perp, p_z)} \sum_{s,u=(0)}^{\infty} \frac{e^{-|\gamma k_0 \Delta z|}}{(1 + \delta_{so} + \delta_{uo})|\gamma|\left[\left(\beta^2 - \chi^2\right)\left(\beta_z^2 + \Gamma^2\right)\right]^2}$$
$$\cdot \cos \chi_{sx} x_1 \cos \chi_{sx} x_2 \cos \chi_{uy} y_1 \cos \chi_{uy} y_2$$
$$\cdot \{p_\perp J_0(p_\perp) J_1(\chi r_0) - \chi r_0 J_1(p_\perp) J_0(\chi r_0)\}^2$$
$$\cdot \begin{bmatrix} p_z \sin p_z ch\Gamma L/2 + \Gamma \cos p_z sh\Gamma L/2 \\ \Gamma \sin p_z ch\Gamma L/2 - \beta_z \cos p_z sh\Gamma L/2 \end{bmatrix}^2, \tag{3.17}$$

where $\gamma = \Gamma/k_0$; $u_{sx} = \left[\chi_{sx}^2 - k_0^2\right]^{1/2}$; $u_{uy} = \left[\chi_{uy}^2 - k_0^2\right]^{1/2}$; $\chi = \left[\chi_{sx}^2 + \chi_{uy}^2\right]^{1/2}$. (x_s, y_s, z_s) coordinates of the center of the Sth DR.

Mutual coupling coefficients can be varied over a wide range not only by choosing the relative sizes and varying the relative distance between the centers of the DR, but also by changing the relative orientation of their axes in the waveguide. Thus, mutual coupling coefficients can be increased by placing one or both DR coaxially (Fig. 3.2a, d) and significantly reduced by using DR positions with orthogonal arrangement of their axes (Fig. 3.2b, c).

From (3.15) it follows that mutual coupling coefficient of the DR with H_{101}^+ oscillations in positions ab vanishes if the resonators are located in one of the symmetry planes of the rectangular waveguide: $x_1 = x_2 = a/2$; or $y_1 = y_2 = a/2$. In position ac (3.16), the coefficients of mutual coupling of even oscillations of the DR are equal to zero in the plane of symmetry waveguide at $x_1 = x_2 = a/2$.

A comparison of the measured and theoretical values of the coupling coefficients of cylindrical DR with fundamental oscillations is shown in Fig. 3.2e–h.

3.3 Coupled Oscillations of Spherical Dielectric Resonators in Circular and Coaxial Waveguides

The mutual coupling coefficients of identical spherical DR located on the axis of a circular waveguide have the form [3]:

when magnetic oscillations are excited in both resonators H_{nml}:

$$k = \alpha_n^{\mathrm{H}}(p, q) F_n^m(k_0 R, k_0 \Delta z); \tag{3.18}$$

when electrical oscillations are excited in both resonators E_{nml}:

$$k = \alpha_n^{E}(p, q) G_n^m(k_0 R, k_0 \Delta z); \tag{3.19}$$

when magnetic H_{nml} oscillations are excited in one resonator and electrical E_{nml} oscillations in the other

$$k = \alpha_n^{H,E}(p, q) S_n^m(k_0 R, k_0 \Delta z). \tag{3.20}$$

The $\alpha_n^{H,E}(p, q)$ functions also coincide with those found earlier (2.20), (3.5). The functions that determine the dependence of the coupling on the waveguide parameters have the form:

$$\begin{aligned} F_n^m(k_0 R, k_0 \Delta z) &= -4 \frac{2n+1}{n(n+1)} \frac{(n-m)!}{(n+m)!} \\ &\cdot \sum_{s=1}^{\infty} \left\{ \frac{1}{|\gamma'|} \left[(1-\gamma'^2) \frac{dP_n^m(\gamma')}{d\gamma'} \right]^2 \cdot \frac{e^{-|\gamma' k_0 \Delta z|}}{(j'^2_{m,s} - m^2) J_m^2(j'_{m,s})} \right. \\ &\left. + \frac{1}{|\gamma|} \left[m P_n^m(\gamma) \right]^2 \cdot \frac{e^{-|\gamma k_0 \Delta z|}}{\left[j_{m,s} J'_m(j_{m,s}) \right]^2} \right\} \end{aligned} \tag{3.21}$$

$$\begin{aligned} G_n^m(k_0 R, k_0 \Delta z) &= -4 \frac{2n+1}{n(n+1)} \frac{(n-m)!}{(n+m)!} \\ &\cdot \sum_{s=1}^{\infty} \left\{ \frac{1}{|\gamma'|} \left[m P_n^m(\gamma') \right]^2 \cdot \frac{e^{-|\gamma' k_0 \Delta z|}}{(j'^2_{m,s} - m^2) J_m^2(j'_{m,s})} \right. \\ &\left. + \frac{1}{|\gamma|} \left[(1-\gamma^2) \frac{dP_n^m(\gamma)}{d\gamma} \right]^2 \cdot \frac{e^{-|\gamma k_0 \Delta z|}}{j^2_{m,s} J'_{m,s}(j_{m,s})} \right\}; \end{aligned} \tag{3.22}$$

$$\begin{aligned} S_n^m(k_0 R, k_0 \Delta z) &= -4m \tfrac{2n+1}{n(n+1)} \tfrac{(n-m)!}{(n+m)!} \cdot \\ &\cdot \sum_{s=1}^{\infty} \left\{ \tfrac{1}{|\gamma'|} (1-\gamma'^2) \tfrac{dP_n^m(\gamma')}{d\gamma'} P_n^m(\gamma') \cdot \tfrac{e^{-|\gamma' k_0 \Delta z|}}{(j'^2_{m,s} - m^2) J_m^2(j'_{m,s})} + \right. \\ &\left. + \tfrac{1}{|\gamma|} (1-\gamma^2) \tfrac{dP_n^m(\gamma)}{d\gamma} P_n^m(\gamma) \cdot \tfrac{e^{-|\gamma k_0 \Delta z|}}{j^2_{m,s} J'^2_m(j_{m,s})} \right\} \end{aligned} \tag{3.23}$$

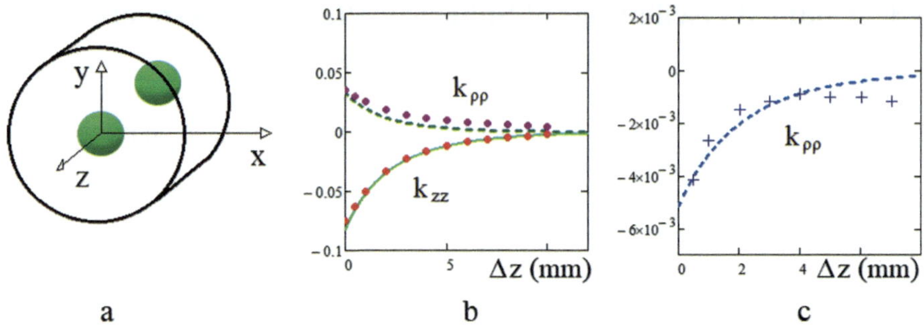

Fig. 3.3 Theoretical dependencies and measurement data of the mutual coupling coefficients of spherical DR ($\varepsilon_{1r} = 81$) with oscillations H_{1m1} (**b**); E_{111} (**c**) in a circular waveguide (**a**) of radius $R = 12, 5$ mm

where $\gamma' = i\left[\left(\frac{j'_{m,s}}{k_0 R}\right)^2 - 1\right]^{1/2}$; $\gamma = i\left[\left(\frac{j_{m,s}}{k_0 R}\right)^2 - 1\right]^{1/2}$.

The results of measurements of the mutual coupling coefficients of spherical DRs in such waveguides showed that the calculation error using formulas (3.21)–(3.23) approximately corresponds to the doubled error in calculating the characteristic parameters of the DR and is within 10% (Fig. 3.3).

Let us study the convergence of series (3.21)–(3.23) for the main values of the angular indices. We assume that the exact summation of functions is carried out up to a certain value M, and subsequent terms of the series are discarded. Let us estimate the size of the discarded residues. For this purpose, taking into account that $s \geq M \gg 1$, we will use the main terms of the asymptotic expansions of the roots and Bessel functions [2]

$$
\begin{aligned}
&j_{\nu,s} \approx (s + \nu/2 - 1/4)\pi;\ j'_{\nu,s} \approx (s + \nu/2 - 3/4)\pi;\\
&J_\nu(x) \approx \sqrt{2/(\pi x)}\cos(x - \nu\pi/2 - \pi/4);\\
&J'_\nu(x) \approx \sqrt{2/(\pi x)}\sin(x - \nu\pi/2 - \pi/4).
\end{aligned}
\tag{3.24}
$$

Substituting (3.24) into (3.21)–(3.23), and summing from M to ∞, we get

$$
\begin{aligned}
\delta_M F_1^0 &\approx 3\left(\frac{\pi}{k_0 R}\right)^3 \cdot \frac{e^{-\left(M-\frac{3}{4}\right)\pi \Delta z/R}}{1-e^{-\pi \Delta z/R}} \\
&\cdot \left\{\left[M-\frac{3}{4}+\frac{e^{-\pi \Delta z/R}}{1-e^{-\pi \Delta z/R}}\right]^2+\frac{e^{-\pi \Delta z/R}}{(1-e^{-\pi \Delta z/R})^2}\right\}; \\
\delta_M F_1^1 &\approx \frac{3}{2}\left(\frac{\pi}{k_0 R}\right)^3 \cdot \frac{e^{-M\pi \Delta z/R}}{1-e^{-\pi \Delta z/R}} \\
&\cdot \left\{e^{+\frac{\pi}{4}\Delta z/R}\left[\left(M-\frac{1}{4}+\frac{e^{-\pi \Delta z/R}}{1-e^{-\pi \Delta z/R}}\right)^2+\frac{e^{-\pi \Delta z/R}}{(1-e^{-\pi \Delta z/R})^2}\right]-\left(\frac{k_0 R}{\pi}\right)^2 e^{-\frac{\pi}{4}\Delta z/R}\right\}; \\
\delta_M G_1^0 &\approx 3\left(\frac{\pi}{k_0 R}\right)^3 \cdot \frac{e^{-\left(M-\frac{1}{4}\right)\pi \Delta z/R}}{1-e^{-\pi \Delta z/R}} \\
&\cdot \left\{\left[M-\frac{1}{4}+\frac{e^{-\pi \Delta z/R}}{1-e^{-\pi \Delta z/R}}\right]^2+\frac{e^{-\pi \Delta z/R}}{(1-e^{-\pi \Delta z/R})^2}\right\}; \\
\delta_M G_1^1 &\approx \frac{3}{2}\left(\frac{\pi}{k_0 R}\right)^3 \cdot \frac{e^{-M\pi \Delta z/R}}{1-e^{-\pi \Delta z/R}} \\
&\cdot \left\{\left(\frac{k_0 R}{\pi}\right)^2 e^{+\frac{\pi}{4}\Delta z/R}-e^{-\frac{\pi}{4}\Delta z/R}\left[\left(M+\frac{1}{4}+\frac{e^{-\pi \Delta z/R}}{1-e^{-\pi \Delta z/R}}\right)^2+\frac{e^{-\pi \Delta z/R}}{(1-e^{-\pi \Delta z/R})}\right]\right\}; \\
\delta_M S_1^1 &\approx \frac{3}{2}\left(\frac{\pi}{k_0 R}\right)^2 \cdot \frac{e^{-M\pi \Delta z/R}}{1-e^{-\pi \Delta z/R}} \\
&\cdot \left\{e^{+\frac{\pi}{4}\Delta z/R}\left[M-\frac{1}{4}+\frac{e^{-\pi \Delta z/R}}{1-e^{-\pi \Delta z/R}}\right]+e^{-\frac{\pi}{4}\Delta z/R}\left[M+\frac{1}{4}+\frac{e^{-\pi \Delta z/R}}{1-e^{-\pi \Delta z/R}}\right]\right\}.
\end{aligned}
\tag{3.25}
$$

The found expressions show that the values of the remainders change with increasing M according to the law: $M^2 e^{-M\pi\Delta z/R}$, or $Me^{-M\pi\Delta z/R}$. These functions decrease monotonically if $M > M_{\max}$, where $M_{\max} = 2R/\pi\Delta z$ in the first case and $M_{\max} = R/\pi\Delta z$ in the second. Thus, we find that the number of taken into account terms of the series decreases with increasing relative distance between the centers of the resonators, or with decreasing waveguide radius. Relations (3.25) can also be used to calculate the remainders of the expansions (3.21)–(3.23).

General expressions for the mutual coupling coefficients for any transverse coordinates of spherical DR in circular (Fig. 3.3a), coaxial (Fig. 3.4a) and sectoral (Fig. 3.4b) waveguides [3, 8] are found using the theorem Graph addition [2] for Bessel functions. For example, when placing resonators in a coaxial waveguide with the coordinates of the centers (ρ_1, α_1, z_1) and (ρ_2, α_2, z_2) (Fig. 3.4a):

in the case of magnetic non-degenerate H_{nml} oscillations in each DR:

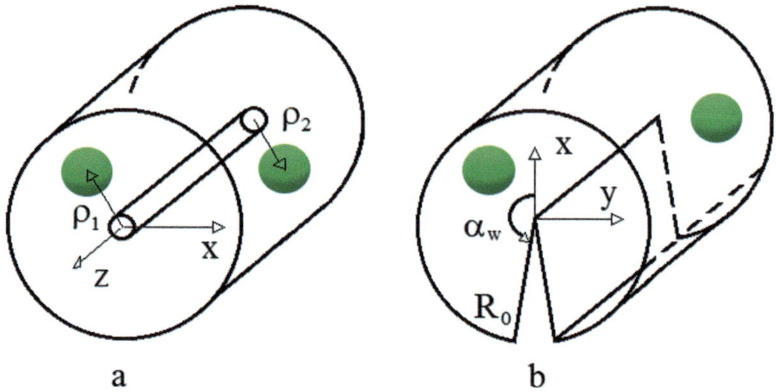

Fig. 3.4 Spherical DR in a coaxial (**a**) and sectoral (**b**) waveguides

$$
\begin{aligned}
k = {} & \frac{-4}{1+\delta_{mo}} \cdot \frac{2n+1}{n(n+1)} \cdot \frac{(n-m)!}{(n+m)!} \cdot \alpha_n^H(p,q) \\
& \cdot \sum_{u=0}^{\infty} \frac{\cos u(\alpha_1-\alpha_2)}{1+\delta_{uo}} \cdot \sum_{s=1}^{\infty} \left\{ \frac{1}{|\gamma'|} \left[\left(1-\gamma'^2\right) \frac{dP_n^m(\gamma')}{d\gamma'} \right]^2 \right. \\
& \cdot \frac{e^{-|\gamma' k_0 \Delta z|}}{C_u^2(\chi R_2)\left[(\chi R_2)^2 - u^2\right] - C_u^2(\chi R_1)\left[(\chi R_1)^2 - u^2\right]} \\
& \cdot \left[C_{m+u}(\chi\rho_1) \mp (-1)^m C_{-m+u}(\chi\rho_1)\right]\left[C_{m+u}(\chi\rho_2) \mp (-1)^m C_{-m+u}(\chi\rho_2)\right] \\
& + \frac{1}{|\gamma|}\left[P_n^m(\gamma)\right]^2 \cdot \frac{m^2 e^{-|\gamma k_0 \Delta z|}}{\left[\chi R_2 D_{u-1}(\chi R_2)\right]^2 - \left[\chi R_1 D_{u-1}(\chi R_1)\right]^2} \\
& \left. \cdot \left[D_{m+u}(\chi\rho_1) \pm (-1)^m D_{-m+u}(\chi\rho_1)\right]\left[D_{m+u}(\chi\rho_2) \pm (-1)^m D_{-m+u}(\chi\rho_2)\right] \right\};
\end{aligned}
\tag{3.26}
$$

for odd H_{nm_1l} oscillations ($h_\theta \sim \sin m_1\varphi$) of the first and even H_{nm_2l} oscillations ($h_\theta \sim \cos m_2\varphi$) of the second DR:

$$
\begin{aligned}
k = & -4 \cdot \frac{2n+1}{n(n+1)} \cdot \left[\prod_{\nu=1}^{2} \frac{1}{1+\delta_{m_\nu o}} \cdot \frac{(n-m_\nu)!}{(n+m_\nu)!} \right]^{1/2} \cdot \alpha_n^H(p,q) \\
& \cdot \sum_{u=1}^{\infty} \sin u(\alpha_2 - \alpha_1) \cdot \sum_{s=1}^{\infty} \left\{ \frac{1}{|\gamma'|} \left(1-\gamma'^2\right)^2 \cdot \frac{dP_n^{m_1}(\gamma')}{d\gamma'} \cdot \frac{dP_n^{m_2}(\gamma')}{d\gamma'} \right. \\
& \cdot \frac{e^{-|\gamma' k_0 \Delta z|}}{C_u^2(\chi R_2)\left[(\chi R_2)^2 - u^2\right] - C_u^2(\chi R_1)\left[(\chi R_1)^2 - u^2\right]} \\
& \cdot \left[C_{m_1+u}(\chi \rho_1) - (-1)^{m_1} C_{-m_1+u}(\chi \rho_1)\right]\left[C_{m_2+u}(\chi \rho_2) + (-1)^{m_2} C_{-m_2+u}(\chi \rho_2)\right] \\
& + \frac{1}{|\gamma|} P_n^{m_1}(\gamma) P_n^{m_2}(\gamma) \cdot \frac{m_1 m_2 e^{-|\gamma k_0 \Delta z|}}{\left[\chi R_2 D_{u-1}(\chi R_2)\right]^2 - \left[\chi R_1 D_{u-1}(\chi R_1)\right]^2} \\
& \left. \cdot \left[D_{m_1+u}(\chi \rho_1) + (-1)^{m_1} D_{-m_1+u}(\chi \rho_1)\right]\left[D_{m_2+u}(\chi \rho_2) - (-1)^{m_2} D_{-m_2+u}(\chi \rho_2)\right] \right\};
\end{aligned}
\tag{3.27}
$$

in the case of electrical non-degenerate E_{nml} oscillations in each DR:

$$
\begin{aligned}
k = & \frac{-4}{1+\delta_{mo}} \cdot \frac{2n+1}{n(n+1)} \cdot \frac{(n-m)!}{(n+m)!} \cdot \alpha_n^E(p,q) \\
& \cdot \sum_{u=0}^{\infty} \frac{\cos u(\alpha_1 - \alpha_2)}{1+\delta_{uo}} \cdot \sum_{s=1}^{\infty} \left\{ \frac{1}{|\gamma'|} \left[P_n^m(\gamma')\right]^2 \right. \\
& \cdot \frac{m^2 e^{-|\gamma' k_0 \Delta z|}}{C_u^2(\chi R_2)\left[(\chi R_2)^2 - u^2\right] - C_u^2(\chi R_1)\left[(\chi R_1)^2 - u^2\right]} \\
& \cdot \left[C_{m+u}(\chi \rho_1) \pm (-1)^m C_{-m+u}(\chi \rho_1)\right]\left[C_{m+u}(\chi \rho_2) \pm (-1)^m C_{-m+u}(\chi \rho_2)\right] \\
& + \frac{1}{|\gamma|} \left[\left(1-\gamma^2\right) \frac{dP_n^m(\gamma)}{d\gamma}\right]^2 \cdot \frac{e^{-|\gamma k_0 \Delta z|}}{\left[\chi R_2 D_{u-1}(\chi R_2)\right]^2 - \left[\chi R_1 D_{u-1}(\chi R_1)\right]^2} \\
& \left. \cdot \left[D_{m+u}(\chi \rho_1) \mp (-1)^m D_{-m+u}(\chi \rho_1)\right]\left[D_{m+u}(\chi \rho_2) \mp (-1)^m D_{-m+u}(\chi \rho_2)\right] \right\};
\end{aligned}
\tag{3.28}
$$

for odd in φ E_{nm_1l} oscillations of the first and even E_{nm_2l} oscillations of the second DR:

$$
\begin{aligned}
k = -4 \cdot \frac{2n+1}{n(n+1)} \cdot \left[\prod_{\nu=1}^{2} \frac{1}{\left(1+\delta_{m_\nu o}\right)} \cdot \frac{(n-m_\nu)!}{(n+m_\nu)!} \right]^{1/2} \cdot \alpha_n^E(p,q) \\
\cdot \sum_{u=1}^{\infty} \sin u(\alpha_2 - \alpha_1) \cdot \sum_{s=1}^{\infty} \left\{ \frac{1}{|\gamma'|} \cdot P_n^{m_1}(\gamma') \cdot P_n^{m_2}(\gamma') \right. \\
\cdot \frac{m_1 m_2 e^{-|\gamma' k_0 \Delta z|}}{C_u^2(\chi R_2)\left[(\chi R_2)^2 - u^2\right] - C_u^2(\chi R_1)\left[(\chi R_1)^2 - u^2\right]} \\
\cdot \left[C_{m_1+u}(\chi\rho_1) + (-1)^{m_1} C_{-m_1+u}(\chi\rho_1)\right]\left[C_{m_2+u}(\chi\rho_2) - (-1)^{m_2} C_{-m_2+u}(\chi\rho_2)\right] \\
+ \frac{1}{|\gamma|}\left(1-\gamma^2\right)^2 \cdot \frac{dP_n^{m_1}(\gamma)}{d\gamma} \cdot \frac{dP_n^{m_2}(\gamma)}{d\gamma} \\
\cdot \frac{e^{-|\gamma k_0 \Delta z|}}{\left[\chi R_2 D_{u-1}(\chi R_2)\right]^2 - \left[\chi R_1 D_{u-1}(\chi R_1)\right]^2} \\
\cdot \left. \left[D_{m_1+u}(\chi\rho_1) - (-1)^{m_1} D_{-m_1+u}(\chi\rho_1)\right]\left[D_{m_2+u}(\chi\rho_2) + (-1)^{m_2} D_{-m_2+u}(\chi\rho_2)\right] \right\}.
\end{aligned}
\tag{3.29}
$$

Oscillations of DR of magnetic and electrical types can also be coupled with each other. In particular, for odd magnetic H_{nml} and even electric E_{nml} (upper sign), as well as even magnetic and odd electric (lower sign):

$$
\begin{aligned}
k = \pm 4m \frac{2n+1}{n(n+1)} \cdot \frac{(n-m)!}{(n+m)!} \cdot \alpha_n^{EH}(p,q) \\
\cdot \sum_{u=0}^{\infty} \frac{\cos u(\alpha_2 - \alpha_1)}{1+\delta_{uo}} \sum_{s=1}^{\infty} \left\{ \frac{1}{|\gamma'|}\left(1-\gamma'^2\right) \frac{dP_n^m(\gamma')}{d\gamma'} P_n^m(\gamma') \right. \\
\cdot \frac{e^{-|\gamma' k_0 \Delta z|}}{C_u^2(\chi R_2)\left[(\chi R_2)^2 - u^2\right] - C_u^2(\chi R_1)\left[(\chi R_1)^2 - u^2\right]} \\
\cdot \left[C_{m+u}(\chi\rho_1) \mp (-1)^m C_{-m+u}(\chi\rho_1)\right]\left[C_{m+u}(\chi\rho_2) \mp (-1)^m C_{-m+u}(\chi\rho_2)\right] \\
+ \frac{1}{|\gamma|}\left(1-\gamma^2\right) \frac{dP_n^m(\gamma)}{d\gamma} P_n^m(\gamma) \cdot \frac{e^{-|\gamma k_0 \Delta z|}}{\left[\chi R_2 D_{u-1}(\chi R_2)\right]^2 - \left[\chi R_1 D_{u-1}(\chi R_1)\right]^2} \\
\cdot \left. \left[D_{m+u}(\chi\rho_1) \pm (-1)^m D_{-m+u}(\chi\rho_1)\right]\left[D_{m+u}(\chi\rho_2) \pm (-1)^m D_{-m+u}(\chi\rho_2)\right] \right\};
\end{aligned}
\tag{3.30}
$$

for odd in φ (upper sign), or even (lower sign) magnetic and electrical oscillations:

$$k = -4m\frac{2n+1}{n(n+1)} \cdot \frac{(n-m)!}{(n+m)!} \cdot \alpha_n^{EH}(p,q)$$
$$\cdot \sum_{u=0}^{\infty} \sin u(\alpha_2 - \alpha_1) \cdot \sum_{s=1}^{\infty} \left\{ \frac{1}{|\gamma'|}(1-\gamma'^2)\frac{dP_n^m(\gamma')}{d\gamma'}P_n^m(\gamma') \right.$$
$$\cdot \frac{e^{-|\gamma' k_0 \Delta z|}}{C_u^2(\chi R_2)\left[(\chi R_2)^2 - u^2\right] - C_u^2(\chi R_1)\left[(\chi R_1)^2 - u^2\right]}$$
$$\cdot \left[C_{m+u}(\chi\rho_1) \mp (-1)^m C_{-m+u}(\chi\rho_1)\right]\left[C_{m+u}(\chi\rho_2) \pm (-1)^m C_{-m+u}(\chi\rho_2)\right]$$
$$+ \frac{1}{|\gamma|}(1-\gamma^2)\frac{dP_n^m(\gamma)}{d\gamma}P_n^m(\gamma) \cdot \frac{e^{-|\gamma k_0 \Delta z|}}{\left[\chi R_2 D_{u-1}(\chi R_2)\right]^2 - \left[\chi R_1 D_{u-1}(\chi R_1)\right]^2}$$
$$\left. \cdot\left[D_{m+u}(\chi\rho_1) \pm (-1)^m D_{-m+u}(\chi\rho_1)\right]\left[D_{m+u}(\chi\rho_2) \mp (-1)^m D_{-m+u}(\chi\rho_2)\right]\right\}, \quad (3.31)$$

where also $\gamma' = i\Gamma_H/k_0$ for magnetic waves; $\gamma = i\Gamma_E/k_0$ for electric waves of the coaxial waveguide.

In the case of a circular waveguide, functions (A.4) transform into Bessel functions, and the parameters take on the values: $\chi R_2 = j'_{u,s}$—for magnetic waves; $\chi R_2 = j_{u,s}$—for electric waves. If one of the resonators is located on the axis of the waveguide, then in the sums (3.26)–(3.30) the terms with $u = m$. In this case, the dependence of the coupling on the angles between the directions of field polarizations at the centers of the DR becomes harmonic: $k(\alpha_2 - \alpha_1) = k(0) \cdot \sin$ or $\cos u(\alpha_2 - \alpha_1)$. A similar effect occurs in the case of cylindrical DRs with H_{101} oscillations rotated relative to each other in a circular waveguide axis [4]. In the general case, the dependence of the coupling on coordinates is more complex.

3.4 Mutual Coupling Coefficients of Various Forms of Dielectric resonators

The spectrum of parasitic oscillations of bandpass filters can be rarefied by using resonators of various types, manufactured in such a way that their main oscillations coincide in frequency, forming a working band, and the frequencies of parasitic oscillations differ from each other. To create such filters, it is necessary to know the coupling parameters of DRs of various forms.

Let us consider a pair of spherical and cylindrical DRs located in one of the symmetry planes of a rectangular waveguide (Fig. 3.5a, b). We assume that the frequency of one of the natural oscillations of the resonators, for example the main, coincides. In this case, the calculation of the mutual coupling coefficients, carried out on the basis of relations (3.1) (3.9)–(3.10), and (2.11)–(2.14), leads to the following expressions:

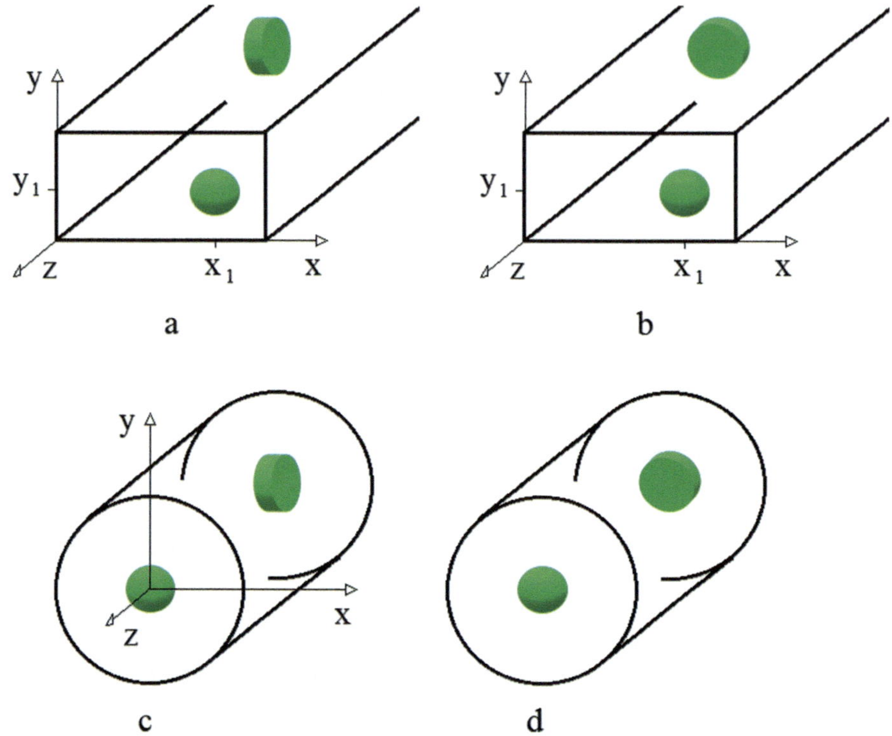

Fig. 3.5 Standard positions of spherical and cylindrical DR in a rectangular (**a, b**) and circular (**c–d**) waveguides

for magnetic azimuthally uniform oscillations in resonators and
in the case of the coupling of the spherical DR with cylindrical DR and
a—position of the cylindrical DR and the location of their centers in the plane: $x_1 = x_2 = a/2$ (Fig. 3.5a):

$$k = \frac{32\pi\left(k_1^2 - k_0^2\right)}{\beta abv(p_\perp, p_z)} \cdot \left|\frac{j_1(p)}{y_1(q)}\right| \cdot \sum_{s,u=(0)}^{\infty}\left[\frac{\chi_{sx}^2|\gamma|}{(1+\delta_{uo})} - \frac{\chi_{uy}^2}{|\gamma|}\right]$$
$$\cdot \frac{e^{-|\gamma k_0 \Delta z|}}{u_{sx}\chi^2\left(\beta^2 + u_{sx}^2\right)\left(\beta_z^2 - \chi_{sx}^2\right)} \cdot \sin^2 s\frac{\pi}{2} \cos\chi_{uy}y_1 \cos\chi_{uy}y_2$$
$$\cdot \{\beta J_0(p_\perp)I_1(u_{sx}r_1) - u_{sx}J_1(p_\perp)I_0(u_{sx}r_1)\}$$
$$\cdot (\beta_z \sin p_z \cos\chi_{sx}L/2 - \chi_{sx}\cos p_z \sin\chi_{sx}L/2); \quad (3.32)$$

in the case of a—position and for magnetic azimuthally uniform oscillations in the cylindrical DR and electrical E_{11l} oscillation in the spherical DR in the plane: $y_1 = y_2 = b/2$:

$$k = \frac{32\pi\left(k_1^2 - k_0^2\right)\varepsilon_{2r}}{\beta ab\upsilon(p_\perp, p_z)}\left|\frac{j_1(p)}{y_1(q)}\right| \cdot \sum_{s,u=(0)}^{\infty} \frac{1}{(1+\delta_{uo})}$$
$$\cdot \frac{e^{-|\gamma k_0 \Delta z|}}{u_{sx}\left(\beta^2 + u_{sx}^2\right)\left(\beta_z^2 - \chi_{sx}^2\right)} \cdot \sin\chi_{sx}x_1 \sin\chi_{sx}x_2 \cos^2 u\frac{\pi}{2}$$
$$\cdot \{\beta J_0(p_\perp)I_1(u_{sx}r_1) - u_{sx}J_1(p_\perp)I_0(u_{sx}r_1)\}$$
$$\cdot (\beta_z \sin p_z \cos\chi_{sx}L/2 - \chi_{sx}\cos p_z \sin\chi_{sx}L/2); \tag{3.33}$$

with a coaxial arrangement (Fig. 3.5b) of a cylindrical DR on the axis of a rectangular waveguide in the case of azimuthally uniform magnetic oscillations:

$$k = \frac{32\pi\left(k_1^2 - k_0^2\right)}{p_\perp k_0^2 ab\upsilon(p_\perp, p_z)}\left|\frac{j_1(p)}{y_1(q)}\right| \cdot \sum_{s,u=(0)}^{\infty} \frac{1}{(1+\delta_{uo}+\delta_{so})}$$
$$\cdot \frac{e^{-|\gamma k_0 \Delta z|}}{|\gamma|} \cdot \frac{\chi}{\left(\beta^2 - \chi^2\right)\left(\beta_z^2 + \Gamma^2\right)} \cdot \cos^2 s\frac{\pi}{2}\cos^2 u\frac{\pi}{2}$$
$$\cdot \{p_\perp J_0(p_\perp)J_1(\chi r_1) - \chi r_1 J_1(p_\perp)J_0(\chi r_1)\}$$
$$\cdot (\beta_z \sin p_z ch\Gamma L/2 + \Gamma\cos p_z sh\Gamma L/2), \tag{3.34}$$

Here r_1, $L/2$—dimensions of the cylindrical, and r_2—radius of the spherical DR; $p_\perp = \beta r_1$; $p_z = \beta_z L/2$—characteristic parameters of a cylindrical resonator; $p = k_1 r_2$; $q = k_0 r_2$—characteristic parameters of a spherical resonator; $\gamma = \Gamma/k_0$; $\chi = \left(\chi_{sx}^2 + \chi_{uy}^2\right)^{1/2}$; $u_{sx} = \left(\chi_{sx}^2 + k_0^2\right)^{1/2}$.

$$\upsilon(p_\perp, p_z) = q_z/p_z\Big\{\left[(k_1/\beta)^2\left\{J_1^2(p_\perp) - J_0(p_\perp)J_2(p_\perp)\right\} + J_0^2(p_\perp) + J_1^2(p_\perp)\right]$$
$$\cdot(2p_z + \sin 2p_z) + (\beta_z/\beta)^2\left\{J_1^2(p_\perp) - J_0(p_\perp)J_2(p_\perp)\right\}\cdot(2p_z - \sin 2p_z)\Big\}; \tag{3.35}$$

for the coupling coefficient of the spherical DR with cylindrical (3.1).

In the case of coupling of a cylindrical DR with the main magnetic type of oscillations of a spherical DR:

$$\upsilon(p_\perp, p_z) \to \upsilon(p, q) = q/3\left\{\left(p^2 - 2\right)j_1^2(p) + \left[j_1(p) - pj_0(p)\right]^2\right\};$$

As well as in the case of coupling of the cylindrical DR with the main electrical type of spherical oscillations:

$$\upsilon(p_\perp, p_z) \to \upsilon(p, q) = \varepsilon_{2r}q/3\left\{\left(p^2 - 2\right)j_1^2(p) + \left[j_1(p) - pj_0(p)\right]^2\right\}.$$

The mutual coupling coefficients of cylindrical and spherical DR located coaxially in a circular waveguide (Fig. 3.5d) have the form:

for magnetic azimuthally homogeneous oscillations

$$
\begin{aligned}
k = {} & \frac{8(k_1^2-k_0^2)}{\beta R^3 k_0^3 \upsilon(p_\perp,p_z)}\left|\frac{j_1(p)}{y_1(q)}\right| \cdot \sum_{s=1}^{\infty}\left\{\frac{j'_{0,s}}{J_0^2(j'_{0,s})}\right. \\
& \cdot \frac{e^{-|\gamma k_0 \Delta z|}}{|\gamma|}\cdot\frac{1}{(\beta^2-\chi^2)(\beta_z^2+\Gamma^2)} \\
& \cdot\{\chi J_0(\chi r_1)J_1(p_\perp)-\beta J_0(p_\perp)J_1(\chi r_1)\} \\
& \left.\cdot(\Gamma\cos p_z sh\Gamma L/2+\beta_z\sin p_z ch\Gamma L/2)\right\};
\end{aligned}
\tag{3.36}
$$

where $\Delta z = |z_1 - z_2|$; $j_{0,s}\left(j'_{0,s}\right)$—root of the (derivative) Bessel function $J_0(x)$; $\upsilon(p_\perp, p_z)$—function (3.35); $\chi = j_{0,s}/R$—for electrical, $\chi = j'_{0,s}/R$ for magnetic waves circular waveguide.

Coupled oscillations can also be realized in a DR system of the same shape, but made of different dielectrics.

Obviously, in the case of spherical DR, the general form of expressions for the coupling coefficients will remain the same as in (3.2)–(3.4), as well as (3.18)–(3.20). Only the type of coupling multipliers that determine the dependence on material parameters will change:

For magnetic or electric types of oscillations in both DRs:

$$
\begin{aligned}
\alpha_n^H(p_1,q_1;p_2,q_2) = {} & \frac{j_n(p_1)}{y_n(q_1)}\cdot\frac{j_n(p_2)}{q_2 y_n(q_2)} \\
& \cdot\frac{1}{\left\{\left[p_2^2-n(n+1)\right]j_n^2(p_2)+\left[n j_n(p_2)-p_2 j_{n-1}(p_2)\right]^2\right\}} \\
\alpha_n^E(p_1,q_1;p_2,q_2) = {} & \frac{\varepsilon_{1r} j_n(p_1)}{y_n(q_1)}\cdot\frac{j_n(p_2)}{q_2 y_n(q_2)} \\
& \cdot\frac{1}{\left\{\left[p_2^2-n(n+1)\right]j_n^2(p_2)+\left[n j_n(p_2)-p_2 j_{n-1}(p_2)\right]^2\right\}}
\end{aligned}
\tag{3.37}
$$

Here $j_n(z)$; $y_n(z)$ are the spherical Bessel and the Neumann functions, respectively [2]. The characteristic parameters $p_t = k_t r_t$, $q_t = k_0 r_t$ can be obtained from the equations for natural oscillations of the tth spherical DR (1.7) or (1.8) (or (1.12), (1.13)). Here r_t—is the radius of the tth DR.

Suppose we have two DRs of cylindrical shape with radius r_1 and r_2, height of L_1 and L_2, respectively. Assume that each resonator is excited in the fundamental magnetic oscillation H_{101}. In this case, the coupling coefficient in the rectangulare waveguide is of the form:

$$k_{12} = \frac{64\pi\beta_{2z}}{v_2} \cdot \frac{r_1}{r_2} \cdot \frac{\beta_2}{\beta_1} \cdot \frac{(k_1^2 - k_0^2)(k_2^2 - k_0^2)}{k_2^2 ab}$$
$$\cdot \{ \sum_{s,u=0}^{\infty} (\frac{|\Gamma|\chi_{sx}^2}{1+\delta_{u0}} - \frac{k_0^2\chi_{uy}^2}{|\Gamma|}) e^{-|\Gamma\Delta z|} \cdot \frac{1}{\chi^2(\chi_{uy}^2 - \Gamma^2)}$$
$$\cdot \prod_{t=1}^{2} \frac{\sin\chi_{sx}x_t \cos\chi_{uy}y_t}{(\beta_t^2 - \chi_{uy}^2 + \Gamma^2)} \Big[\beta_t J_0(p_{t\perp}) J_1(r_t\sqrt{\chi_{uy}^2 - \Gamma^2})$$
$$-\sqrt{\chi_{uy}^2 - \Gamma^2} J_1(p_{t\perp}) J_0(r_t\sqrt{\chi_{uy}^2 - \Gamma^2})\Big]$$
$$\cdot \frac{1}{\beta_{tz}^2 - \chi_{sx}^2} \Big[\beta_{tz} \sin p_{tz} \cos\chi_{sx}\frac{L_t}{2} - \chi_{sx}\cos p_{tz} \sin\chi_{sx}\frac{L_t}{2}\Big]\}. \tag{3.38}$$

Where $v_2 = v(p_{2\perp}, p_{2z})$; $p_{t\perp} = \beta_t r_t$; $p_{tz} = \beta_{tz}L_t/2$ $(t = 1, 2)$;

$$v(p_\perp, p_z) = \{J_1^2(p_\perp) - J_0(p_\perp)J_2(p_\perp)\} \cdot (2p_z + \sin 2p_z).$$

On Fig. 3.6 shows the dependences of the mutual coupling coefficients calculated using formulas (3.21), (3.37), (3.38) for the main magnetic oscillations of the resonators. It can be seen that the difference between the coefficients of mutual coupling k_{12} and k_{21} can be significant.

3.5 Coupled Oscillations of Dielectric Resonators in a Planar Waveguide

The application of a transformation of the form (2.62)–(2.63), carried out by passing to a continuous spectrum, allows us to find new useful expressions for the mutual coupling coefficients of DR in a plane waveguide (Figs. 3.7 and 3.8a). In this case, it is more convenient to determine the complex coefficient of mutual coupling between two DRs in a more general form:

$$\kappa_{12} = k_{12} + i\tilde{k}_{12}, \tag{3.39}$$

where

$$\tilde{k}_{12} = \frac{1}{\omega_0 w_2} \cdot \sum_{t<t_M} (c_t^{1\pm})_0 (c_t^{2\pm})_0^* e^{-i\Gamma|\Delta z|}. \tag{3.40}$$

In (3.40), the t multi-index runs through values characteristic only of propagating waves of the line, and k_{12} is calculated according to (3.1) by expanding the field of

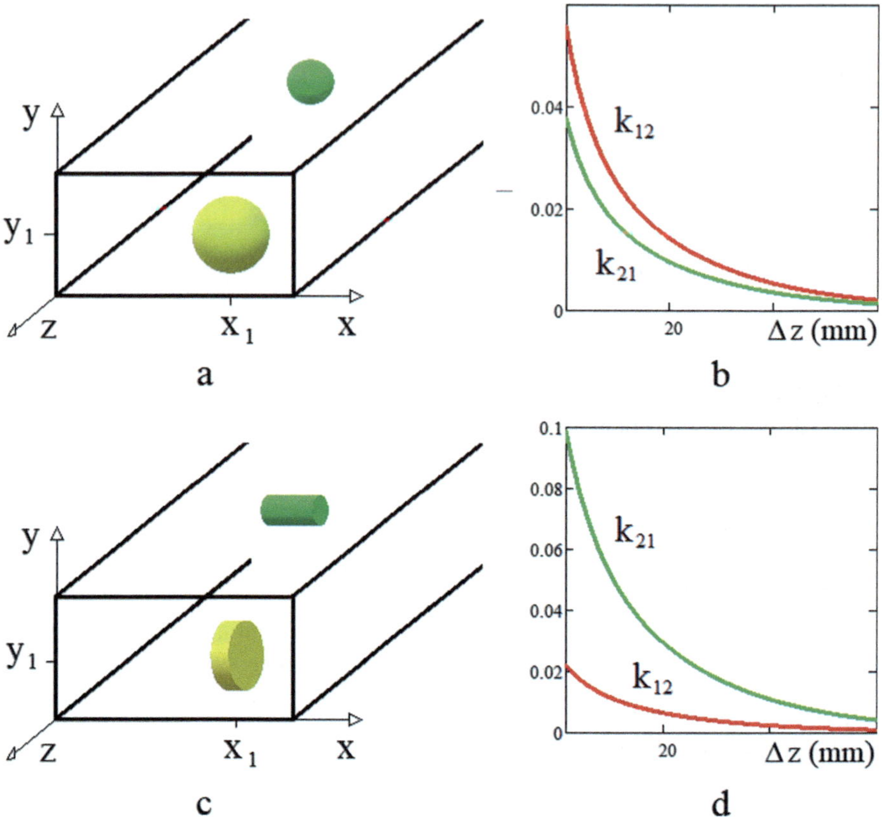

Fig. 3.6 Different spherical (**a**); cylindrical (**c**), DRs in a rectangular waveguide. Dependence of mutual coupling coefficients versus distance between DR centers (**b**, **d**): $\varepsilon_{1r} = 36$; $\varepsilon_{2r} = 82$; $x_1 = x_2 = a/2$; $y_1 = y_2 = b/2$; $\Delta_1 = L_1/2r_1 = 0,3$; $\Delta_2 = L_2/2r_2 = 1,5$; $a = 25\,\text{mm}$; $b = 25\,\text{mm}$; $f_0 = 4\,\text{GHz}$

one of the DRs only into damped waves ($\Delta z = z_2 - z_1$). Coupling coefficients (3.39) can be calculated without dividing it into real and imaginary parts, based on general expressions for the coefficients of expansion of the DR field into transmission line waves.

As an example, let us consider the main H_{1m1} magnetic oscillations of spherical DR located in a rectangular waveguide. Let us go to the limit $a \to \infty$ in (3.1), (3.39), using (2.11), (2.12) for $n = 1$; $m = 0$; 1. When moving to integrals, we will also use a corollary of the well-known Riemann-Lebesgue Lemma [5], when the coordinates of the resonators in the cross section of the waveguide $x_s \to \infty$ ($s = 1, 2$) for finite $\Delta x = x_2 - x_1$.

The resulting integrals can be calculated exactly (D.14) and after the necessary algebraic transformations, we present (3.39) in the form:

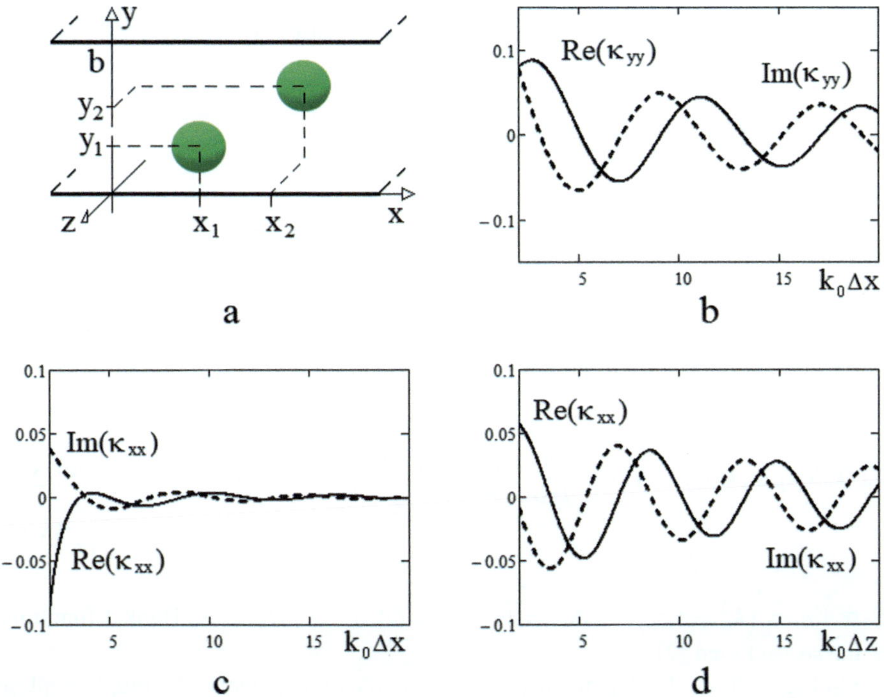

Fig. 3.7 Dependences of the real and imaginary parts of mutual coupling coefficients of spherical DR with H_{1m1} oscillations in a plane waveguide (**a**); $k_0b = 5$; $\varepsilon_{1r} = 9$; $y_1 = y_2 = b/2$. Polarization of the magnetic field at the center of the DR, parallel to the y axis (**b**); x (**c, d**)

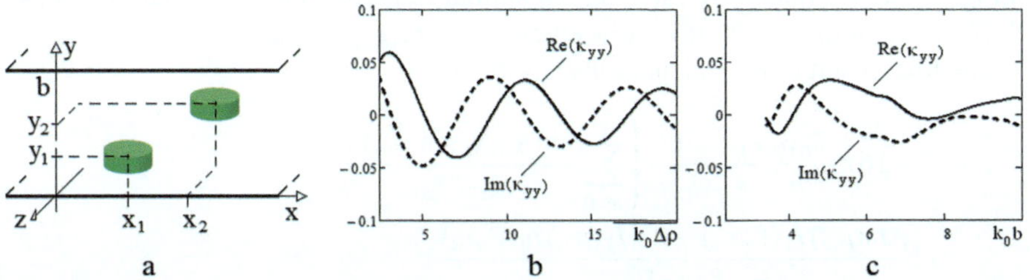

Fig. 3.8 Dependences of the real and imaginary parts of the mutual coupling coefficients (**b, c**) of cylindrical DRs with H_{101}^{+} oscillations in a plane waveguide (**a**); $f_0 = 7$ GHz; $\varepsilon_{1r} = 9$; $\Delta = 0, 4$; $y_1 = y_2 = b/2$; $k_0b = 5$ (**b**); $k_0\Delta z = 11$ (**c**)

$$\kappa_{12} = \alpha_1^{\mathrm{H}}(p, q)\Im_1^m(k_0b; k_0\Delta x, k_0\Delta y, k_0\Delta z) \tag{3.41}$$

The resulting expressions for the multipliers $\Im_1^m$ of mutual coupling coefficients will be further distinguished by the direction of the magnetic field at the centers of the resonators:

for H_{11l} oscillations with magnetic field polarization at the centers of the DR parallel to the axis x:

$$\Im_{xx} = \frac{6\pi i}{k_0 b} \cdot \left(\sum_{u=0}^{\infty} \frac{\cos\chi_{uy}y_1 \cos\chi_{uy}y_2}{1+\delta_{uo}} \right. \\ \left. \cdot \left\{ \left[\left(\frac{\Gamma_{II}\Delta z}{k_0\Delta\rho} \right)^2 + \left(\frac{\chi_{uy}}{k_0} \right)^2 \right] H_0^{(2)}(\Gamma_{II}\Delta\rho) - \frac{\Gamma_{II}}{k_0} \cdot \frac{k_0^2(\Delta z^2 - \Delta x^2)}{(k_0\Delta\rho)^3} H_1^{(2)}(\Gamma_{II}\Delta\rho) \right\} \tag{3.42}$$

for H_{11l} oscillations with magnetic field polarization at the center of the DR parallel to the axis y:

$$\Im_{yy}(k_0 b; k_0 x_0, k_0 y_0, k_0\Delta z) = \frac{6\pi i}{k_0 b} \cdot \sum_{u=0}^{\infty} \frac{\sin\chi_{uy}y_1 \sin\chi_{uy}y_2}{1+\delta_{uo}} \cdot \left(\frac{\Gamma_{II}}{k_0} \right)^2 \cdot H_0^{(2)}(\Gamma_{II}\Delta\rho). \tag{3.43}$$

Here $\Gamma_{II} = \left(k_0^2 - \chi_{uy}^2\right)^{1/2}$; $\Delta\rho = \left[(\Delta x)^2 + (\Delta z)^2\right]^{1/2}$; $H_n^{(2)}(z)$—Hankel functions of the second kind [2].

On Fig. 3.7b–d, the dependences of the complex coefficients of mutual coupling are given for varying the coordinates of the resonators in the plane of symmetry of the waveguide.

Expressions of the mutual coupling coefficients of azimuthally homogeneous magnetic oscillations of cylindrical DR in a plane waveguide in the position shown in Fig. 3.8, a, we obtain by carrying out similar transformations of relations (3.14), (2.27) taking into account the replacement: $a \rightleftarrows b$; $\chi_{sx} \rightleftarrows \chi_{uy}$.

The final result of the calculations has the form:

$$\kappa_{yy} = 16\pi i \frac{k_0^2\beta_z}{b} \frac{\varepsilon_{1r} - 1}{\varepsilon_{1r} v_0} \cdot \left\{ \sum_{u=0}^{\infty} \frac{\sin\chi_{uy}y_1 \sin\chi_{uy}y_2}{1+\delta_{u0}} \right. \\ \cdot \left[\frac{\beta J_0(p_\perp) J_1(\Gamma_u r_0) - \Gamma_u J_1(p_\perp) J_0(\Gamma_u r_0)}{\beta^2 - \Gamma_u^2} \right]^2 \\ \left. \cdot \left[\frac{(\beta_z \sin p_z \cos\chi_{uy}L/2 - \chi_{uy}\cos p_z \sin\chi_{uy}L/2)}{\beta_z^2 - \chi_{uy}^2} \right]^2 H_0^{(2)}(\Gamma_u\Delta\rho) \right\} \tag{3.44}$$

Here

$$v_0 = v(p_\perp, p_z) = [J_1^2(p_\perp) - J_0(p_\perp)J_2(p_\perp)](2p_z + \sin 2p_z). \tag{3.45}$$

On Fig. 3.8b shows the dependence of the mutual coupling of identical DRs made of dielectric with $\varepsilon_{1r} = 9$, calculated using formulas (3.44), on the distance between their centers (b) and the height of the waveguide (c). The above-considered coupled DR in a flat waveguide is convenient to use for calculating and analyzing the characteristics of screened gratings excited by the main types of natural oscillations.

3.6 Mutual Coupling Coefficients of Dielectric Resonators in a Rectangular Waveguide During Rotation of the Axes

The improvement of filter characteristics can be achieved by using less traditional structures, in particular, dielectric resonators with rotation of their axes relative to each other and the transmission line [6, 7]. To calculate and analyze the characteristics of such filters, it is required to knowledge the coupling coefficients of the DR located at an arbitrary angle with respect to the waveguide axis.

We investigated the values of the maximum variation coupling coefficients when changing the angles of resonator rotation. The dimensions of the DR corresponded to the condition of excitation of the main magnetic oscillations in them H_{111}.

To calculate (3.1), we represented the waveguide field (A.2), (A.3) [6] in the local coordinate system of the dielectric resonator (x', y', z'). After that, for simplicity, we rewrite obtained result in the form:

$$k_{12} = k_{12}^0 \sum_{t \geq t_M} f_t^1(\mp i\Gamma)(f_t^2(\pm i\Gamma))^* e^{-\Gamma|\Delta z|} \tag{3.46}$$

where

$$k_{12}^0 = \frac{16}{w_0}(\varepsilon_{1r} - 1)^2 \frac{p_x p_y p_z q_x q_y q_z}{\upsilon H};$$
$$\upsilon H = \varepsilon_{1r} k_0^2 [\beta_y^2 \pi_x^\mp \pi_y^\pm \pi_z^\mp + \beta_x^2 \pi_x^\pm \pi_y^\mp \pi_z^\mp] + \beta_x^2 \beta_z^2 \cdot \pi_x^\pm \pi_y^\mp \pi_z^\pm + \beta_y^2 \beta_z^2 \cdot \pi_x^\mp \pi_y^\pm \pi_z^\pm + +(\beta_x^2 + \beta_y^2)^2 \cdot \pi_x^\mp \pi_y^\mp \pi_z^\mp$$

Here

$$\pi_v^\pm = p_v \pm \sin p_v \cos p_v;$$

$v = (x, y, z)$. The upper and lower signs of $\pi_v^\pm$ correspond to the distributions of the DR field shown in brackets (1.32). $\varepsilon_{1r} = \varepsilon_1/\varepsilon_0$; $w_0 = (\mu_0/\varepsilon_0)^{1/2}$; $p_x = \beta_x a_0/2$; $p_y = \beta_y b_0/2$; $p_z = \beta_z L/2$ as well as $q_x = k_0 a_0/2$; $q_y = k_0 b_0/2$; $q_z = k_0 L/2$—is the resonator characteristic parameters.

(a) In the case of rotation of the resonator relatively to the x-axis in the waveguide coordinate system (Fig. 3.9a), the functions $f_t^n(\mp i\Gamma)$ of (3.46) take the form for $n = 1, 2$:

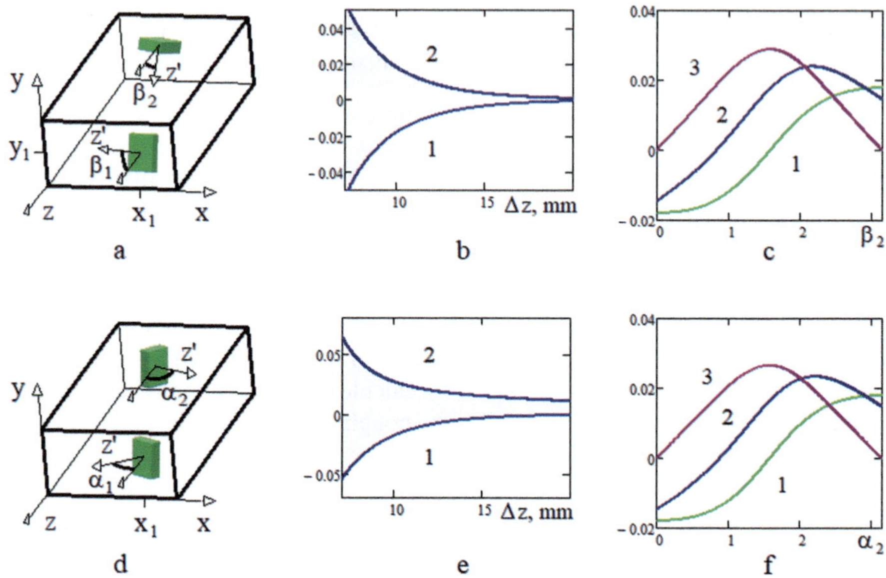

Fig. 3.9 Rectangular dielectric resonators rotating relatively the x (**a**) and y axis (**d**) in the rectangular waveguide. Variation of the mutual coupling coefficients at different longitudinal distances between the centers of rotating dielectric resonators with H_{111} oscillations ($\varepsilon_{1r} = 36$; $\Delta = 0, 4$; $B = 1$) at $f_0 = 7$ GHz in the waveguide $a = 20$ mm; $b = 15$ mm (**b**, **e**). Dependence of the mutual coupling coefficient on the angle of rotation of the second resonator (**c**, **f**)

$$
\begin{aligned}
f_t^n(\mp i\Gamma) = & \begin{pmatrix} i\sin\chi_{sx}x_n \\ \cos\chi_{sx}x_n \end{pmatrix} \cdot \{-E_{x0}^{*}\,\beta_y\omega_x(\xi_{sx}) \\
& \cdot\Big[e^{-i\chi_{uy}y_n}\,\overline{\omega}_y(-(\eta_{uy}\cos\beta_n \mp i\gamma\sin\beta_n))\omega_z(-(\eta_{uy}\sin\beta_n \pm i\gamma\cos\beta_n)) \\
& -e^{i\chi_{uy}y_n}\overline{\omega}_y(\eta_{uy}\cos\beta_n \pm i\gamma\sin\beta_n)\omega_z(\eta_{uy}\sin\beta_n \mp i\gamma\cos\beta_n)\Big] \\
& -E_{y0}^{*}\beta_y\cos\beta_n\overline{\omega}_x(\xi_{sx})\cdot\Big[e^{-i\chi_{uy}y_n}\,\omega_y(-(\eta_{uy}\cos\beta_n \mp i\gamma\sin\beta_n))\omega_z(-(\eta_{uy}\sin\beta_n \pm i\gamma\cos\beta_n)) \\
& +e^{i\chi_{uy}y_n}\,\omega_y(\eta_{uy}\cos\beta_n \pm i\gamma\sin\beta_n)\omega_z(\eta_{uy}\sin\beta_n \mp i\gamma\cos\beta_n)\Big] - iE_{z0}^{*}\beta_x\sin\beta_n\overline{\omega}_x(\xi_{sx}) \\
& \cdot\Big[e^{-i\chi_{uy}y_n}\,\omega_y(-(\eta_{uy}\cos\beta_n \mp i\gamma\sin\beta_n))\omega_z(-(\eta_{uy}\sin\beta_n \pm i\gamma\cos\beta_n)) \\
& -e^{i\chi_{uy}y_n}\,\omega_y(\eta_{uy}\cos\beta_n \pm i\gamma\sin\beta_n)\omega_z(\eta_{uy}\sin\beta_n \mp i\gamma\cos\beta_n)\Big]\Big\}.
\end{aligned}
\tag{3.47}
$$

Here $\chi_{sx} = s\pi/a$; $\chi_{uy} = u\pi/b$; $\xi_{sx} = \chi_{sx}/k_0$; $\eta_{uy} = \chi_{uy}/k_0$; $\Gamma = [(\chi_{sx})^2 + (\chi_{uy})^2 - k_0^2]^{1/2}$; $\gamma = \Gamma/k_0$—is the wave numbers of a rectangular cut-off waveguide with a cross section $a \times b$.

h_{su}^0; e_{su}^0 are from the (A.3); (x_n, y_n, z_n) is the coordinates of the center of the nth resonator in the waveguide; β_n—is the angle between the z'—axes of the local coordinate system of the resonator and the z—axes waveguide (Fig. 3.9a); $a_0 \times b_0 \times L$—dimensions of the Rectangular DR (Fig. 1.8a);

$$\omega_v(\xi) = \frac{1}{p_v^2 - (q_v\xi)^2} \cdot \begin{pmatrix} -i[p_v \cos p_v \sin(q_v\xi) - q_v\xi \sin p_v \cos(q_v\xi)] \\ p_v \sin p_v \cos(q_v\xi) - q_v\xi \cos p_v \sin(q_v\xi) \end{pmatrix};$$

$$\overline{\omega}_v(\xi) = \frac{1}{p_v^2 - (q_v\xi)^2} \cdot \begin{pmatrix} p_v \sin p_v \cos(q_v\xi) - q_v\xi \cos p_v \sin(q_v\xi) \\ i[p_v \cos p_v \sin(q_v\xi) - q_v\xi \sin p_v \cos(q_v\xi)] \end{pmatrix},$$

for $v = (x, y, z)$.

(b) The coupling coefficient of the dielectric resonators rotated relatively the y-axis determines by the functions (see Fig. 3.9d):

$$\begin{aligned} f_t^n(\mp i\Gamma) = &\begin{pmatrix} i \sin \chi_{uy} y_n \\ \cos \chi_{uy} y_n \end{pmatrix} \cdot \Big\{ E_{y0}^* \beta_x \omega_y(\eta_{uy}) \\ &\cdot \big[e^{-i\chi_{uy}x_n} \overline{\omega}_x(-(\xi_{sx} \cos \alpha_n \mp i\gamma \sin \alpha_n))\omega_z(-(\xi_{sx} \sin \alpha_n \pm i\gamma \cos \alpha_n)) \\ &- e^{i\chi_{sx}x_n} \overline{\omega}_x(\xi_{sx} \cos \alpha_n \pm i\gamma \sin \alpha_n)\omega_z(\xi_{sx} \sin \alpha_n \mp i\gamma \cos \alpha_n) \big] \\ &+ E_{x0}^* \beta_y \cos \alpha_n \overline{\omega}_y(\eta_{uy}) \\ &\cdot \big[e^{-i\chi_{sx}x_n} \omega_x(-(\xi_{sx} \cos \alpha_n \mp i\gamma \sin \alpha_n))\omega_z(-(\xi_{sx} \sin \alpha_n \pm i\gamma \cos \alpha_n)) \\ &+ e^{i\chi_{sx}x_n} \omega_x(\xi_{sx} \cos \alpha_n \pm i\gamma \sin \alpha_n)\omega_z(\xi_{sx} \sin \alpha_n \mp i\gamma \cos \alpha_n) \big] \\ &+ iE_{z0}^* \beta_y \sin \alpha_n \overline{\omega}_y(\eta_{uy}) \\ &\cdot \big[e^{-i\chi_{sx}x_n} \omega_x(-(\xi_{sx} \cos \alpha_n \mp i\gamma \sin \alpha_n))\omega_z(-(\xi_{sx} \sin \alpha_n \pm i\gamma \cos \alpha_n)) \\ &- e^{i\chi_{sx}x_n} \omega_x(\xi_{sx} \cos \alpha_n \pm i\gamma \sin \alpha_n)\omega_z(\xi_{sx} \sin \alpha_n \mp i\gamma \cos \alpha_n) \big] \Big\}. \end{aligned} \tag{3.48}$$

Here α_n—is the angle between the z'—axes of the local coordinate system of the resonator and the z—axes of the waveguide (Fig. 3.9d).

Despite the fact that the functions $f_t^s(\pm i\Gamma)$ are complex, the mutual coupling coefficients in most cases remain real when the resonator axes rotate (see, for example, Fig. 3.9c).

In Fig. 3.9b. e the dependences of the coupling coefficients on the longitudinal distance between the resonator centers are shown for $\alpha_1, \beta_1 = 0$ and $\alpha_2, \beta_2 =$ 0—1; $\alpha_2, \beta_2 = \pi$—2.

In Fig. 3.9c, f show the dependences of the coupling on the angle of rotation of the second resonator α_2, β_2 for $\alpha_1, \beta_1 = 0$—1; $\alpha_1, \beta_1 = \pi/4$—2; $\alpha_1, \beta_1 = \pi/2$—3; $\Delta z = 10mm$.

(c) In the case of rotation of the dielectric resonator relatively z-axis at the initial position of the resonator axis parallel to the waveguide axis (Fig. 3.10a), the coupling coefficient defined by functions:

$$\begin{aligned} f_t^n(\mp i\Gamma) = &\omega_z(\mp i\gamma) \cdot \{ -E_{x0}^* \beta_y \cos \alpha_n \cdot [\omega_x(\xi_{sx} \cos \alpha_n - \eta_{uy} \sin \alpha_n) \\ &\overline{\omega}_y(\xi_{sx} \sin \alpha_n + \eta_{uy} \cos \alpha_n) \cos(\chi_{sx}x_n + \chi_{uy}y_n) \\ &- \omega_x(\chi_{sx} \cos \alpha_n + \eta_{uy} \sin \alpha_n)\overline{\omega}_y(\xi_{sx} \sin \alpha_n - \eta_{uy} \cos \alpha_n) \\ &\cos(\chi_{sx}x_n - \chi_{uy}y_n)] + E_{y0}^* \beta_y \sin \alpha_n [\omega_x(\xi_{sx} \cos \alpha_n - \eta_{uy} \sin \alpha_n) \\ &\overline{\omega}_y(\xi_{sx} \sin \alpha_n + \eta_{uy} \cos \alpha_n) \cos(\chi_{sx}x_n + \chi_{uy}y_n) + \omega_x(\xi_{sx} \cos \alpha_n + \eta_{uy} \sin \alpha_n) \\ &\overline{\omega}_y(\xi_{sx} \sin \alpha_n - \eta_{uy} \cos \alpha_n) \cos(\chi_{sx}x_n - \chi_{uy}y_n)] + E_{x0}^* \beta_x \sin \alpha_n \\ &\cdot [\overline{\omega}_x(\xi_{sx} \cos \alpha_n - \eta_{uy} \sin \alpha_n)(\omega_y(\xi_{sx} \sin \alpha_n + \eta_{uy} \cos \alpha_n) \cos(\chi_{sx}x_n + \chi_{uy}y_n) \end{aligned}$$

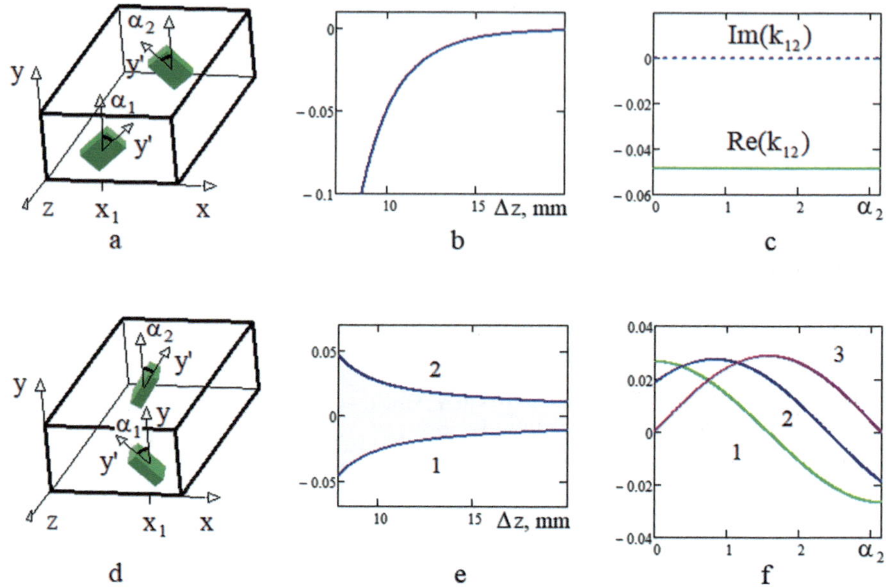

Fig. 3.10 Rectangular DRs rotating relatively the z axis for different initial positions in the rectangular waveguide (**a, d**). Variation of the mutual coupling coefficients at different longitudinal distances between the centers of resonators in the waveguide (**b, e**). Dependence of the mutual coupling coefficient on the angle of rotation of the second resonator

$$
\begin{aligned}
&- \overline{\omega}_x(\xi_{sx}\cos\alpha_n + \eta_{uy}\sin\alpha_n)\omega_y(\xi_{sx}\sin\alpha_n - \eta_{uy}\cos\alpha_n)\cos(\chi_{sx}x_n - \chi_{uy}y_n)] \\
&+ E^*_{y0}\beta_x\cos\alpha_n \cdot [\overline{\omega}_x(\xi_{sx}\cos\alpha_n - \eta_{uy}\sin\alpha_n)\omega_y(\xi_{sx}\sin\alpha_n + \eta_{uy}\cos\alpha_n) \\
&\cos(\chi_{sx}x_n + \chi_{uy}y_n) + \overline{\omega}_x(\xi_{sx}\cos\alpha_n + \eta_{uy}\sin\alpha_n) \\
&\omega_y\xi_{sx}\sin\alpha_n - \eta_{uy}\cos\alpha_n)\cos(\chi_{sx}x_n - \chi_{uy}y_n)]\}. \qquad (3.49)
\end{aligned}
$$

Here α_n is the angle between the y—axis of the waveguide and the y'—axis of the n—th resonator.

As can be seen from the data obtained, the maximum variation of the coupling coefficients is determined by the distribution of the natural oscillation field of the resonators in the transverse plane (x', y') of local coordinate system. For natural oscillations of a rectangular DR H_{111}, the spatial distribution of the field is close to the dipole and uniform in the plane (x', y'), therefore, the rotation of the resonator in this plane has little effect on the coupling coefficients. A good example of this effect is the rotation of resonators about the longitudinal axis of the waveguide shown in Fig. 3.10c.

(d) In the case of rotation of the dielectric resonator relative z-axis at the initial position of the resonator y'-axis parallel to the waveguide y-axis (Fig. 3.10d):

$$
\begin{gathered}
f_t^n(\mp i\Gamma) = i \cdot \{-E_{x0}^* \beta_x \sin\alpha_n \overline{\omega}_x(\mp i\gamma)\cdot \\
[\omega_z(\xi_{sx}\cos\alpha_n - \eta_{uy}\sin\alpha_n)\omega_y(\xi_{sx}\sin\alpha_n + \eta_{uy}\cos\alpha_n)\sin(\chi_{sx}x_n + \chi_{uy}y_n) - \\
-\omega_z(\xi_{sx}\cos\alpha_n + \eta_{uy}\sin\alpha_n)\omega_y(\xi_{sx}\sin\alpha_n - \eta_{uy}\cos\alpha_n)\sin(\chi_{sx}x_n - \chi_{uy}y_n)] - \\
-E_{y0}^* \beta_x \cos\alpha_n \overline{\omega}_x(\mp i\gamma)\cdot \\
\cdot[\omega_z(\xi_{sx}\cos\alpha_n - \eta_{uy}\sin\alpha_n)\omega_y(\xi_{sx}\sin\alpha_n + \eta_{uy}\cos\alpha_n)\sin(\chi_{sx}x_n + \chi_{uy}y_n) + \cdot \\
+\omega_z(\xi_{sx}\cos\alpha_n + \eta_{uy}\sin\alpha_n)\omega_y(\xi_{sx}\sin\alpha_n - \eta_{uy}\cos\alpha_n)\sin(\chi_{sx}x_n - \chi_{uy}y_n)] - \\
-iE_{z0}^* \beta_y \omega_x(\mp i\gamma)\cdot \\
[\omega_z(\xi_{sx}\cos\alpha_n - \eta_{uy}\sin\alpha_n)\overline{\omega}_y(\xi_{sx}\sin\alpha_n + \eta_{uy}\cos\alpha_n)\sin(\chi_{sx}x_n + \chi_{uy}y_n) - \\
-\omega_z(\xi_{sx}\cos\alpha_n + \eta_{uy}\sin\alpha_n)\overline{\omega}_y(\xi_{sx}\sin\alpha_n - \eta_{uy}\cos\alpha_n)\sin(\chi_{sx}x_n - \chi_{uy}y_n)]\}
\end{gathered}
\tag{3.50}
$$

In Figs. 3.9 and 3.10b, e depending on the coupling coefficients on the longitudinal distance between resonator centers with; $a_0 = b_0$; $L/a_0 = 0, 4$. The cross section of the waveguide was $a \times b = 20 \times 15$ mm; frequency of the fundamental magnetic oscillation of the resonator H_{111} ($h_z = h_1 \cos\beta_x x' \cos\beta_y y' \cos\beta_z z'$ in (1.32)) $f_0 =$ 7 GHz.

As follows from the indicated dependencies, the coupling coefficients can take both positive and negative values. This can be due to physical and mathematical reasons. If co-directional z'-axes of the resonators with H_{111} oscillations lie on the z-axis of the waveguide and $\Delta z > L$, the mutual coupling coefficient is negative. If the co-directed z'-axes are parallel and the resonators are coupled along the side wall ($z'_n < L$), the coupling coefficients are usually positive for H_{111} oscillations. If there is continuous rotation between these two positions of the DR, it leads to a change in the sign of the coupling (see Fig. 3.9c–f). However, the rotation of one of the resonators at an angle π can also lead to a change in the sign of the coupling. This change in coupling purely mathematically compensates for the change in the direction of the natural oscillation field of one of the resonators (see, for example, Fig. 3.9b or Fig. 3.10b).

As can be seen from the results of calculations, in spite of the fact that the coupling functions are complex, all coupling coefficients of the resonators in the rectangular cut-off waveguide are purely real.

Calculation of the mutual coupling coefficients of identical cylindrical DRs with H_{101} oscillations (1.17) in a cut-off rectangular waveguide leads to the expressions (3.46) in the coordinate system (x, y, z) (Fig. 3.11a–c). here for H_{101} [7]:

$$
k_{12}^0 = -2\pi \frac{q_\perp^2 p_z}{k_0 w_0} \frac{(\varepsilon_{1r} - 1)^2}{\varepsilon_{1r}} \frac{L}{v_0};
\tag{3.51}
$$

$p_\perp = \beta r_0$; $p_z = \beta_z L/2$; $q_\perp = k_0 L/2$; $q_z = k_0 L/2$; as in(3.45):

$$
v_0 = [J_1^2(p_\perp) - J_0(p_\perp)J_2(p_\perp)](2p_z + \sin 2p_z).
$$

(a) In the case of rotation of the resonator axis about the x axis (Fig. 3.11a):

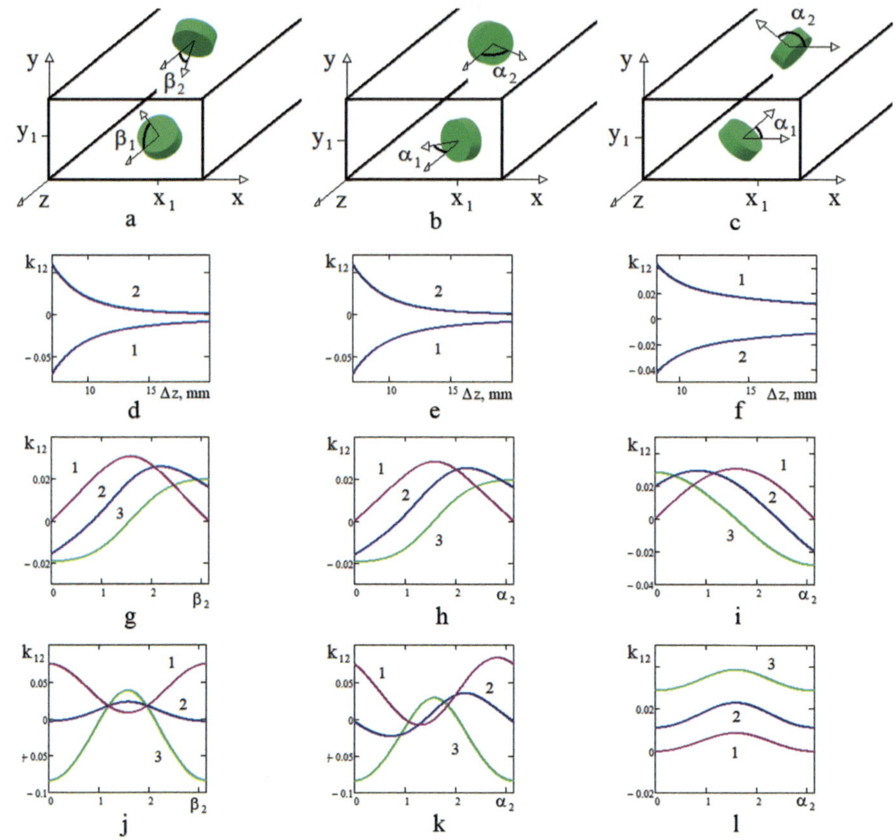

Fig. 3.11 Simple rotations of the cylindrical DRs relatively the x-axis (**a**); y-axis (**b**); z-axis (**c**) of the rectangular waveguide. Dependence of mutual coupling coefficients on the DR longitudinal coordinate $\Delta z = |z_1 - z_2|$ for $y_1 = y_2 = 0,5b$; (**d**) $\beta_1 = 0$;1—$\beta_2 = 0$; 2—$\beta_2 = \pi$; (**e, f**) $\alpha_1 = 0$;1—$\alpha_2 = 0$; 2—$\alpha_2 = \pi$. Dependence of the coupling coefficients on the rotation angle of the 2 DR for $x_1 = x_2 = 0,5a$; $y_1 = y_2 = 0,5b$; $\Delta z = 10$ mm: (g–i) 1—$\beta_1, \alpha_1 = 0$; 2—$\beta_1, \alpha_1 = \pi/4$;3—$\beta_1, \alpha_1 = \pi/2$. Dependence of the coupling coefficients on the rotation angle ((j) $\beta_1 = \beta_2$; (k – l) $\alpha_1 = \alpha_2$) of the DRs for $x_1 = 0,25a$; $y_1 = y_2 = 0,5b$; $\Delta z = 10$ mm, and (j – l) 1—$x_2 = 0,25a$; 2—$x_2 = 0,5a$; 3—$x_2 = 0,75a$

$$
\begin{aligned}
f_t^n(\mp i\Gamma) = \cos(\chi_{1x}x_n)\Big\{&[E_{x0}^*(\chi_{uy}\cos\beta_n \mp i\Gamma\sin\beta_n) - E_{y0}^*\chi_{sx}\cos\beta_n + iE_{z0}^*\chi_{sx}\sin\beta_n] \\
&\cdot e^{-i\chi_{uy}y_n}\upsilon(\chi_{sx}, (\chi_{uy}\cos\beta_n \mp i\Gamma\sin\beta_n), (\chi_{uy}\sin\beta_n \pm i\Gamma\cos\beta_n)) \\
&+ [E_{x0}^*(\chi_{uy}\cos\beta_n \pm i\Gamma\sin\beta_n) - E_{y0}^*\chi_{sx}\cos\beta_n + iE_{z0}^*\chi_{sx}\sin\beta_n] \\
&\cdot e^{i\chi_{uy}y_n}\upsilon(\chi_{sx}, (\chi_{uy}\cos\beta_n \pm i\Gamma\sin\beta_n), -(\chi_{uy}\sin\beta_n \mp i\Gamma\cos\beta_n))\Big\};
\end{aligned}
\tag{3.52}
$$

(b) Rotation of the resonator relative to the y axis of the waveguide (Fig. 3.11b):

$$
f_t^n(\mp i\Gamma) = \cos(\chi_{uy}y_n)\Big\{[E_{x0}^*\chi_{uy}\cos\alpha_n - E_{y0}^*(\chi_{sx}\cos\alpha_n \mp i\Gamma\sin\alpha_n)
$$

$$
\begin{aligned}
&- iE_{z0}^{*}\chi_{uy}\sin\alpha_n]\cdot e^{-i\chi_{sx}x_n}\upsilon((\chi_{sx}\cos\alpha_n \mp i\Gamma\sin\alpha_n),\chi_{uy},\\
&(\chi_{sx}\sin\alpha_n \pm i\Gamma\cos\alpha_n)) + [E_{x0}^{*}\chi_{uy}\cos\alpha_n - E_{y0}^{*}(\chi_{sx}\cos\alpha_n \pm i\Gamma\sin\alpha_n)\\
&+ iE_{z0}^{*}\chi_{uy}\sin\alpha_n]\cdot e^{i\chi_{sx}x_n}\upsilon((\chi_{sx}\cos\alpha_n \pm i\Gamma\sin\alpha_n),\chi_{uy},\\
&-(\chi_{sx}\sin\alpha_n \mp i\Gamma\cos\alpha_n))\}; \qquad (3.53)
\end{aligned}
$$

(c) Rotation of the resonator relative to the z axis of the waveguide (Fig. 3.11c) on the angle α_n:

$$
\begin{aligned}
&f_t^n(\mp i\Gamma) = [\pm i\Gamma(E_{x0}^{*}\sin\alpha_n + E_{y0}^{*}\cos\alpha_n) - i(\chi_{sx}\sin\alpha_n + \chi_{uy}\cos\alpha_n)E_{z0}^{*}]\\
&\cdot\sin(\chi_{sx}x_n + \chi_{uy}y_n)\upsilon(-i\Gamma,(\chi_{sx}\sin\alpha_n + \chi_{uy}\cos\alpha_n),(\chi_{sx}\cos\alpha_n - \chi_{uy}\sin\alpha_n))\\
&+ [\pm i\Gamma(E_{x0}^{*}\sin\alpha_n - E_{y0}^{*}\cos\alpha_n) - i(\chi_{sx}\sin\alpha_n - \chi_{uy}\cos\alpha_n)E_{z0}^{*}]\\
&\cdot\sin(\chi_{sx}x_n - \chi_{uy}y_n)\upsilon(-i\Gamma,(\chi_{sx}\sin\alpha_n - \chi_{uy}\cos\alpha_n),(\chi_{sx}\cos\alpha_n + \chi_{uy}\sin\alpha_n)). \qquad (3.54)
\end{aligned}
$$

where $\chi = [\chi_{sx}^2 + \chi_{uy}^2]^{1/2}$; (x_n, y_n, z_n) is the coordinates of the center of the nth resonator in the waveguide coordinate system; β_n, α_n—is the angles between the z'-axes of the local coordinate system of the resonator and the z-axes waveguide (Fig. 3.11a, b), or x-axes waveguide (Fig. 3.11c) ($n = 1, 2$).

If the axis of one of the resonators is rotated by an angle π, then the sign of the mutual coupling coefficient changes due to a change in the direction of the field (Fig. 3.11d–f).

In Fig. 3.11g–l shows the calculated dependences of the coupling on the mutual orientations of the axes of the DR and waveguide. The effect of the independence of the value of the mutual coupling from the transverse coordinate of the second resonator at a certain angle of inclination of their axes, shown in Fig. 3.11j, in the case of rotation relative the x axis.

As also follows from the data shown in Fig. 3.11g–k, for each resonator 1 position, there is an optimal resonator 2 position at which the coupling coefficient between them reaches a maximum.

References

1. A.A. Trubin, *Mutual Coupling Coefficients of Spherical Dielectric Resonators in a Rectangular Waveguide*, vol. 28 (1991), pp. 27–32. (in Russian)
2. M. Abramowitz, I.A. Stegun (eds.), *Handbook of Mathematical Functions with Formulas, Graphs and Mathematical Tables. Applied Mathematics Series*, vol. 55 (National Bureau of Standarts, 1964), 830 p
3. A.A. Trubin, Dielectric ball in a circular waveguide. Bull. NTUU "KPI" Ser. Radiotechnique **27**, 31–34 (1990). (in Russian)
4. A.A. Trubin, Coupling coefficients of orthogonally oriented cylindrical dielectric resonators in a circular waveguide. Radioelectronics **6**, 10–15 (1997)

5. F.W.J. Olver, *Introduction to Asymptotics and Special Functions* (Academic Press, New York and London, 1974), 592 p
6. A.A. Trubin, Coupling coefficients of rotating rectangular dielectric resonators in cut-off rectangular waveguide. Inf. Telecommun. Sci. **12**, **1**(22), 48–54 (2021)
7. A.A. Trubin, Rotations of cylindrical dielectric resonators in a rectangular waveguide. Visnyk NTUU KPI Seriia - Radiotekhnika Radioaparatobuduvannia **86**, 22–28 (2021)
8. A.A. Trubin, Dielectric ball in a coaxial waveguide. Bull. NTUU "KPI" Ser. Radiotechnique **30**, 43–50 (1993). (in Russian)

Chapter 4
Coupling Coefficients of Dielectric Resonators in Microstrip Lines

Ease of combination with a variety of strip structures, improved quality factor and thermal stability have led to the fact that dielectric resonators have found wide application in various hybrid-integrated devices of almost all wavelength ranges [1–5]. At the same time, the lack of simple analytical relationships for the fields of both the DRs themselves and strip lines significantly complicates the task of studying their interaction. The coupling coefficients can be calculated approximately using the representation of the field in the DR material as the field of its aigen oscillations without taking into account the influence of current-carrying conductors, using the decomposition of stripline waves along a continuous spectrum, for example (B.1), and in the case of TEM waves—in the form of the Poisson integral (B.6). This method makes it possible to obtain the most general analytical expressions of coupling coefficients, which, in the future, can be used for calculations even in cases where the line field is specified implicitly, for example, in the form of conformal mappings.

This chapter discusses the coupling coefficients of cylindrical and spherical DR with slot, asymmetric strip and coplanar lines. Analytical expressions are also considered for the coupling coefficients of a spherical DR with multiwire lines, and in addition, the coupling coefficients of orthogonally oriented cylindrical DR in a circular waveguide.

4.1 Coupling Parameters of Dielectric Resonators and Open Waveguide Structures

If a spherical resonator is placed near a transmission line formed by flat interfaces between media, its natural oscillations will be characterized by a given field parity in the center of the ball relative to one of the symmetry planes of the structure. In this case, it is convenient to represent the line field in the form of an expansion in plane waves, for example, H-type (B.1). Since if the result of the expansion of the

A. Trubin, *Introduction to the Theory of Dielectric Resonators*, Springer Series in Advanced Microelectronics 65, https://doi.org/10.1007/978-3-031-65396-4_4

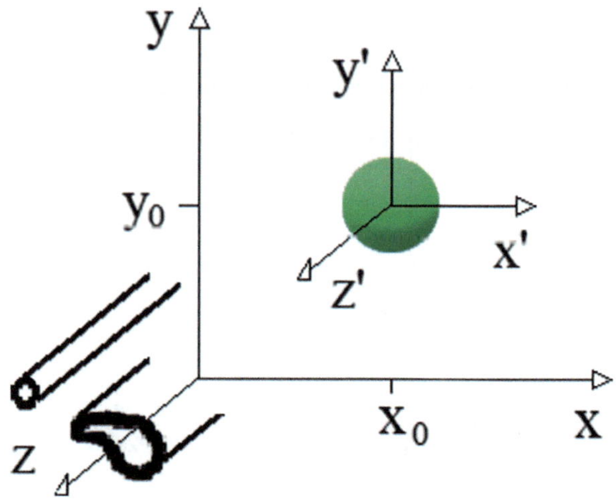

Fig. 4.1 Spherical DR in a multiwire transmission line

field of a spherical DR natural oscillations in plane waves is known (Chapter 2), we can assume that the final expression will take the form of at least an integral transformation. In fact, as will be shown below, the indicated expansion coefficients can often be expressed in the most general form—through the field of the line at the center of the DR.

Let us denote by (x_0, y_0, z_0) the coordinates of the DR center in the coordinate system of an open multiwire transmission line (Fig. 4.1). To expand the field of natural oscillations of the ball (1.2, 1.3) into line waves, we use the bilinearity property (2. 4) with respect to the DR field and the field of the transmission line. First, let's calculate the expansion coefficients for plane waves included in the integrands (B.1) using (D.13). After this, we substitute the obtained result into integrals of the form (B.1), taking into account the expressions for the initial amplitudes $a(\eta, \xi)$, $b(\eta, \xi)$, respectively, for magnetic and electric types of waves. The expansion coefficients obtained as a result of this transformation will take the form:

for magnetic types H_{nml} of DR oscillations and magnetic waves of the line:

$$c_H^{nml\pm} = -2\pi \frac{w_0}{k_0} \cdot \frac{j_n(p)}{h_n^{(2)}(q)} \cdot e^{in\frac{\pi}{2} \pm i\gamma k_0 z_0} \frac{dP_n^m(\pm\gamma)}{d(\pm\gamma)} \cdot \int_{-\infty}^{\infty} a(\eta, \xi)^* \cdot \begin{bmatrix} \sin m\tilde{\varphi}_0 \\ \cos m\tilde{\varphi}_0 \end{bmatrix} \cdot e^{i\eta k_0 \left(\xi x_0 + \sqrt{1-\xi^2} y_0\right)} d\xi; \quad (4.1)$$

for H_{nml} and electric waves of the line:

$$c_E^{nml\pm} = -2\pi \frac{1}{k_0} \cdot \frac{j_n(p)}{h_n^{(2)}(q)} e^{in\frac{\pi}{2} \pm i\gamma k_0 z_0} m \frac{P_n^m(\pm\gamma)}{1-\gamma^2} \cdot \int_{-\infty}^{\infty} b(\eta, \xi)^* \cdot \begin{bmatrix} \cos m\varphi_0 \\ -\sin m\varphi_0 \end{bmatrix} \cdot e^{i\eta k_0\left(\xi x_0 + \sqrt{1-\xi^2} y_0\right)} d\xi. \tag{4.2}$$

Similarly, we find for electrical types of oscillations of DR E_{nml} and magnetic waves:

$$c_H^{nml\pm} = -2\pi \frac{\varepsilon_{1r}}{k_0} \cdot \frac{j_n(p)}{h_n^{(2)}(q)} e^{in\frac{\pi}{2} \pm i\gamma k_0 z_0} m \frac{P_n^m(\pm\gamma)}{1-\gamma^2} \cdot \int_{-\infty}^{\infty} a(\eta, \xi)^* \cdot \begin{bmatrix} \cos m\tilde{\varphi}_0 \\ -\sin m\tilde{\varphi}_0 \end{bmatrix} \cdot e^{i\eta k_0\left(\xi x_0 + \sqrt{1-\xi^2} y_0\right)} d\xi; \tag{4.3}$$

for electric E_{nml} oscillations and electric waves lines:

$$c_E^{nml\pm} = -2\pi \frac{\varepsilon_{1r}}{w_0 k_0} \cdot \frac{j_n(p)}{h_n^{(2)}(q)} e^{in\frac{\pi}{2} \pm i\gamma k_0 z_0} \frac{dP_n^m(\pm\gamma)}{d(\pm\gamma)} \cdot \int_{-\infty}^{\infty} b(\eta, \xi)^* \cdot \begin{bmatrix} \sin m\tilde{\varphi}_0 \\ \cos m\tilde{\varphi}_0 \end{bmatrix} \cdot e^{i\eta k_0\left(\xi x_0 + \sqrt{1-\xi^2} y_0\right)} d\xi, \tag{4.4}$$

where $\eta = \sqrt{1-\gamma^2}$; $\cos\tilde{\varphi}_0 = \xi$. Here we assumed that the z' axis of the local coordinate system, associated with the center of the ball, is parallel to the z axis of the line (Fig. 4.1) as well as the field of the electric wave of the line is represented similarly to (B.1) with $H_z = 0$; $E_z \neq 0$ and with a known spectral function $b(\eta, \xi)$ instead $a(\eta, \xi)$.

In a particular case $m = 0, 1$, taking into account the original representation of the field (B.1), we find [6] in explicit form:

for H_{nml}

$$c_H^{nol\pm} = -2\pi e^{in\frac{\pi}{2}} \frac{w_0}{k_0} \cdot \frac{j_n(p)}{h_n^{(2)}(q)} \cdot \frac{dP_n(\pm\gamma)}{d(\pm\gamma)} \cdot H_z(x_0, y_0, z_0)^*;$$

$$c^{n1l\pm}_{\begin{pmatrix} E \\ H \end{pmatrix}} = 2\pi e^{in\frac{\pi}{2}} \frac{w_0}{k_0} \cdot \frac{j_n(p)}{h_n^{(2)}(q)} \cdot \begin{pmatrix} \frac{P_n^1(\pm\gamma)}{\eta} \\ \frac{\eta}{\pm\gamma} \cdot \frac{dP_n^1(\pm\gamma)}{d(\pm\gamma)} \end{pmatrix} \cdot \begin{bmatrix} H_y(x_0, y_0, z_0)^* \\ H_x(x_0, y_0, z_0)^* \end{bmatrix}; \tag{4.5}$$

for E_{nml}:

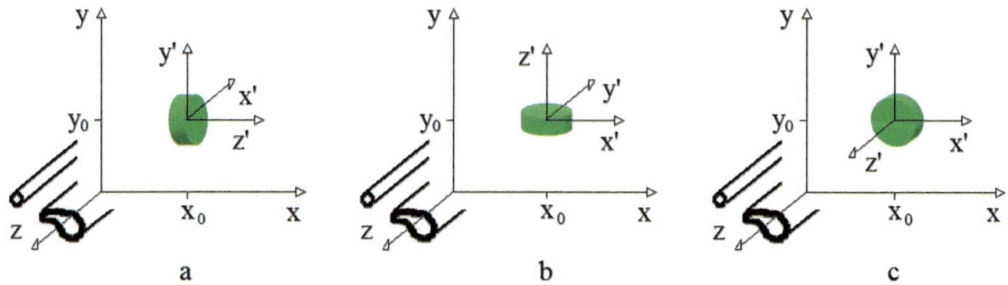

Fig. 4.2 Standard positions of a cylindrical DR in an open multi-wire transmission line

$$c_E^{nol\pm} = -2\pi e^{in\frac{\pi}{2}} \frac{\varepsilon_{1r}}{w_0 k_0} \cdot \frac{j_n(p)}{h_n^{(2)}(q)} \cdot \frac{dP_n(\pm\gamma)}{d(\pm\gamma)} \cdot E_z(x_0, y_0, z_0)^*;$$

$$c^{n1l\pm}_{\begin{pmatrix} E \\ H \end{pmatrix}} = 2\pi e^{in\frac{\pi}{2}} \frac{\varepsilon_{1r}}{w_0 k_0} \cdot \frac{j_n(p)}{h_n^{(2)}(q)} \cdot \begin{pmatrix} \frac{\eta}{\pm\gamma} \cdot \frac{dP_n^1(\pm\gamma)}{d(\pm\gamma)} \\ \frac{P_n^1(\pm\gamma)}{\eta} \end{pmatrix} \cdot \begin{bmatrix} E_y(x_0, y_0, z_0)^* \\ E_x(x_0, y_0, z_0)^* \end{bmatrix}. \quad (4.6)$$

From this we obtain that for the main types of oscillations, characterized by a maximum magnetic (for H_{nml}) or electric (for E_{nml}) field at the center of the spherical DR, the indicated expansion coefficients (4.5, 4.6) turn out to be proportional to the corresponding components of the magnetic or electric field of the transmission line.

To calculate the power emitted by a cylindrical DR into an open transmission line, we also expand in general form the field of its own oscillations into the natural waves of the line. In this case, it is no longer possible to present the expansion coefficients (2.4) in a compact form, except for the coaxial arrangement of the cylindrical DR and the line (Fig. 4.2c). In the general case, this expansion takes the form of a series containing partial derivatives of one of the line field components.

Let us also fix three standard positions of the cylindrical DR relative to the rectangular coordinate system of the line: A—z' axis of the local coordinate system associated with the axis of the dielectric cylinder is parallel to the x axis (Fig. 4.2a); B—z' axis of the dielectric cylinder is parallel to the y axis (Fig. 4.2b) and C—z' axis of the dielectric cylinder is parallel to the z axis of the line (Fig. 4.2c). As usual, we represent the field of natural oscillations of a cylindrical DR near one of its resonant frequencies in the form (1.17). We find the expansion coefficients by integrating the plane waves included in (B.1) over the surface of the resonator (2.1.4), and then presenting the obtained result in a superposition corresponding to the general expansion of the line field in plane waves (B.1). The found analytical expressions in the case of azimuthally homogeneous magnetic oscillations $H_{nol}^{\pm}$ along the magnetic waves of the line take the form:

for A position (Fig. 4.2a):

$$c_H^{nol\pm} = \mp\frac{\pi i}{2} r_0^3 L w_0 h_1 \left(k_1^2 - k_0^2\right) \cdot \left(\frac{\gamma}{\eta}\right) e^{\pm i\gamma k_0 z_0} q_\perp$$
$$\cdot \int_{-\infty}^{\infty} a(\eta, \xi)^* \cdot e^{i\eta k_0 \left(\xi x_0 + \sqrt{1-\xi^2} y_0\right)} f(\eta\xi) \cdot \xi d\xi; \tag{4.7}$$

for B position (Fig. 4.2b):

$$c_H^{nol\pm} = \mp\frac{\pi i}{2} r_0^3 L w_0 h_1 \left(k_1^2 - k_0^2\right) \cdot \left(\frac{\gamma}{\eta}\right) e^{\pm i\gamma k_0 z_0} q_\perp$$
$$\cdot \int_{-\infty}^{\infty} a(\eta, \xi)^* e^{i\eta k_0 \left(\xi x_0 + \sqrt{1-\xi^2} y_0\right)} f\left(\eta\sqrt{1-\xi^2}\right) \sqrt{1-\xi^2} d\xi; \tag{4.8}$$

for C position (Fig. 4.2c):

$$c_H^{nol\pm} = \frac{\pi i}{2} r_0^3 L w_0 h_1 \left(k_1^2 - k_0^2\right) \cdot \left(\frac{\gamma}{\eta}\right) e^{\pm i k_0 \gamma z_0} q_\perp$$
$$\cdot f(\gamma) \int_{-\infty}^{\infty} a(\eta, \xi)^* e^{i\eta k_0 \left(\xi x_0 + \sqrt{1-\xi^2} y_0\right)} d\xi, \tag{4.9}$$

where (x_0, y_0, z_0)—coordinates of the DR center; $k_1 = \varepsilon_{1r}^{1/2} k_0$;

$$f(x) = \frac{J_0(p_\perp) J_2\left(q_\perp\sqrt{1-x^2}\right) - J_2(p_\perp) J_0\left(q_\perp\sqrt{1-x^2}\right)}{p_\perp^2 - q_\perp^2\left[1-x^2\right]}$$
$$\cdot \frac{1}{p_z^2 - (q_z x)^2} \cdot \begin{bmatrix} p_z \sin p_z \cos q_z x - q_z x \cos p_z \sin q_z x \\ i(p_z \cos p_z \sin q_z x - q_z x \sin p_z \cos q_z x) \end{bmatrix}. \tag{4.10}$$

Comparing (4.10) with the original representation (B.1), we note that the expansion coefficient of the DR field at position C, as for a spherical resonator, turns out to be proportional to the z component of the magnetic field of the open waveguide structure. Next, we note that function (4.10) has no poles for any finite values of the argument. In this case, in view of the exponential decrease of the integrand expression for big ξ, we represent function (4.10) in the form of a Taylor series at a point $\xi = 0$ and integrate the resulting terms taking into account the original representation of the line field (B.1) (in position A (upper); B (lower)):

$$c_H^{nol\pm} = \frac{\pi i}{2}(w_0 h_1) r_0^3 L\left(k_1^2 - k_0^2\right) q_\perp \cdot \sum_{s=0}^{\infty} \frac{f^{(s)}(0)}{s!} \left(\frac{-i}{\eta k_0}\right)^s \cdot \begin{pmatrix} \frac{\partial^s H_x^{\pm}(x_0, y_0, z_0)^*}{\partial x_0^s} \\ \frac{\partial^s H_y^{\pm}(x_0, y_0, z_0)^*}{\partial y_0^s} \end{pmatrix}. \tag{4.11}$$

It follows that the convergence of expansion (4.11) is determined by the rate of change of the field in the transverse plane of the waveguide structure. Note also that for even oscillations H_{n0l} $f^{(1)}(0) = 0$, therefore the correction to the first term of expansion (4.13) turns out to be proportional to the value of the second derivative of the line field. Taking into account this remark, we find in general form approximate expansions of the field of the main magnetic oscillations of the cylindrical DR in position A (upper); B (average); C (lower) along the field (B.1) of the open waveguide structure

$$c_H^{nol\pm} = \frac{\pi i}{2}(w_0 h_1) r_0^3 L q_\perp \left(k_1^2 - k_0^2\right) \cdot \begin{pmatrix} f(0) H_x^{\pm}(x_0, y_0, z_0)^* \\ f(0) H_y^{\pm}(x_0, y_0, z_0)^* \\ f(\gamma) H_z^{\pm}(x_0, y_0, z_0)^* \end{pmatrix}. \tag{4.12}$$

The resulting expressions significantly simplify the calculation of the external field of the DR in microstrip lines of complex cross sections.

The relationships found also indicate the possibility of using high-quality DRs made of high dielectric constant material as local indicators of the moduli of line field components, determined by the magnitude of their coupling coefficients.

4.2 Coupling Coefficients of Dielectric Resonators with a Slot Line

We use relations (4.1, 4.5) to calculate the coupling coefficients of the spherical DR with the slot line (Fig. 4.3a). It is known [7] that $a(\eta, \xi)$ the slot wave function in the quasi-static approximation ($k_0 w \ll 1$) has the form (B.2, B.3). Replacing the Bessel function in (B.2) with the zero term of the expansion and using (D.14), we find:

for H_{nml}:

$$\tilde{k} = 2\pi \frac{Z}{w_0} \cdot \left[(1-\gamma^2) \frac{dP_n^m(\gamma)}{d\gamma}\right]^2 \cdot \frac{1}{1+\delta_{m0}} \cdot \frac{2n+1}{n(n+1)} \cdot \frac{(n-m)!}{(n+m)!} \cdot \alpha_n^H(p, q) \cdot \begin{bmatrix} \sin^2 m\psi_0 \\ \cos^2 m\psi_0 \end{bmatrix} \cdot \left| H_m^{(2)}\left(\eta k_0 \sqrt{x_0^2 + y_0^2}\right) \right|^2; \tag{4.13}$$

for E_{nml}:

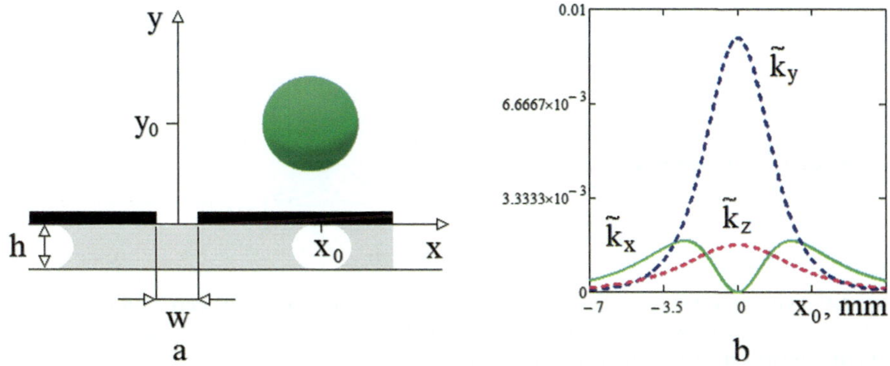

Fig. 4.3 Spherical DR in a slot microstrip line (**a**). Coupling coefficients of the main magnetic oscillations of a spherical DR with the slot line: $H_{111} - \tilde{k}_x, \tilde{k}_y$; $H_{101} - \tilde{k}_z$ **b** ($\varepsilon_{1r} = 81$)

$$\tilde{k} = 2\pi\frac{Z}{w_0}\cdot\left[mP_n^m(\gamma)\right]^2\cdot\frac{2n+1}{n(n+1)}\cdot\frac{(n-m)!}{(n+m)!}\cdot\alpha_n^E(p,q)$$
$$\cdot\begin{bmatrix}\cos^2 m\psi_0\\ \sin^2 m\psi_0\end{bmatrix}\cdot\left|H_m^{(2)}\left(\eta k_0\sqrt{x_0^2+y_0^2}\right)\right|^2. \qquad (4.14)$$

Here Z—characteristic impedance of the slot line; $\gamma^2 = \varepsilon_{eff}$—effective dielectric constant [7]; $\cos\psi_0 = x_0/\sqrt{x_0^2+y_0^2}$; $H_m^{(2)}(z)$—Hankel function of the second kind [8].

On Fig. 4.3b shows the dependences of the coupling coefficients of the main magnetic oscillations of the ball at a frequency of 7 GHz for $H_{101} - \tilde{k}_z$ and $H_{111} - \tilde{k}_{x,y}$ with $\varepsilon_{1r} = 81$; on the transverse coordinate of the DR at $y_0 = 3$ mm with a slot line with $\varepsilon_\Lambda = 3, 8$; $w = 0, 1$ mm. It can be seen that the slot line field can be coupled with oscillations of the ball of all three types of polarization. In the plane of symmetry $x_0 = 0$ the most significant coupling is for y oscillations. The field of the slot transmission line in the transverse plane is concentrated mainly on a small part of the surface adjacent to the edges of the current-carrying conductors. The dimensions of the DR, even for the main types of oscillations, are much larger than the characteristic transverse dimensions of the line, so the main difficulty in achieving an acceptable coupling arises from the difficulty of implementing a design with a sufficiently large value of the line field at the center of the ball. This coupling can be increased by widening the gap or by using other forms of DRs.

We now use the relations obtained above to calculate the coupling coefficients of the main oscillations of the cylindrical DR with the slot line (Fig. 4.4a). The coupling coefficient of the cylindrical DR, taking into account (4.12), as well as (B.1–B.4), in the region $y > 0$, is found in the form: in position A (upper); B (average); C (bottom):

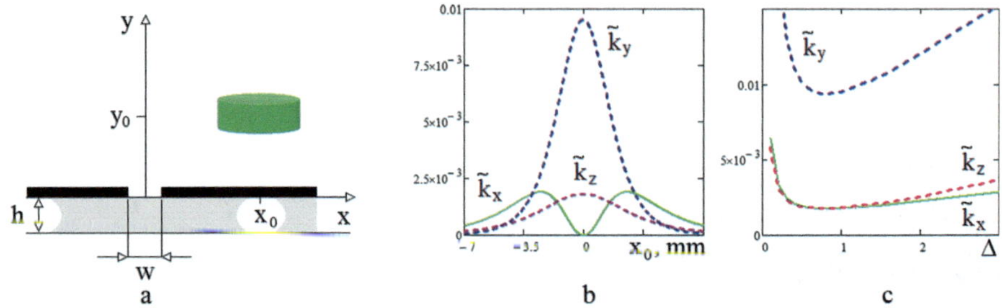

Fig. 4.4 Cylindrical DR in a slotted microstrip line (**a**). Coupling coefficients with the slot line of the main magnetic H_{101}^{+} oscillations for different DR orientations—$\tilde{k}_x$, $\tilde{k}_y$; $\tilde{k}_z$ on transverse coordinates (**b**); on relative sizes $\Delta = L/2r_0$ (**c**)

$$\tilde{k} = \frac{4}{\pi} \cdot \frac{Z}{w_0} \cdot \frac{(\varepsilon_{1r}-1)^2}{\varepsilon_{1r}} \cdot \frac{p_\perp^2 p_z q_\perp^4 q_z}{[J_1^2(p_\perp) - J_0(p_\perp)J_2(p_\perp)] \cdot (2p_z + \sin 2p_z)} \cdot \begin{pmatrix} f(0) \cdot \begin{Bmatrix} \cos\varphi_0 \\ \sin\varphi_0 \end{Bmatrix} \cdot |\eta|\gamma K_1(|\eta|k_0\rho_0) \\ f(\gamma)\eta^2 K_0(|\eta|k_0\rho_0) \end{pmatrix}^2. \quad (4.15)$$

Here $\sin\varphi_0 = y_0/\rho_0$; $\rho_0 = \left(x_0^2 + y_0^2\right)^{1/2}$; $K_n(x)$—modified Bessel function [8].

In Fig. 4.4c shows the relationship between the main magnetic oscillations H_{101}^{+} at the $f_0 = 7$ GHz frequency of the DR ($\varepsilon_{1r} = 81$) with the slot line $\varepsilon_\Lambda = 3, 8$; $w = 0, 1$ mm; at $y_0 = 3$ mm from the compression ratio of the DR, calculated using formulas (4.15). for $x_0 = 0$ in the position B, C and $x_0 = 2$ mm in the position A. The coupling between the resonator and the line in this case can be increased by varying its relative sizes (Fig. 4.4c).

A more accurate value of the coupling can be obtained using a second-order correction arising from expansion (4.11) for position A:

$$\frac{\delta^2 c^{\pm}}{c_0^{\pm}} = \frac{f^{(2)}(0)}{2f(0)}\left[\frac{1}{|\eta|k_0\rho_0}\left(4\cos^2\varphi_0 - 3\right)\frac{K_0(|\eta|k_0\rho_0)}{K_1(|\eta|k_0\rho_0)} + \cos^2\varphi_0\left(1 + \frac{8}{(|\eta|k_0\rho_0)^2}\right) - \frac{6}{(|\eta|k_0\rho_0)^2}\right] \quad (4.16)$$

(in the B position $\cos\varphi_0$ needs to be replaced with $\sin\varphi_0 = y_0/\rho_0$; $\rho_0 = \left(x_0^2 + y_0^2\right)^{1/2}$). The magnitude of the correction quickly decreases as the center of the resonator moves away from the slot line.

4.3 Coupling Coefficients of a Spherical Dielectric Resonator with TEM Waves of Strip Lines

When expanding the field of a spherical DR (1.2) in terms of TEM waves, it is convenient to use the representation of the transverse components of the field of a strip line in the form of the Poisson integral (B.6). Let us assume that the dielectric ball is entirely located in the harmonic region Ω of the function $\varphi(\tilde{z})$ (Fig. 4.5). Let us denote by $\tilde{z}_0 = x_0 + iy_0$—the coordinate of the center of the ball (for simplicity we will accept: $\tilde{z}_0 = 0$), and through R—the radius of a circle with a center in a plane $z = const$ such that $r_0 \sin\theta < R$ it lies entirely in Ω. Let us also denote by $\tilde{z} = re^{i\tau}$ the coordinates of the inner point of this circle. Applying the Poisson formula, we substitute the line field (B.5, B.6), and the field of natural oscillations of the spherical DR in (2.4). Changing the order of integration taking into account (D.12), orthogonality of trigonometric functions, after a series of algebraic transformations, we obtain:

for magnetic types of ball oscillations H_{nml}:

$$c_{TEM}^{nml\pm} = m\frac{(n+m)!}{(n-m)!}e^{i(n-m+1)\frac{\pi}{2}\pm ik_0 z_0}\frac{j_n(p)}{y_n(q)} \cdot \frac{1}{k_0^m}\int\limits_0^{2\pi}\begin{bmatrix}\cos m\tau \\ -\sin m\tau\end{bmatrix}\frac{\varphi\left(\mathrm{R}e^{i\tau}\right)^*}{R^m}d\tau; \tag{4.17}$$

for electrical types of ball oscillations E_{nml}:

$$c_{TEM}^{nml\pm} = m\frac{(n+m)!}{(n-m)!}e^{i(n-m+1)\frac{\pi}{2}\pm ik_0 z_0}\cdot\frac{\varepsilon_{1r}}{w_0}\cdot\frac{j_n(p)}{y_n(q)} \cdot \frac{1}{k_0^m}\int\limits_0^{2\pi}\begin{bmatrix}\sin m\tau \\ \cos m\tau\end{bmatrix}\frac{\varphi\left(\mathrm{R}e^{i\tau}\right)^*}{R^m}d\tau; \tag{4.18}$$

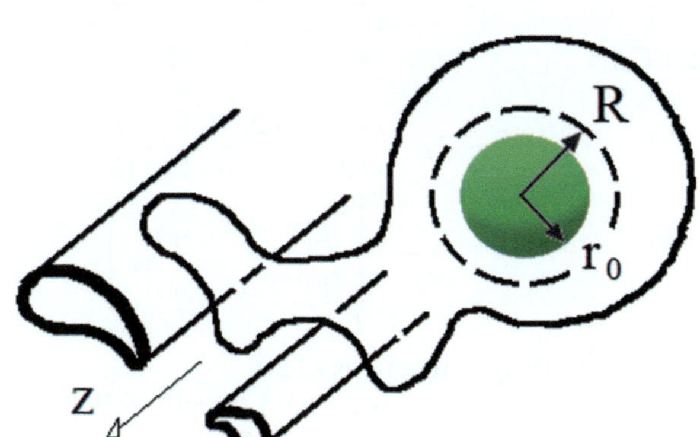

Fig. 4.5 Spherical DR in a multiwire line

The last integrals are easily calculated using Cauchy's formula [9]. Returning to an arbitrary coordinate of the center of the DR, we find the final expressions of the expansions (4.17, 4.18):

for H_{nml}:

$$c_{TEM}^{nml\pm} = -\pi w_0 \cdot \frac{m}{m!} \cdot \frac{(n+m)!}{(n-m)!} e^{i(n-m+1)\frac{\pi}{2}} \cdot \frac{1}{k_0^m} \cdot \frac{j_n(p)}{y_n(q)} \cdot \begin{bmatrix} H_y^{(m-1)}(\tilde{z}_0, z_0)^* \\ H_x^{(m-1)}(\tilde{z}_0, z_0)^* \end{bmatrix}; \tag{4.19}$$

for E_{nml}:

$$c_{TEM}^{nml\pm} = \frac{\pi}{w_0} \cdot \frac{m}{m!} \cdot \frac{(n+m)!}{(n-m)!} e^{i(n-m+1)\frac{\pi}{2}} \cdot \frac{\varepsilon_{1r}}{k_0^m} \cdot \frac{j_n(p)}{y_n(q)} \cdot \begin{bmatrix} E_y^{(m-1)}(\tilde{z}_0, z_0)^* \\ E_x^{(m-1)}(\tilde{z}_0, z_0)^* \end{bmatrix}; \tag{4.20}$$

Here $A^{(m)}(\tilde{z}_0, z_0) = \partial_m / \partial \tilde{z}_0^m A(\tilde{z}_0, z_0)$; $\tilde{z}_0 = x_0 + i y_0$; $y_n(x)$—spherical Neumann function [8]. The resulting relations are convenient for calculations, since they directly express the expansion coefficients through the field of a multiwire transmission line. Harmonic functions have partial derivatives of any order [9], therefore the expressions for the right-hand sides of equalities (4.19, 4.20) always exist. Using them, we find in general form expressions for the coupling coefficients of spherical DRs with a multiwire transmission line [10]:

for H_{nml}:

$$\tilde{k} = 2\pi \cdot \left(\frac{m}{m!}\right)^2 \cdot \frac{2n+1}{n(n=1)} \cdot \frac{(n+m)!}{(n-m)!} \cdot \frac{1}{w_0} \cdot \alpha_n^H(p,q) \cdot \frac{1}{k_0^{2m}} \cdot \begin{bmatrix} \left|E_x^{(m-1)}(x_0, y_0)\right|^2 \\ \left|E_y^{(m-1)}(x_0, y_0)\right|^2 \end{bmatrix}; \tag{4.21}$$

for E_{nml}:

$$\tilde{k} = 2\pi \cdot \left(\frac{m}{m!}\right)^2 \cdot \frac{2n+1}{n(n+1)} \cdot \frac{(n+m)!}{(n-m)!} \cdot \frac{1}{w_0} \cdot \alpha_n^E(p,q) \cdot \frac{1}{k_0^{2m}} \cdot \begin{bmatrix} \left|E_y^{(m-1)}(x_0, y_0)\right|^2 \\ \left|E_x^{(m-1)}(x_0, y_0)\right|^2 \end{bmatrix}, \tag{4.22}$$

where $\alpha_n^{H,E}(p,q)$—functions (2.20).

From the relationships found it directly follows that azimuthally homogeneous oscillations of a dielectric ball are not coupled with TEM waves at all, and the

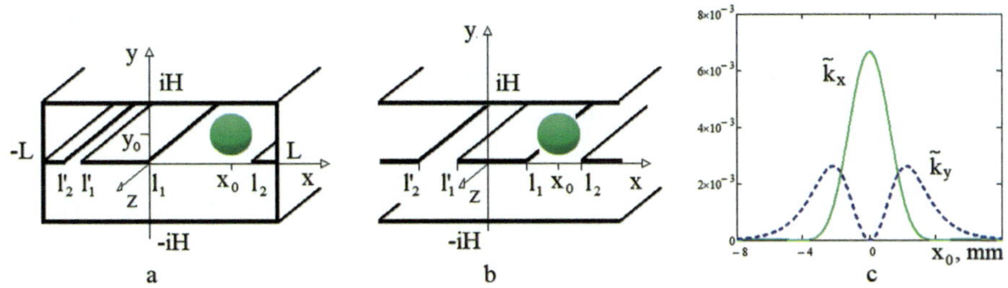

Fig. 4.6 Spherical resonator in a shielded fin strip lines (**a, b**). Coefficients of coupling of the main magnetic H_{111} oscillations of the DR with a ribbed strip line (**b**) $f_0 = 7$ GHz; $\varepsilon_{1r} = 81$; $l_1 = -l'_1 = 1,5$ mm; $l_2 = -l'_2 = 1,7$ mm; $H = 10$ mm; $y_0 = 3$ mm

coupling coefficients of oscillations with the transverse polarization of the field ($m = 1$) at the center of the ball are proportional to the square of the amplitudes of the corresponding line field.

We use the obtained expressions to calculate the coupling coefficients of a spherical DR with various types of multiwire lines. As follows from (4.21, 4.22), to obtain analytical relations, it is enough to calculate the normalized components of the field of TEM waves of the line.

Let us first calculate the coupling coefficients of the main magnetic oscillations of a spherical DR with a screened ribbed strip line (Fig. 4.6a). The characteristics of such a transmission line are determined using three sequential conformal mappings [11]. In this case, direct calculation of the coupling between the DR and the line is extremely difficult. Using relations (4.21, 4.22) directly leads to the desired expressions:

For example for H_{1ml}:

$$\begin{pmatrix} \tilde{k}_x \\ \tilde{k}_y \end{pmatrix} = \frac{\alpha_1^H(p,q)}{80} \cdot \frac{Z_0}{\kappa \cdot (k_0 L)^2} \cdot \left[\frac{K(k)}{K'(k)}\right]^2 \cdot \left[s(l_1) - s(l'_1)\right] \cdot \left[s(l_2) - s(l'_2)\right] \cdot \left(\begin{matrix} \mathrm{Re} \\ Im \end{matrix} \sqrt{\frac{\left(1 - k^2\omega_1^2\right)\cdot\left(1-\omega_1^2\right)}{[\omega_1 - s(l_1)]\cdot[\omega_1 - s(l_2)]\cdot\left[\omega_1 - s(l'_1)\right]\cdot\left[\omega_1 - s(l'_2)\right]}}\right)^2, \tag{4.23}$$

where, for brevity, denoted: $\omega_1 = s(\tilde{z}_0) = sn\left(K(k)\frac{\tilde{z}_0}{L}, k\right)$; $s(l_s) = sn\left(K(k)\frac{l_s}{L}, k\right)$ $(s = 1, 2)$; $sn(u, k)$—elliptic Jacobi function [8].

As follows from (4.21, 4.22), in the case of electrical E_{nml} oscillations of the DR, the coupling coefficients $\tilde{k}_x$ and $\tilde{k}_y$ formally change places.

$$\begin{pmatrix} \tilde{k}_x \\ \tilde{k}_y \end{pmatrix} = \frac{\alpha_1^E(p,q)}{80} \cdot \frac{Z_0}{\kappa \cdot (k_0 L)^2} \cdot \left[\frac{K(k)}{K'(k)}\right]^2 \cdot \left[s(l_1) - s(l'_1)\right] \cdot \left[s(l_2) - s(l'_2)\right] \cdot \left(\begin{matrix} Im \\ \mathrm{Re} \end{matrix} \sqrt{\frac{\left(1 - k^2\omega_1^2\right)\cdot\left(1-\omega_1^2\right)}{[\omega_1 - s(l_1)]\cdot[\omega_1 - s(l_2)]\cdot\left[\omega_1 - s(l'_1)\right]\cdot\left[\omega_1 - s(l'_2)\right]}}\right)^2.$$

In some limiting cases, Jacobi functions are expressed in terms of elementary:

(1) $L = \infty$, and H—is finite (B.14) (Fig. 4.6b). In this case, the coupling coefficients (4.23) are transformed to the form:

$$\begin{pmatrix} \tilde{k}_x \\ \tilde{k}_y \end{pmatrix} = \frac{\alpha_1^H(p,q)}{80} \cdot \frac{Z_0}{\kappa} \cdot \left[\frac{\pi}{2k_0 H K^{\cdot}(\kappa)}\right]^2 \cdot \left[t(l_1) - t(l_1')\right] \cdot \left[t(l_2) - t(l_2')\right]$$
$$\cdot \begin{pmatrix} \mathrm{Re} \\ Im \end{pmatrix} \frac{1-\omega_1^2}{\sqrt{[\omega_1 - t(l_1)]\cdot[\omega_1 - t(l_2)]\cdot[\omega_1 - t(l_1')]\cdot[\omega_1 - t(l_2')]}}\Bigg)^2; \tag{4.24}$$

Here $\omega_1 = th(\pi\tilde{z}_0/2H)$; $t(l) = th(\pi l/2H)$.

(2) $H = \infty$, and L—is finite (B.16) (Fig. 4.7a)

$$\begin{pmatrix} \tilde{k}_x \\ \tilde{k}_y \end{pmatrix} = \frac{\alpha_1^H(p,q)}{80} \cdot \frac{Z_0}{\kappa} \cdot \left[\frac{\pi}{2k_0 L K^{\cdot}(\kappa)}\right]^2 \cdot \left[s(l_1) - s(l_1')\right] \cdot \left[s(l_2) - s(l_2')\right]$$
$$\cdot \begin{pmatrix} \mathrm{Re} \\ Im \end{pmatrix} \sqrt{\frac{1-\omega_1^2}{[\omega_1 - s(l_1)]\cdot[\omega_1 - s(l_2)]\cdot[\omega_1 - s(l_1')]\cdot[\omega_1 - s(l_2')]}}\Bigg)^2; \tag{4.25}$$

where $s(\tilde{z}_v) \equiv sn(K(k)\tilde{z}_v/L, 0) = \sin \pi\tilde{z}_v/2L$ $(v = 1, 2)$.

(3) Using (4.24), we obtain a simple expression for the coupling coefficients of a spherical DR with a ribbed line in open space. Indeed, in the limit $H = \infty$ taking into account: $\lim H\omega_1 = \pi\tilde{z}_0/2$ and $\lim Hth(l_v) = \pi l_v/2$:

$$\begin{pmatrix} \tilde{k}_x \\ \tilde{k}_y \end{pmatrix} = \frac{\alpha_1^H(p,q)}{80} \cdot \frac{Z_0}{\kappa} \cdot \frac{(l_1 - l_1')\cdot(l_2 - l_2')}{[k_0 K'(\kappa)]^2}$$
$$\cdot \begin{pmatrix} \mathrm{Re} \\ Im \end{pmatrix} \frac{1}{\sqrt{(\tilde{z}_0 - l_1)\cdot(\tilde{z}_0 - l_1')\cdot(\tilde{z}_0 - l_2)\cdot(\tilde{z}_0 - l_2')}}\Bigg)^2. \tag{4.26}$$

The last expression can be used to calculate the coupling coefficients of a spherical DR with a coplanar strip line, if we take into account the change in its wave resistance Z_0 through the value of the effective permeability [11].

General expressions (4.21, 4.22) are also convenient to use for calculating the coupling coefficients of a spherical DR with lines of simpler cross sections. In this case, the final formulas are obtained explicitly, for example, for a symmetric strip line (see. Figure 4.8):

$$\begin{pmatrix} \tilde{k}_x \\ \tilde{k}_y \end{pmatrix} = 6\pi \cdot \frac{Z_0}{w_0} \cdot \alpha_1^H(p,q) \cdot \begin{pmatrix} \cos\theta \\ \sin\theta \end{pmatrix}^2 / (k_0 h\rho)^2, \tag{4.27}$$

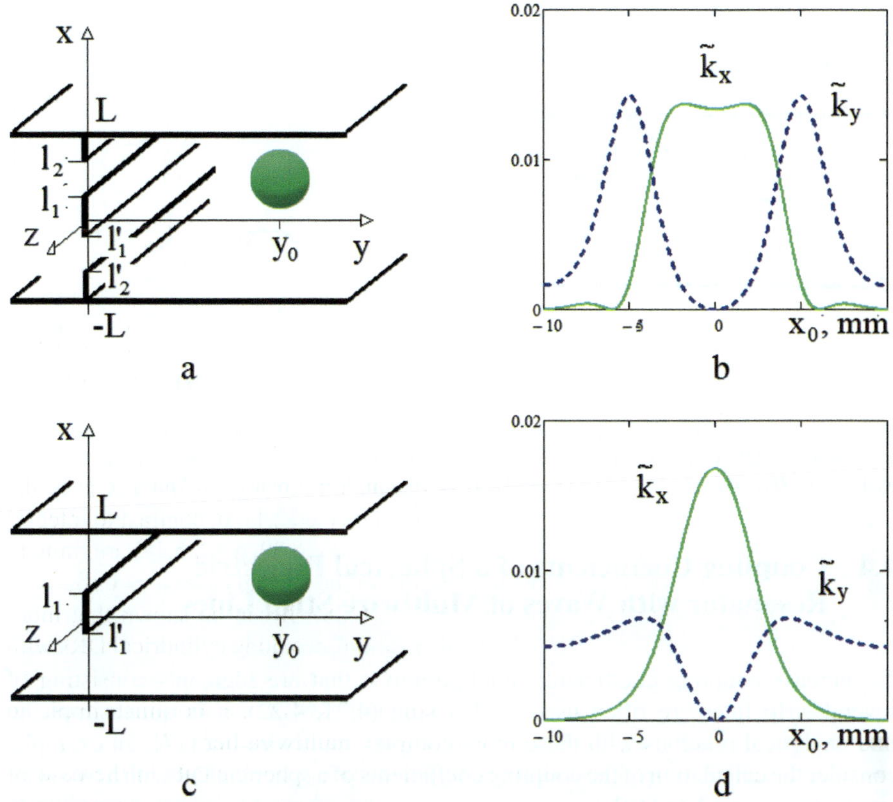

Fig. 4.7 Spherical DR in a vertical ribbed microstrip line (**a**, **c**). Dependences of coupling coefficients on transverse coordinates for the main magnetic oscillations H_{111} ($f_0 = 7$ GHz; $\varepsilon_{1r} = 81$; **b** $l_1 = 4$ mm; $l_2 = 6$ mm; $l'_1 = -4$ mm; $l'_2 = -6$ mm; $L = 10$ mm; $y_0 = 2, 3$ mm; **d** $l_1 = 2$ mm; $l'_1 = -2$ mm; $L = 10$ mm; $y_0 = 3$ mm)

where $\rho = (1 - 2e^{-\mu}\cos\nu + e^{-2\mu})^{1/4}$; $\theta = 0,5 \cdot \text{arctg}[e^{-\mu}\sin\nu/(1 - e^{-\mu}\cos\nu)]$.

In Figs. 4.6, 4.7, and 4.8 show typical dependences of the coupling coefficients of a spherical DR with $\varepsilon_{1r} = 81$ and strip lines of the types considered on the horizontal coordinate, calculated using formulas (4.23–4.27). It can be seen that the maximum coupling between the main magnetic oscillations of the ball of vertical polarization $\tilde{k}_y$ of its field at the center does not always fall into the slit region, is insignificant and is shifted to the outer region of the line. The coupling coefficients of the main magnetic oscillations with polarization parallel to plane of the strip are usually larger and take maximum values in the plane of symmetry ($x = 0$).

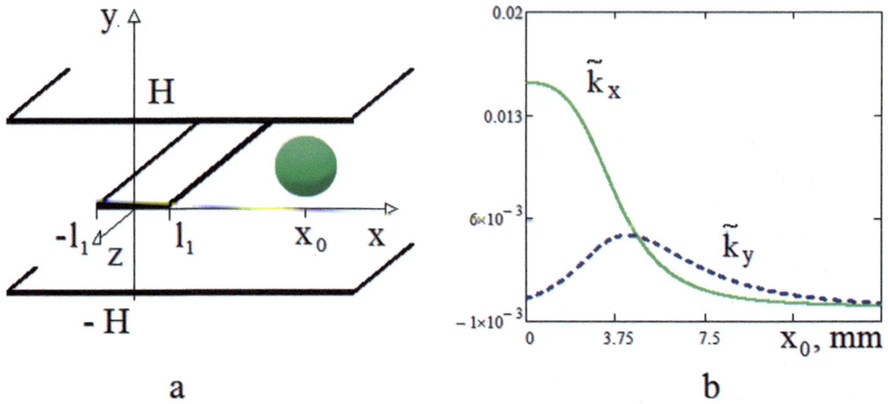

Fig. 4.8 Spherical DR in a symmetrical microstrip line (**a**). Dependences of the coupling coefficients of the DR with a symmetrical microstrip line on the coordinates for the main magnetic oscillations H_{111} (**b**); ($f_0 = 7$ GHz; $\varepsilon_{1r} = 81$; $H = 10$ mm; $l_1 = 3$ mm; $y_0 = 3$ mm)

4.4 Coupling Coefficients of a Spherical Dielectric Resonator with Waves of Multiwire Strip Lines

To increase coupling coefficients, configurations that are segments consisting of several strip lines are often used [11]. Using (4.21, 4.22), it is quite simple to find analytical relations with these more complex multiwire lines. As an example, consider the calculation of the coupling coefficients of a spherical DR with a two-wire symmetrical strip line [12].

Let us denote by H—the distance between the screen and the conductors of a symmetrical edge coupled strip line; the width of the strips through w, and the distance between their edges through s (Fig. 4.9a). The coupling coefficients of the main magnetic oscillations of the ball, characterized by the polarization of the magnetic field at its center (x_0, y_0) parallel to the x axes or y, as above, will be denoted, respectively, by $\tilde{k}_x$, $\tilde{k}_y$. In this case, the parameters related to waves of an even type (characterized by the same value of the instantaneous charges of the strips in the transverse plane) will be denoted by the e index, and the parameters related to waves of an odd type (characterized by opposite values of the instantaneous charges of the strips) will be denoted by the o index.

Using (4.21, 4.22), and (B.18, B.19), we find analytical expressions for the coupling coefficients of the main magnetic oscillations of a spherical DR with the edge-coupled symmetric strip line for its even waves:

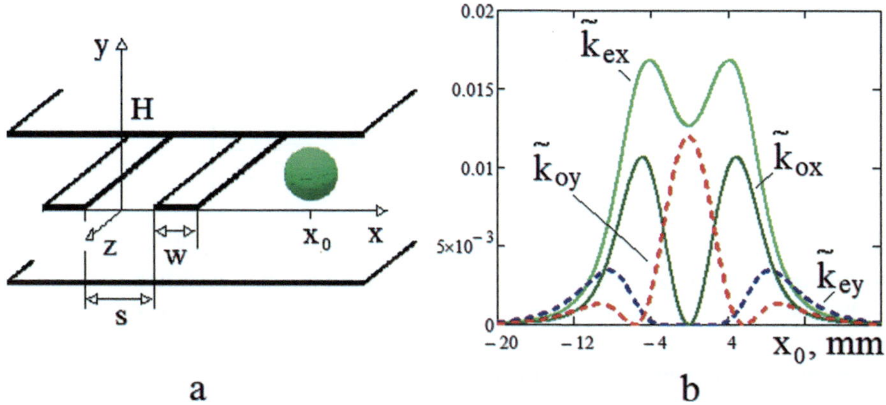

Fig. 4.9 Spherical DR in a two-wire edge coupled strip line (**a**). Dependences of coupling coefficients with even ($\tilde{k}_{ex}$, $\tilde{k}_{ey}$); odd ($\tilde{k}_{ox}$, $\tilde{k}_{oy}$) waves from the transverse coordinates of the DR with H_{111} oscillations ($f_0 = 7$ GHz; $\varepsilon_{1r} = 81$; $w = 4$ mm; $s = 5$ mm; $H = 10$ mm; $y_0 = 4$ mm) (**b**)

$$\begin{pmatrix} \tilde{k}_{ex} \\ \tilde{k}_{ey} \end{pmatrix} = \alpha_1^H(p,q) \cdot \left[\frac{\pi}{k_0 b K'}\right]^2 \cdot \frac{Z_{0e}}{5p_e(1-\kappa_e^2)} \cdot \left[\begin{matrix}\mathrm{Re}\\ \mathrm{Im}\end{matrix} \frac{sh(\pi\tilde{z}_0/b)}{\sqrt{\left[ch^2\frac{\pi\tilde{z}_0}{b} - ch^2\frac{\pi s}{2b}\right]\cdot\left[ch^2\frac{\pi\tilde{z}_0}{b} - ch^2\frac{\pi s}{2b}(2w+s)\right]}}\right]^2; \tag{4.28}$$

for its odd waves:

$$\begin{pmatrix} \tilde{k}_{ox} \\ \tilde{k}_{oy} \end{pmatrix} = \alpha_1^H(p,q) \cdot \left[\frac{\pi}{k_0 b K'}\right]^2 \cdot \frac{Z_{0o}}{5p_o(1-\kappa_o^2)} \cdot \left[\begin{matrix}\mathrm{Re}\\ \mathrm{Im}\end{matrix} \frac{ch(\pi\tilde{z}_0/b)}{\sqrt{\left[ch^2\frac{\pi\tilde{z}_0}{b} - ch^2\frac{\pi s}{2b}\right]\cdot\left[ch^2\frac{\pi\tilde{z}_0}{b} - ch^2\frac{\pi s}{2b}(2w+s)\right]}}\right]^2; \tag{4.29}$$

where Z_{0e} (Z_{0o})—wave impedance of even (odd) type of waves and $p_e, \kappa_e; p_o, \kappa_o$—are parameters of (B.17, B.20). Here $b = 2H$; $K' = K(\kappa')$—complete elliptic integral [8], corresponding to the additional module $\kappa' = \left(1-\kappa^2\right)^{1/2}$. The calculated dependences of the coupling coefficients between a spherical DR and a two-wire edge-coupled symmetric strip line are shown in Fig. 4.9b.

When the strips move away from each other, expressions (4.28, 4.29) coincide exactly with the formulas for the coupling coefficients of the main magnetic oscillations of a dielectric ball with an isolated symmetrical strip line:

$$\lim_{s=\infty}\begin{pmatrix}\tilde{k}_x^{e,o}\\ \tilde{k}_y^{e,o}\end{pmatrix} = \frac{\alpha_1^H(p,q)}{20}\cdot Z_0\left[\frac{\pi}{k_0 bK'}\right]^2 \cdot\left[\begin{matrix}Re\\ Im\end{matrix}\ \frac{1}{ch\frac{\pi\tilde{z}}{b}\sqrt{(th^2\frac{\pi\tilde{z}}{b}-th^2\frac{\pi w}{2b})}}\right]^2.$$

Here $Z_0 = \lim\limits_{s=\infty} Z_{0e}$; $\tilde{z} = \tilde{z}_0 - (s+w)/2$—new transverse complex coordinates of the ball center related to the plane of symmetry of an isolated strip line.

To calculate the coupling coefficients between a DR and a high-quality symmetrical coupled strip line (Fig. 4.10a), we use the results of [11], which approximately determine the parameters of TEM waves for wide strips of finite thickness. Keeping the notation [11]: where $2h_2$—minimum distance between stripes; $h_1 = h_2 + t$; t—the thickness of the strips we obtained from (B.7–B.9):

$$\begin{pmatrix}\tilde{k}_x\\ \tilde{k}_y\end{pmatrix} = \frac{\alpha_1^H(p,q)}{20}\cdot\frac{Z_0}{(k_0H)^2} \cdot\left[\begin{matrix}\mathrm{Re}\\ Im\end{matrix}\ \frac{\sqrt{(\omega+a_1)\cdot(\omega+a_2)}}{\frac{\sqrt{(a_1+1)\cdot(a_2+1)}}{1-\omega}+\frac{h_2}{H}\cdot\frac{\sqrt{a_1a_2}}{\omega}+\frac{h_1}{H}-1}\cdot\varphi(\omega)\right]^2, \tag{4.30}$$

where $Z_0 = Z_{0e}$—for even; Z_{0o}—for odd waves of the line and

$\varphi(\omega) = \left(\frac{1}{\sqrt{\omega(\omega-1)}}\right)$—for even, as well as

$$\varphi(\omega) = \left(\frac{1}{\omega}\right)\text{-for odd waves.} \tag{4.31}$$

For thin strips, relations (4.30, 4.31) are significantly simplified.

in $h = h_1 = h_2$ we find:

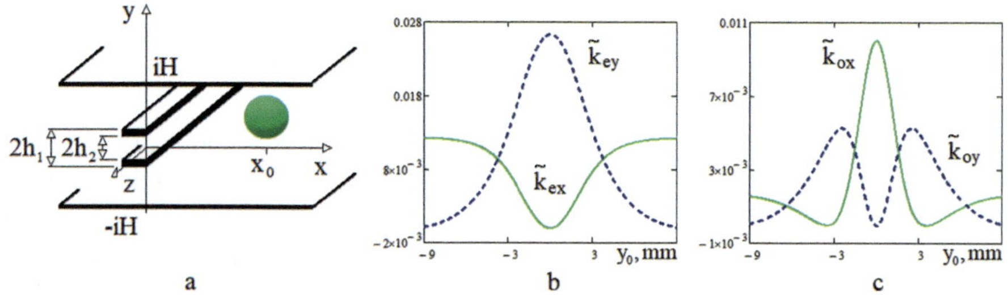

Fig. 4.10 Spherical DR in a high-quality symmetrical coupled strip line (**a**). Dependences of coupling coefficients with even ($\tilde{k}_{ex}$, $\tilde{k}_{ey}$); odd ($\tilde{k}_{ox}$, $\tilde{k}_{oy}$) waves on the transverse coordinates of the DR with H_{111} oscillations ($f_0 = 7$ GHz; $\varepsilon_{1r} = 81$; $w = 1$ mm; $h_1 = 1, 1$ mm; $h_2 = 1$ mm; $H = 10$ mm; $x_0 = 3$ mm)

$$\begin{pmatrix} \tilde{k}_x \\ \tilde{k}_y \end{pmatrix} = \alpha_1^H(p,q) \cdot \frac{Z_0}{20k_0^2(H-h)^2} \cdot \begin{pmatrix} \mathrm{Re} \\ \mathrm{Im} \end{pmatrix} f(\omega) \Big), \tag{4.32}$$

where
$f(\omega) = \left(\frac{\sqrt{\omega(\omega-1)}}{\omega+a}\right)$—for even waves, and

$$f(\omega) = \left(\frac{\omega-1}{\omega+a}\right)\text{-for odd waves;} \tag{4.33}$$

$a = h/(H-h)$. The transformation $\tilde{z}_0(\omega)$ has the form (B.11).

Calculations of the coupling coefficients of DR with two-wire lines in comparison with a single-wire line show that the maximum values of the coupling coefficients approximately double when an additional strip appears.

Let's consider a special case: $H \gg h$. In this case, the relations for the coupling coefficients of the DR with the odd wave of a high-quality coupled strip line can be expressed in a simpler form:

$$\begin{pmatrix} \tilde{k}_x \\ \tilde{k}_y \end{pmatrix} = \frac{1}{20k_0^2} Z_0 \alpha_1^H(p,q) \cdot \frac{1}{(hr)^2} \cdot \begin{pmatrix} 1 + e^u \cos v \\ e^u \sin v \end{pmatrix}^2, \tag{4.34}$$

where $r = 1 + 2e^u \cos v + e^{2u}$, and the parameters u, v are determined by the coordinates of the dielectric ball in the transverse plane by the relations:

$$\frac{\pi x_0}{h} = e^u \cos v + u + 1; \quad \frac{\pi y_0}{h} = e^u \sin v + v. \tag{4.35}$$

In this case, such a line is equivalent to an asymmetric microstrip line (Fig. 4.11a). In Fig. 4.11b shows the dependences of the coupling coefficients of the main oscillations of the DR with the TEM wave of the microstrip line, calculated using the formulas (4.34, 4.35) for $Z_0 = 50$ Ohm; $\varepsilon_\Lambda = 9, 6$; $w = h = 1$ mm. The experimental measurements carried out showed good agreement between the calculated and experimental values of the coupling coefficients for H_{111}^y in the region remote from the edge of the conductor. In the region above the current-carrying conductor, the experimental coupling values are less than the calculated ones. This is due to the inaccuracy of the representation of the microstrip line field in the conductor region (the corresponding expressions for the line field components were obtained for an infinitely wide strip) as well as the approximate representation of the resonator field in the line. In the case of a finite width, the strip is a component of the magnetic field of the microstrip line, and along with it the coupling coefficient $\tilde{k}_y$, in the symmetry plane $x = -w/2$ becomes zero. However, in the practically used coordinate region DR ($x_0 > 0$), the obtained relations are characterized by acceptable accuracy and can be used for calculations. Note also that the $H_{111}^{x,y}$ oscillations are degenerate, so accurate measurements of the coupling are impossible due to their mutual influence on each other.

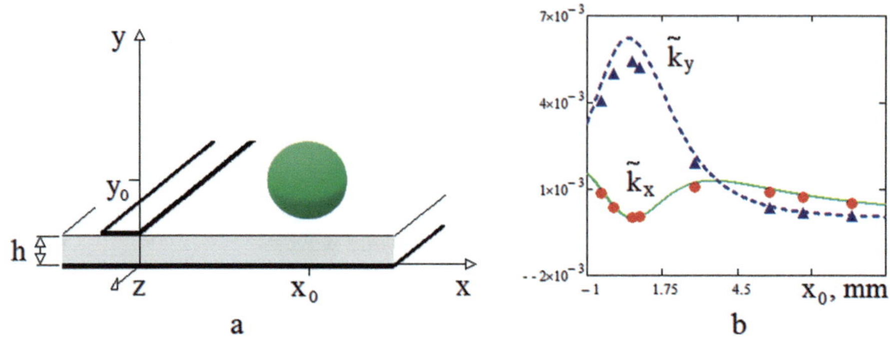

Fig. 4.11 Spherical DR in a asymmetric microstrip line (**a**). Dependences of microstrip line coupling coefficients on coordinates for fundamental magnetic oscillations H_{111} (**b**); (points—measurement results; $f_0 = 7$ GHz; $\varepsilon_{1r} = 81$; $w = 1,25$; $h = 1$ mm; $\varepsilon_L = 9,4$; $y_0 = 3,25$ mm)

4.5 Coupling of a Cylindrical Dielectric Resonator and a Microstrip Line

TEM waves also can be considered as a limiting case of waves H or E type, where the longitudinal field components and transverse wave numbers vanish [13]. Indeed, let us replace the variables $\wp = \eta\xi$ in (B.1) and go to the limit $\eta = 0$, now assuming that $\eta^2 + \gamma^2 = \varepsilon_{eff}$. Taking into account the condition $\sqrt{-\wp^2} = -i\wp$ for the field to decrease at: $y \to \infty$ for waves characterized by spectral density $a(\eta, \wp/\eta) \sim O(\eta^3)$, we obtain a relation of the form (B.5), where the potential function $\varphi(x, y)$ determines the field of the microstrip line:

$$\varphi(x, y) = \frac{i}{k_0} \int\limits_{-\infty}^{\infty} e^{-ik_0(x-iy)\wp} \frac{\varphi(\wp)}{\wp} d\wp, \tag{4.36}$$

and the function $\varphi(\wp)$ is determined from the relation:

$$\lim_{\eta=0,\eta\xi=\wp} \frac{a(\eta, \wp/\eta)}{\eta^3} = i\frac{\varphi(\wp)}{\wp}. \tag{4.37}$$

The application of transformations (4.36, 4.37) allows us to directly calculate from (4.7–4.9) the expansion coefficients of the main H_{101} oscillation of the field of a cylindrical DR from the TEM waves of the strip line [14].

For A position (Fig. 4.2a):

$$c_{TEM}^{nol\pm} = \mp\frac{\pi}{2} h_1 r_0^3 L q_\perp \left(k_1^2 - k_0^2\right) \cdot \sqrt{\varepsilon_{eff}} \cdot e^{\pm i\sqrt{\varepsilon_{eff}} k_0 z_0}$$

$$\cdot \int_{-\infty}^{\infty} \varphi(\wp)^* \cdot f(\wp) \cdot e^{ik_0(x_0 - iy_0)\wp} d\wp; \tag{4.38}$$

For B position (Fig. 4.2b):

$$c_{TEM}^{nol\pm} = \pm i \frac{\pi}{2} h_1 r_0^3 L q_\perp \left(k_1^2 - k_0^2\right) \cdot \sqrt{\varepsilon_{eff}} \cdot e^{\pm i\sqrt{\varepsilon_{eff}} k_0 z_0}$$
$$\cdot \int_{-\infty}^{\infty} \varphi(\wp)^* \cdot f(-i\wp) \cdot e^{ik_0(x_0 - iy_0)\wp} d\wp, \tag{4.39}$$

Here the function $f(x)$ is defined by the equality (4.10), and

$$f(-i\wp) = \frac{J_0(p_\perp)J_2\left(q_\perp\sqrt{1+\wp^2}\right) - J_2(p_\perp)J_0\left(q_\perp\sqrt{1+\wp^2}\right)}{p_\perp^2 - q_\perp^2\left[1+\wp^2\right]}$$
$$\cdot \frac{1}{p_z^2 + (q_z\wp)^2} \cdot \begin{bmatrix} p_z \sin p_z ch(q_z\wp) - q_z\wp \cos p_z sh(q_z\wp) \\ p_z \cos p_z sh(q_z\wp) - q_z\wp \sin p_z ch(q_z\wp) \end{bmatrix} \tag{4.40}$$

Let us now expand the functions $f(\wp)$ and $f(-i\wp)$ in the Taylor series:

$$f(\wp) = \sum_{s=0}^{\infty} \frac{f^{(s)}(0)}{s!} \cdot \wp^s$$

and we will use the expressions for the field of the microstrip line (B.1) as well as the expansion (4.36). From where we also find that for positions A, B (Fig. 4.2a, b) the indicated coefficients also take the form (4.11), where formally we should put $\eta = 1$.

In the particular case, if the field of the line changes exponentially, for example in the direction of the x axis at $|x| > \breve{x}$:

$$H_x^{\pm} \approx H_0 e^{-\chi x} \psi(y) e^{\mp i\sqrt{\varepsilon_{eff}} k_0 z},$$

then

$$\partial^{(s)}/\partial x^s H_x^{\pm} = (-\chi)^s H_x^{\pm}. \tag{4.41}$$

Substituting (4.41) into (4.11), we obtain for the A position:

$$c_{TEM}^{nol\pm} = -i\frac{\pi}{2}(w_0 h_1) r_0^3 L q_\perp \left(k_1^2 - k_0^2\right) \cdot f(i\frac{\chi^*}{k_0}) H_x^{\pm}(x_0, y_0, z_0)^*.$$

In this case, again the expansion coefficient of the fundamental H_{101} oscillation of the resonator turns out to be proportional to the field of the transmission line at its center.

In position C (Fig. 4.2c) the use of a transformation of the form (4.37), or direct calculation according to (B.6) based on the representation of the Poisson integral, for natural oscillations EH_{nml} or HE_{nml} leads to the relation:

$$c_{TEM}^{nml\pm} = \pi\frac{m}{m!}e^{im\pi}r_0^m\left(k_1^2 - k_0^2\right)\frac{J_m(p_\perp)}{k_0\beta}\left(e_1\frac{\beta_z}{\beta} + i\omega\mu_0\frac{h_1}{\beta}\right) \cdot \frac{\partial^{m-1}}{\partial x_0^{m-1}}\left\{\begin{matrix} H_y^\pm(x_0, y_0, z_0) \\ H_x^\pm(x_0, y_0, z_0) \end{matrix}\right\}\frac{1}{\beta_z^2 - k_0^2}\left[\begin{matrix} \beta_z \sin p_z \cos q_z - k_0 \cos p_z \sin q_z \\ \mp i(k_0 \sin p_z \cos q_z - \beta_z \cos p_z \sin q_z) \end{matrix}\right]. \tag{4.42}$$

From which we also come to the conclusion that in the case of a cylindrical DR in position C, the field expansion coefficients in TEM waves are proportional to the partial derivatives of the magnetic (or electric) field components of this line in the center of the resonator. In the particular case $m = 1$, within the accepted assumptions, the indicated coefficients turn out to be proportional to the corresponding components of the transmission line field. Azimuthally homogeneous oscillations with this arrangement of the DR are not coupled with TEM waves. The resulting expressions make it possible to directly determine the power emitted by the resonator into a multiwire line, the field of which is usually calculated in a complex way.

As an example, let us show the result of calculating the coupling coefficients of azimuthally homogeneous oscillations of a cylindrical DR in position B with the asymmetric shielded microstrip line (Fig. 4.12a); the influence of dielectric plates and the screen should be taken into account using the representation of longitudinal dependencies (1.27, 1.28). In this case, the expansion coefficient of the resonator field over the line field, calculated on the basis of (4.11), takes the form:

$$c_{TEM}^{nol\pm} = \frac{\pi i}{2} \cdot (w_0 h_1) \cdot r_0^3 L q_\perp \left(k_1^2 - k_0^2\right) \cdot \left[\frac{J_0(p_\perp)J_2(q_\perp) - J_2(p_\perp)J_0(q_\perp)}{p_\perp^2 - q_\perp^2}\right] \cdot \left\{a\frac{\sin p_z}{p_z}H_y^\pm(x_0, y_0, z_0) - b\frac{q_z}{p_z^2}(p_z \cos p_z - \sin p_z)\frac{\partial H_y^\pm(x_0, y_0, z_0)}{k_0 \partial y_0}\right\}. \tag{4.43}$$

From here we get [12]

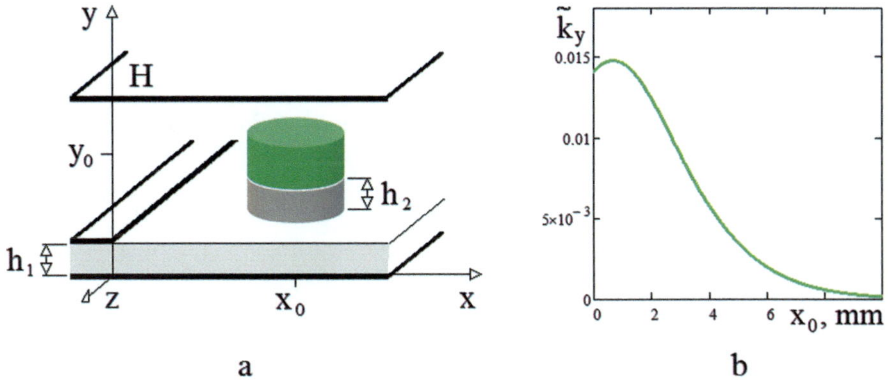

Fig. 4.12 Dependence of the coupling coefficient with an asymmetric shielded microstrip line (**a**) of a cylindrical DR with H_{101} oscillations on the transverse coordinate (**b**) for $f_0 = 6$ GHz; $\varepsilon_{1r} = 81$; $\Delta = 0,44$; $Z_0 = 50$ Ohm; $\varepsilon_L = 9,6$; $\varepsilon_P = 2,2$; $h_1 = 1$ mm; $h_2 = 3$ mm; $H = 10$ mm; $y_0 = h_1 + h_2 + L/2$

$$\tilde{k}_y = 4\pi\frac{Z_0}{w_0}\cdot\frac{(\varepsilon_{1r}-1)^2}{\varepsilon_{1r}}\cdot\frac{p_\perp^2 p_z q_\perp^4 q_z}{[J_1^2(p_\perp) - J_0(p_\perp)J_2(p_\perp)]\cdot(2p_z + \cos 2\delta\sin 2p_z)}$$
$$\cdot\left[\frac{J_0(p_\perp)J_2(q_\perp) - J_2(p_\perp)J_0(q_\perp)}{p_\perp^2 - q_\perp^2}\right]^2\cdot\frac{1}{[k_0(H-h_1)]^2}\cdot\left[\frac{(a_\Lambda+1)\rho\sin\tau}{\rho^2 + a_\Lambda^2 + 2a_\Lambda\rho\cos\tau}\right]^2$$
$$\cdot\left\{\cos\delta\frac{\sin p_z}{p_z} - \frac{\pi}{k_0(H-h_1)}\cdot\sin\delta\frac{q_z}{p_z^2}(p_z\cos p_z - \sin p_z)\right.$$
$$\left.\cdot\frac{2a_\Lambda\rho\left(\rho^2 - a_\Lambda\right) + \left(\rho^3 - a_\Lambda^3 + 3a_\Lambda\rho(a_\Lambda - 1)\right)\cos\tau + \rho\left(a_\Lambda^3 - \rho^2\right)\cos 2\tau}{\sin\tau\left(\rho^2 + a_\Lambda^2 + 2a_\Lambda\rho\cos\tau\right)^2}\right\}^2. \tag{4.44}$$

Here $\cos\delta = a/\sqrt{a^2+b^2}$; $\omega = \rho e^{i\tau}$; $a_\Lambda = h_1/(H-h_1)$.

When deriving relations (4.43, 4.44), we took into account the influence of metal screens as well as dielectric plates of the stand and microstrip line; the field inside the DR is distributed asymmetrically along the axis relative to its center.

The dependence of the coupling coefficient, calculated using formula (4.44), is shown in Fig. 4.12b.

Let us apply relations (4.41) to calculate the coupling coefficients of hybrid oscillations EH_{111}^{+} of a cylindrical DR with asymmetric microstrip line:

$$\begin{pmatrix} \tilde{k}_x \\ \tilde{k}_y \end{pmatrix} = 8\pi \frac{Z_0}{w_0} \cdot \frac{c_{11}(\varepsilon_{1r}, \Delta)}{[k_0(H - h_1)]^2} \cdot \frac{1}{(\rho^2 + a_\Lambda^2 + 2a_\Lambda \rho \cos \tau)^2} \cdot \begin{pmatrix} \rho^2 - a_\Lambda + \rho(a_\Lambda - 1)\cos\tau \\ (a_\Lambda + 1)\rho \sin\tau \end{pmatrix}^2, \quad (4.45)$$

where

$$c_{11}(\varepsilon_{1r}, \Delta) = \frac{(\varepsilon_{1r} - 1)^2}{\varepsilon_{1r}} \cdot \left(\frac{J_1(p_\perp)}{\Delta p_\perp}\right)^2 |e_0 - h_0|^2 \Big/ \left[\frac{1}{2}\{(|e_0| - |h_0|)^2 \left[J_0^2(p_\perp) + J_1^2(p_\perp)\right] + (|e_0| + |h_0|)^2 \cdot \left[J_2^2(p_\perp) - J_1(p_\perp)J_3(p_\perp)\right]\} \cdot (2p_z \pm \sin 2p_z) + |\Delta p_\perp / p_z e_0|^2 \cdot \left[J_1^2(p_\perp) - J_0(p_\perp)J_2(p_\perp)\right] \cdot (2p_z \mp \sin 2p_z)\right] \cdot \frac{\beta_z}{k_0}\left(\frac{k_0^2}{\beta_z^2 - k_0^2}\right)^2 \cdot \left[\frac{p_z \sin p_z \cos q_z - q_z \cos p_z \sin q_z}{q_z \sin p_z \cos q_z - p_z \cos p_z \sin q_z}\right]^2; \quad (4.46)$$

and e_0, h_0—amplitudes (2.31).

In cases where the centers of the resonators need to be located at the maximum distance from the metal plates, which leads to minimal deterioration in the quality factor associated with losses in the metal, it is convenient to use a symmetrical strip line. The characteristics of symmetrical strip line are well studied [11], and in the case of an infinitely thin central conductor, a simple exact solution is known. The coupling coefficients with a symmetrical strip line of a cylindrical DR c located coaxially (Fig. 4.13a) are found in the form:

$$\begin{pmatrix} \tilde{k}_x \\ \tilde{k}_y \end{pmatrix} = \frac{\pi^3}{2K(\kappa)K'(\kappa)} \cdot \frac{c_{11}(\varepsilon_{1r}, \Delta)}{(k_0 H)^2} \cdot \begin{pmatrix} \mathrm{Re} \\ Jm \end{pmatrix} \frac{i}{\sqrt{sh^2 \frac{\pi \tilde{z}_0}{2H} - \kappa^2 ch^2 \frac{\pi \tilde{z}_0}{2H}}} \Bigg)^2, \quad (4.47)$$

where H—screen height; w—strip width; $\kappa = th[\pi w/(4H)]$; $K(\kappa)$, $K'(\kappa)$—complete elliptic integrals of the first kind for κ and $\kappa' = (1 - \kappa^2)^{1/2}$, respectively [8].

The magnitude of the coupling coefficients quickly decreases as the center of the DR moves away from the edge of the strip (Fig. 4.13b).

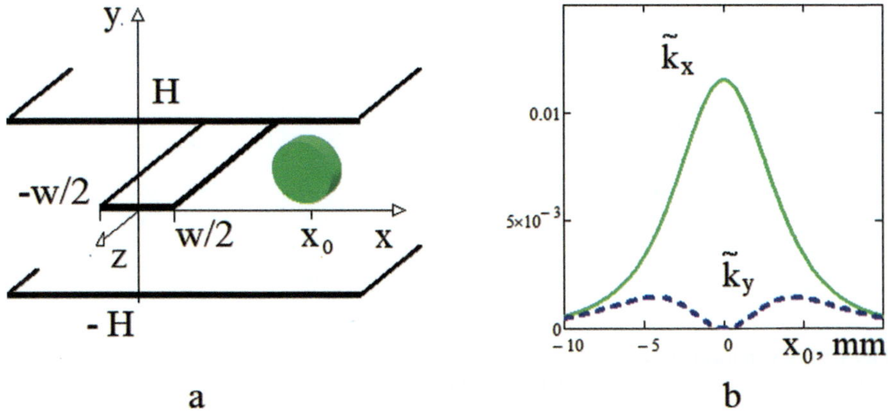

Fig. 4.13 Cylindrical DR in a symmetrical strip line (**a**). Coupling coefficients with an symmetrical strip line (**a**) of a cylindrical DR with EH_{111} oscillations on the transverse coordinate (**b**) for $f_0 = 7$ GHz; $\varepsilon_{1r} = 81$; $\Delta = 0, 4$; $H = 10$ mm; $w = 2$ mm; $y_0 = 5$ mm

4.6 Coupling Coefficients of an Orthogonally Oriented Cylindrical Dielectric Resonator in a Circular Waveguide

When designing filters and phase shifters, it is convenient to use cylindrical DRs, the axes of which are oriented orthogonally to the axis of the circular waveguide. Calculation of the DR coupling coefficients in such structures is much more complicated due to the non-coordinated boundary surfaces between the media. However, obtaining analytical relations is possible if we use repeated expansion of the field of a circular waveguide in plane waves, as well as previously obtained results of the form (4.7–4.10).

Let us denote by α_0—the angle of rotation around the z axis defined in Fig. 4.14 (x', y', z') coordinate systems relative to the (x, y, z) waveguide coordinate system. Denoting in (D.13) by $z_1 = \eta k_0 x'$; $z_2 = \eta k_0 y'$; $m = u$, let us express the field of a circular waveguide in the (x, y, z) system in the form of an expansion in plane waves H- or E-type, but now through integrals over finite limits. Comparing the result of transformations (D.13) with the original representation (B.1), we obtain expressions for the corresponding spectral functions:

$$\begin{pmatrix} a(\xi,\eta) \\ b(\xi,\eta) \end{pmatrix} = \begin{pmatrix} -N^{su}/\mu_0 \\ \pm K^{su}/\varepsilon_0 \end{pmatrix} \frac{(\eta k_0)^2}{2\pi\omega} e^{iu\frac{\pi}{2}} \left[\begin{matrix} \sin \\ \cos \end{matrix} u(\arccos\xi + \alpha_0) \right] / \sqrt{1-\xi^2},$$

and from (4.7–4.10), we find the required expansion coefficients [15] , for example, for the position of DR A:

by magnetic waves H_{su}

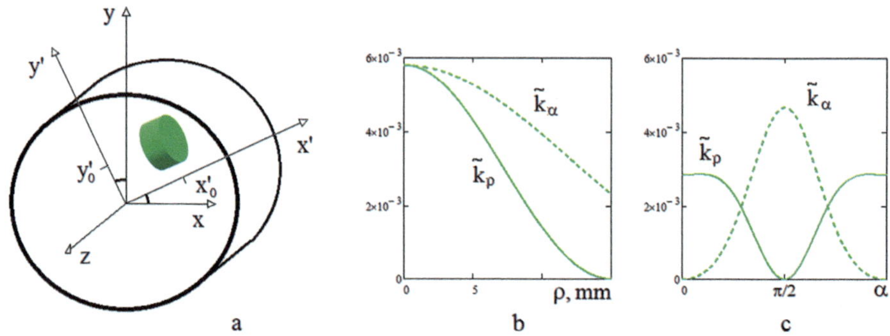

Fig. 4.14 Cylindrical DR in a circular metal waveguide (**a**). Dependences of coupling coefficients on the transverse coordinates of the DR ($f_0 = 7$ GHz; $\varepsilon_{1r} = 81$; $\Delta = 1$;$R = 15$ mm)

$$c_H^{n0l\pm} = c_0^{\pm}\frac{N^{su}w_0}{\omega\mu_0}\gamma\int\limits_{-\pi}^{\pi}\left[\begin{matrix}\sin\\ \cos\end{matrix}u(\tau+\alpha_0)\right]$$

$$\cdot f(\cos\tau)\cdot e^{i\eta k_0\left(x_0'\cos\tau+y_0'\sin\tau\right)}\cos\tau d\tau; \tag{4.48}$$

by electric waves E_{su}

$$c_E^{n0l\pm} = c_0^{\pm}\frac{K^{su}}{\omega\varepsilon_0}\int\limits_{-\pi}^{\pi}\left[\begin{matrix}\sin\\ \cos\end{matrix}u(\tau+\alpha_0)\right]$$

$$\cdot f(\cos\tau)\cdot e^{i\eta k_0\left(x_0'\cos\tau+y_0'\sin\tau\right)}\sin\tau d\tau, \tag{4.49}$$

where

$$c_0^{\pm} = \mp h_1/(4\beta)Lp_{\perp}q_{\perp}^3\left(k_1^2-k_0^2\right)\cdot\eta e^{i\left[(u+1)\frac{\pi}{2}\pm\gamma k_0 z_0\right]};$$

$\eta = \chi/k_0$—transverse, and $\gamma = \Gamma/k_0$—longitudinal wave numbers of a circular waveguide; $f(\xi)$—function (4.10).

The expressions found make it possible to relatively simply calculate the coupling coefficients of magnetic azimuthally homogeneous oscillations of a cylindrical DR in position A with the waves of a circular waveguide:

For H_{su}

$$\tilde{k} = \frac{2}{\pi^2(1+\delta_{u0})(j_{us}^{\cdot 2} - u^2)|J_u(j_{us}^{\cdot})|^2} \cdot \frac{(\varepsilon_{1r}-1)^2}{\varepsilon_{1r}}$$
$$\cdot \frac{q_\perp^4 q_z p_\perp^2 p_z}{\{J_1^2(p_\perp) - J_0(p_\perp)J_2(p_\perp)\} \cdot [2p_z \pm \sin 2p_z]}$$
$$\cdot |\gamma|\eta^2 \left| \int_{-\pi}^{\pi} \left[\begin{matrix} \sin \\ \cos \end{matrix} u(\tau+\alpha_0) \right] \cdot e^{i\eta k_0 (x_0' \cos\tau + y_0' \sin\tau)} f(\cos\tau) \cdot \cos\tau d\tau \right|^2; \quad (4.50)$$

For E_{su}

$$\tilde{k} = \frac{2}{\pi^2(1+\delta_{u0})|j_{us} J_u^{\cdot}(j_{us})|^2} \cdot \frac{(\varepsilon_{1r}-1)^2}{\varepsilon_{1r}}$$
$$\cdot \frac{q_\perp^4 q_z p_\perp^2 p_z}{\{J_1^2(p_\perp) - J_0(p_\perp)J_2(p_\perp)\} \cdot [2p_z \pm \sin 2p_z]}$$
$$\cdot \frac{\eta^2}{|\gamma|} \left| \int_{-\pi}^{\pi} \left[\begin{matrix} \sin \\ \cos \end{matrix} u(\tau+\alpha_0) \right] \cdot e^{i\eta k_0 (x_0' \cos\tau + y_0' \sin\tau)} f(\cos\tau) \cdot \sin\tau d\tau \right|^2. \quad (4.51)$$

In position B, the function $f(\cos\tau)\cos\tau$ from (4.50) is replaced by $f(\sin\tau)\sin\tau$ and $f(\cos\tau)\sin\tau$ from (4.51) by $f(\sin\tau)\cos\tau$.

The resulting relations in general describe the coupling of the DR with a circular waveguide with an arbitrary orientation of its axis in the transverse plane. In practice, simpler standard provisions are used: $\tilde{k}_\rho = \tilde{k}_A$ and $\tilde{k}_\alpha = \tilde{k}_B$ at $y_0' = 0$; $x_0' = \rho_0$, where (ρ_0, α_0)—polar coordinates of the DR center. In Fig. 4.14, the dependences of the coupling coefficients of the main H_{101} oscillations with the H_{11} wave of a circular waveguide of radius $R = 15$ mm are shown for the two above positions: $\tilde{k}_\alpha$ (for $\alpha_0 = \pi/2$). It was assumed that the remaining parameters were fixed: resonant frequency—$f_0 = 7$ GHz; $\varepsilon_{1r} = 81$. The meaning of the obtained dependencies becomes clear if we imagine the H_{11} wave polarization and the orientation of the DR axis as constant. The coefficient $\tilde{k}_\rho$ in this case corresponds to the change in coupling when the center of the DR is displaced in the radial direction along the magnetic field of the H_{11} wave, and the coefficient $\tilde{k}_\alpha$ in the orthogonal direction.

References

1. D. Kajfez, P. Guillon, *Dielectric Resonators* (Artech House, 1986), 201 p
2. *Dielectric Resonators*, ed. by M.E. Ilchenko (Radio I Sviaz, 1989), 327 p (in Russian)
3. M.E. Ilchenko, A.A. Trubin, *Electrodynamics' of Dielectric Resonators* (Naukova Dumka, 2004), 265 p (in Russian)

4. *Optical Microcavities*, ed. by K. Vahala (California Institute of Technology, USA, 2004), 516 p
5. I. Chremmos, O. Schwelb, N. Uzunoglu, *Photonic Microresonator Research and Applications*. Series in Optical Sciences—156 (Springer International Publishing, Switzerland, 2010), 517 p
6. A. Trubin, Calculation of the coupling between a spherical dielectric resonator and a slot transmission line. Radioelectronics **38**(2), 61–66 (1995) (in Russian)
7. E.I Nefedov, A.T. Fialcovsky, *Strip Transmission Lines* (Nauka, 1980), 312 p (in Russian)
8. *Handbook of Mathematical Functions with Formulas, Graphs and Mathematical Tables*, ed. by M. Abramowitz, I.A. Stegun, Applied Mathematics Series—55 (National Bureau of Standarts. 1964), 830 p
9. G.A. Korn, T.M. Korn, *Mathematical Handbook* (McGraw-Hill Book Company, 1968), 943 p
10. A. Trubin, Coupling coefficients of a spherical dielectric resonator with TEM waves of strip lines, Radioelectronics **38**(11), 31–38 (1995) (in Russian)
11. *Striplines and Microwave Devices*, ed. by V.M. Sedyh (Harkov. Visha shkola, 1974), 275 p (in Russian)
12. A. Trubin, Parameters of coupling of a spherical dielectric resonator with multiwire strip structures, Radioelectronics **39**(3), 66–70 (1996) (in Russian)
13. V.V. Nikolsky, *Electrodynamics and Radio Wave Propagation* (Nauka, 1973), 606 p (in Russian)
14. A. Trubin, Coupling coefficients of asymmetrical oscillations of cylindrical dielectric resonators with TEM waves of a microstrip transmission line, Radioelectronics **39**(6), 25–30 (1996) (in Russian)
15. A. Trubin, Coupling coefficients of orthogonally oriented cylindrical dielectric resonators in a circular waveguide, Radioelectronics **6**, 10–15 (1997) (in Russian)

Chapter 5
Dielectric Resonators in Open Space

Dielectric resonators are open structures, they always radiate part of the electromagnetic energy into the surrounding space. This serves as an objective prerequisite for using them as antenna elements [1–3]. Antennas built on the basis of DR, along with the formation of radiation patterns, can combine the functions of input filters. To calculate the characteristics of antennas on DR, you must be able to calculate their fields and coupling coefficients with open space. In this chapter we will consider the fields of various DRs in the wave zone, as well as the coupling coefficients caused by their radiation into open space.

The DR field in the wave zone can be found using the Kirchhoff integral, or by passing to the continuous spectrum (Sect. 3.5). In both cases, it is expressed through integrals either from the field on the surface of the DR, or in a secondary expansion through the field of the line when removing its walls. Direct calculation of these integrals makes it possible to find in analytical form the dependence of the pattern and coupling coefficients on the resonator parameters. The obtained relationships make it possible to establish the conditions for narrowing the radiation pattern for main azimuthally homogeneous and hybrid types of oscillations of the DR different shapes: spherical, cylindrical and rectangular. The directional characteristics of main dipole and quadruple types of oscillations are calculated. Below we also consider the possibility of unidirectional radiation from a cylindrical and spherical DR upon excitation of degenerate magnetic and electrical oscillations in it.

The radiation patterns of spherical DR located near a metal wedge, as well as near a metal cylinder were studied. The coupling coefficients of DR located in open space near a metal wedge, cylinder and near a metal plane also are considered.

The influence of metallic and dielectric inhomogeneities on the coupling coefficients of a spherical DR with open space has been studied. The possibility of increasing coupling by creating an air cavity in the resonator material is noted.

A. Trubin, *Introduction to the Theory of Dielectric Resonators*, Springer Series in Advanced Microelectronics 65, https://doi.org/10.1007/978-3-031-65396-4_5

5.1 Radiation Characteristics of Spherical Dielectric Resonators in Open Space

The angular distribution of the field of spherical DR is known exactly (1.2, 1.3), so its analysis is rather of methodological interest.

Note that for $|z| \to \infty$, up to $O(1/|z|^2)$ spherical Hankel functions of the second kind:

$$h_n^{(2)}(z) \approx i^{n+1} e^{-iz}/z; \quad 1/z \cdot d/dz\{zh_n^{(2)}(z)\} \approx i^n e^{-iz}/z, \tag{5.1}$$

Therefore, from (1.2, 1.3) and (5.1) we conclude that the angular dependences of the field of a spherical DR in the wave zone $\left|\mathbf{e}^0\right| = (\left|e_\theta^0\right|^2 + \left|e_\varphi^0\right|^2)^{1/2}$ for magnetic and electrical non-degenerate oscillations with the same indices m and n coincide. Magnetic and electric fields differ from each other only in polarization.

In addition, in the case of excitation of a spherical DR by an external source (see, for example, comments on 1.12, 1.13), the characteristic parameters p, q become real and its field in the wave zone satisfies the radiation conditions (Fig. 1.1).

On Fig. 5.1 shows the wave zone field distributions for the main types of oscillations. We note that the lowest natural oscillations have a dipole field distribution, magnetic for H_{1m1}, electric for E_{1m1} (Fig. 5.1a–c). These distributions differ from each other by rotation $\pi/2$ by angles θ, or φ.

With simultaneous excitation in the DR of magnetic and electric type oscillations degenerate in frequency, it is possible to obtain a radiation pattern in a shape close to cardioid. Let us determine the conditions for implementing the cardioid radiation characteristic for a spherical DR. When a dielectric ball is filled with a metallic inhomogeneity of a spherical shape, for example, for $\varepsilon_{1r} = 81$ at $r_1/r_2 \approx 0,4$ the frequencies of its main magnetic and electrical oscillations coincide (Fig. 1.15b). So-called super degenerate oscillations arise [4]. The total field in the far zone of such a resonator can be represented as:

$$e_\theta^\infty = \lim_{r\to\infty} (e_\theta^0) \approx -i\left(e^{-ik_0 r}/k_0 r\right)[w_0 h_0 + e_0 \cos\theta] \cos\varphi;$$

$$e_\varphi^\infty = \lim_{r\to\infty} (e_\varphi^0) \approx i\left(e^{-ik_0 r}/k_0 r\right)[w_0 h_0 \cos\theta + e_0] \sin\varphi. \tag{5.2}$$

From here we find the condition for excitation of unidirectional radiation:

$$w_0 h_0 = \pm e_0, \tag{5.3}$$

where $w_0 = (\mu_0/\varepsilon_0)^{1/2}$. Indeed, substituting (5.3) into (5.2), we find:

$$e_\theta^\infty \approx -ie_0\left(e^{-ik_0 r}/k_0 r\right) \cdot [\pm 1 + \cos\theta] \cos\varphi;$$

$$e_\varphi^\infty \approx ie_0\left(e^{-ik_0 r}/k_0 r\right) \cdot [\pm 1 + \cos\theta] \sin\varphi, \tag{5.4}$$

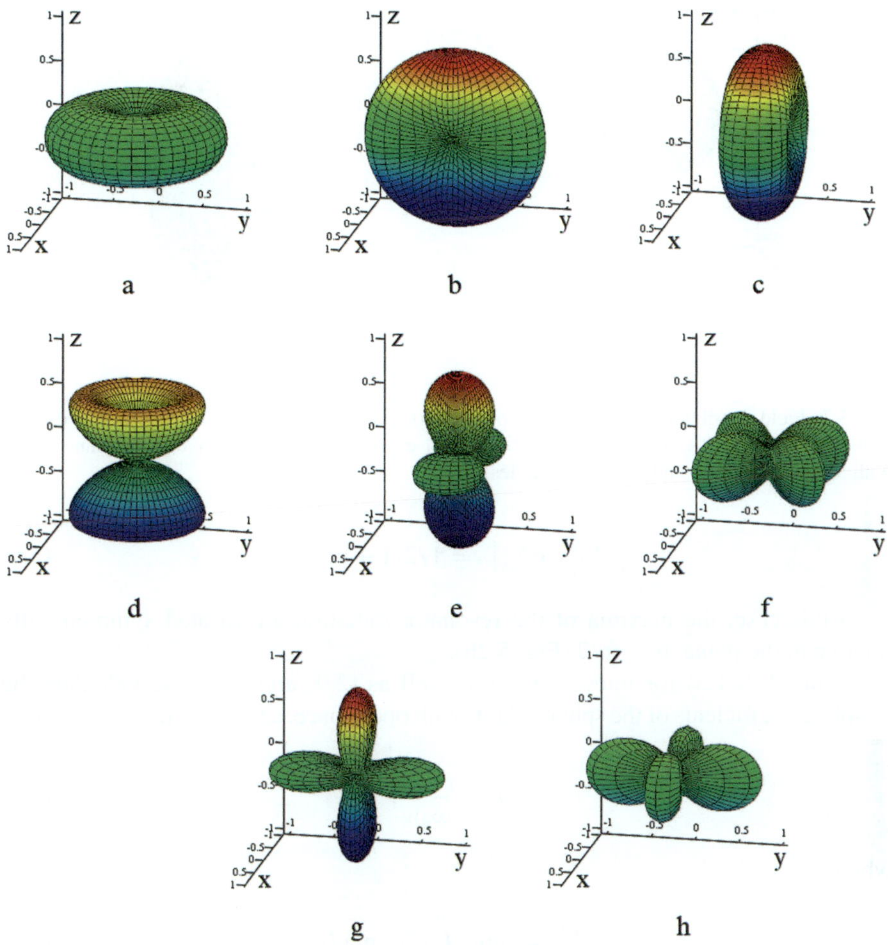

Fig. 5.1 Field distribution in the far zone of a spherical DR with oscillations H_{101} (**a**), H_{111} (**b**), (**c**), H_{201} (**d**), H_{211} (**e, g**), H_{221} (**f, h**). $e_\theta \sim \sin(2\varphi)$ (**e, f**); $e_\theta \sim \cos(2\varphi)$ (**g, h**)

or

$$\left|\mathbf{e}^{\infty}\right|^2 / \left|\mathbf{e}^{\infty}_{\max}\right|^2 = |\pm 1 + \cos\theta|^2/4, \tag{5.5}$$

the pattern of such a resonator exactly takes the form of a cardioid. In this case, the " + " sign corresponds to the maximum radiation in the direction of the axis z, and the "−" sign—in the direction of the $-z$. (Fig. 5.2a). In the plane $\theta = \pi/2$, the radiation pattern of the ball has the shape of a circle.

When excitation of degenerate oscillations of the magnetic type H_{111}, even and odd relative to the selected plane $\varphi = 0$, with a phase shift $\pi/2$ in the case of equal amplitudes, the resonator pattern has the form:

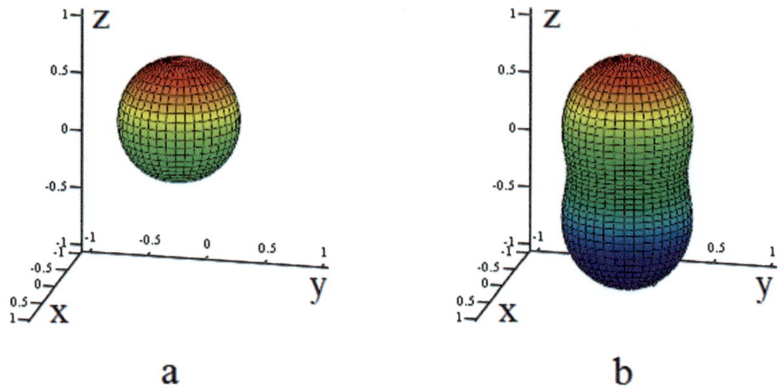

Fig. 5.2 Field distribution in the far zone of a spherical DR upon excitation of degenerate oscillations H_{111} and E_{111} (**a**); magnetic oscillations H_{111} even and odd type relative to the plane $\varphi = 0$ with a phase shift $\pi/2$ and equal amplitudes (**b**)

$$\left|\mathbf{e}^{\infty}\right|^2 / \left|\mathbf{e}^{\infty}_{\max}\right|^2 = 1/2(1 + \cos^2\theta). \tag{5.6}$$

In this case, the maxima of the resonator radiation are located symmetrically relative to the plane: $\theta = \pi/2$ (Fig. 5.2b).

Using (1.2, 1.3) for integers m, n as well as (2.9) and (5.1), we calculate the coupling coefficients of the spherical DR with open space, according to the definition:

$$\tilde{k} = \frac{P^{\Sigma}}{\omega_0 w}, \tag{5.7}$$

where

$$P^{\Sigma} = \lim_{r\to\infty} \int\limits_{\Omega} (\Pi \cdot \mathbf{n}_r) ds, \tag{5.8}$$

power emitted by DR, w—energy stored in a dielectric; $\mathbf{\Pi} = 1/2\mathrm{Re}[\mathbf{e}, \mathbf{h}^*]$—Poynting vector; $\mathbf{n}_r$—unit vector in the radial direction of a spherical coordinate system (r, θ, φ), with the beginning at the center of the resonator. Integration is carried out over solid angles: $\Omega : \theta \in [0, \pi]; \varphi \in [0, 2\pi]$; where $ds = r^2 \sin\theta \cdot d\theta d\varphi$.

As a result of the transformations we find exactly:

for magnetic oscillations of a spherical DR:

$$\tilde{k} = 2\alpha_n^H(p, q); \tag{5.9}$$

For electrical oscillations

$$\tilde{k} = 2\alpha_n^E(p, q) \tag{5.10}$$

here $\alpha_n^{H,E}(p, q)$ functions (2.20).

Let us now investigate the influence exerted on the coupling coefficients by various types of spherically symmetric metallic and dielectric inhomogeneities formed in the DR material.

It is easy to show that the dependence of the coupling on the parameters of layered spherical DRs are also expressed by relations of the form (5.9, 5.10) through the functions $\alpha_n^{H,E}$.

In the case of a spherical DR containing a metal ideally conducting ball at the center (Fig. 1.15a), the indicated coupling functions take the form:

For magnetic H_{nml} (top), electrical E_{nml} (lower) resonator oscillations:

$$\begin{pmatrix} \alpha_n^H(s, p; q) \\ \alpha_n^E(s, p; q) \end{pmatrix} = \begin{pmatrix} \varepsilon_{1r}^{1/2} \\ \varepsilon_{1r}^{3/2} \end{pmatrix} \left| \frac{\psi_n(p)}{y_n(q)} \right|^2 / \Big[x\{[x^2 - n(n+1)]\psi_n^2(x) + \big[n\psi_n(x) - x\psi_{n-1}(x)\big]^2\} \Big]_s^p. \tag{5.11}$$

here $[f(x)]_s^p = f(p) - f(s)$; $\psi_n(x) = f_n^{(1)}(x) = a_1 j_n(x) + b_1 y_n(x)$—linear combination of spherical functions with coefficients (1.74), or (1.75).

Despite the formal similarity of relations (5.11) for oscillations of magnetic and electric types, the functions and characteristic parameters included in it are different. As the size of the metallic inhomogeneity decreases, in the limit $r_1/r_2 = 0$ (5.11) it transforms into (2.20), obtained for a homogeneous DR. An increase in the size of the metallic inhomogeneity leads to an increase in the coupling functions for oscillations of magnetic types (Fig. 5.3b) and a slight decrease for oscillations of electric. The coupling functions of a homogeneous ball for the main oscillations of electrical types are somewhat smaller compared to the functions of magnetic types for $\varepsilon_{1r}^{1/2} > 6$ (Fig. 5.3a).

The coupling functions of two-layer spherical DR (Fig. 1.16a) have the form:

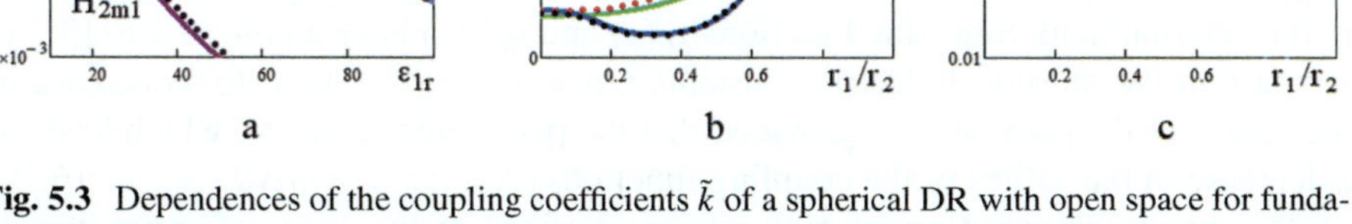

Fig. 5.3 Dependences of the coupling coefficients $\tilde{k}$ of a spherical DR with open space for fundamental oscillations on the relative dielectric constant of the material (**a**). Dependences of the coupling coefficients of the spherical DR with metallic inhomogeneity ($\varepsilon_{1r} = 36$) (**b**); hollow (**c**) ($\varepsilon_{2r} = 36$); on the relative sizes of heterogeneity. Solid curves—calculation using complex characteristic parameters; points—calculation using formulas (5.10–5.13)

for magnetic types H_{nml} of oscillations:

$$\alpha_n^H(p,s,t,q) = \left|\frac{\psi_n(t)}{y_n(q)}\right|^2 \Big/ \left\{\varepsilon_{1r}^{-1/2} p \left\{\left[p^2 - n(n+1)\right] j_n^2(p) + \left[n j_n(p) - p j_{n-1}(p)\right]^2\right\} + \varepsilon_{2r}^{-1/2}\left[x\left\{\left[x^2 - n(n+1)\right]\psi_n^2(x) + \left[n\psi_n(x) - x\psi_{n-1}(x)\right]^2\right\}\right]_s^t\right\}; \tag{5.12}$$

for electric types E_{nml}:

$$\alpha_n^E(p,s,t,q) = \varepsilon_{2r}^2 \left|\frac{\psi_n(t)}{y_n(q)}\right|^2 \Big/ \left\{\varepsilon_{1r}^{1/2} p \left\{\left[p^2 - n(n+1)\right] j_n^2(p) + \left[n j_n(p) - p j_{n-1}(p)\right]^2\right\} + \varepsilon_{2r}^{1/2}\left[x\left\{\left[x^2 - n(n+1)\right]\psi_n^2(x) + \left[n\psi_n(x) - x\psi_{n-1}(x)\right]^2\right\}\right]_s^t\right\}; \tag{5.13}$$

here $\psi_n(x) = f_n^{(2)}(x)$—linear combination of spherical functions (1.78): $f_n^{(2)}(x) = a_2 j_n(x) + b_2 y_n(x)$ with coefficients (1.79), or (1.80).

On Fig. 5.3 shows comparative dependences of the coupling coefficients of the main oscillations of spherical DR with open space. It can be seen that the found relations (5.9, 5.10) correlate well for large ε_{1r} ones with the values obtained from the formula $\tilde{k} = 2p_J/p_R$ based on solving the problem of natural oscillations (sec. 1. 1). For layered resonators, the discrepancy between the formulas increases for fundamental magnetic oscillations as the size of the inhomogeneity increases (Fig. 5.3b, c). Let us pay attention to the possibility of increasing the coupling coefficients during the formation of an air cavity in the volume of the DR.

The generalization of relations (5.12, 5.13) to the case of an arbitrary number of dielectric layers is obvious: in the denominators of the resulting formulas, the terms containing $\psi_n(x)$, should be replaced by the sum of the parameters of each layer, with the exception of the central inhomogeneity. In this case, the index t has the meaning of a characteristic parameter of the outer layer of the dielectric. The ratio $|\psi_n(t)/y_n(q)|$ is proportional to the ratio of the field of the outer layer of the resonator to the external field. Since the functions $y_n(x)$ and $h_n^{(2)}(x)$ have a pole at zero [5], an increase in the ratio of dielectric constants $t/q = (\varepsilon_N/\varepsilon_0)^{1/2}$ leads to a decrease in the values of the parameter q, provided that the parameter t is finite, which leads to a decrease in the values of the coupling functions.

The results obtained are also simply generalized to the cases of expansion of the field of spherically inhomogeneous resonators in waveguides or other structures, calculated without taking into account the influence of external inhomogeneities on their natural oscillations, since the same expressions for $\alpha_n^{H,E}$ are included in them in the form of factors.

5.2 Radiation Characteristics of a Cylindrical Dielectric Resonator

In analytical form, a simple one-wave approximation for the natural oscillations of a cylindrical DR allows one to express the field quite accurately only in the volume of the dielectric. The field of natural oscillations in the far zone is not correct, since it does not satisfy the radiation conditions (Chapter 1). It can be assumed that for internal oscillations, the field stored in the dielectric serves as a source of emitted waves, therefore its representation in the far zone of the resonator can be obtained using the Kirchhoff integral [6]. Let us denote by **r**—the radius vector directed from the origin of the local coordinate system associated with the DR to the observation point, and also $r = |\mathbf{r}|$; $\mathbf{n}_0 = \mathbf{r}/|\mathbf{r}|$. Kirchhoff's theorem connects the field at a point and on any closed surface containing it. If all sources are contained in a region of small compared to r size, and the observation point is located at a distance significantly greater than the wavelength in free space ($k_0 r >> 1$), then the field in the far zone can be represented as a known approximate integral over the surface covering these sources, in our case DR [7, 8]:

$$\mathbf{e}(\mathbf{r}) \approx \frac{k_0^2}{4\pi\varepsilon_0} \cdot \frac{e^{-ik_0 r}}{r} \left\{ [[\mathbf{n}_0, \boldsymbol{\nu}]\mathbf{n}_0] - \frac{1}{w_0}[\mathbf{n}_0, \boldsymbol{\eta}] \right\}; \tag{5.14}$$

$$\mathbf{h}(\mathbf{r}) \approx \frac{k_0^2}{4\pi\mu_0} \cdot \frac{e^{-ik_0 r}}{r} \{[[\mathbf{n}_0, \boldsymbol{\eta}]\mathbf{n}_0] + w_0[\mathbf{n}_0, \boldsymbol{\nu}]\}$$

where

$$\boldsymbol{\nu} = \frac{-i}{\omega} \oint_{s_p} [\mathbf{n}, \mathbf{h}] \cdot e^{ik_0(\mathbf{n}_0, \mathbf{x}')} ds'; \tag{5.15}$$

$$\boldsymbol{\eta} = \frac{-i}{\omega} \oint_{s_p} [\mathbf{e}, \mathbf{n}] \cdot e^{ik_0(\mathbf{n}_0, \mathbf{x}')} ds'$$

Here w_0—characteristic impedance of free space, and **n**—normal to the resonator surface: s_p. It is easy to see that field (5.14) satisfies the radiation condition (1.1), as well as the relation:

$$\mathbf{h}(\mathbf{r}) = \frac{1}{w_0} \cdot [\mathbf{n}_0, \mathbf{e}(\mathbf{r})]. \tag{5.16}$$

In a spherical coordinate system (r, θ, φ), expression (5.14) takes on a simpler form:

$$e_\theta \approx \frac{k_0^2}{4\pi\varepsilon_0} \cdot \frac{e^{-ik_0 r}}{r} \left\{ \nu_\theta + \frac{1}{w_0} \eta_\varphi \right\}; \tag{5.17}$$

$$e_\varphi \approx \frac{k_0^2}{4\pi\varepsilon_0} \cdot \frac{e^{-ik_0 r}}{r} \left\{ \nu_\varphi - \frac{1}{w_0} \eta_\theta \right\}$$

Let us represent the power emitted by the DR into open space in the form:

$$P^\Sigma = \frac{(\omega k_0)^2}{32\pi^2} \int\limits_\Omega \left\{ w_0 \left(|\nu_\theta|^2 + |\nu_\varphi|^2 \right) + w_0^{-1} \left(|\eta_\theta|^2 + |\eta_\varphi|^2 \right) \right. \\ \left. + 2\mathrm{Re}\left(\nu_\theta \eta_\varphi^* - \nu_\varphi \eta_\theta^* \right) \right\} do, \tag{5.18}$$

where $do = \sin\theta \, d\theta \, d\varphi$. We also carry out integration according to Ω: $\theta \in [0, \pi]$; $\varphi \in [0, 2\pi]$.

Let us now substitute the expressions for the natural oscillation field of the dielectric cylinder (1.17) into (5.15) and integrate them over the resonator surface using (D.9). In the found field components, we highlight the dependence on the angular variables:

$$e_\theta = \frac{e^{-ik_0 r}}{k_0 r} \cdot e^{-i(m-1)\frac{\pi}{2}} f_\theta(\varepsilon_{1r}, \Delta; \cos\theta) \left\{ \begin{matrix} \sin m\varphi \\ \cos m\varphi \end{matrix} \right\}; \tag{5.19}$$

$$e_\varphi = \frac{e^{-ik_0 r}}{k_0 r} \cdot e^{+i(m-1)\frac{\pi}{2}} f_\varphi(\varepsilon_{1r}, \Delta; \cos\theta) \left\{ \begin{matrix} \cos m\varphi \\ -\sin m\varphi \end{matrix} \right\},$$

where

$$f_\theta(\varepsilon_{1r}, \Delta; \cos\theta) = \frac{e_0}{\beta_z} \left[\frac{k_0^2}{\beta^2 - (k_0 n_{0\perp})^2} \left\{ p_\perp J_m(p_\perp) J_m'(q_\perp n_{0\perp}) - q_\perp n_{0\perp} J_m'(p_\perp) J_m(q_\perp n_{0\perp}) \right\} \right. \\ \cdot \left[\begin{matrix} \varepsilon_{1r} k_0 n_{0z} \sin p_z \cos q_z n_{0z} - \beta_z \cos p_z \sin q_z n_{0z} \\ i(\varepsilon_{1r} k_0 n_{0z} \cos p_z \sin q_z n_{0z} - \beta_z \sin p_z \cos q_z n_{0z}) \end{matrix} \right] \\ + \left\{ p_\perp J_m(p_\perp) J_m'(q_\perp n_{0\perp}) - \varepsilon_{1r} q_\perp n_{0\perp} J_m'(p_\perp) J_m(q_\perp n_{0\perp}) \right\} \\ \left. \cdot \frac{k_0^2}{\beta_z^2 - (k_0 n_{0z})^2} \cdot \left[\begin{matrix} k_0 n_{0z} \sin p_z \cos q_z n_{0z} - \beta_z \cos p_z \sin q_z n_{0z} \\ i(k_0 n_{0z} \cos p_z \sin q_z n_{0z} - \beta_z \sin p_z \cos q_z n_{0z}) \end{matrix} \right] \right] \\ - \frac{h_0}{\beta} m J_m(p_\perp) J_m(q_\perp n_{0\perp}) \cdot \\ \cdot \frac{k_1^2 - k_0^2}{\beta_z^2 - (k_0 n_{0z})^2} \cdot \frac{n_{0z}}{n_{0\perp}} \cdot \left[\begin{matrix} \beta_z \sin p_z \cos q_z n_{0z} - k_0 n_{0z} \cos p_z \sin q_z n_{0z} \\ i(\beta_z \cos p_z \sin q_z n_{0z} - k_0 n_{0z} \sin p_z \cos q_z n_{0z}) \end{matrix} \right]; \tag{5.20}$$

$$f_\varphi(\varepsilon_{1r},\Delta;\cos\theta)=\left[\frac{e_0}{\beta}\cdot\frac{m}{n_{0\perp}}\cdot J_m(p_\perp)J_m(q_\perp n_{0\perp})-h_0\cdot\frac{k_0}{\beta^2-(k_0n_{0\perp})^2}\right.$$
$$\left.\cdot\left\{p_\perp J_m(p_\perp)J_m'(q_\perp n_{0\perp})-q_\perp n_{0\perp}J_m'(p_\perp)J_m(q_\perp n_{0\perp})\right\}\right]$$
$$\cdot\frac{k_1^2-k_0^2}{\beta_z^2-(k_0n_{0z})^2}\cdot\begin{bmatrix}\beta_z\sin p_z\cos q_zn_{0z}-k_0n_{0z}\cos p_z\sin q_zn_{0z}\\ i(\beta_z\cos p_z\sin q_zn_{0z}-k_0n_{0z}\sin p_z\cos q_zn_{0z})\end{bmatrix}.$$

f_θ, f_φ—functions that determine the dependence of the DR field components in the azimuthally plane ($\varphi = const$); $(n_{0\perp}, n_{0z}) = (\sin\theta, \cos\theta)$; e_0, h_0—amplitudes (2.31).

In the particular case of azimuthally homogeneous oscillations $m = 0$ in relations (5.19, 5.20) one should put formally $e_0 = 0$ for the $H^{\pm}_{nol}$; $h_0 = 0$ for $E^{\pm}_{nol}$. In this case, expressions (5.20) take on a simpler form:

for $H^{\pm}_{nol}$:

$$f_\theta(\varepsilon_{1r},\Delta;\cos\theta)=0; \tag{5.21}$$

$$f_\varphi(\varepsilon_{1r},\Delta;\cos\theta)=h_0\cdot\frac{k_0}{\beta^2-(k_0n_{0\perp})^2}$$
$$\cdot\left\{p_\perp J_0(p_\perp)J_1(q_\perp n_{0\perp})-q_\perp n_{0\perp}J_1(p_\perp)J_0(q_\perp n_{0\perp})\right\}$$
$$\cdot\frac{k_1^2-k_0^2}{\beta_z^2-(k_0n_{0z})^2}$$
$$\cdot\begin{bmatrix}\beta_z\sin p_z\cos q_zn_{0z}-k_0n_{0z}\cos p_z\sin q_zn_{0z}\\ i(\beta_z\cos p_z\sin q_zn_{0z}-k_0n_{0z}\sin p_z\cos q_zn_{0z})\end{bmatrix}$$

for $E^{\pm}_{nol}$:

$$f_\theta(\varepsilon_{1r},\Delta;\cos\theta)=\frac{e_0}{\beta_z}\left[\frac{k_0^2}{\beta^2-(k_0n_{0\perp})^2}\left\{q_\perp n_{0\perp}J_1(p_\perp)J_0(q_\perp n_{0\perp})-p_\perp J_0(p_\perp)J_1(q_\perp n_{0\perp})\right\}\right.$$
$$\cdot\begin{bmatrix}\varepsilon_{1r}k_0n_{0z}\sin p_z\cos q_zn_{0z}-\beta_z\cos p_z\sin q_zn_{0z}\\ i(\varepsilon_{1r}k_0n_{0z}\cos p_z\sin q_zn_{0z}-\beta_z\sin p_z\cos q_zn_{0z})\end{bmatrix}$$
$$+\left\{\varepsilon_{1r}q_\perp n_{0\perp}J_1(p_\perp)J_0(q_\perp n_{0\perp})-p_\perp J_0(p_\perp)J_1(q_\perp n_{0\perp})\right\}$$
$$\left.\cdot\frac{k_0^2}{\beta_z^2-(k_0n_{0z})^2}\cdot\begin{bmatrix}k_0n_{0z}\sin p_z\cos q_zn_{0z}-\beta_z\cos p_z\sin q_zn_{0z}\\ i(k_0n_{0z}\cos p_z\sin q_zn_{0z}-\beta_z\sin p_z\cos q_zn_{0z})\end{bmatrix}\right];$$

$$f_\varphi(\varepsilon_{1r},\Delta;\cos\theta)=0. \tag{5.22}$$

The resulting expressions in general establish the angular distribution of the field in the wave zone from the parameters of natural oscillations of an isolated cylindrical DR in open space.

$|\mathbf{e}(\mathbf{r})|^2/|\mathbf{e}_{\max}|^2$ dependences for the main oscillations of the DR, calculated according to (5.19), are shown in Fig. 5.4. As can be seen, the field structure for the main types of oscillations in the far zone is similar to the field of magnetic or electric dipoles. At that, an increase in the height of the cylinder leads to a noticeable narrowing of the radiation pattern of azimuthally homogeneous oscillations. In the case of hybrid oscillations, the radiation pattern toroid turns out to be somewhat flattened in the "equator" region: $\theta = \pi/2$ (Fig. 5.4c, d).

Having integrated the power flow over the remote sphere using (5.18–5.20) and (2.9), we find an expression for the coupling coefficients of the cylindrical DR with free space:

$$\tilde{k} = \frac{8\beta_z}{\varepsilon_{1r} k_0^3 r_0^2} \cdot \frac{i(p_\perp, p_z; q_\perp, q_z)}{v_m(p_\perp, p_z)}, \tag{5.23}$$

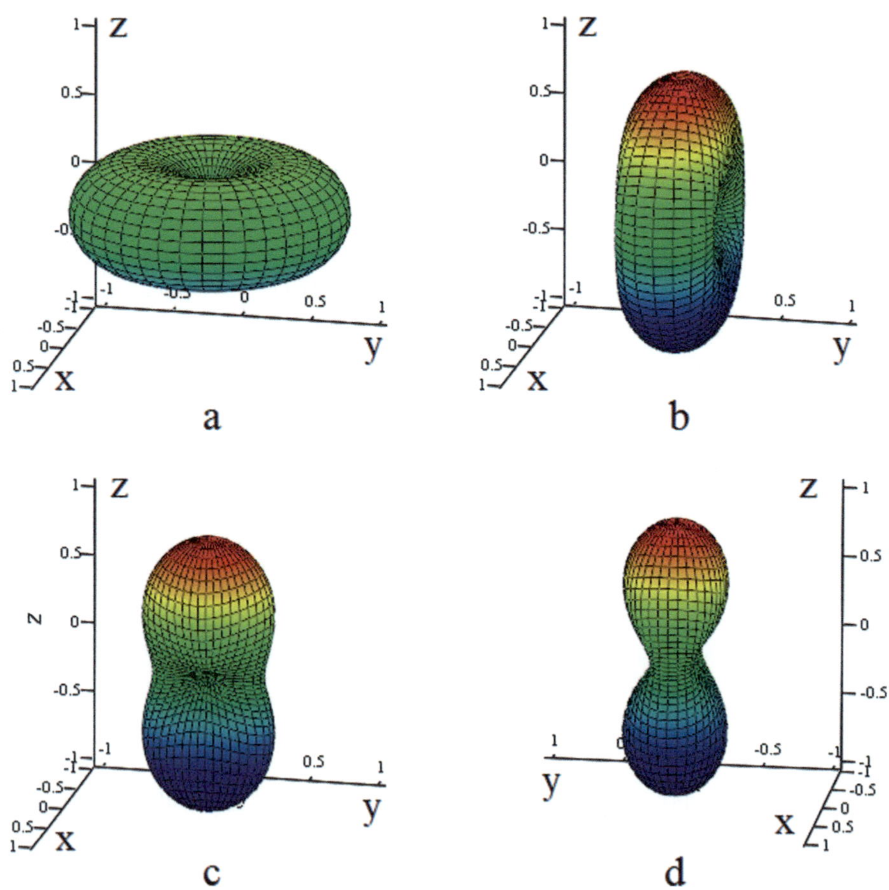

Fig. 5.4 Field distribution in the far zone of a cylindrical DR with oscillations H_{101} (**a**), EH_{111} (**b**), HE_{111} (**c**), (**d**)

where

$$i(p_\perp, p_z; q_\perp, q_z) = \int_0^1 \left\{ |f_\theta|^2 + |f_\varphi|^2 \right\} d\xi; \tag{5.24}$$

and $v_m(p_\perp, p_z)$—is the function (2.30).

In the special case $m = 0$, integral (5.24) is simplified:

for $H^{\pm}_{nol}$:

$$\begin{aligned} i(p_\perp, p_z; q_\perp, q_z) &= \left[h_0 q_\perp^2 q_z (\varepsilon_{1r} - 1)\right]^2 \\ &\cdot \int_0^1 \left(\frac{p_\perp J_0(p_\perp) J_1(q_\perp \eta) - q_\perp \eta J_1(p_\perp) J_0(q_\perp \eta)}{p_\perp^2 - (q_\perp \eta)^2} \right. \\ &\left. \cdot \frac{\left[p_z \begin{matrix}\sin\\ \cos\end{matrix} p_z \begin{matrix}\cos\\ \sin\end{matrix} q_z \xi - q_z \xi \begin{matrix}\cos\\ \sin\end{matrix} p_z \begin{matrix}\sin\\ \cos\end{matrix} q_z \xi \right]}{p_z^2 - (q_z \xi)^2} \right)^2 d\xi; \end{aligned} \tag{5.25}$$

for $E^{\mp}_{nol}$:

$$\begin{aligned} i(p_\perp, p_z; q_\perp, q_z) &= \left(\frac{e_0}{p_z} \right)^2 \\ &\cdot \int_0^1 \left| \frac{q_\perp^2}{p_\perp^2 - (q_\perp \eta)^2} \cdot \{ q_\perp \eta J_1(p_\perp) J_0(q_\perp \eta) - p_\perp J_0(p_\perp) J_1(q_\perp \eta) \} \right. \\ &\cdot \left[\varepsilon_{1r} q_z \xi \begin{matrix}\sin\\ \cos\end{matrix} p_z \begin{matrix}\cos\\ \sin\end{matrix} q_z \xi - p_z \begin{matrix}\cos\\ \sin\end{matrix} p_z \begin{matrix}\sin\\ \cos\end{matrix} q_z \xi \right] \\ &+ \{ \varepsilon_{1r} q_\perp \eta J_1(p_\perp) J_0(p_\perp \eta) - p_\perp J_0(p_\perp) J_1(q_\perp \eta) \} \\ &\cdot \frac{q_z^2}{p_z^2 - (q_z \xi)^2} \cdot \left[q_z \xi \begin{matrix}\sin\\ \cos\end{matrix} p_z \begin{matrix}\cos\\ \sin\end{matrix} q_z \xi - p_z \begin{matrix}\cos\\ \sin\end{matrix} p_z \begin{matrix}\sin\\ \cos\end{matrix} q_z \xi \right] \Bigg)^2 d\xi \end{aligned} \tag{5.26}$$

oscillations. Here $\eta = \left(1 - \xi^2\right)^{1/2}$.

The dependence of the coupling coefficients of the main oscillations of a cylindrical DR with $\varepsilon_{1r} = 81$, and a fixed frequency of natural oscillations on $\Delta = L/2r_0$ is shown in Fig. 5.5. The values of the coupling coefficients calculated using the formula: $\tilde{k} = 1/Q^\Sigma$, where Q^Σ—the radiation quality factor was determined from [9], are shown by dots.

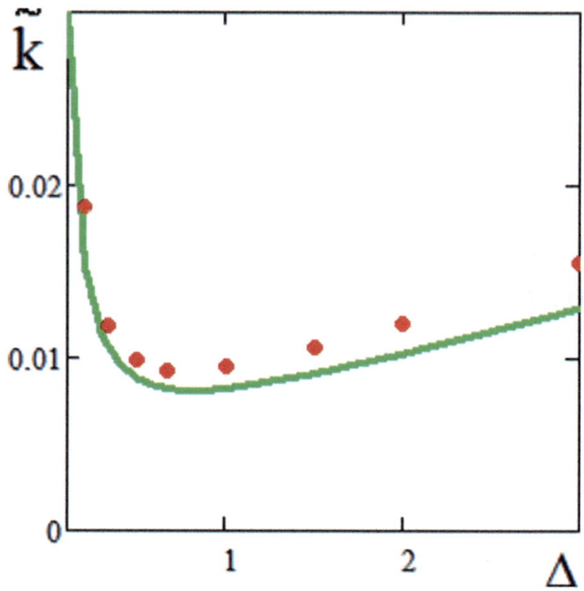

Fig. 5.5 Dependence of the coupling coefficient with the open space of a cylindrical DR for the main magnetic type of oscillations H_{101} ($\varepsilon_{1r} = 81$) on $\Delta = L/2r_0$

As follows from the results obtained, increasing the height of the dielectric cylinder not only narrows the width of the radiation pattern, but also leads to an increase in the coupling coefficient of the magnetic azimuthally uniform oscillation.

5.3 Radiation Characteristics of a Rectangular Dielectric Resonator

We find the field of a rectangular DR in the wave zone in the same way as a cylindrical resonator. To do this, we use expressions (1.32, 1.33) for the field in the dielectric, which we substitute into relations (5.14, 5.15). After integration we find:

$$
\begin{aligned}
\nu_\theta + \frac{1}{w_0}\mu_\varphi = \frac{2}{\omega_0}\{ & ib_0L\big[(k_1^2 - \beta_z^2)\omega_y(n_{0y})\omega_z(n_{0z})\cos\theta\sin\varphi \\
& +\beta_y\beta_z\overline{\omega}_y(n_{0y})\overline{\omega}_z(n_{0z})\sin\theta\big] \\
& \cdot\begin{pmatrix} \sin p_x\cos(q_xn_{0x}) \\ i\cos p_x\sin(q_xn_{0x}) \end{pmatrix} \\
& - ia_0L\big[(k_1^2 - \beta_z^2)\omega_x(n_{0x})\omega_z(n_{0z})\cos\theta\cos\varphi \\
& +\beta_x\beta_z\overline{\omega}_x(n_{0x})\overline{\omega}_z(n_{0z})\sin\theta] \\
& \cdot\begin{pmatrix} \sin p_y\cos(q_yn_{0y}) \\ i\cos p_y\sin(q_yn_{0y}) \end{pmatrix} \\
& + a_0b_0[\beta_y\omega_x(n_{0x})\overline{\omega}_y(n_{0y})\cos\varphi - \beta_x\overline{\omega}_x(n_{0x})\omega_y(n_{0y})\sin\varphi] \\
& \cdot\left\{\begin{matrix} k_0\sin p_z\cos(q_zn_{0z}) - \beta_zn_{0z}\cos p_z\sin(q_zn_{0z}) \\ i[k_0\cos p_z\sin(q_zn_{0z}) - \beta_zn_{0z}\sin p_z\cos(q_zn_{0z})] \end{matrix}\right\}\Bigg\};
\end{aligned}
\tag{5.27}
$$

$$
\begin{aligned}
\nu_\varphi - \frac{1}{w_0}\mu_\theta = \frac{2}{\omega_0} & L\omega_z(n_{0z}) \\
& \cdot\left\{i(k_1^2 - \beta_z^2)\left[a_0\omega_x(n_{0x})\sin\varphi\begin{bmatrix} \sin p_y\cos(q_yn_{0y}) \\ i\cos p_y\sin(q_yn_{0y}) \end{bmatrix}\right.\right. \\
& \left. +b_0\omega_y(n_{0y})\cos\varphi\begin{pmatrix} \sin p_x\cos(q_xn_{0x}) \\ i\cos p_x\sin(q_xn_{0x}) \end{pmatrix}\right] \\
& + i\frac{a_0b_0}{2}[\beta_z^2 - (k_0n_{0z})^2][\beta_x\overline{\omega}_x(n_{0x})\omega_y(n_{0y})\cos\varphi \\
& + \beta_y\omega_x(n_{0x})\overline{\omega}_y(n_{0y})\sin\varphi] \\
& - k_0\left[a_0\beta_y\omega_x(n_{0x})\begin{bmatrix} i\cos p_y\sin(q_yn_{0y}) \\ -\sin p_y\cos(q_yn_{0y}) \end{bmatrix}\right. \\
& \left.\left. +b_0\beta_x\omega_y(n_{0y})\cos\varphi\begin{pmatrix} i\cos p_x\sin(q_xn_{0x}) \\ -\sin p_x\cos(q_xn_{0x}) \end{pmatrix}\right]\sin\theta\right\}.
\end{aligned}
\tag{5.28}
$$

From (5.17, 5.18, 5.27, 5.28) and also (2.39) we find the expressions for the coupling coefficients of the rectangular resonator with open space:

$$
\tilde{k} = 1/(2\omega_0w_0w)\cdot\int_0^\pi\int_0^{2\pi}(|e_\theta|^2 + |e_\varphi|^2)r^2\sin\theta d\theta d\varphi \tag{5.29}
$$

The result of calculating the coupling coefficient from the relative sizes of the rectangular DR using formula (5.29) is shown in Fig. 5.6b, c, for the main type of magnetic oscillations H_{111} with $\varepsilon_{1r} = 21$. The dots indicate the results of calculations using the numerical method [10]. Let us note here that the discrepancy between these data increases as one of the relative sizes of the cuboid decreases.

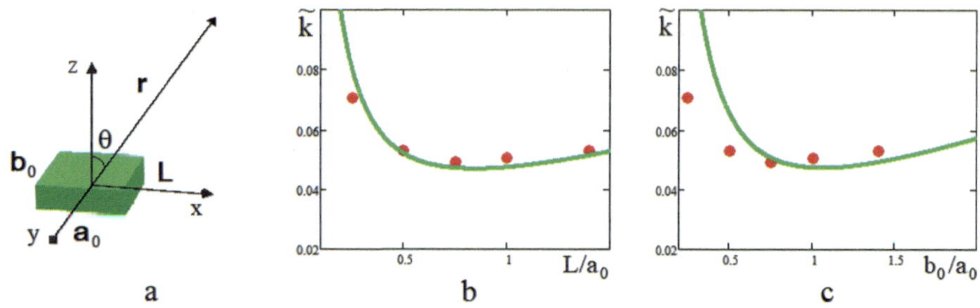

Fig. 5.6 Dependences of coupling coefficients with open space on the relative sizes of a rectangular DR with $\varepsilon_{1r} = 21$ for H_{111} oscillation. $b_0/a_0 = 1$ (**b**); $L/a_0 = 1$ (**c**). Points are the result of numerical simulation [10]

Using expressions (5.27, 5.28), and also (5.17), we find the angular dependences of the field in the wave zone. The result of calculating the field for the main magnetic oscillations of a rectangular DR with $\varepsilon_{1r} = 36$ shown in Fig. 5.7. We see that the field of the wave zone of a rectangular DR for the main types of oscillations with an odd distribution of one or several components (1.32, 1.33) relative to one of the symmetry planes approximately corresponds to the angular field distribution characteristic of "quadrupole" ($n = 2$) oscillations of spherical DR (Fig. 5.1d–h).

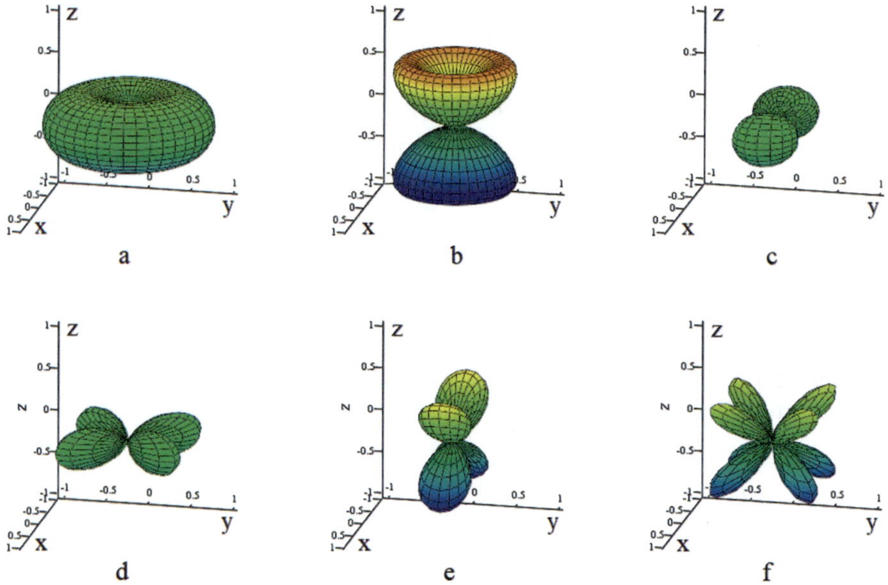

Fig. 5.7 Field distribution in the far zone of a rectangular DR with oscillations H_{111}^{ccc} (**a**), H_{111}^{ccs} (**b**), H_{111}^{scc} (**c**), H_{111}^{ssc} (**d**), H_{111}^{scs} (**e**), H_{111}^{sss} (**f**) ($\varepsilon_{1r} = 36$; $L/a_0 = 0, 2$; $b_0/a_0 = 1$)

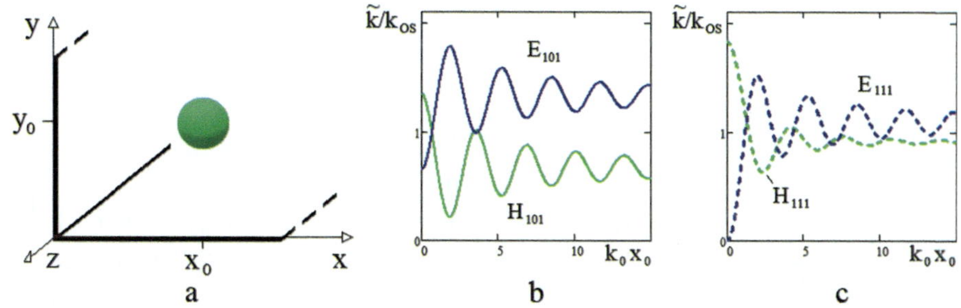

Fig. 5.8 Dependences of the coupling coefficients $\tilde{k}$ with the space of a rectangular metal wedge on the coordinates of the Spherical DR (k_{OS}—coupling coefficient of an isolated DR with open space obtained from [5.9, 5.10])

5.4 Radiation of Dielectric Resonators Above a Metal Plane

When used as antenna elements, DRs are usually located near metal surfaces of various shapes. These surfaces also form the radiation pattern of the structure, and in addition, change the coupling coefficients. To take into account the influence of metal surfaces on the characteristics of DR in open space, let us consider several useful examples that allow us to obtain fairly simple relationships for the field and coupling coefficients.

Let us consider a spherical DR located in the space of a rectangular metal wedge (Fig. 5.8a). To calculate the field, we use transformations of the found expansions of the form (2.64) and move to the limit: $b = \infty$. Thus, using integrals (D.24–D.26), from (2.64) we find the field components in the Cartesian coordinate system.

For fundamental magnetic oscillations H_{10l}:

$$\mathbf{e}_m^0 = \begin{pmatrix} e_x^0 \\ e_y^0 \\ e_z^0 \end{pmatrix} = \omega\mu_0 \left[\frac{j_1(p)}{y_1(q)}\right] \cdot \begin{pmatrix} (y - y_0) \sum\limits_{s=1,4} \frac{1}{|\mathbf{r}_s|} h_1^{(2)}(k_0|\mathbf{r}_s|) + (y + y_0) \sum\limits_{s=2,3} \frac{1}{|\mathbf{r}_s|} h_1^{(2)}(k_0|\mathbf{r}_s|) \\ -(x - x_0) \sum\limits_{s=1,2} \frac{1}{|\mathbf{r}_s|} h_1^{(2)}(k_0|\mathbf{r}_s|) - (x + x_0) \sum\limits_{s=3,4} \frac{1}{|\mathbf{r}_s|} h_1^{(2)}(k_0|\mathbf{r}_s|) \\ 0 \end{pmatrix}, \quad (5.30)$$

For fundamental oscillations H_{11l}^x with magnetic field polarization at the center of the DR (1.2), parallel to the x axis (Fig. 5.8a):

$$\mathbf{e}_m^0 = \begin{pmatrix} e_x^0 \\ e_y^0 \\ e_z^0 \end{pmatrix} = \omega\mu_0 \left[\frac{j_1(p)}{y_1(q)}\right] \cdot \begin{pmatrix} 0 \\ (z-z_0)\left\{-\frac{1}{|\mathbf{r}_1|}h_1^{(2)}(k_0|\mathbf{r}_1|) - \frac{1}{|\mathbf{r}_2|}h_1^{(2)}(k_0|\mathbf{r}_2|) + \frac{1}{|\mathbf{r}_3|}h_1^{(2)}(k_0|\mathbf{r}_3|) + \frac{1}{|\mathbf{r}_4|}h_1^{(2)}(k_0|\mathbf{r}_4|)\right\} \\ -\frac{(y-y_1)}{|\mathbf{r}_1|}h_1^{(2)}(k_0|\mathbf{r}_1|) - \frac{(y-y_2)}{|\mathbf{r}_2|}h_1^{(2)}(k_0|\mathbf{r}_2|) + \frac{(y-y_3)}{|\mathbf{r}_3|}h_1^{(2)}(k_0|\mathbf{r}_3|) + \frac{(y-y_4)}{|\mathbf{r}_4|}h_1^{(2)}(k_0|\mathbf{r}_4|) \end{pmatrix} \tag{5.31}$$

For fundamental oscillations H_{11l}^y with $e_x^0 \sim \sin\varphi$ and magnetic field polarization at the center of the DR (1.2), parallel to the y axis (Fig. 5.8a):

$$\mathbf{e}_m^0 = \begin{pmatrix} e_x^0 \\ e_y^0 \\ e_z^0 \end{pmatrix} = \omega\mu_0 \left[\frac{j_1(p)}{y_1(q)}\right] \cdot \begin{pmatrix} (z-z_0)\left\{\frac{1}{|\mathbf{r}_1|}h_1^{(2)}(k_0|\mathbf{r}_1|) - \frac{1}{|\mathbf{r}_2|}h_1^{(2)}(k_0|\mathbf{r}_2|) - \frac{1}{|\mathbf{r}_3|}h_1^{(2)}(k_0|\mathbf{r}_3|) + \frac{1}{|\mathbf{r}_4|}h_1^{(2)}(k_0|\mathbf{r}_4|)\right\} \\ 0 \\ \frac{(x-x_1)}{|\mathbf{r}_1|}h_1^{(2)}(k_0|\mathbf{r}_1|) - \frac{(x-x_2)}{|\mathbf{r}_2|}h_1^{(2)}(k_0|\mathbf{r}_2|) - \frac{(x-x_3)}{|\mathbf{r}_3|}h_1^{(2)}(k_0|\mathbf{r}_3|) + \frac{(x-x_4)}{|\mathbf{r}_4|}h_1^{(2)}(k_0|\mathbf{r}_4|) \end{pmatrix} \tag{5.32}$$

where $|\mathbf{r}_s| = \left[(x \pm x_0)^2 + (y \pm y_0)^2 + (z - z_0)^2\right]^{1/2}$—distance from the observation point to the center of the DR (Fig. 5.8a) and its images located symmetrically relative to the tip of the metal wedge in the plane: $z = z_0$ ($s = 1, 2, 3, 4$). At that, the centers of the mirror images of the DR are numbered clockwise. It is easy to verify that obtained expressions (5.30–5.32) satisfy the boundary conditions on the wedge surface.

Thus, the fields of a spherical DR in the space of a rectangular wedge take the form of a sum consisting of the resonator field in open space and three of its images. The result obtained is not surprising, since when deriving the field expansion coefficients (2.11–2.14) we excluded the inverse influence of the waveguide walls on the resonator field. It is interesting to note here that the singularity of expansions outside the region occupied by the dielectric disappears completely. Fields and oscillations of other types take a similar form. The passage to the limit in the obtained formulas to the metal plane is obvious: in relations (5.30–5.32) it is enough to x_0 direct to ∞ under the condition of finiteness $x - x_0$. In this case, the components of the electric field of the DR take the form:

For the azimuthally homogeneous oscillations H_{10l}:

$$\mathbf{e}_m^0 = \omega\mu_0 \left[\frac{j_1(p)}{y_1(q)}\right] \cdot \begin{pmatrix} \frac{(y-y_0)}{|\mathbf{r}_1|}h_1^{(2)}(k_0|\mathbf{r}_1|) + \frac{(y+y_0)}{|\mathbf{r}_2|}h_1^{2}(k_0|\mathbf{r}_2|) \\ -\frac{(x-x_0)}{|\mathbf{r}_1|}h_1^{(2)}(k_0|\mathbf{r}_1|) - \frac{(x-x_0)}{|\mathbf{r}_2|}h_1^{(2)}(k_0|\mathbf{r}_2|) \\ 0 \end{pmatrix}. \tag{5.33}$$

For H^{y}_{11l} oscillations with magnetic field polarization at the center of the DR, orthogonal to the metal plane:

$$\mathbf{e}^0_m = \omega\mu_0\left[\frac{j_1(p)}{y_1(q)}\right]\cdot\begin{pmatrix}(z-z_0)\left\{\frac{1}{|\mathbf{r}_1|}h_1^{(2)}(k_0|\mathbf{r}_1|)-\frac{1}{|\mathbf{r}_2|}h_1^{(2)}(k_0|\mathbf{r}_2|)\right\}\\ 0\\ \frac{(x-x_1)}{|\mathbf{r}_1|}h_1^{(2)}(k_0|\mathbf{r}_1|)-\frac{(x-x_2)}{|\mathbf{r}_2|}h_1^{(2)}(k_0|\mathbf{r}_2|)\end{pmatrix}. \tag{5.34}$$

Now y_0 directing to ∞ at a finite value $y-y_0$, we move on to free space. Having rewritten the expressions for the field components in a spherical coordinate system associated with the center of the ball,

$(x-x_0)/|\mathbf{r}_1| = \sin\theta\cos\varphi$; $(y-y_0)/|\mathbf{r}_1| = \sin\theta\sin\varphi$ and $(z-z_0)/|\mathbf{r}_1| = \cos\theta$ as well as remembering the replacement of functions $y_1(q)\approx ih_1^{(2)}(q)$ performed in 1.1, we are convinced that the found limit expression ultimately coincides exactly with the field of a spherical DR in open space.

Let us now go to the limit: $b=\infty$ in (2.65, 2.66). Using (D.18) we find the coupling coefficients between the spherical DR and the space of a rectangular metal wedge:

$$\begin{aligned}\tilde{k} = 3\alpha_1^{H,E}(p,q)\cdot\Bigg[&\frac{2}{3}\pm j_0(2k_0x_0)\mp\frac{1}{2k_0x_0}j_1(2k_0x_0)\\ &\pm j_0(2k_0y_0)\mp\frac{1}{2k_0y_0}j_1(2k_0y_0)+j_0\left(2k_0\sqrt{x_0^2+y_0^2}\right)\\ &-\frac{1}{2k_0\sqrt{x_0^2+y_0^2}}j_1\left(2k_0\sqrt{x_0^2+y_0^2}\right)\Bigg],\end{aligned} \tag{5.35}$$

where the upper sign corresponds to magnetic H_{10l} oscillations, and the lower sign to electrical E_{10l} oscillations, characterized by the polarization of the magnetic (in the second case, electric) field in the center of the dielectric ball, parallel to the edge of the wedge.

For H_{11l} oscillations (upper sign),, E_{11l} (lower sign) with field polarization at the center of the DR, parallel to the y axis, we obtain:

$$\begin{aligned}\tilde{k} = 3\alpha_1^{H,E}(p,q)\cdot\Bigg[&\frac{2}{3}\pm j_0(2k_0x_0)\mp\frac{1}{2k_0x_0}j_1(2k_0x_0)\\ &\mp\frac{1}{k_0y_0}j_1(2k_0y_0)-\frac{x_0^2}{x_0^2+y_0^2}j_0\left(2k_0\sqrt{x_0^2+y_0^2}\right)\\ &+\frac{(2k_0x_0)^2-2(2k_0y_0)^2}{\left[(2k_0x_0)^2+(2k_0y_0)^2\right]^{3/2}}j_1\left(2k_0\sqrt{x_0^2+y_0^2}\right)\Bigg].\end{aligned} \tag{5.36}$$

The relations for oscillations of the second polarization, parallel to the x axis, are obtained from (5.36), replacing $x_0 \Leftrightarrow y_0$.

From here we also find the coupling coefficients of a spherical DR above a metal plane at a height y_0 in the case of H_{11l} oscillations or E_{11l} with vertical (upper ratio) and horizontal (lower ratio) polarization of the field at the center of the resonator:

$$\tilde{k} = 3\alpha_1^{H,E}(p,q) \cdot \left[\frac{\frac{2}{3} \mp \frac{1}{k_0 y_0} j_1(2k_0 y_0)}{\frac{2}{3} \pm j_0(2k_0 y_0) \mp \frac{1}{2k_0 y_0} j_1(2k_0 y_0)} \right]. \tag{5.37}$$

When the metal plane is removed, ($y_0 \to \infty$) the coupling coefficients (5.37) transform into coupling coefficients of the spherical DR with free space: $\tilde{k} = \tilde{k}_{os} = 2\alpha_1^{H,E}(p,q)$ (see 5.9, 5.10).

It is obvious that the oscillatory nature of the coupling is determined by the interference of emitted and re-reflected waves of the resonator plane (Fig. 5.8). Near the metal plane, the coupling coefficients for H_{101} oscillations with parallel polarization of the magnetic field increase and formally double at $y_0 = 0$, and for H_{11l} oscillations of orthogonal polarization they decrease to zero. In the case of electrical oscillations, the opposite pattern is observed: the coupling coefficients of oscillations with orthogonal polarization of the electric field at the center of the ball—$E_{11l}^{\perp}$, increase as the DR approaches the metal plane, and the coupling coefficients of oscillations E_{11l}^{II}, with polarization parallel to the plane decrease.

Transformation of the form (2.61–2.63) can also be used to calculate the coupling coefficients of a cylindrical DR in the space of a rectangular metal wedge, or above a metal plane (Fig. 5.9).

Let us first consider the coupling coefficients between cylindrical DR and the space of a rectangular metal wedge (Fig. 5.9a, b). To derive the necessary relationships, we will use the expressions found on the basis of (2.32, 2.34) for the coupling coefficients of cylindrical DR with a rectangular waveguide, or semi-infinite trench.

Going to the limit at $a \to \infty$ and $b \to \infty$ for the $H_{n0l}^{\pm}$ we obtain:in case the resonator axis is located orthogonal to the wedge edge (Fig. 5.9a):

$$\begin{aligned}
\tilde{k} = {} & \frac{16p_z}{\varepsilon_{1r} q_z} \cdot \frac{(\varepsilon_{1r}-1)^2}{[J_1^2(p_\perp) - J_0(p_\perp)J_2(p_\perp)]} \Big/ \begin{bmatrix} 2p_z + \sin 2p_z \\ 2p_z - \sin 2p_z \end{bmatrix} \\
& \cdot \int_0^1 \frac{1 + J_0(2k_0 y_0 \eta)}{\left[(p_\perp/q_\perp)^2 - \eta^2\right]^2 \cdot \left[(p_z/q_z)^2 - \xi^2\right]^2} \\
& \cdot \{p_\perp/q_\perp J_0(p_\perp) J_1(q_\perp \eta) - \eta J_1(p_\perp) J_0(q_\perp \eta)\}^2 \\
& \cdot \begin{bmatrix} \sin^2(k_0 x_0 \xi) \cdot (p_z/q_z \sin p_z \cos q_z \xi - \xi \cos p_z \sin q_z \xi)^2 \\ \cos^2(k_0 x_0 \xi) \cdot (p_z/q_z \cos p_z \sin q_z \xi - \xi \sin p_z \cos q_z \xi)^2 \end{bmatrix} \cdot d\xi;
\end{aligned} \tag{5.38}$$

in case the resonator axis is located parallel to the wedge edge (Fig. 5.9b)

$$\tilde{k} = \frac{8p_z}{\varepsilon_{1r}q_z} \cdot \frac{(\varepsilon_{1r}-1)^2}{[J_1^2(p_\perp) - J_0(p_\perp)J_1(p_\perp)]} \Big/ \begin{bmatrix} 2p_z + \sin 2p_z \\ 2p_z - \sin 2p_z \end{bmatrix}$$
$$\cdot \int_0^1 \frac{1 + J_0(2k_0x_0\eta) + J_0(2k_0y_0\eta) + J_0\left(2k_0\sqrt{x_0^2+y_0^2}\eta\right)}{\left[(p_\perp/q_\perp)^2 - \eta^2\right]^2 \cdot \left[(p_z/q_z)^2 - \xi^2\right]^2}$$
$$\cdot \{p_\perp/q_\perp J_0(p_\perp)J_1(q_\perp\eta) - \eta J_1(p_\perp)J_0(q_\perp\eta)\}^2$$
$$\cdot \begin{bmatrix} p_z/q_z \sin p_z \cos q_z\xi - \xi \cos p_z \sin q_z\xi \\ p_z/q_z \cos p_z \sin q_z\xi - \xi \sin p_z \cos q_z\xi \end{bmatrix}^2 \cdot d\xi, \quad (5.39)$$

where $\eta = \left(1-\xi^2\right)^{1/2}$. When deriving the last relation, we used the expansion of the field of a cylindrical DR in terms of the propagating waves of a rectangular waveguide, followed by transformation to the limit at $a \to \infty$ and $b \to \infty$ in cylindrical coordinates on the plane $(\chi_x/k_0, \chi_y/k_0)$. In this case, integration was carried out using (D.3) in the region $\chi/k_0 \le 1$.

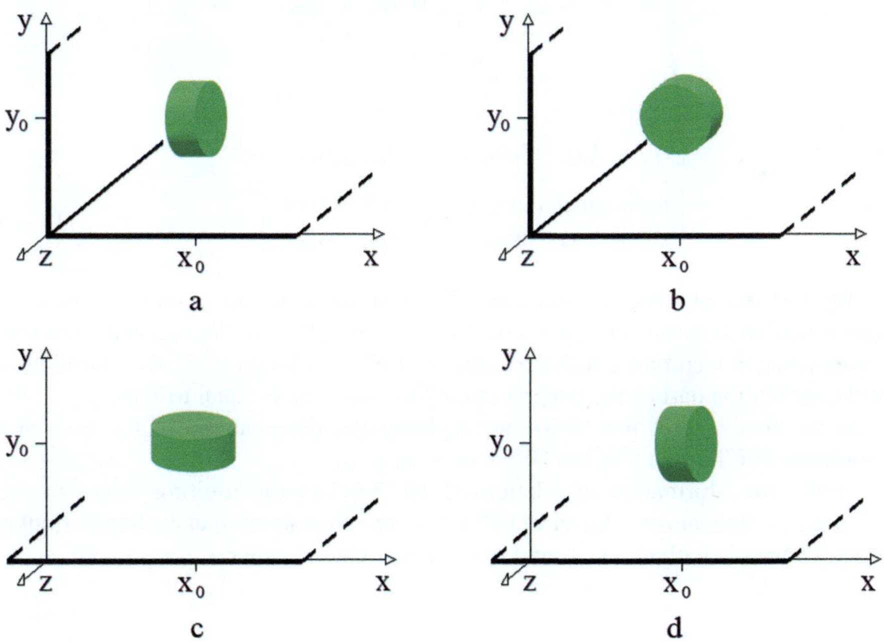

Fig. 5.9 Standard positions of a cylindrical DR in the space of a metal wedge (**a**, **b**); above the metal plane (**c**, **d**)

Transition in (5.38, 5.39) to the limit at $y_0 \to \infty$, $x_0 \to \infty$, respectively, leads us to expressions for the coupling coefficients of cylindrical DR located above the metal plane:

for $H^{\pm}_{n0l}$ and orthogonal to the plane of the resonator axis (Fig. 5.9c):

$$\tilde{k} = \frac{16p_z}{\varepsilon_{1r}q_z} \cdot \frac{(\varepsilon_{1r}-1)^2}{[J_1^2(p_\perp) - J_0(p_\perp)J_2(p_\perp)]} \bigg/ \begin{bmatrix} 2p_z + \sin 2p_z \\ 2p_z - \sin 2p_z \end{bmatrix}$$
$$\cdot \int\limits_0^1 \frac{1}{\left[(p_\perp/q_\perp)^2 - \eta^2\right]^2 \cdot \left[(p_z/q_z)^2 - \xi^2\right]^2}$$
$$\cdot \{p_\perp/q_\perp J_0(p_\perp)J_1(q_\perp\eta) - \eta J_1(p_\perp)J_0(q_\perp\eta)\}^2$$
$$\cdot \begin{bmatrix} \sin(k_0y_0\xi)(p_z/q_z \sin p_z \cos q_z\xi - \xi \cos p_z \sin q_z\xi) \\ \cos(k_0y_0\xi)(p_z/q_z \cos p_z \sin q_z\xi - \xi \sin p_z \cos q_z\xi) \end{bmatrix}^2 \cdot d\xi; \tag{5.40}$$

with a parallel arrangement of the DR axis and the metal plane (Fig. 5.9d):

$$\tilde{k} = \frac{8p_z}{\varepsilon_{1r}q_z} \cdot \frac{(\varepsilon_{1r}-1)^2}{[J_1^2(p_\perp) - J_0(p_\perp)J_1(p_\perp)]} \bigg/ \begin{bmatrix} 2p_z + \sin 2p_z \\ 2p_z - \sin 2p_z \end{bmatrix}$$
$$\cdot \int\limits_0^1 \frac{1 + J_0(2k_0y_0\eta)}{\left[(p_\perp/q_\perp)^2 - \eta^2\right]^2 \cdot \left[(p_z/q_z)^2 - \xi^2\right]^2}$$
$$\cdot \{p_\perp/q_\perp J_0(p_\perp)J_1(q_\perp\eta) - \eta J_1(p_\perp)J_0(q_\perp\eta)\}^2$$
$$\cdot \begin{bmatrix} p_z/q_z \sin p_z \cos q_z\xi - \xi \cos p_z \sin q_z\xi \\ p_z/q_z \cos p_z \sin q_z\xi - \xi \sin p_z \cos q_z\xi \end{bmatrix}^2 d\xi; \tag{5.41}$$

The last relation also follows from (5.40), if the dependence on coordinates is represented in the form $\sin^2 k_0x_0\xi\left(\cos^2 k_0x_0\xi\right) = 1/2(1 \mp \cos 2k_0x_0\xi)$ and taken into account that, in accordance with the Riemann–Lebesgue lemma [11], the contribution of the oscillating part of the integral in the limit $x_0 = \infty$ is equal to zero.

In the final expression (5.40) we replaced the designations of the axes and coordinates of the center of the DR: $x \Leftrightarrow y$; $x_0 \Leftrightarrow y_0$.

Further transformation of relations (5.40, 5.41) to the limiting values of the coupling coefficients of cylindrical DR with open space shows that the final formulas exactly coincide with (5.23), found using the Kirchhoff integral.

5.5 Radiation Parameters of a Spherical Dielectric Resonator Near a Round Metal Cylinder

Let us consider the coupling coefficients of a spherical DR in the presence of an infinite metal cylinder near its surface. As above, we will assume that external inhomogeneities have little effect on the distribution of the natural oscillation field in the DR material. Let us show that in this case the field of the system as a whole is represented as a superposition of the field of an isolated DR in free space and diffraction terms associated with the re-emission of the resonator field by a metal cylinder. As the inhomogeneity moves away from the surface of the ball, these terms tend to zero.

To determine the field components of a spherical DR near a metal cylinder, we use the previously found expansions in terms of the waves of a coaxial waveguide (Sec. 2.3) and go to the limit $R_2 = \infty$, where R_2—coax screen radius. Let us denote $\eta = \chi/k_0$ the relative transverse wave number, then for a regular waveguide $\Gamma/k_0 = \gamma = \sqrt{1-\eta^2}$. Considering that in each of the terms the u value is fixed, and the contribution to the sums is determined by large values of the arguments, we use the asymptotic expansions of cylindrical functions and their roots [5]. Then for magnetic and (electric) waves the lines:

$$d\eta = \lim_{R_2,s\to\infty}\left[\frac{j'_{u,s+1}-j'_{u,s}}{k_0R_2}\right] \sim \left[\frac{j_{u,s+1}-j_{u,s}}{k_0R_2}\right] \sim \frac{\pi}{k_0R_2}, \text{ and}$$

$$\begin{gathered}\lim_{R_2,s\to\infty}\left\{C_u^2(\chi R_2)\left[(\chi R_2)^2-u^2\right]-C_u^2(\chi R_1)\left[(\chi R_1)^2-u^2\right]\right\}^{-1}\\ \sim \lim_{R_2,s\to\infty}\left\{\left[\chi R_2 D_{u-1}(\chi R_2)\right]^2-\left[\chi R_1 D_{u-1}(\chi R_1)\right]\right\}^{-1} \sim d\eta/2\eta.\end{gathered} \quad (5.42)$$

We also use (2.3), (A.5–A.7), (2.62) and (5.42). As a result of transformation to the limit $R_2 = \infty$, the expressions for the electric field components of the DR in a cylindrical coordinate system take the form:

For magnetic oscillations H_{nml}:

$$\begin{aligned} e_\rho^\pm &= \mp i^{n-m-1}\omega\mu_0\frac{j_n(p)}{y_n(q)}\sum_{u=0}^{\infty}\frac{1}{1+\delta_{uo}}\\ &\cdot\int_0^\infty\left\{\frac{dP_n^m(\gamma)}{d\gamma}\cdot\frac{u}{k_0\rho}C_u(k_0\rho\eta)\right.\\ &\cdot\begin{bmatrix}-\left[\mathrm{C}_{m+u}(k_0\rho_0\eta)-(-1)^m\mathrm{C}_{-m+u}(k_0\rho_0\eta)\right]\cos u\Delta\alpha\\ \left[\mathrm{C}_{m+u}(k_0\rho_0\eta)+(-1)^m\mathrm{C}_{-m+u}(k_0\rho_0\eta)\right]\sin u\Delta\alpha\end{bmatrix}\\ &+m\frac{\gamma}{\eta}P_n^m(\gamma)D'_u(k_0\rho\eta)\\ &\left.\cdot\begin{bmatrix}-\left[\mathrm{D}_{m+u}(k_0\rho_0\eta)+(-1)^m\mathrm{D}_{-m+u}(k_0\rho_0\eta)\right]\cos u\Delta\alpha\\ \left[\mathrm{D}_{m+u}(k_0\rho_0\eta)-(-1)^m\mathrm{D}_{-m+u}(k_0\rho_0\eta)\right]\sin u\Delta\alpha\end{bmatrix}\right\}\end{aligned}$$

$$\cdot\, e^{-i\gamma k_0 \Delta z}\eta/\gamma\, d\eta;$$

$$e_\alpha^\pm = \mp i^{n-m-1}\omega\mu_0 \frac{j_n(p)}{y_n(q)} \sum_{u=0}^{\infty} \frac{1}{1+\delta_{uo}}$$
$$\cdot \int_0^\infty \left\{ \eta \frac{dP_n^m(\gamma)}{d\gamma} \cdot C_u'(k_0\rho\eta) \right.$$
$$\cdot \begin{bmatrix} \left[\mathrm{C}_{\mathrm{m+u}}(k_0\rho_0\eta) - (-1)^m \mathrm{C}_{\mathrm{-m+u}}(k_0\rho_0\eta)\right] \sin u\Delta\alpha \\ \left[\mathrm{C}_{\mathrm{m+u}}(k_0\rho_0\eta) + (-1)^m \mathrm{C}_{\mathrm{-m+u}}(k_0\rho_0\eta)\right] \cos u\Delta\alpha \end{bmatrix}$$
$$+ m\frac{\gamma}{\eta} P_n^m(\gamma) \frac{u}{k_0\rho\eta} D_u(k_0\rho\eta)$$
$$\left. \cdot \begin{bmatrix} \left[\mathrm{D}_{\mathrm{m+u}}(k_0\rho_0\eta) + (-1)^m \mathrm{D}_{\mathrm{-m+u}}(k_0\rho_0\eta)\right] \sin u\Delta\alpha \\ \left[\mathrm{D}_{\mathrm{m+u}}(k_0\rho_0\eta) - (-1)^m \mathrm{D}_{\mathrm{-m+u}}(k_0\rho_0\eta)\right] \cos u\Delta\alpha \end{bmatrix} \right\}$$
$$\cdot\, e^{-i\gamma k_0 \Delta z}\eta/\gamma\, d\eta;$$

$$e_z^\pm = \pm i^{n-m}\omega\mu_0 \frac{j_n(p)}{y_n(q)} m \sum_{u=0}^{\infty} \frac{1}{1+\delta_{uo}}$$
$$\cdot \int_0^\infty P_n^m(\gamma) D_u(k_0\rho\eta)$$
$$\cdot \begin{bmatrix} -\left[\mathrm{D}_{\mathrm{m+u}}(k_0\rho_0\eta) + (-1)^m \mathrm{D}_{\mathrm{-m+u}}(k_0\rho_0\eta)\right] \cos u\Delta\alpha \\ \left[\mathrm{D}_{\mathrm{m+u}}(k_0\rho_0\eta) - (-1)^m \mathrm{D}_{\mathrm{-m+u}}(k_0\rho_0\eta)\right] \sin u\Delta\alpha \end{bmatrix}$$
$$\cdot\, e^{-i\gamma k_0 \Delta z}\eta/\gamma\, d\eta; \tag{5.43}$$

For electrical oscillations E_{nml}

$$e_\rho^\pm = \mp i^{n-m+1} k_0\varepsilon_{1r} \frac{j_n(p)}{y_n(q)} \sum_{u=0}^{\infty} \frac{1}{1+\delta_{uo}}$$
$$\cdot \int_0^\infty \left\{ \frac{m}{\eta} P_n^m(\gamma) \frac{u}{\eta k_0\rho} \cdot C_u'(k_0\rho\eta) \right.$$
$$\cdot \begin{bmatrix} \left[\mathrm{C}_{\mathrm{m+u}}(k_0\rho_0\eta) + (-1)^m \mathrm{C}_{\mathrm{-m+u}}(k_0\rho_0\eta)\right] \sin u\Delta\alpha \\ \left[\mathrm{C}_{\mathrm{m+u}}(k_0\rho_0\eta) - (-1)^m \mathrm{C}_{\mathrm{-m+u}}(k_0\rho_0\eta)\right] \cos u\Delta\alpha \end{bmatrix}$$
$$+ \gamma\eta \frac{dP_n^m(\gamma)}{d\gamma} D_u'(k_0\rho\eta)$$
$$\left. \cdot \begin{bmatrix} \left[\mathrm{D}_{\mathrm{m+u}}(k_0\rho_0\eta) - (-1)^m \mathrm{D}_{\mathrm{-m+u}}(k_0\rho_0\eta)\right] \sin u\Delta\alpha \\ \left[\mathrm{D}_{\mathrm{m+u}}(k_0\rho_0\eta) + (-1)^m \mathrm{D}_{\mathrm{-m+u}}(k_0\rho_0\eta)\right] \cos u\Delta\alpha \end{bmatrix} \right\}$$
$$\cdot\, e^{-i\gamma k_0 \Delta z}\eta d\eta/\gamma;$$

$$
\begin{aligned}
e_\alpha^{\pm} = \mp i^{n-m+1} k_0 \varepsilon_{1r} \frac{j_n(p)}{y_n(q)} \sum_{u=0}^{\infty} \frac{1}{1+\delta_{uo}} \\
\cdot \int_0^{\infty} \left\{ \frac{m}{\eta} P_n^m(\gamma) \cdot C_u'(k_0 \rho \eta) \right. \\
\cdot \begin{bmatrix} \left[C_{m+u}(k_0\rho_0\eta) + (-1)^m C_{-m+u}(k_0\rho_0\eta)\right] \cos u\Delta\alpha \\ -\left[C_{m+u}(k_0\rho_0\eta) - (-1)^m C_{-m+u}(k_0\rho_0\eta)\right] \sin u\Delta\alpha \end{bmatrix} \\
+ \eta\gamma \frac{dP_n^m(\gamma)}{d\gamma} \cdot \frac{u}{k_0\rho\eta} D_u(k_0\rho\eta) \\
\left. \cdot \begin{bmatrix} \left[D_{m+u}(k_0\rho_0\eta) - (-1)^m D_{-m+u}(k_0\rho_0\eta)\right] \cos u\Delta\alpha \\ -\left[D_{m+u}(k_0\rho_0\eta) + (-1)^m D_{-m+u}(k_0\rho_0\eta)\right] \sin u\Delta\alpha \end{bmatrix} \right\} \\
\cdot e^{-i\gamma k_0 \Delta z} \eta d\eta / \gamma;
\end{aligned}
$$

$$
\begin{aligned}
e_z^{\pm} = \pm i^{n-m} k_0 \varepsilon_{1r} \frac{j_n(p)}{y_n(q)} \sum_{u=0}^{\infty} \frac{1}{1+\delta_{uo}} \\
\cdot \int_0^{\infty} \frac{dP_n^m(\gamma)}{d\gamma} \cdot D_u(k_0\rho\eta) \\
\cdot \begin{bmatrix} \left[D_{m+u}(k_0\rho_0\eta) - (-1)^m D_{-m+u}(k_0\rho_0\eta)\right] \sin u\Delta\alpha \\ \left[D_{m+u}(k_0\rho_0\eta) + (-1)^m D_{-m+u}(k_0\rho_0\eta)\right] \cos u\Delta\alpha \end{bmatrix} \\
\cdot e^{-i\gamma k_0 \Delta z} \eta^3 d\eta / \gamma
\end{aligned}
\tag{5.44}
$$

where (ρ_0, α_0, z_0)—coordinates of the ball's center; (ρ, α, z)—observation point coordinates; $\Delta\alpha = \alpha - \alpha_0$; $\Delta z = |z - z_0|$; $\gamma = \left(1 - \eta^2\right)^{1/2}$ (Fig. 5.10a).

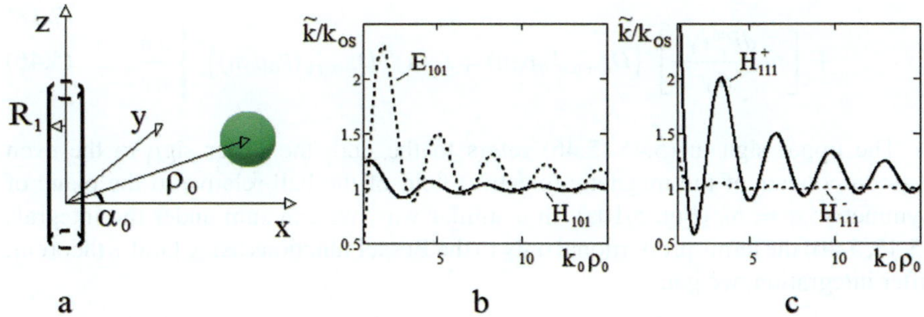

Fig. 5.10 Dependences of coupling coefficients $\tilde{k}$ with open space near a metal cylinder on the coordinates of the spherical DR (k_{OS}—coupling coefficient between DR and open space; $k_0 R_1 = 0, 5$; $f_0 = 7$ GHz; $\varepsilon_{1r} = 36$)

Let us select in $C_\nu(x)$ and $D_\nu(x)$ the terms containing only the Bessel functions and sum the series using the addition theorem [12]. We integrate the obtained relations using (D.16) and move to a spherical coordinate system associated with the center of the DR. The resulting expressions exactly coincide with the natural oscillation field of a spherical DR in open space (1.2, 1.3).

Thus, we are convinced that terms (5.43, 5.44), containing only Bessel functions, determine the field of natural oscillations of an isolated spherical DR in open space. The remaining terms containing the Neumann functions are due to diffraction of the ball field on the metal cylinder. Taking into account the data on the behavior of cylindrical functions at zero and at infinity [13], we conclude that these terms vanish at $R_1 = 0$, or $\rho_0 \to \infty$.

Using (2.57–2.60), in the limit $R_2 = \infty$, we also find:

for H_{nml} oscillations:

$$\tilde{k} = \frac{2}{1+\delta_{m0}} \cdot \frac{2n+1}{n(n+1)} \cdot \frac{(n-m)!}{(n+m)!} \cdot \alpha_n^H(p,q) \sum_{u=0}^{\infty} \frac{1}{1+\delta_{u0}}$$
$$\cdot \int_0^1 \left\{ \left[\eta^2 \frac{dP_n^m(\gamma)}{d\gamma} \right]^2 \left\{ C_{m+u}(k_0\rho_0\eta) \mp (-1)^m C_{-m+u}(k_0\rho_0\eta) \right\}^2 \right.$$
$$\left. + \left[mP_n^m(\gamma) \right]^2 \left\{ D_{m+u}(k_0\rho_0\eta) \pm (-1)^m D_{-m+u}(k_0\rho_0\eta) \right\}^2 \right\} \frac{d\eta}{\eta\gamma} \tag{5.45}$$

For E_{nml} oscillations:

$$\tilde{k} = \frac{2}{1+\delta_{m0}} \cdot \frac{2n+1}{n(n+1)} \cdot \frac{(n-m)!}{(n+m)!} \cdot \alpha_n^E(p,q) \sum_{u=0}^{\infty} \frac{1}{1+\delta_{u0}}$$
$$\cdot \int_0^1 \left\{ \left[mP_n^m(\gamma) \right]^2 \left\{ C_{m+u}(k_0\rho_0\eta) \pm (-1)^m C_{-m+u}(k_0\rho_0\eta) \right\}^2 \right.$$
$$\left. + \left[\eta^2 \frac{dP_n^m(\gamma)}{d\gamma} \right]^2 \left\{ D_{m+u}(k_0\rho_0\eta) \mp (-1)^m D_{-m+u}(k_0\rho_0\eta) \right\}^2 \right\} \frac{d\eta}{\eta\gamma}. \tag{5.46}$$

The upper sign in (5.45, 5.46) refers to the odd, the lower sign to the even θ—components of the magnetic (electric) field of the ball relative to the plane of symmetry: $\alpha = \alpha_0$ (Fig. 5.10a). In a similar way, we can sum under the integrals (5.45, 5.46) the terms proportional only to the Bessel functions using Graf's theorem; after integration, we get:

for H_{nml}:

$$\tilde{k} = (1 \mp \delta_{m0}) \cdot \alpha_n^H(p, q) + \frac{2}{1 + \delta_{m0}} \cdot \frac{2n+1}{n(n+1)} \cdot \frac{(n-m)!}{(n+m)!} \cdot \alpha_n^H(p, q)$$
$$\cdot \int_0^1 \left\{ \left[\eta^2 \frac{dP_n^m(\gamma)}{d\gamma} \right]^2 S_m^{H\mp}(k_0 \rho_0 \eta) + \left[m P_n^m(\gamma) \right]^2 S_m^{E\pm}(k_0 \rho_0 \eta) \right\} \frac{d\eta}{\eta \gamma}, \tag{5.47}$$

for E_{nml}:

$$\tilde{k} = (1 \mp \delta_{m0}) \cdot \alpha_n^E(p, q) + \frac{2}{1 + \delta_{m0}} \cdot \frac{2n+1}{n(n+1)} \cdot \frac{(n-m)!}{(n+m)!} \cdot \alpha_n^E(p, q)$$
$$\cdot \int_0^1 \left\{ \left[m P_n^m(\gamma) \right]^2 S_m^{H\pm}(k_0 \rho_0 \eta) + \left[\eta^2 \frac{dP_n^m(\gamma)}{d\gamma} \right]^2 S_m^{E\mp}(k_0 \rho_0 \eta) \right\} \frac{d\eta}{\eta \gamma}. \tag{5.48}$$

Here

$$S_m^{HE\pm}(x) = \sum_{u=-\infty}^{\infty} W_u^{HE} \left\{ \frac{1}{2} W_u^{HE} \left[Y_{m+u}(x) \pm (-I)^m Y_{-m+u}(x) \right] \right.$$
$$\left. - \left[J_{m+u}(x) \pm (-I)^m J_{-m+u}(x) \right] \right\} \cdot \left[Y_{m+u}(x) \pm (-I)^m Y_{-m+u}(x) \right] \tag{5.49}$$

$$W_u^H = J_u'(k_0 R_1 \eta)/Y_u'(k_0 R_1 \eta); \quad W_u^E = J_u(k_0 R_1 \eta)/Y_u(k_0 R_1 \eta). \tag{5.50}$$

From (5.47, 5.48) we see that for $R_1 \to 0$, $\rho_0 \to \infty$, the found relations also coincide with the coupling coefficients of the dielectric ball with free space: $\tilde{k} = \tilde{k}_{os} = 2\alpha_1^{H,E}(p, q)$. The influence of the rod most significantly affects the types of oscillations that induce longitudinal currents in the metal rod (Fig. 5.10b, c).

The resulting integral expressions for the field and coupling coefficients contain singularities due to the requirement that boundary conditions be met on the surface of the metal cylinder. As follows from (5.50), integrals (5.45–5.48) diverge if $k_0 R_1 > y_{0,1}$, $y_{0,1}$—first zero of the Neumann function $Y_0(z)$ [5]. From here we derive the condition for their direct application: $k_0 R_1 < y_{0,1} = 0, 89358$.

5.6 Characteristics of a Spherical Dielectric Resonator Near a Metal Wedge

In practical applications in antenna devices, dielectric resonators are often placed near metal surfaces, parts of which are various modifications of a metal wedge. Placing metal surfaces in close proximity to the DR leads to a change in its quality factor, and in addition, the spatial distribution of the radiation field. To calculate the radiation characteristics of such structures, we apply the previously obtained results of decomposition of the DR field into the waves of a sector waveguide (2.44–2.47); (A.5, A.6); (A.12, A.13). Let us move to the limit $R = \infty$, taking into account for the coupling coefficients the contribution of only propagating types of waves. The final result takes the form in the case of magnetic oscillations H_{nml} (odd h_θ—upper sign, even h_θ—lower sign relative to the plane: $\alpha = \alpha_0$):

$$\begin{aligned}\tilde{k} = &\frac{2\pi}{\alpha_w} \cdot \frac{2n+1}{n(n+1)} \cdot \frac{(n-m)!}{(n+m)!} \cdot \alpha_n^H(p,q) \cdot \frac{1}{1+\delta_{m0}} \cdot \\ &\cdot \sum_{s=0}^{\infty} \int_0^1 \left\{ \left[\eta^2 \frac{dP_n^m(\gamma)}{d\gamma} \right]^2 \right. \\ &\cdot \left[J_{m+u(s)}(k_0\rho_0\eta) \mp (-1)^m J_{-m+u(s)}(k_0\rho_0\eta) \right]^2 + \left[m P_n^m(\gamma) \right]^2 \\ &\left. \cdot \left[J_{m+u(s)}(k_0\rho_0\eta) \pm (-1)^m J_{-m+u(s)}(k_0\rho_0\eta) \right]^2 \right\} \begin{matrix} \sin^2 \\ \cos^2 \end{matrix} \left[\left(\frac{\alpha_0}{\alpha_w} + 1 \right) \frac{\pi}{2} s \right] \cdot \frac{d\eta}{\eta\gamma},\end{aligned} \tag{5.51}$$

where $\gamma = \left(1-\eta^2\right)^{1/2}$; (ρ_0, α_0, z_0)—coordinates of the DR center; α_w—wedge opening angle (Fig. 5.11a); $u(s) = (\pi/2\alpha_w) \cdot s$. The expression for coupling coefficients in the case of electrical type oscillations E_{nml} also has the form (5.51) taking into account the replacement $\alpha_n^H \Leftrightarrow \alpha_n^E$ и $\sin^2 \Leftrightarrow \cos^2$.

Let us consider some particular values of the wedge opening angles, when the summation over s can be carried out under the integral sign.

When $\alpha_w = \pi/2$ the index $u(s)$ takes only integer values and you can use the Graph addition theorem for Bessel functions [12] (in the general case, such summation is possible if $\alpha_w = \pi/s$ where $s = 2, 3, \ldots$):

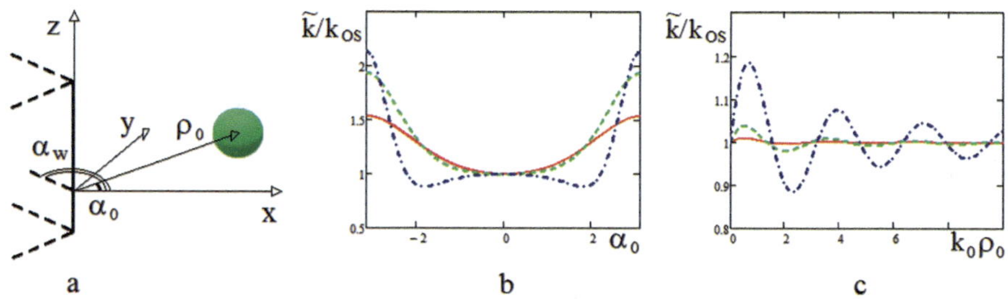

Fig. 5.11 Spherical DR near a metal wedge (**a**). Dependences of coupling coefficients $\tilde{k}$ with open space near a metal half-plane ($\alpha_w = \pi$) on the coordinates of the DR (k_{OS}—coupling coefficient between isolated DR and open space) for $\varepsilon_{1r} = 36$; **b** $k_0\rho_0 = 0, 5$—solid line; $k_0\rho_0 = 1$ dotted curve; $k_0\rho_0 = 2$—dash-dotted curve; **c** $\alpha_0 = \pi/8$—solid line; $\alpha_0 = \pi/4$—dotted line; $\alpha_0 = \pi/2$—dash-dotted line

$$\tilde{k} = \frac{1 \mp \delta_{m0}}{1 + \delta_{m0}} \cdot 2\alpha_n^{H,E}(p,q) + \frac{(\pm 2)}{1 + \delta_{m0}} \frac{2n+1}{n(n+1)} \cdot \frac{(n-m)!}{(n+m)!} \cdot \alpha_n^{H,E}(p,q)$$
$$\cdot \left\{ \int_0^1 \left\{ \left[\eta^2 \frac{dP_n^m(\gamma)}{d\gamma} \right]^2 - \left[mP_n^m(\gamma) \right]^2 \right\} J_{2m}(2k_0\rho_0 \cos\alpha_0 \cdot \eta) \frac{d\gamma}{\eta^2} \right.$$
$$\mp (-1)^m \cos 2m\alpha_0 \int_0^1 \left\{ \left[\eta^2 \frac{dP_n^m(\gamma)}{d\gamma} \right]^2 \right.$$
$$\left. + \left[mP_n^m(\gamma) \right]^2 \right\} J_0(2k_0\rho_0 \cos\alpha_0 \cdot \eta) \frac{d\gamma}{\eta^2}, \tag{5.52}$$

where the (+) sign before the second term in parentheses is taken for magnetic type oscillations, and (−) for electric ones. The first term (5.52) determines the coupling coefficient of the DR with open space, the second is due to the contribution of the re-emitted field and depends on the height of the center of the ball above the metal plane $\rho_0 \cos \alpha_0$ (Fig. 5.11a). Relation (5.52) generalizes the result obtained in Sect. 5.4 to the case of natural oscillations of a ball "polarized" at an angle α_0 relative to the normal. When $\alpha_0 = 0$ the found expressions coincide.

If the dielectric ball is located above the metal half-plane: $\alpha_w = \pi$, then for H_{nml}:

$$\tilde{k} = \frac{1 \mp \delta_{m0}}{1 + \delta_{m0}} \cdot \alpha_n^H(p,q)(\pm) \frac{2n+1}{n(n+1)} \cdot \frac{(n-m)!}{(n+m)!} \cdot \alpha_n^H(p,q) \cdot \frac{1}{1 + \delta_{m0}}$$
$$\cdot \left(\int_0^1 \left\{ \left[\eta^2 \frac{dP_n^m(\gamma)}{d\gamma} \right]^2 - \left[mP_n^m(\gamma) \right]^2 \right\} J_{2m}(2k_0\rho_0 \sin\alpha_0 \cdot \eta) \frac{d\eta}{\gamma\eta} \right.$$
$$\mp \cos 2m\alpha_0 \int_0^1 \left\{ \left[\eta^2 \frac{dP_n^m(\gamma)}{d\gamma} \right]^2 + \left[mP_n^m(\gamma) \right]^2 \right\} J_0(2k_0\rho_0 \sin\alpha_0 \cdot \eta) \frac{d\eta}{\gamma\eta}$$
$$+ \frac{2}{1 + \delta_{m0}} \cdot \frac{2n+1}{n(n+1)} \cdot \frac{(n-m)!}{(n+m)!} \cdot \alpha_n^H(p,q)$$
$$\cdot \int_0^1 \left(\left[\eta^2 \frac{dP_n^m(\gamma)}{d\gamma} \right]^2 \sum_{t=0}^{\infty} \left\{ J_{m+t+1/2}(k_0\rho_0\eta) \mp (-1)^m J_{-m+t+1/2}(k_0\rho_0\eta) \right\}^2 \right.$$
$$+ \left[mP_n^m(\gamma) \right]^2 \sum_{t=0}^{\infty} \left\{ J_{m+t+1/2}(k_0\rho_0\eta) \right.$$
$$\left. \pm (-1)^m J_{-m+t+1/2}(k_0\rho_0\eta) \right\}^2 \Big) \frac{\cos^2}{\sin^2} \frac{\alpha_0}{2} (2t+1) \frac{d\eta}{\gamma\eta}. \tag{5.53}$$

For electrical oscillations, as in the previous case, it is necessary to replace with α_n^H; on α_n^E, $\cos^2$ on $\sin^2$, as well as signs in (+)—for H_{nml} on (−) for E_{nml}. Comparing the obtained result with (5.52), we find that the coupling coefficient of a dielectric ball over a metal half-plane is represented as a superposition of half values of the coupling coefficients with open space and the metal plane, as well as a term due to the edge field. The summation of expansions (5.53) under the integral sign for the fundamental oscillations can be carried out using (D.32)

for H_{10l} (top sign); E_{10l} (lower):

$$\tilde{k} = \alpha_1^{H,E}(p,q)\Bigg(1 \pm \frac{3}{2}\left[j_0(2k_0\rho_0\sin\alpha_0) - \frac{1}{2k_0\rho_0\sin\alpha_0} j_1(2k_0\rho_0\cos\alpha_0)\right] + \frac{3}{\pi}\int\limits_0^1 [Si(2k_0\rho_0\cdot\eta) \mp S(2k_0\rho_0\sin\alpha_0\cdot\eta,\, 2k_0\rho_0\cos\alpha_0\cdot\eta)]\eta^2 d\gamma\Bigg), \quad (5.54)$$

for H_{11l} with $h_\theta \sim \dfrac{\sin}{\cos}\alpha$:

$$\tilde{k} = \alpha_1^{H}(p,q)\Bigg\{1 + \frac{3}{4}\Big[\mp\cos 2\alpha_0[\frac{1}{2k_0\rho_0\sin\alpha_0} j_1(2k_0\rho_0\sin\alpha_0) + j_0(2k_0\rho_0\cos\alpha_0)] + j_2(2k_0\rho_0\sin\alpha_0)\Big] + \frac{3}{4}\int\limits_0^1 \left[\Phi_1^{\mp}(k_0\rho_0\eta, 0) \pm \Phi_1^{\mp}(k_0\rho_0\eta, \alpha_0)\right]\cdot(\gamma + \frac{1}{\gamma})\eta d\eta\Bigg\} \quad (5.55)$$

where

$$\Phi_1^{\pm}(x,\alpha) = \frac{2}{\pi}\Big\{(\cos 2\alpha \mp 1)S(2x\sin\alpha, 2x\cos\alpha) \mp \frac{1}{x^2\sin 2\alpha}\left[\frac{d}{d\alpha}S(2x\sin\alpha, 2x\cos\alpha) + \sin\alpha\sin 2x\right] + \frac{1}{x}\cos\alpha\cos 2x\Big\}$$

and

$$\Phi_1^{\pm}(x,0) = \frac{2}{\pi}\left\{Si(2x) \mp j_1(2x) + \frac{1}{x}\cos 2x\right\}$$

$Si(x)$—integral sine function [3]; $S(a,x)$—generalized integral sine function (D.27).

Using (D.34, D.35), as well as asymptotic representations of hypergeometric functions of large arguments [14], we find by the stationary phase method from (5.54), (5.55) at $k_0\rho_0 \to \infty$ with accuracy $(k_0\rho_0)^{-3/2}$.

for H_{10l}—top sign (E_{10l}—lower):

$$\tilde{k} \approx 2\alpha_1^{H,E}(p,q)\cdot\left[1 \pm \frac{3}{4}\left(1-\frac{\pi}{4}\right)\frac{\sin(2k_0\rho_0\sin\alpha_0)}{2k_0\rho_0\sin\alpha_0}+\ldots\right]$$

for H_{11l} oscillations:

$$\tilde{k} \approx 2\alpha_1^{H}(p,q)\cdot\left[1 - \frac{3}{4}\cdot\frac{\sin(2k_0\rho_0\sin\alpha_0)}{2k_0\rho_0\sin\alpha_0}+\ldots\right]$$

From where we see that the coupling coefficients at $k_0\rho_0 \to \infty$ also tend to the coupling coefficients of the ball with open space.

On Fig. 5.11 the dependence normalized to the magnitude of the coupling coefficients between the spherical DR and open space is given for the case of H_{10l} oscillations of the ball over a metal half-plane. The physical meaning of the obtained relations is that the indicated inhomogeneity has the main influence on this type of oscillations in the region adjacent to the surface, where the coupling coefficient of the DR with open space approximately doubles. If the DR falls into the plane: $\alpha = 0$, the coupling coefficient is set equal to its coupling with open space (Fig. 5.11c).

In special cases of basic oscillations of the DR, expansion (5.51) can be integrated term by term. So for H_{10l} we find:

$$\tilde{k} = 3\frac{\pi^{3/2}}{\alpha_w}\cdot\alpha_1^{H}(p,q)\cdot\binom{0}{1}\cdot\sum_{s=0}^{\infty}\frac{1}{1+\delta_{u0}}\cdot\left(\frac{k_0\rho_0}{2}\right)^{2u}\cos^2\left(\frac{\alpha_0}{\alpha_w}+1\right)\frac{\pi}{2}s$$
$$\cdot\Gamma\left[\begin{matrix}u+2\\ u+5/2,\,u+1,\,u+1\end{matrix}\right]{}_2F_3\left(\left.\begin{matrix}u+1/2,\,u+2\\ u+1,\,u+5/2,\,2u+1\end{matrix}\right| -(k_0\rho_0)^2\right). \quad (5.56)$$

Here $u = u(s) = (\pi/2\alpha_b)s$; $\Gamma\left[\begin{matrix}v\\ x,y,z\end{matrix}\right] = \Gamma(v)/\Gamma(x)/\Gamma(y)/\Gamma(z)$; ${}_2F_3(\alpha,\beta;\gamma,\delta,\varepsilon|x)$—This is a hypergeometric function [15].

Expansion (5.56) is suitable for calculating coupling coefficients for any values of wedge opening angles α_w.

In a particular case $\alpha_w = \pi/2$, relation (5.56) coincides exactly with (5.37).

The field of natural oscillations of a spherical DR, taking into account the influence of a metal wedge, is found similarly. In a case $\alpha_w = \pi/2$, such expansions can also be summed under the integral sign by applying Graph's theorem and integrating the result using the generalized Sommerfeld integral (D.16). In this case, the final expression is also represented as a superposition of the DR field and its image in open space. For example, the electric field z component of magnetic oscillations H_{nml} takes the form:

$$e_z = i\omega\mu_0 m\frac{j_n(p)}{h_n^{(2)}(q)}\cdot$$

$$\left\{\begin{matrix}\cos\\ -\sin\end{matrix}\, m\varphi^{+}P_n^m(\cos\theta^{+})h_n^{(2)}(k_0r^{+}) - \begin{matrix}\cos\\ -\sin\end{matrix}\, m\varphi^{-}P_n^m(\cos\theta^{-})h_n^{(2)}(k_0r^{-})\right\},$$

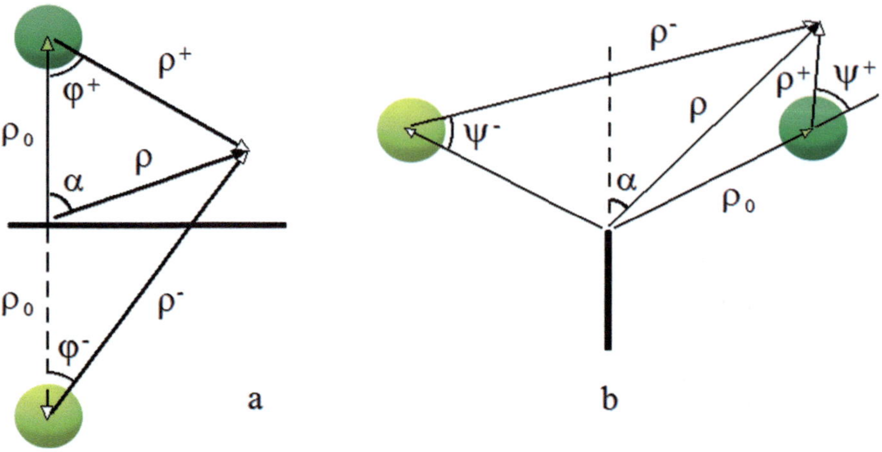

Fig. 5.12 Spherical DR near flat metal surfaces

where $\varphi^{\pm} = \arcsin\left[\rho/\rho^{\pm}\sin(\alpha \mp \alpha_0)\right]$; $r^{\pm} = \left[\left(\rho^{\pm}\right)^2 + (z - z_0)^2\right]^{1/2}$; $\cos\theta^{\pm} = (z - z_0)/r^{\pm}$; $\rho^{\pm} = \left[\rho^2 + \rho_0^2 \mp 2\rho\rho_0\cos(\alpha \mp \alpha_0)\right]^{1/2}$.

It is easy to verify that when $\alpha = \pi/2$ the field satisfies the boundary conditions on the surface of the conductor. In the coordinate system with $\alpha_0 = 0$, angles $\varphi^{\pm}$ are measured from the normal (Fig. 5.12a). By transforming the field components to a spherical coordinate system and passing to the limit ρ, $\rho_0 \to \infty$ so that $r^- = \infty$, and r^+ is finite, we make sure that the resulting expression exactly coincides with the field of the ball in free space.

We use expressions (D.32) to calculate the field of a spherical DR in the case of its placement near a metal half-plane: $\alpha_w = \pi$ (Fig. 5.12b). In this case, the total field is represented as a superposition of half the field of the ball above the plane and the term due to the influence of the edge. For example, for magnetic azimuthally uniform oscillations H_{10l} and $|z - z_0| > r_0$:

$$
\begin{aligned}
e_\rho &= \frac{1}{2} i\omega\mu_0 \frac{j_1(p)}{h_1^{(2)}(q)} \Big[-\sin\left(\alpha - \alpha_0 + \psi^+\right)\sin\tau^+ h_1^{(2)}\left(k_0 r^+\right) \\
&\quad - \sin\left(\alpha + \alpha_0 + \psi^-\right)\sin\tau^- h_1^{(2)}\left(k_0 r^-\right) + \frac{2}{k_0\rho}\frac{\partial}{\partial\alpha}\psi_0(\alpha, \alpha_0; \ldots) \Big]; \\
e_\alpha &= -\frac{1}{2} i\omega\mu_0 \frac{j_1(p)}{h_1^{(2)}(q)} \Big[\cos\left(\alpha - \alpha_0 + \psi^+\right)\sin\tau^+ h_1^{(2)}\left(k_0 r^+\right) \\
&\quad + \cos\left(\alpha + \alpha_0 + \psi^-\right)\sin\tau^- h_1^{(2)}\left(k_0 r^-\right) + 2\frac{\partial}{\partial(k_0\rho)}\psi_0(\alpha, \alpha_0; \ldots) \Big]; \\
e_z &= 0
\end{aligned}
\tag{5.57}
$$

where $r^{\pm} = \left[\left(\rho^{\pm} \right)^2 + |z - z_0|^2 \right]^{1/2}$; $\cos \tau^{\pm} = |z - z_0|/r^{\pm}$;

$$\psi_0(\alpha, \ldots) = \frac{1}{\pi} \int_0^{\infty} \left[S\left(k_0 \rho^{+} \eta, 2k_0 \sqrt{\rho \rho_0} \cos \frac{\alpha - \alpha_0}{2} \eta \right) \right.$$
$$\left. - S\left(k_0 \rho^{-} \eta, 2k_0 \sqrt{\rho \rho_0} \cos \frac{\alpha + \alpha_0}{2} \eta \right) \right] \cdot e^{-ik_0 \gamma |z - z_0|} \eta d\eta / \gamma;$$

$\rho^{\pm} = \left[\rho^2 + \rho_0^2 - 2\rho\rho_0 \cos(\alpha \mp \alpha_0) \right]^{1/2}$; $\psi^{\pm} = \arcsin\left[\rho \sin(\alpha \mp \alpha_0)/\rho^{\pm} \right]$—variables that also have the meaning of coordinates of the center of the DR and its image (Fig. 5.12b), symmetrical relative to the plane $\alpha = 0, \pi$. To interpret the result obtained in the wave zone, you can also use the method of saddle point and ratio (D.37), with the help of which you can isolate the field caused by the diffraction terms (5.57).

References

1. M.E. Ilchenko, Study of an antenna with an open dielectric resonator. Radioelectronics **21**(1), 15–18 (1978)
2. K.M. Luk, K.W. Leung (eds.), *Dielectric Resonator Antennas* (Research Studies Press Ltd, UK, 2003), 388 p
3. A. Petosa, *Dielectric Resonator Antenna Handbook* (Artech House Inc, UK, 2007), 308 p
4. A.A. Trubin, On the influence of metallic inhomogeneity on the resonant oscillations of a dielectric ball. Bulletin of NTUU "KPI" Ser. Radiotechnique **21**, 81–82 (1984) (in Russian)
5. M. Abramowitz, I.A. Stegun (eds.), *Handbook of Mathematical Functions with Formulas, Graphs and Mathematical Tables*, vol. 55 (National Bureau of Standarts, Applied Mathematics Ser, 1964), 830 p
6. J.D. Jackson, *Classical Electrodynamics* (John Willey and Sons, Inc, 1962), 656 p
7. A. Trubin, Study of the radiation characteristics of a disk dielectric resonator. Bulletin of NTUU "KPI" Ser. Radiotechnique **21**, 29–33 (1984) (in Russian)
8. A. Trubin, Directional patterns of axisymmetric oscillations of a disk dielectric resonator. Bulletin of NTUU "KPI" Ser. Radiotechnique **22**, 37–40 (1985) (in Russian)
9. M. Verplanke, J. Van Bladel, The magnetic-dipol resonances of ring resonators of very high permittivity. IEEE Trans. Microwave Theory Techn. **27**(4), 328–333 (1979)
10. A.S. Kupriianov, V.I. Fesenko, A.A. Trubin, V.R. Tuz, Coupling coefficients for dielectric cuboids located in free space. Appl. Opt. OSA **59**(23/10), 6918–6924 (2020)
11. F.W.J. Olver, *Introduction to Asymptotics and Special Functions* (Academic Press, New York and London, 1974), 592 p
12. H. Bateman, A. Erdelyi, *Higher Transcendental Functions*, vol. 2 (McGraw-Hill Book Company, Inc., New York, 1953), p.365
13. G.N. Watson, *A Treatise on the Theory of Bessel Functions* (University Press, Cambridge, 1945), p.804
14. H. Bateman, A. Erdelyi, *Higher Transcendental Functions*, vol. 1 (McGraw-Hill Book Company, Inc., New York, 1953), p.302
15. Y.L. Luke, *Mathematical Functions and their Approximations* (Academic Press Inc., New York, San Francisco, London, 1975), 586 p

Chapter 6
Coupling Coefficients of Dielectric Resonators in Open Space

In this chapter we consider the complex coefficients of DR mutual coupling in open space. The mutual influence of spherical, cylindrical and rectangular DRs on each other is studied. It is shown that in the general case, mutual coupling coefficients can be represented in the form of integral expressions, which can then be calculated numerically, or using various analytical methods. In special cases of spherical DR, the resulting integral expressions can be resolved in quadratures. At the same time, the found formulas serve as the basis for obtaining various approximate expressions for the coupling coefficients of DRs other forms, or in more complex cases of their mutual arrangement. In this way we obtain approximate analytical formulas for rotating cylindrical and rectangular DR.

6.1 Natural Oscillations of Dielectric Resonators in Open Space

The influence of DR on each other can be taken into account using the concept of mutual coupling coefficients. If dielectric resonators are placed in open structures, they can influence each other through reactive fields as well as by emitting propagating waves. In this case, the coefficients of mutual coupling become complex functions of the parameters. In Chapter 3 we have already considered the complex coefficients of mutual coupling of DRs in plane waveguides, found in the form of integral expansions. Let us apply a similar method to obtain the coupling coefficients of DR in open space.

Let there be 2 DRs, which we denote by the indices s, t located in some external structure, in such a way that the field of one resonator can influence the other resonator and vice versa. Let us also assume that the frequency of any natural oscillation of the first resonator (s) coincides with the frequency of natural oscillation of the second resonator (t) and is equal to ω_0. We denote by $(\mathbf{e}_s, \mathbf{h}_s)$—the field of natural oscillations

A. Trubin, *Introduction to the Theory of Dielectric Resonators*, Springer Series in Advanced Microelectronics 65, https://doi.org/10.1007/978-3-031-65396-4_6

in the given external structure of the s–th, and $(\mathbf{e}_t, \mathbf{h}_t)$—the field of the t–th DR. Let us determine the coefficient of mutual coupling between the resonators in the form of a surface integral:

$$\kappa_{st} = \frac{i}{2\omega_0 w_t(1+\delta_{st})} \oint\limits_{S_t} \left\{\left[\mathbf{e}_s, \mathbf{h}_t^*\right] + \left[\mathbf{e}_t^*, \mathbf{h}_s\right]\right\}\mathbf{n}ds \tag{6.1}$$

where; $\mathbf{n}$—normal to surface s_t of the t-th DR, and w_t—energy stored in a dielectric of the t—th resonator; δ_{st}—Kronecker symbol, $i = \sqrt{-1}$; $*$—complex conjugate symbol.

From (6.1) we see that in the general case the mutual coupling coefficient is a complex function determined by the natural oscillation field of each of the resonators, as well as their location in the external containing structure. Usually, direct calculation of the integral using formula (6.1) is impossible, with the exception of a few special cases. However, if we know the expansion of the resonator field in the external structure in terms of any complete system of eigenfunctions, we can transform (6.1) thus to significantly simplify it, or present it in a form convenient for further transformations.

For example, suppose that both resonators are located in some metal waveguide. Then we can expand the field of one of them into the eigenwaves (2.3), and then again use the definition of the coefficients of such expansion (2.4). In this case, it is convenient to represent the mutual coupling coefficient in general form as (see 3.39, 3.40):

$$\kappa_{st} = k_{st} + i\tilde{k}_{st} \tag{6.2}$$

where k_{st}—coupling coefficient expressed in the form of expansion in terms of non-propagating waves, and $\tilde{k}_{st}$—mutual coupling coefficient expressed through the expansion in terms of the set M of all propagating waves of the line.

The function k_{st} is real, but $\tilde{k}_{st}$ complex, therefore (6.2) is not a partition of the function κ_{st} into real and imaginary parts.

It is easy to show that in this case:

$$\tilde{k}_{st} = \frac{1}{\omega_0 w_t(1+\delta_{st})} \cdot \left[\sum_{\nu\in M} (c_\nu^{s\pm})_0 (c_\nu^{t\pm})_0^* e^{-i\Gamma|\Delta z|}\right] \tag{6.3}$$

Here $(c_\nu^{s\pm})_0$—expansion coefficient of the field s-th DR over propagating waves of the line without taking into account the dependence on its longitudinal coordinate of the center z_s (2.7); $\Delta z = z_s - z_t$.

Expansions (6.3) serve as the basis for obtaining analytical expressions for the mutual coupling coefficients of DR in open space by removing the metal conductors of the waveguide to infinity and using transformations of the form (2.62, 2.63). Indeed, the removal of the metal walls of the waveguides leads to the fact that all critical frequencies of natural waves become equal to zero; and in formula (6.2) only the second term remains. Thus, we find, for two DR in open space:

$$\kappa_{st} = \frac{i}{\omega_0 w_t (1 + \delta_{st})} \cdot \lim_{\Sigma \to \infty} \left[\sum_{\nu \in M} (c_\nu^{s\pm})_0 (c_\nu^{t\pm})_0^* e^{-i\Gamma|\Delta z|} \right] \tag{6.4}$$

Here Σ—a set of transverse geometric parameters of the waveguide used in the transition to an open structure. The result of such a transformation is an integral representation, which in some cases can be found in quadratures. As follows from the data in Chapter 2, expansion (6.4) is valid in the region $|\Delta z| > 1/2(d_{sz} + d_{tz})$ (d_z—maximum longitudinal size of the resonator along z—axis).

6.2 Mutual Coupling Coefficients of Spherical Dielectric Resonators in Open Space

To obtain complex coefficients of mutual coupling of the main magnetic types of oscillations H_{1m1} of spherical DR in open space (Fig. 6.1a), we use relations (6.4) and previously found expressions for the expansion coefficients of the field of a spherical DR in the propagating waves of a rectangular waveguide (2.11, 2.12). Assuming $m_1, m_2 = 0$, or 1, in the resulting expressions we move to the limit: $a, b \to \infty$ for $\Delta z > 2r_0$:

$$\kappa_{xx} = i\frac{2 \cdot 3}{\pi}\alpha_1^H(p,q) \cdot \int_0^\infty \int_0^\infty e^{-i\gamma k_0|\Delta z|} \cos(k_0 \Delta x \xi) \cos(k_0 \Delta y \eta)(1 - \xi^2)\frac{d\xi d\eta}{\gamma};$$

$$\kappa_{yy} = i\frac{2 \cdot 3}{\pi}\alpha_1^H(p,q) \cdot \int_0^\infty \int_0^\infty e^{-i\gamma k_0|\Delta z|} \cos(k_0 \Delta x \xi) \cos(k_0 \Delta y \eta)(1 - \eta^2)\frac{d\xi d\eta}{\gamma};$$

$$\kappa_{zz} = i\frac{2 \cdot 3}{\pi}\alpha_1^H(p,q) \cdot \int_0^\infty \int_0^\infty e^{-i\gamma k_0|\Delta z|} \cos(k_0 \Delta x \xi) \cos(k_0 \Delta y \eta)(\xi^2 + \eta^2)\frac{d\xi d\eta}{\gamma};$$

$$\kappa_{xy} = i\frac{2 \cdot 3}{\pi}\alpha_1^H(p,q) \cdot \int_0^\infty \int_0^\infty e^{-i\gamma k_0|\Delta z|} \sin(k_0 \Delta x \xi) \sin(k_0 \Delta y \eta) \cdot \xi\eta \cdot \frac{d\xi d\eta}{\gamma};$$

$$\tag{6.5}$$

$$\kappa_{xz} = -\frac{2 \cdot 3}{\pi}\alpha_1^H(p,q) \cdot \int_0^\infty \int_0^\infty e^{-i\gamma k_0|\Delta z|} \sin(k_0 \Delta x \xi) \cos(k_0 \Delta y \eta) \cdot \xi \cdot d\xi d\eta;$$

$$\kappa_{yz} = -\frac{2 \cdot 3}{\pi}\alpha_1^H(p,q) \cdot \int_0^\infty \int_0^\infty e^{-i\gamma k_0|\Delta z|} \sin(k_0 \Delta x \xi) \cos(k_0 \Delta y \eta) \cdot \eta \cdot d\xi d\eta.$$

Here $\xi^2 + \eta^2 + \gamma^2 = 1$; $\Delta x = x_1 - x_2$; $\Delta y = y_1 - y_2$; $\Delta z = z_1 - z_2$; $\alpha_1^H(p,q)$—is the function (2.20).

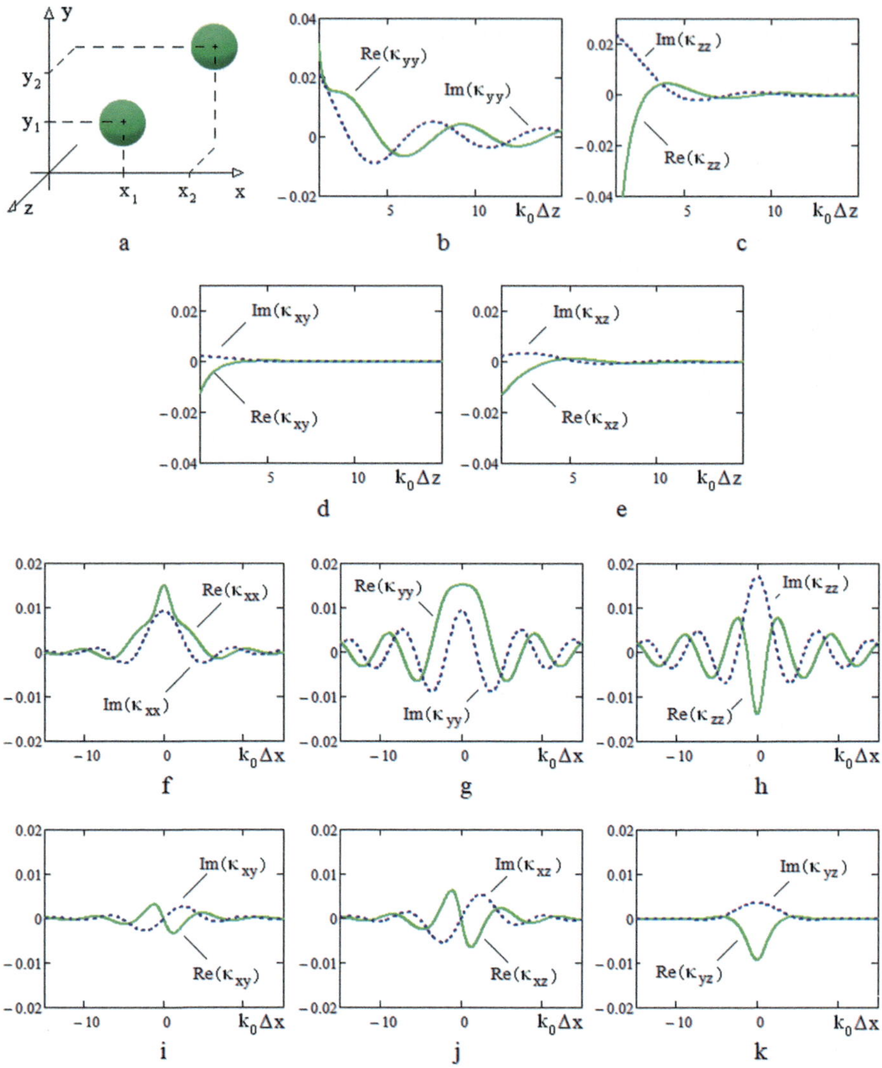

Fig. 6.1 Two spherical DR in a given rectangular coordinate system (**a**). Dependences of the coupling coefficients of the magnetic oscillations H_{1m1} of a spherical DRs on the mutual distance between there centers for $\varepsilon_{1r} = 36$; $k_0\Delta x = k_0\Delta y = 0$ (**b**, **c**); $k_0\Delta x = k_0\Delta y = 1$ (**d**, **e**); $k_0\Delta y = 0$; $k_0\Delta z = 2$ (**f**–**h**); $k_0\Delta y = 1$; $k_0\Delta z = 2$ (**i**–**k**)

For H_{1m1} we denoted κ_{uv} by u, v—directions of the magnetic field in the DR centers ($u, v = x, y, z$) of the rectangular coordinate system (Fig. 6.1a).

From (6.5), using integral relations (D.1–D.8), as well as (D.16), we have obtained simple formulas for the most general cases of excitation of magnetic oscillations H_{1m1}, which we will give a more symmetrical form:

for $\Delta z > 2r_0$:

$$\kappa_{xx} = i\alpha_1^H(p,q)\cdot\left\{2h_0^{(2)}(k_0\Delta r) + \left[\frac{2\Delta x^2 - \Delta y^2 - \Delta z^2}{\Delta r^2}\right]h_2^{(2)}(k_0\Delta r)\right\};$$

$$\kappa_{yy} = i\alpha_1^H(p,q)\cdot\left\{2h_0^{(2)}(k_0\Delta r) + \left[\frac{2\Delta y^2 - \Delta x^2 - \Delta z^2}{\Delta r^2}\right]h_2^{(2)}(k_0\Delta r)\right\};$$

$$\kappa_{zz} = i\alpha_1^H(p,q)\cdot\left\{2h_0^{(2)}(k_0\Delta r) + \left[\frac{2\Delta z^2 - \Delta x^2 - \Delta y^2}{\Delta r^2}\right]h_2^{(2)}(k_0\Delta r)\right\}; \quad (6.6)$$

$$\kappa_{xy} = 3i\alpha_1^H(p,q)\cdot\frac{\Delta x\Delta y}{\Delta r^2}h_2^{(2)}(k_0\Delta r);$$

$$\kappa_{xz} = 3i\alpha_1^H(p,q)\cdot\frac{\Delta x\Delta z}{\Delta r^2}h_2^{(2)}(k_0\Delta r);$$

$$\kappa_{yz} = 3i\alpha_1^H(p,q)\cdot\frac{\Delta y\Delta z}{\Delta r^2}h_2^{(2)}(k_0\Delta r).$$

where $\Delta r = \sqrt{\Delta x^2 + \Delta y^2 + \Delta z^2}$—distance between resonator centers.

As follows from (6.6), the formulas actually coincide with each other with an appropriate permutation of the coordinates. For example:

$$\kappa_{yy}(\Delta x, \Delta y, \Delta z) = \kappa_{xx}(\Delta y, \Delta x, \Delta z);$$

$$\kappa_{zz}(\Delta x, \Delta y, \Delta z) = \kappa_{xx}(\Delta z, \Delta y, \Delta x);$$

$$\kappa_{xz}(\Delta x, \Delta y, \Delta z) = \kappa_{xy}(\Delta x, \Delta z, \Delta y);$$

$$\kappa_{yz}(\Delta x, \Delta y, \Delta z) = \kappa_{xy}(\Delta y, \Delta z, \Delta x).$$

This becomes obvious if we take into account the maximum spatial symmetry of the shape of both resonators.

On Fig. 6.1b–k shows the dependences of the complex coefficients of mutual coupling for identical spherical DR in open space on the distance between their centers for $\varepsilon_{1r} = 36$, calculated using formulas (6.6).

Let us note here that the mutual coupling coefficients are even functions of the transverse coordinates of the DR for the main magnetic oscillations of the same "polarization" of the magnetic field at their centers. The values of the real and imaginary parts of the functions are comparable, but in open space they decrease quite quickly as the resonators move away from each other.

6.3 Mutual Coupling Coefficients of Cylindrical Dielectric Resonators in Open Space

We again use relations (6.4) and the expansion coefficients of the field of a cylindrical DR over the propagating waves of a rectangular waveguide. In the resulting expressions we also go to the limit $a, b \to \infty$. As a result, we obtain for the main magnetic oscillations H_{10l} in each resonator:

In the case of orientation of the DR along the coordinate x axis (Fig. 6.2a), for $\Delta z > 2r_0$:

$$\kappa_{xx} = 2i\kappa_0 \int_0^{\infty} \int_0^{\infty} \cos[\xi k_0 \Delta x] \cdot \cos[\eta k_0 \Delta y] \cdot \upsilon(k_0\gamma, k_0\eta, k_0\xi)^2 \cdot (1 - \xi^2) \cdot \frac{e^{-i\gamma k_0 |\Delta z|}}{\gamma} d\xi d\eta \tag{6.7}$$

In the case of orientation of 1 DR along the x axis, and 2 DR along the y axis of the coordinate system (Fig. 6.2b), for $|\Delta z| > 2r_0$:

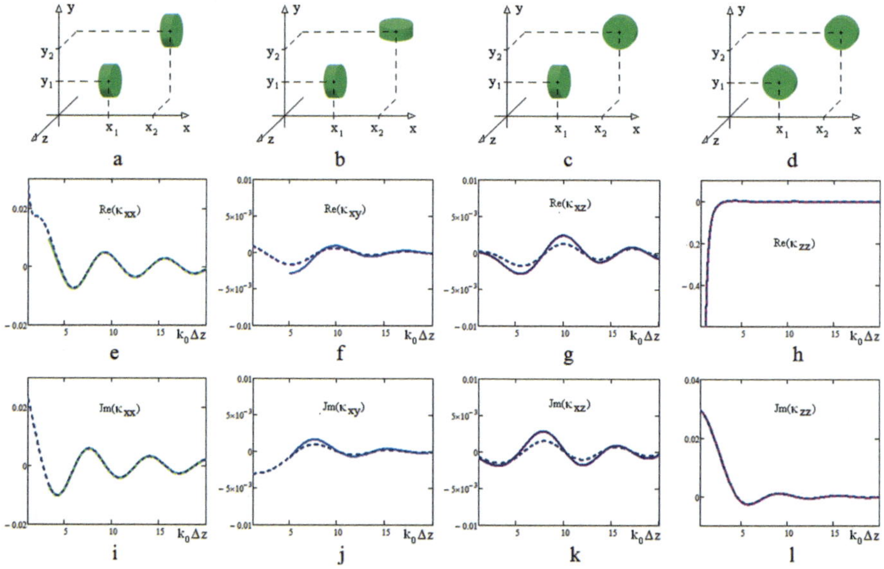

Fig. 6.2 Dependences of the real and imaginary parts of the mutual coupling coefficients for standard positions of cylindrical DR ($\varepsilon_{1r} = 36$; $\Delta = 0, 4$) in a rectangular coordinate: **a**—κ_{xx}, e, i ($k_0\Delta x = k_0\Delta y = 0$); **b**—$\kappa_{xy}(k_0\Delta x = k_0\Delta y = 5)$; **c**—$\kappa_{xz}(k_0\Delta x = k_0\Delta y = 5)$; **d**—$\kappa_{zz}(k_0\Delta x = k_0\Delta y = 0)$. Solid curves are the result of numerical calculation of integral expressions (6.7–6.10); dotted curves—calculations taking into account approximation (6.12)

$$\kappa_{xy} = 2i\kappa_0 \int_0^\infty \int_0^\infty \sin[\xi k_0 \Delta x] \cdot \sin[\eta k_0 \Delta y] \cdot \upsilon(k_0\gamma, k_0\eta, k_0\xi) \cdot \upsilon(k_0\gamma, k_0\xi, k_0\eta) \cdot \xi\eta \cdot \frac{e^{-i\gamma k_0|\Delta z|}}{\gamma} d\xi d\eta; \quad (6.8)$$

In the case of orientation of 1 DR along the x axis, and 2 DR along the z axis в случае (Fig. 6.2c), for $|\Delta z| > r_0 + L/2$:

$$\kappa_{xz} = -2\kappa_0 \int_0^\infty \int_0^\infty \sin[\xi k_0 \Delta x] \cdot \cos[\eta k_0 \Delta y] \cdot \upsilon(k_0\gamma, k_0\eta, k_0\xi) \cdot \upsilon(k_0\xi, k_0\eta, k_0\gamma) \cdot e^{-i\gamma k_0|\Delta z|} \cdot \xi \cdot d\xi d\eta; \quad (6.9)$$

In the case of orientation of the DR along the axis z (Fig. 6.2d), for $|\Delta z| > L$:

$$\kappa_{zz} = 2i\kappa_0 \int_0^\infty \int_0^\infty \cos[\xi k_0 \Delta x] \cdot \cos[\eta k_0 \Delta y] \cdot \upsilon(k_0\xi, k_0\eta, k_0\gamma)^2 \cdot \left(\xi^2 + \eta^2\right) \cdot \frac{e^{-i\gamma k_0|\Delta z|}}{\gamma} d\xi d\eta. \quad (6.10)$$

Here $\xi^2 + \eta^2 + \gamma^2 = 1$; $\Delta x = x_1 - x_2$; $\Delta y = y_1 - y_2$; $\Delta z = z_1 - z_2$;

$$\kappa_0 = 8 \cdot \frac{(\varepsilon_{1r} - 1)^2}{\varepsilon_{1r}} \frac{q_\perp^2 q_z p_z}{\pi v_0} \cdot k_0^2. \quad (6.11)$$

For H_{101} oscillations $v_0 = \left[J_1^2(p_\perp) - J_0(p_\perp)J_2(p_\perp)\right](2p_z + \sin 2p_z)$; $\upsilon(s_x, s_y, s_z)$—is the real function (2.73): $\upsilon(s_x, s_y, s_z) = \upsilon(s_x, s_y, s_z)^*$; $\varepsilon_{1r} = \varepsilon_1/\varepsilon_0$; $p_\perp = \beta r_0$; $p_z = \beta_z L/2$ and $q_\perp = k_0 r_0$; $q_z = k_0 L/2$—is the characteristic parameters; r_0—radius; L—height of the resonators.

The resulting relations (6.7–6.10) are complex functions of the parameters. Their convergence in the region $\xi^2 + \eta^2 > 1$ is ensured by the choice of the sign of the function γ in the exponential: $\gamma = -i\sqrt{\xi^2 + \eta^2 - 1}$.

A good approximation when calculating integrals (6.7–6.10) is the condition: $s_x^2 + s_y^2 \approx 1$ for functions:

$$\upsilon(k_0 s_x, k_0 s_y, k_0 s_z) \approx \upsilon_1 = \frac{\left[p_\perp J_1(q_\perp) J_0(q_\perp) - q_\perp J_0(q_\perp) J_1(q_\perp)\right]}{q_\perp^2 - p_\perp^2} \cdot \frac{\omega_z(0)}{k_0}. \quad (6.12)$$

In this case, expressions (6.7–6.10) coincide with (6.6) up to amplitude, in which the factors $3 \cdot \alpha_n^H(p, q)$ must be replaced by:

$$\kappa_1 = \pi \cdot \kappa_0 \cdot \upsilon_1^2 \quad (6.13)$$

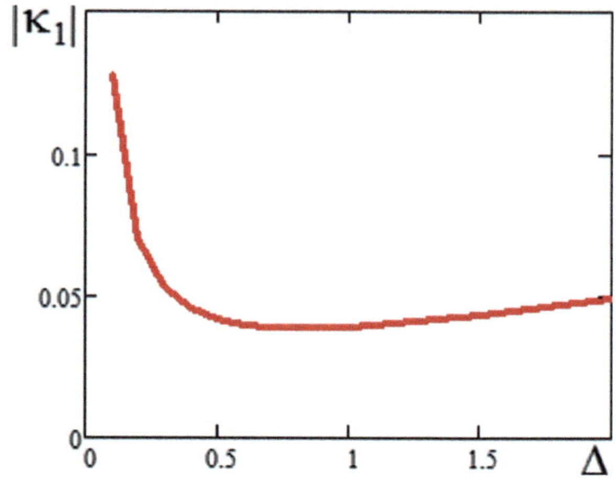

Fig. 6.3 Dependence of the multiplier module on the DR compression ratio: $\Delta = L/2r_0$ for $\varepsilon_{1r} = 36$

In Fig. 6.2 shows the results of calculating coupling coefficients when varying the distance between the centers of the DR. Solid curves depict the dependences obtained using the numerical calculation of integral expressions (6.7–6.10); dotted curves—the result of calculations taking into account the approximation (6.12). It can be seen that the approximate formulas coincide well with the numerical solutions for the integrals for the same orientation of the resonator axes.

The dependence of the coupling coefficient multipliers on the relative sizes of the resonators is shown in Fig. 6.3. Taking into account approximation (6.6), such a dependence corresponds to a relative change in the coupling with variation in the shape of the DR.

Similarly, it is possible to calculate the mutual coupling coefficients of DRs of various shapes, for example, spherical and cylindrical, as well as DRs with different types of natural oscillations.

6.4 Mutual Coupling Coefficients of Rectangular Dielectric Resonators in Open Space

Let us now consider the calculation of the mutual coupling coefficients of identical rectangular DRs with H_{nml} magnetic types of oscillations in open space (1.32). To do this, we again use the previously found expressions for the expansion coefficients of the resonator field over the propagating waves of a rectangular waveguide (2.35–2.38).

Using (6.4) and also (2.39), we find expressions for the mutual coupling coefficients of two identical rectangular DRs in open space:

for the disposition shown in Fig. 6.4a; for $\Delta z > a_0$:

$$\kappa_{xx} = i\kappa_0 \cdot \int_0^\infty \int_0^\infty e^{-ik_0|\Delta z|\gamma} \Big\{ \beta_x^2 \xi^2 |\overline{\omega}_x(\gamma)|^2 |\omega_y(\eta)|^2$$
$$+ \big|\beta_y(\xi^2 + \eta^2)\omega_x(\gamma)\overline{\omega}_y(\eta) + \beta_x \gamma \eta \overline{\omega}_x(\gamma)\omega_y(\eta)\big|^2 \Big\}$$
$$\cdot |\omega_z(\xi)|^2 \cos(k_0 \Delta x \xi) \cos(k_0 \Delta y \eta) \frac{d\xi d\eta}{\gamma(\xi^2 + \eta^2)} \tag{6.14}$$

for the disposition shown in Fig. 6.4b; for $\Delta z > a_0$:

$$\kappa_{xy} = i\kappa_0 \cdot \int_0^\infty \int_0^\infty e^{-ik_0|\Delta z|\gamma} \{ \beta_x^2 \xi \eta |\overline{\omega}_x(\gamma)|^2 \omega_y(\xi)\omega_y(\eta)$$
$$+ \left[\beta_y(\xi^2 + \eta^2)\omega_x(\gamma)\overline{\omega}_y(\eta) + \beta_x \gamma \eta \overline{\omega}_x(\gamma)\omega_y(\eta)\right]$$
$$\cdot \left[\beta_y(\xi^2 + \eta^2)\omega_x(\gamma)\overline{\omega}_y(\xi) + \beta_x \gamma \xi \overline{\omega}_x(\gamma)\omega_y(\xi)\right]^*$$
$$\cdot \omega_z(\xi)\omega_z(\eta)^* \sin(k_0 \Delta x \xi) \sin(k_0 \Delta y \eta) \frac{d\xi d\eta}{\gamma(\xi^2 + \eta^2)}; \tag{6.15}$$

for the disposition shown in Fig. 6.4c; for $\Delta z > a_0/2 + L/2$:

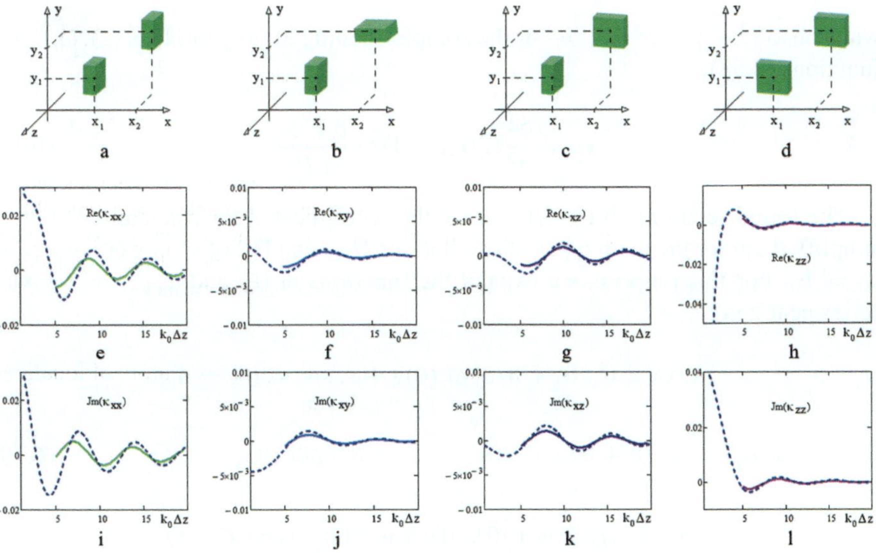

Fig. 6.4 Dependences of mutual coupling coefficients for standard positions of rectangular DRs ($\varepsilon_{1r} = 36$; $a_0 = b_0$; $\Delta = 0, 4$): **a**—κ_{xx}, e, i ($k_0\Delta x = k_0\Delta y = 0$); **b**—$\kappa_{xy}$ ($k_0\Delta x = k_0\Delta y = 5$); **c**—$\kappa_{xz}$ ($k_0\Delta x = k_0\Delta y = 5$); **d**—$\kappa_{zz}$($k_0\Delta x = k_0\Delta y = 0$). Solid curves are the result of numerical calculations of integral expressions (6.14–6.17); dotted curves—calculations using formulas (6.6) taking into account the formal replacement (6.18)

$$\kappa_{xz} = -\kappa_0 \cdot \int_0^\infty \int_0^\infty e^{-ik_0|\Delta z|\gamma} \omega_z(\xi)\omega_z(\gamma)^*$$
$$\cdot \{\beta_x \xi \overline{\omega}_x(\gamma)\omega_y(\eta) \cdot [\beta_y \eta \omega_x(\xi)\overline{\omega}_y(\eta) + \beta_x \xi \overline{\omega}_x(\xi)\omega_y(\eta)]^*$$
$$- \gamma^*[\beta_y(\xi^2 + \eta^2)\omega_x(\gamma)\overline{\omega}_y(\eta)$$
$$-\beta_x \gamma \eta \overline{\omega}_x(\gamma)\omega_y(\eta)][\beta_y \xi \omega_x(\xi)\overline{\omega}_y(\eta) - \beta_x \eta \overline{\omega}_x(\xi)\omega_y(\eta)]^*\}$$
$$\cdot \sin(k_0 \Delta x \xi)\cos(k_0 \Delta y \eta)\frac{d\xi d\eta}{\gamma(\xi^2 + \eta^2)}; \tag{6.16}$$

for the disposition shown in Fig. 6.4d; for $\Delta z > L$:

$$\kappa_{zz} = i\kappa_0 \cdot \int_0^\infty \int_0^\infty e^{-ik_0|\Delta z|\gamma} |\omega_z(\gamma)|^2$$
$$\cdot \Big\{(1/\gamma) \cdot \big|[\beta_y \eta \omega_x(\xi)\overline{\omega}_y(\eta) + \beta_x \xi \overline{\omega}_x(\xi)\omega_y(\eta)]\big|^2$$
$$+\gamma \cdot \big|[\beta_y \xi \omega_x(\xi)\overline{\omega}_y(\eta) - \beta_x \eta \overline{\omega}_x(\xi)\omega_y(\eta)]\big|^2\Big\}$$
$$\cdot \cos(k_0 \Delta x \xi)\cos(k_0 \Delta y \eta)\frac{d\xi d\eta}{(\xi^2 + \eta^2)}; \tag{6.17}$$

where also $\xi^2 + \eta^2 + \gamma^2 = 1$; $*$—is the complex conjugation symbol; $\omega_\nu(\xi)$, $\overline{\omega}_\nu(\xi)$—functions (2.38);

$$\kappa_0 = \frac{64}{\pi^2} k_0^5 (\varepsilon_{1r} - 1)^2 \cdot \frac{\beta_x \beta_y \beta_z}{\upsilon H} \tag{6.18}$$

The found relations (6.14–6.17) are rather complicated for calculations. Let us simplify them for the main types of oscillations H_{111} and DR of square cross section $a_0 = b_0$. For this purpose, we expand the functions $\omega_\nu(\xi)$ and $\overline{\omega}_\nu(\xi)$ in a Taylor series near zero:

$$\omega_\nu(\xi) = \omega_\nu(0) + \partial/\partial\xi\,\omega_\nu(0)\xi + \ldots \approx \omega_\nu(0) = A_\nu;$$

$$\overline{\omega}_\nu(\xi) = \overline{\omega}_\nu(0) + \partial/\partial\xi\,\overline{\omega}_\nu(0)\xi + \ldots \approx \partial/\partial\xi\,\overline{\omega}_\nu(0)\xi = C_\nu \xi; \tag{6.19}$$

$$\omega_\nu(\gamma(\xi, \eta)) \approx \omega_\nu(\gamma(0, 0)) = \omega_\nu(1); \quad (\nu = x, y, z).$$

and limit ourselves to the first terms of such an expansion.

After integration (6.14–6.17), using Somerfield's integral (D.16) we obtain approximate expressions coinciding with (6.6) up to a constant factor. Thus, we come to the conclusion that for the main magnetic oscillations relations (6.14–6.17)

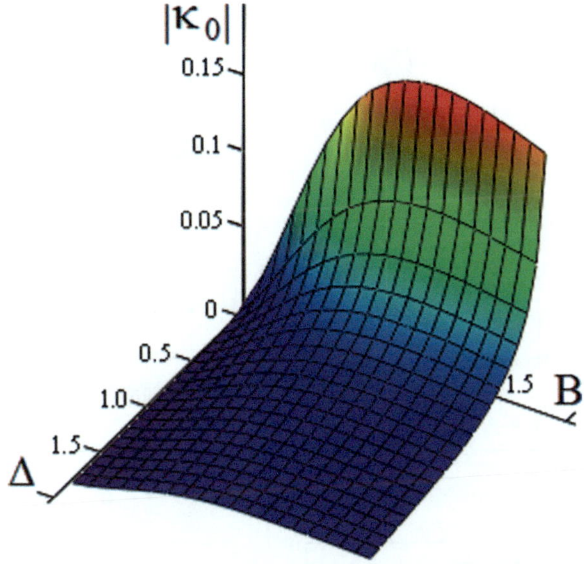

Fig. 6.5 Dependence of the coupling multiplier on the shape $\Delta = a_0/L, B = b_0/a_0$ of the rectangular DRs for $\varepsilon_{1r} = 36$

are similar functions of coordinates, as for spherical DR (6.6). In this case, it is also sufficient for us to replace the constant factors $3i \cdot \alpha_n^H(p, q)$ by κ_0 (6.4.5).

On Fig. 6.4e—l shows the results coupling coefficients calculating for DR of square cross section $a_0 = b_0$, obtained numerically from (6.14–6.17) and on the approximation (6.6). It can be seen that the approximate expressions correspond with good accuracy to numerical calculations at distances $k_0|\Delta z| > 5$.

The dependence of the coupling multipliers on the shape of the resonators is shown in Fig. 6.5. As follows from (6.18), the value $|\kappa_0|$ increases when the resonators are flattened $L << a_0, b_0$, and decreases with a decrease in the ratio b_0/a_0.

6.5 Coupled Oscillations of Rotating Dielectric Resonators in Open Space

As we saw in Sects. 2.6 and 3.6, rotation of the resonators in the waveguide makes it possible to change the mutual coupling coefficients within a wide range without changing their spatial coordinates. Let us find analytical expressions for the coupling coefficients of rotating DR in open space. We will study the simplest cases of rotation of the resonators relative to one of the fixed axes of the rectangular coordinate system (see Fig. 6.6; 6.10). In each concrete case we will specify the directions of the initial positions of the resonator axes.

Let us first consider the rotation of a cylindrical DR.

Let's start by considering rotations about the x axis. We will assume that the initial orientation of the resonator axes z' is parallel to the axis z (Fig. 6.6a). Without loss of

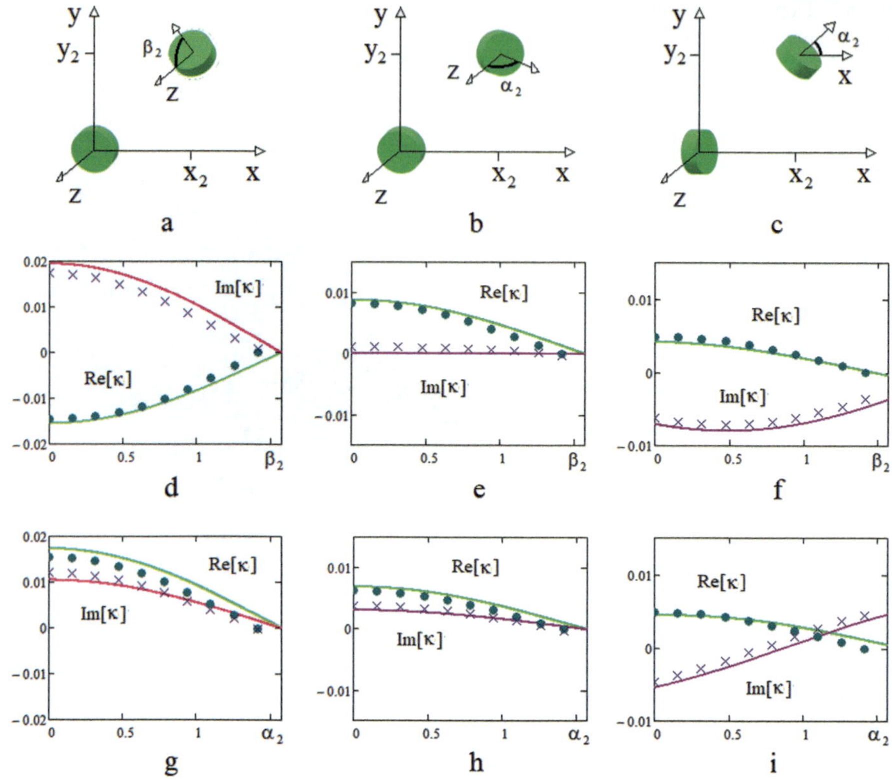

Fig. 6.6 Comparison of calculated coupling dependencies on the relative rotation of the resonator axes for $(x_1, y_1) = (0, 0)$ mm and $\Delta z = 12$ mm: relative to the axis x: (**d–f**); axis z: (**g–i**). $(x_2, y_2) = (0, 0)$ mm (**d, g**); $(x_2, y_2) = (15, 0)$ mm (**e, h**); $(x_2, y_2) = (15, 15)$ mm (**f, i**). Dots and crosses show the calculated values in the package COMSOL Multiphysics electromagnetic solver [1]

generality, we can also assume that the coordinates of the center of the first resonator are equal $(x_1, y_1, z_1) = (0, 0, 0)$, and the coordinates of the second resonator are arbitrary and equal: (x_2, y_2, z_2).

To obtain the necessary analytical formulas, we again use the relations for expanding the field of a rotating DR in a rectangular waveguide (Sect. 3.6) and transform them by directing the cross-sectional dimensions to infinity using (6.4).

In the case of rotation of resonator 2 relative to the x axis (Fig. 6.6a) by an angle β_2:

$$\kappa_{12} = i\kappa_0 \int_0^\infty \int_0^\infty \cos[\xi k_0 \Delta x] \cdot \upsilon(k_0\xi, k_0\eta, k_0\gamma) \cdot \frac{e^{-i\gamma k_0|\Delta z|}}{\gamma}$$
$$\cdot \{[(\xi^2 + \eta^2)\cos\beta_2 + \gamma\eta\sin\beta_2]e^{-i\eta k_0 \Delta y}\upsilon[k_0\xi, k_0(\eta\cos\beta_2$$
$$+\gamma\sin\beta_2), k_0(\eta\sin\beta_2 - \gamma\cos\beta_2)]$$
$$+ [(\xi^2 + \eta^2)\cos\beta_2 - \gamma\eta\sin\beta_2]e^{i\eta k_0 \Delta y}\upsilon[k_0\xi, k_0(\eta\cos\beta_2$$
$$(-\gamma\sin\beta_2), k_0(\eta\sin\beta_2 + \gamma\cos\beta_2)]\} \cdot d\xi d\eta; \tag{6.20}$$

Here $\xi^2 + \eta^2 + \gamma^2 = 1$; and also $\Delta x = x_1 - x_2$; $\Delta y = y_1 - y_2$; $\Delta z = z_1 - z_2$; κ_0 coincides with (6.11).

For H_{101}^+ oscillations (1.17), $v_0 = [J_1^2(p_\perp) - J_0(p_\perp)J_2(p_\perp)](2p_z + \sin 2p_z)$; $\upsilon = \upsilon(k_0 s_x, k_0 s_y, k_0 s_z)$ is the function (2.73); $\varepsilon_{1r} = \varepsilon_1/\varepsilon_0$; $p_\perp = \beta r_0$; $p_z = \beta_z L/2$ and $q_\perp = k_0 r_0$; $q_z = k_0 L/2$—is the characteristic parameters; r_0—radius; L—height of the resonator (Fig. 1.6a); $J_n(z)$—is the Bessel functions.

Using approximation (6.12), as well as Sommerfeld integrals (D.16), we find a simple approximate representation of the integral (6.20):

$$\kappa_{12} \approx \kappa_{ZZ}\cos\beta_2 + \kappa_{YZ}\sin\beta_2 \tag{6.21}$$

where

$$\kappa_{ZZ} = i\kappa_1 \left\{ \frac{2}{3} h_0^{(2)}(k_0\Delta r) + \frac{1}{3}\left[\frac{2\Delta z^2 - \Delta x^2 - \Delta y^2}{\Delta r^2}\right] h_2^{(2)}(k_0\Delta r) \right\} \tag{6.22}$$

$$\kappa_{YZ} = i\kappa_1 \frac{\Delta y \Delta z}{\Delta r^2} h_2^{(2)}(k_0\Delta r) \tag{6.23}$$

κ_1 defined from (6.13); $\Delta r = \sqrt{\Delta x^2 + \Delta y^2 + \Delta z^2}$; $h_n^{(2)}(z)$—is the Hankel spherical functions of the second kind [2].

In the case of rotation of second resonator relative to the y axis (Fig. 6.6b) by an angle α_2 we obtain:

$$\kappa_{12} = i\kappa_0 \int_0^\infty \int_0^\infty \cos[\eta k_0 \Delta y] \cdot \upsilon(k_0\xi, k_0\eta, k_0\gamma) \cdot \frac{e^{-i\gamma k_0|\Delta z|}}{\gamma}$$
$$\cdot \{[(\xi^2 + \eta^2)\cos\alpha_2 + \gamma\xi\sin\alpha_2]e^{-i\xi k_0 \Delta x}\upsilon[k_0(\xi\cos\alpha_2$$
$$+\gamma\sin\alpha_2), k_0\eta, k_0(\xi\sin\alpha_2 - \gamma\cos\alpha_2)]$$
$$+ [(\xi^2 + \eta^2)\cos\alpha_2 - \gamma\xi\sin\alpha_2]e^{i\xi k_0 \Delta x}\upsilon[k_0(\xi\cos\alpha_2$$
$$-\gamma\sin\alpha_2), k_0\eta, k_0(\xi\sin\alpha_2 + \gamma\cos\alpha_2)]\} \cdot d\xi d\eta; \tag{6.24}$$

Using approximation (6.12) again, we find:

$$\kappa_{12} \approx \kappa_{ZZ} \cos\alpha_2 + \kappa_{XZ} \sin\alpha_2 \tag{6.25}$$

Here

$$\kappa_{XZ} = i\kappa_1 \frac{\Delta x \Delta z}{\Delta r^2} h_2^{(2)}(k_0 \Delta r) \tag{6.26}$$

Let us consider an important type of rotation of resonators relative to the axis z. In this case, the initial position of the axes of both resonators will be considered oriented parallel to the axis x (Fig. 6.6c). In the case of rotation of the second resonator relative to the axis z (Fig. 6.6c) by an angle α_2:

$$\begin{aligned}\kappa_{12} = i\kappa_0 \int_0^\infty \int_0^\infty & \upsilon(k_0\gamma, k_0\eta, k_0\xi) \cdot \frac{e^{-i\gamma k_0|\Delta z|}}{\gamma} \\ & \cdot \{[(1-\xi^2)\cos\alpha_2 + \xi\eta\sin\alpha_2]\cos[k_0(\xi\Delta x + \eta\Delta y)] \\ & \cdot \upsilon[k_0\gamma, k_0(\xi\sin\alpha_2 + \eta\cos\alpha_2), k_0(\xi\cos\alpha_2 - \eta\sin\alpha_2)] \\ & + [(1-\xi^2)\cos\alpha_2 - \xi\eta\sin\alpha_2]\cos[k_0(\xi\Delta x - \eta\Delta y)] \\ & \cdot \upsilon[k_0\gamma, k_0(\xi\sin\alpha_2 - \eta\cos\alpha_2), k_0(\xi\cos\alpha_2 + \eta\sin\alpha_2)]\} \\ & \cdot d\xi d\eta;\end{aligned} \tag{6.27}$$

Using (6.12) we again obtain from (6.27):

$$\kappa_{12} \approx \kappa_{XX} \cos\alpha_2 + \kappa_{XY} \sin\alpha_2 \tag{6.28}$$

where

$$\kappa_{XX} = i\kappa_1 \left\{ \frac{2}{3} h_0^{(2)}(k_0\Delta r) + \frac{1}{3}\left[\frac{2\Delta x^2 - \Delta y^2 - \Delta z^2}{\Delta r^2}\right] h_2^{(2)}(k_0\Delta r) \right\} \tag{6.29}$$

$$\kappa_{XY} = i\kappa_1 \frac{\Delta x \Delta y}{\Delta r^2} h_2^{(2)}(k_0 \Delta r) \tag{6.30}$$

The results of calculating the dependences of the coupling coefficients on the coordinates and the angle of rotation of the resonator are shown in Figs. 6.7, 6.8, and 6.9 for H_{101} oscillations and for $\varepsilon_{1r} = 36$; $L/2r_0 = 0,4$; $f_0 = 8$ GHz. On Fig. 6.7 solid curves show dependencies calculated using formulas (6.20, 6.24, 6.27). The dotted curves show the dependences obtained using approximate formulas (6.21, 6.25, 6.28). As can be seen from the data presented, the approximate relationships are in good agreement with the integral representation of the coupling coefficients in the region of resonator coordinates that interests us.

On Figs. 6.8 and 6.9 show the dependences of the coupling coefficients with variations in the transverse coordinates of the second resonator, calculated using formulas (6.21, 6.25, 6.28) for different axis orientations. The reason for the asymmetry of the dependence $\kappa(\Delta x)$ in the case of y—rotation (Fig. 6.8b, e) is due to the different

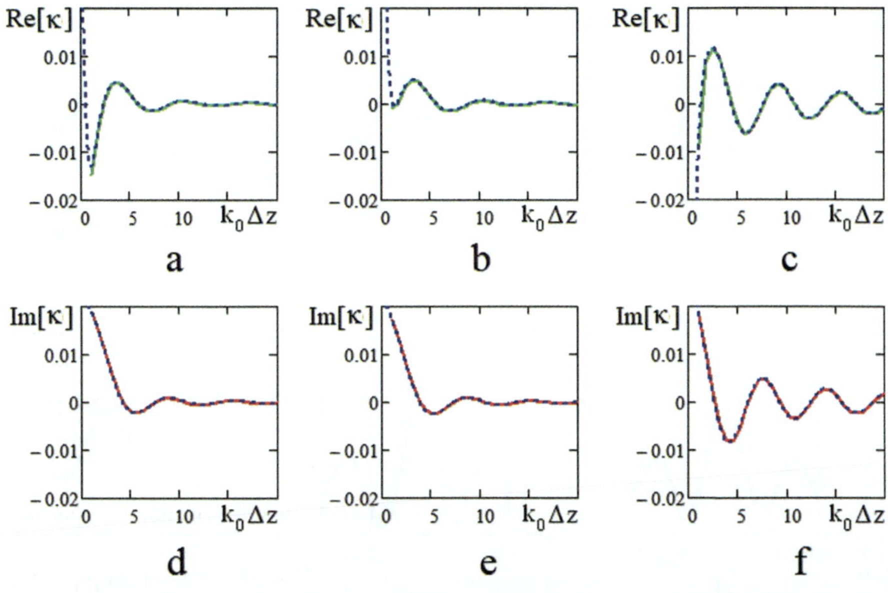

Fig. 6.7 Dependence of rotated DRs about the x-axis (Fig. 6.6a) on the distance $k_0 \Delta z$ (**a**, **d**). Dependence of the DR coupling coefficients rotated relatively the y-axis (Fig. 6.6b) on the distance (**b**, **e**). Dependence of rotated DRs about the z-axis (Fig. 6.6c) on the distance (**c**, **f**) ($k_0 \Delta x = 1$; $k_0 \Delta y = 0$; $\alpha_1, \beta_1 = 0$; $\alpha_2, \beta_2 = 0, 5$)

initial spatial distribution of the vector fields of natural oscillations relative to the plane of symmetry of the first resonator $x = 0$.

On Fig. 6.6, shows the result of comparing the calculation of coupling coefficients using formulas (6.21, 6.25, 6.28)—solid curves; and received with the help COMSOL Multiphysics electromagnetic solver [1]—dots and crosses. The data presented demonstrate the good accuracy of the found approximate expressions.

In a similar way, you can calculate the coupling coefficients in the case of various types of rotations (Fig. 6.10) of rectangular resonators, for example with magnetic types of oscillations H_{111} [3].

In the case of rotation of the resonator axes relative to the x—axis of the selected rectangular coordinate system, we will assume that the initial orientation of the axes of both resonators z' is parallel to the z—axis (Fig. 6.10a) and the direction of the x' and y' axes coincides with the direction of the x and y axes, respectively. Let us also assume that the coordinates of the center of the first resonator are equal (x_1, y_1, z_1), and the coordinates of the second resonator are equal (x_2, y_2, z_2) and the main type of magnetic oscillations is excited in the resonators H_{111}.

Using general expressions for the mutual coupling coefficients of rotating DR in a rectangular waveguide, for the field of which is described by expression (1.32), and also transformations (6.4), we find:

In the case of rotation of 1 resonator relative to the x-axis (Fig. 6.10a) by an angle β_1 and the second resonator relative to the x-axis by an angle β_2, the mutual coupling

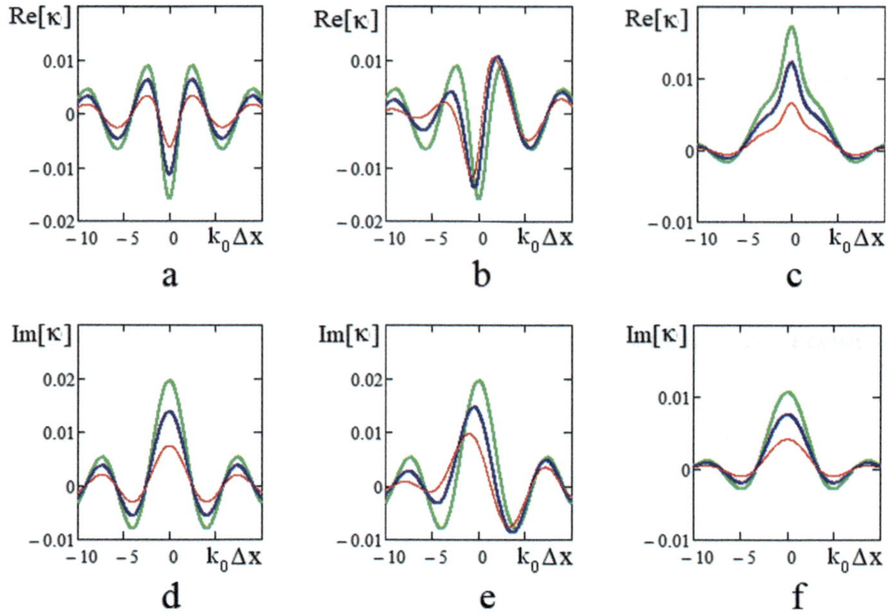

Fig. 6.8 Dependence of rotated DRs about the x-axis (Fig. 6.6a) on the distance $k_0\Delta x$ (**a**, **d**). Dependence of the DR coupling coefficients rotated relatively the y-axis (Fig. 6.6b) on the distance $k_0\Delta x$ (**b**, **e**). Dependence of rotated DRs about the z-axis (Fig. 6.6c) on the distance $k_0\Delta x$ (**c**, **f**) ($k_0\Delta y = 0$; $k_0\Delta z = 2$; $\alpha_1, \beta_1 = 0$; $\alpha_2, \beta_2 = 0$—green curves; $\alpha_2, \beta_2 = \pi/4$—blue curves; $\alpha_2, \beta_2 = 3\pi/8$—red curves)

coefficient takes the form:

$$\begin{aligned}
\kappa_{12} = i\kappa_0 \int_0^\infty \int_0^\infty \cos[\xi k_0 \Delta x] \cdot \frac{e^{-i\gamma k_0|\Delta z|}}{\gamma(\xi^2+\eta^2)} \cdot \\
\cdot \Bigg\{ \prod_{s=1}^{2} \big[\eta\beta_y \omega_x(\xi)\overline{\omega}_y(\eta\cos\beta_s + \gamma\sin\beta_s) \\
+ \xi\beta_x \cos\beta_s \overline{\omega}_x(\xi)\omega_y(\eta\cos\beta_s \\
+\gamma\sin\beta_s)\big]\omega_z(\eta\sin\beta_s - \gamma\cos\beta_s)e^{-i\eta k_0\Delta y} \\
+ \prod_{s=1}^{2} \big[\eta\beta_y \omega_x(\xi)\overline{\omega}_y(\eta\cos\beta_s - \gamma\sin\beta_s) \\
+ \xi\beta_x \cos\beta_s \overline{\omega}_x(\xi)\omega_y(\eta\cos\beta_s \\
-\gamma\sin\beta_s)\big]\omega_z(\eta\cos\beta_s + \gamma\sin\beta_s)e^{i\eta k_0\Delta y} \\
+ \prod_{s=1}^{2} \big[\xi\gamma\beta_y \omega_x(\xi)\overline{\omega}_y(\eta\cos\beta_s + \gamma\sin\beta_s)
\end{aligned}$$

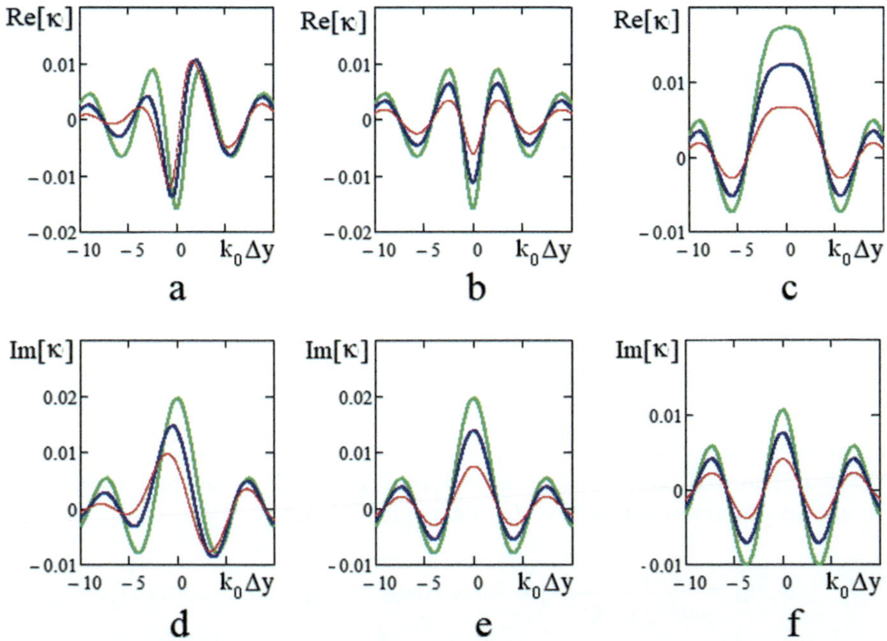

Fig. 6.9 Dependence of rotated DRs about the x-axis (Fig. 6.6a) on the distance $k_0\Delta y$ (**a, d**). Dependence of the DR coupling coefficients rotated relatively the y-axis (Fig. 6.6b) on the distance $k_0\Delta y$ (**b, e**). Dependence of rotated DRs about the z-axis (Fig. 6.6c) on the distance $k_0\Delta y$ (**c, f**) ($\alpha_2, \beta_2 = 0$—green curves; $\alpha_2, \beta_2 = \pi/4$—blue curves; $\alpha_2, \beta_2 = 3\pi/8$—red curves. $k_0\Delta z = 2$; $\alpha_1, \beta_1 = 0$)

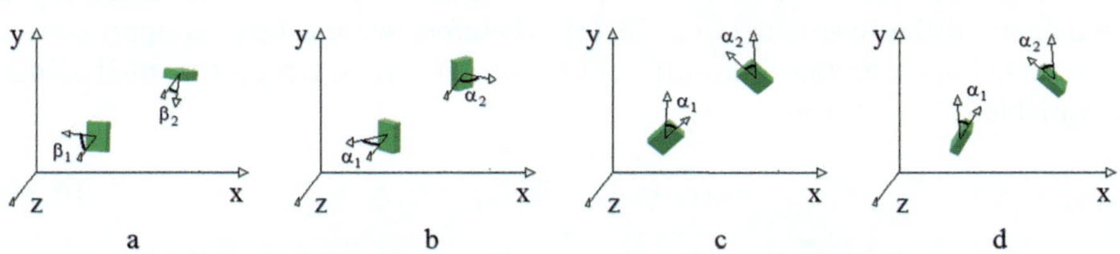

Fig. 6.10 Rectangular resonators rotated about the axis x (**a**); y (**b**); z (**c, d**)

$$
\begin{aligned}
&- \eta\gamma\beta_x \cos\beta_s \overline{\omega}_x(\xi)\omega_y(\eta\cos\beta_s + \gamma\sin\beta_s) \\
&+ \left(\xi^2 + \eta^2\right)\beta_x \sin\beta_s \overline{\omega}_x(\xi)\omega_y(\eta\cos\beta_s \\
&+ \gamma\sin\beta_s)\big]\omega_z(\eta\cos\beta_s - \gamma\sin\beta_s)e^{-i\eta k_0\Delta y} \\
&+ \prod_{s=1}^{2}\big[\xi\gamma\beta_y\omega_x(\xi)\overline{\omega}_y(\eta\cos\beta_s - \gamma\sin\beta_s) \\
&- \eta\gamma\beta_x \cos\beta_s \overline{\omega}_x(\xi)\omega_y(\eta\cos\beta_s - \gamma\sin\beta_s) \\
&+ \left(\xi^2 + \eta^2\right)\beta_x \sin\beta_s \overline{\omega}_x(\xi)\omega_y(\eta\cos\beta_s
\end{aligned}
$$

$$-\gamma \sin \beta_s)\big]\omega_z(\eta \cos \beta_s + \gamma \sin \beta_s)e^{i\eta k_0 \Delta y}\big\}d\xi d\eta; \tag{6.31}$$

where $\xi^2 + \eta^2 + \gamma^2 = 1$;

$$\kappa_0 = \frac{32k_0^2}{\pi^2}(\varepsilon_{1r} - 1)^2\frac{p_x p_y p_z q_x q_y q_z}{\upsilon H}; \tag{6.32}$$

$$\begin{aligned}\upsilon H &= \varepsilon_{1r}k_0^2\Big[\beta_y^2\pi_x^+\pi_y^-\pi_z^+ + \beta_x^2\pi_x^-\pi_y^+\pi_z^+\Big]\\ &+ \beta_x^2\beta_z^2\cdot\pi_x^-\pi_y^+\pi_z^- + \beta_y^2\beta_z^2\cdot\pi_x^+\pi_y^-\pi_z^-\\ &+ \left(\beta_x^2 + \beta_y^2\right)^2\cdot\pi_x^+\pi_y^+\pi_z^+;\end{aligned}$$

$\pi_\nu^\pm = p_\nu \pm \sin p_\nu \cos p_\nu$; $\nu = (x, y, z)$; $\varepsilon_{1r} = \varepsilon_1/\varepsilon_0$; $p_x = \beta_x a_0/2$; $p_y = \beta_y b_0/2$; $p_z = \beta_z L/2$, as well as $q_x = k_0 a_0/2$; $q_y = k_0 b_0/2$; $q_z = k_0 L/2$—are the characteristic parameters of the resonators. For H_{111} oscillations (1.32)

$$\omega_\nu(\xi) = \frac{1}{p_\nu^2 - (q_\nu\xi)^2}\cdot\big[p_\nu \sin p_\nu \cos(q_\nu\xi) - q_\nu\xi \cos p_\nu \sin(q_\nu\xi)\big];$$

$$\overline{\omega}_\nu(\xi) = \frac{i}{p_\nu^2 - (q_\nu\xi)^2}\cdot\big[p_\nu \cos p_\nu \sin(q_\nu\xi) - q_\nu\xi \sin p_\nu \cos(q_\nu\xi\big]. \tag{6.33}$$

Integral (6.31) is not very convenient for calculations, it is characterized by poor convergence, therefore, to calculate it, we will use the approximation.

The essential region of integration (6.31) is in the region of small values of the arguments of the functions $\omega_\nu(z)$, $\overline{\omega}_\nu(z)$. Therefore we use here an approximate representation of the functions $\omega_\nu(z)$, $\overline{\omega}_\nu(z)$, $\nu = (x, y, z)$, typical for small values of variables (ξ, η) (6.4.6):

$$\omega_\nu(z) \approx A_\nu, \quad \overline{\omega}_\nu(z) \approx C_\nu z \tag{6.34}$$

where A_ν, C_ν are constants. Applying (6.34) and Sommerfeld's integrals (D.16), we obtain a simpler formula, expressed in terms of the angles of rotation of the resonators:

$$\kappa_{12} \approx \kappa_{ZZ}\cos\beta_1\cos\beta_2 + \kappa_{YY}\sin\beta_1\sin\beta_2 + \kappa_{YZ}\sin(\beta_1 + \beta_2) \tag{6.35}$$

where

$$\kappa_{ZZ} = i\kappa_1\left\{\frac{\Delta x^2 + \Delta y^2}{\Delta r^2}h_0^{(2)}(k_0\Delta r) - \left[1 - 3\frac{\Delta z^2}{\Delta r^2}\right]\frac{h_1^{(2)}(k_0\Delta r)}{k_0\Delta r}\right\} \tag{6.36}$$

$$\kappa_{YZ} = i\kappa_1\frac{\Delta y\Delta z}{\Delta r^2}h_2^{(2)}(k_0\Delta r) \tag{6.37}$$

(x_s, y_s, z_s) is the coordinates of the s-th resonator in the rectangular coordinate system $(s = 1, 2)$; $\Delta x = x_1 - x_2$; $\Delta y = y_1 - y_2$; $\Delta z = z_1 - z_2$; $\Delta r = \sqrt{\Delta x^2 + \Delta y^2 + \Delta z^2}$; $h_n^{(2)}(z)$—is the Hankel spherical functions of the second kind [2].

Under the conditions: $A_x = A_y$ and $\beta_x C_x = \beta_y C_y$, which occur, for example, for resonators with a square cross-section ($a_0 = b_0$), the coefficient $\kappa_1 = \frac{\pi}{2}\kappa_0 \cdot |A_x\beta_x C_x A_z|^2$ gives unsatisfactory results. The easiest way to choose a numerical value of κ_1 is based on the mean theorem, directly for the given parameters of the resonator from (see 1.3) for a particular orientation, for example, for $\beta_1 = \beta_2 = 0$ [4] or use (6.18).

In the case of rotation of each of the resonators, respectively, at an angle α_1, α_2 (Fig. 6.10b) relative to the y-axis, we obtain similarly:

$$\begin{aligned}
\kappa_{12} = i\kappa_0 \int_0^\infty \int_0^\infty \cos[\eta k_0 \Delta y] \cdot \frac{e^{-i\gamma k_0|\Delta z|}}{\gamma(\xi^2+\eta^2)} \\
\cdot \Bigg\{ \prod_{s=1}^{2} \big[\eta\beta_y \cos\alpha_s \overline{\omega}_y(\eta)\omega_x(\xi\cos\alpha_s + \gamma\sin\alpha_s) \\
+ \xi\beta_x\omega_y(\eta)\overline{\omega}_x(\xi\cos\alpha_s \\
+\gamma\sin\alpha_s)\big]\omega_z(\xi\cos\alpha_s - \gamma\sin\alpha_s)e^{-i\xi k_0\Delta x} \\
+ \prod_{s=1}^{2} \big[\eta\beta_y \cos\alpha_s \overline{\omega}_y(\eta)\omega_x(\xi\cos\alpha_s - \gamma\sin\alpha_s) \\
+ \xi\beta_x\omega_y(\eta)\overline{\omega}_x(\xi\cos\alpha_s \\
-\gamma\sin\alpha_s)\big]\omega_z(\xi\sin\alpha_s + \gamma\cos\alpha_s)e^{i\xi k_0\Delta x} \\
+ \prod_{s=1}^{2} \big[\xi\gamma\beta_y \cos\alpha_s \overline{\omega}_y(\eta)\omega_x(\xi\cos\alpha_s + \gamma\sin\alpha_s) \\
- \eta\gamma\beta_x\omega_y(\eta)\overline{\omega}_x(\xi\cos\alpha_s + \gamma\sin\alpha_s) \\
- (\xi^2+\eta^2)\beta_y\sin\alpha_s\overline{\omega}_y(\eta)\omega_x(\xi\cos\alpha_s \\
+\gamma\sin\alpha_s)\big]\omega_z(\xi\sin\alpha_s - \gamma\cos\alpha_s)e^{-i\xi k_0\Delta x} \\
+ \prod_{s=1}^{2} \big[\xi\gamma\beta_y \cos\alpha_s \overline{\omega}_y(\eta)\omega_x(\xi\cos\alpha_s - \gamma\sin\alpha_s) \\
- \eta\gamma\beta_x\omega_y(\eta)\overline{\omega}_x(\xi\cos\alpha_s - \gamma\sin\alpha_s) \\
+ (\xi^2+\eta^2)\beta_y\sin\alpha_s\overline{\omega}_y(\eta)\omega_x(\xi\cos\alpha_s \\
-\gamma\sin\alpha_s)\big]\omega_z(\xi\sin\alpha_s + \gamma\cos\alpha_s)e^{i\xi k_0\Delta x}\Bigg\} d\xi d\eta;
\end{aligned} \tag{6.38}$$

Using again approximation (6.34), we find:

$$\kappa_{12} \approx \kappa_{ZZ}\cos\alpha_1\cos\alpha_2 + \kappa_{XX}\sin\alpha_1\sin\alpha_2 + \kappa_{XZ}\sin(\alpha_1+\alpha_2) \tag{6.39}$$

Here

$$\kappa_{XX} = i\kappa_1 \left\{ \frac{2}{3} h_0^{(2)}(k_0 \Delta r) + \frac{1}{2} \left[\frac{1}{3} - \frac{\Delta y^2 - \Delta x^2 + \Delta z^2}{\Delta r^2} \right] h_2^{(2)}(k_0 \Delta r) \right\}; \tag{6.40}$$

$$\kappa_{XZ} = i\kappa_1 \frac{\Delta x \Delta z}{\Delta r^2} h_2^{(2)}(k_0 \Delta r)$$

In the case of rotation of 1 and 2 resonators relative to z—axis (Fig. 6.10c), respectively, at an angle α_1, α_2, relative to z—axis, the mutual coupling coefficient takes the form:

$$\begin{aligned}
\kappa_{12} = i\kappa_0 \int_0^\infty \int_0^\infty & \frac{e^{-i\gamma k_0 |\Delta z|}}{\gamma(\xi^2 + \eta^2)} \\
& \cdot \left\{ \prod_{s=1}^{2} \left[\beta_y(\eta \cos\alpha_s + \xi \sin\alpha_s)\omega_x(\xi \cos\alpha_s \right.\right. \\
& -\eta \sin\alpha_s)\overline{\omega}_y(\xi \sin\alpha_s + \eta \cos\alpha_s) \\
& - \beta_x(\eta \sin\alpha_s - \xi \cos\alpha_s)\overline{\omega}_x(\xi \cos\alpha_s \\
& \left. -\eta \sin\alpha_s)\omega_y(\xi \sin\alpha_s + \eta \cos\alpha_s) \right] \cos(\xi k_0 \Delta x + \eta k_0 \Delta y) \\
& + \prod_{s=1}^{2} \left[\beta_y(\eta \cos\alpha_s - \xi \sin\alpha_s)\omega_x(\xi \cos\alpha_s \right. \\
& +\eta \sin\alpha_s)\overline{\omega}_y(\xi \sin\alpha_s - \eta \cos\alpha_s) \\
& - \beta_x(\eta \sin\alpha_s + \xi \cos\alpha_s)\overline{\omega}_x(\xi \cos\alpha_s \\
& \left. +\eta \sin\alpha_s)\omega_y(\xi \sin\alpha_s - \eta \cos\alpha_s) \right] \cos(\xi k_0 \Delta x - \eta k_0 \Delta y) \\
& + \prod_{s=1}^{2} \left[\beta_y \gamma(\xi \cos\alpha_s - \eta \sin\alpha_s)\omega_x(\xi \cos\alpha_s \right. \\
& -\eta \sin\alpha_s)\overline{\omega}_y(\xi \sin\alpha_s + \eta \cos\alpha_s) \\
& - \beta_x \gamma(\xi \sin\alpha_s + \eta \cos\alpha_s)\overline{\omega}_x(\xi \cos\alpha_s \\
& \left. -\eta \sin\alpha_s)\overline{\omega}_y(\xi \sin\alpha_s + \eta \cos\alpha_s) \right] \cos(\xi k_0 \Delta x + \eta k_0 \Delta y) \\
& + \prod_{s=1}^{2} \left[\beta_y \gamma(\xi \cos\alpha_s + \eta \sin\alpha_s)\omega_x(\xi \cos\alpha_s \right. \\
& +\eta \sin\alpha_s)\overline{\omega}_y(\xi \sin\alpha_s - \eta \cos\alpha_s) \\
& - \beta_x \gamma(\xi \sin\alpha_s - \eta \cos\alpha_s)\overline{\omega}_x(\xi \cos\alpha_s \\
& \left.\left. +\eta \sin\alpha_s)\overline{\omega}_y(\xi \sin\alpha_s - \eta \cos\alpha_s) \right] \cos(\xi k_0 \Delta x - \eta k_0 \Delta y) \right\} \\
& \cdot |\omega_z(\gamma)|^2 \cdot d\xi d\eta.
\end{aligned} \tag{6.41}$$

Using approximation (6.34), we obtain from (6.41):

$$\kappa_{12} \approx \kappa_{ZZ} \tag{6.42}$$

Let us consider another important type of resonator rotation about the z axis, in which the initial position of the axes of both resonators is oriented parallel to the x axis (Fig. 6.10d). In this case

$$\begin{aligned}
\kappa_{12} = i\kappa_0 \int_0^\infty \int_0^\infty & \frac{e^{-i\gamma k_0|\Delta z|}}{\gamma(\xi^2+\eta^2)} \cdot \\
& \cdot \left\{ \prod_{s=1}^{2} [\beta_x(\eta \sin\alpha_s - \xi\cos\alpha_s)\overline{\omega}_x(\gamma)\omega_y(\xi\sin\alpha_s \right. \\
& +\eta\cos\alpha_s)\omega_z(\xi\cos\alpha_s - \eta\sin\alpha_s)] \cos(\xi k_0\Delta x + \eta k_0 \Delta y) \\
& + \prod_{s=1}^{2} [\beta_x(\eta\sin\alpha_s + \xi\cos\alpha_s)\overline{\omega}_x(\gamma)\omega_y(\xi\sin\alpha_s \\
& -\eta\cos\alpha_s)\omega_z(\xi\cos\alpha_s + \eta\sin\alpha_s)] \cos(\xi k_0 \Delta x - \eta k_0 \Delta y) \\
& + \prod_{s=1}^{2} [\beta_x\gamma(\xi\sin\alpha_s + \eta\cos\alpha_s)\overline{\omega}_x(\gamma)\omega_y(\xi\sin\alpha_s \\
& +\eta\cos\alpha_s)\beta_y(\xi^2+\eta^2)\omega_x(\gamma)\overline{\omega}_y(\xi\sin\alpha_s + \eta\cos\alpha_s)] \\
& \cdot \omega_z(\xi\cos\alpha_s - \eta\sin\alpha_s)\cos(\xi k_0\Delta x + \eta k_0\Delta y) \\
& + \prod_{s=1}^{2} [\beta_x\gamma(\xi\sin\alpha_s - \eta\cos\alpha_s)\overline{\omega}_x(\gamma)\omega_y(\xi\sin\alpha_s - \eta\cos\alpha_s) \\
& +\beta_y(\xi^2+\eta^2)\omega_x(\gamma)\overline{\omega}_y(\xi\sin\alpha_s - \eta\cos\alpha_s)] \\
& \left. \cdot\omega_z(\xi\cos\alpha_s + \eta\sin\alpha_s)\cos(\xi k_0\Delta x - \eta k_0\Delta y) \right\} d\xi d\eta.
\end{aligned} \tag{6.43}$$

Taking into account (6.34), expression (6.43) takes the form:

$$\kappa_{12} \approx \kappa_{XX}\cos\alpha_1\cos\alpha_2 + \kappa_{YY}\sin\alpha_1\sin\alpha_2 + \kappa_{XY}\sin(\alpha_1+\alpha_2) \tag{6.44}$$

where

$$\kappa_{YY} = i\kappa_1 \left\{ \frac{2}{3} h_0^{(2)}(k_0\Delta r) + \frac{1}{2}\left[\frac{1}{3} - \frac{\Delta x^2 - \Delta y^2 + \Delta z^2}{\Delta r^2}\right] h_2^{(2)}(k_0\Delta r) \right\}; \tag{6.45}$$

$$\kappa_{XY} = i\kappa_1 \frac{\Delta x \Delta y}{\Delta r^2} h_2^{(2)}(k_0\Delta r)$$

Integral relations as well as approximate formulas make it possible to calculate the mutual coupling coefficients of the rectangular DRs in the cases of the considered rotations of their axes. Integral relations (6.31, 6.38, 6.41, 6.43) are valid under

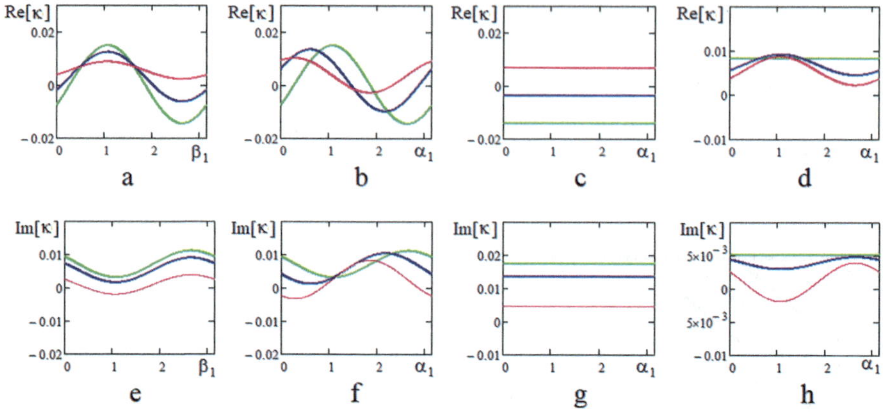

Fig. 6.11 Angular dependence of mutual coupling coefficients during pseudo-rotation ($\beta_2 = \beta_1+1$; $\alpha_2 = \alpha_1 + 1$) resonators relative to the axis x (**a**, **c**); y (**b**, **f**); z (Fig. 6.10c, g); z (Fig. 6.10d, h) for $(k_0\Delta x, k_0\Delta y, k_0\Delta z) = (0, 0, 2)$—green curves; $(k_0\Delta x, k_0\Delta y, k_0\Delta z) = (1, 0, 2)$—blue curves; $(k_0\Delta x, k_0\Delta y, k_0\Delta z) = (2, 0, 2)$—red curves

the condition $|\Delta z| > d_{\max}$, where $d_{\max}$ is the maximum size of the DR is in the direction of the straight line connecting their centers (see 2.1). In contrast to integral representations, the found approximate formulas (6.35, 6.39, 6.42, 6.44) are finite in the entire spatial region of resonator interaction. In the special case $\alpha_1 = 0$ ($\beta_1 = 0$) formulas (6.31, 6.38, 6.44) go into (6.21, 6.25, 6.28). It can be shown that more general expressions (6.31, 6.38, 6.44) are also valid for rotating cylindrical DRs with the main magnetic types of oscillations.

In particular cases of parallelism of the resonator axes of one coordinate axes, the found analytical expressions coincide with those obtained earlier [17].

Figure 6.11 shows the dependences of the coupling coefficients in the case of pseudo-rotation of resonators: $\alpha_2, \beta_2 = \alpha_1, \beta_1 + a$, where $a = const$, according to formulas (6.31, 6.38, 6.41, 6.43) for $\varepsilon_{1r} = 36; a_0 = b_0; L/2a_0 = 0,4; f_0 = 8$ GHz. By pseudo-rotation of resonators we mean rotation in which the relative direction of the axes of each resonator remains fixed. A change in the mutual coupling coefficients during pseudo rotation is associated with a change in the relative coordinates of the DRs. In the latter case (Fig. 6.11d, h), if the centers of the resonators are located on the axis of rotation, then the coefficients of mutual coupling remain constant.

6.6 Coupling Coefficients of Higher Type Oscillations of Spherical Dielectric Resonators in Open Space

In Sects. 6.2–6.5 we considered the coupling coefficients between DR with the main types of magnetic natural oscillations. Let us show that the method proposed in Sect. 6.1 is also suitable for obtaining analytical expressions for the mutual coupling

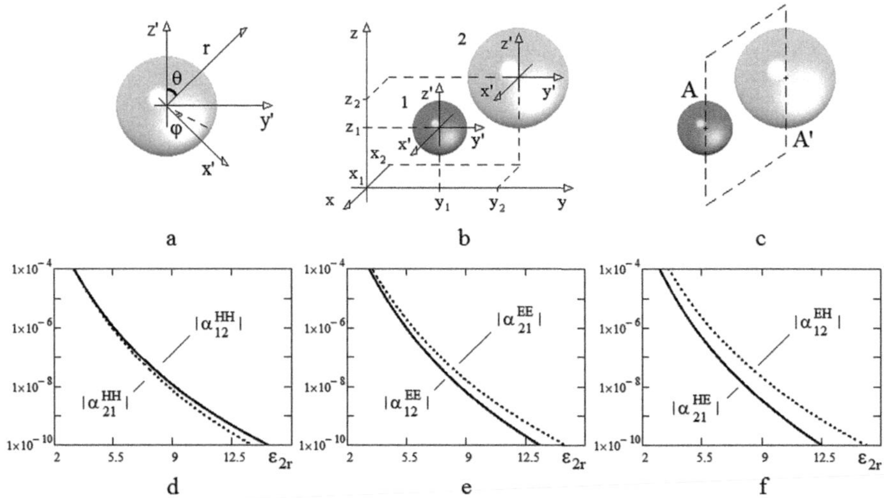

Fig. 6.12 **a**—the spherical resonator in the open space. Two spherical DRs in the rectangular coordinate system (**b**). Plane of symmetry AA' of two different spherical resonators (**c**). Dependence of module of coupling coefficient multipliers $|\alpha|$ as a function of the second resonator dielectric permittivity (**d–f**) ($\varepsilon_{1r} = 2; n_1 = 16; n_2 = 22$)

coefficients of the DR with oscillations of higher types. Let us consider in more detail the coefficients of mutual coupling of various spherical DR with higher types of oscillations.

In local spherical coordinate system (Fig. 6.12a), associated with resonator center, the field of the H_{nml} (or E_{nml}) mode describes by radial magnetic-field (electric-field) component (1.2) (or (1.3)). When considering DR, we will assume that the direction of the axes (x_s', y_s', z_s') of their local coordinates coincides with the direction of the axes of the external coordinate system (Fig. 6.12b) (x, y, z). We will classify the parity of the field of natural oscillations according to the angular coordinate φ_s' (1.1.1) of each of the resonators relative to the plane $x_s' = 0$ ($s = 1, 2$).

The coupling coefficients we obtain from already known expressions (6.4); (2.11–2.14). recorded for different spherical resonators. As a result of integration, next relationships has been obtained: with magnetic mode $H_{n_1m_1l_1}$ in the first resonator, the magnetic mode in the second $H_{n_2m_2l_2}$ and for the same parity with respect to m_1, m_2 and for the same parity to φ:

$$\begin{aligned}\kappa_{12} = {} & \kappa_0 \alpha^{HH}_{n_1 n_2}(p_1, q_1; p_2, q_2) \\ & \cdot \int_0^{\infty} e^{-i\gamma k_0 |\Delta z|} \Big\{ \big(1-\gamma^2\big)^2 \frac{dP^{m_1}_{n_1}(\gamma)}{d\gamma} \cdot \frac{dP^{m_2}_{n_2}(\gamma)}{d\gamma} \\ & \cdot \big[\cos(m_1 - m_2)\Delta\alpha \cdot J_{m_1-m_2}(k_0 \Delta\rho\kappa) \\ & \mp (-1)^{m_1 m_2} \cos(m_1 + m_2)\Delta\alpha \cdot J_{m_1+m_2}(k_0 \Delta\rho\kappa)\big] \\ & + m_1 m_2 P^{m_1}_{n_1}(\gamma) P^{m_2}_{n_2}(\gamma) \cdot \big[\cos(m_1 - m_2)\Delta\alpha \cdot J_{m_1-m_2}(k_0 \Delta\rho\kappa) \\ & \pm (-1)^{m_1 m_2} \cos(m_1 + m_2)\Delta\alpha \cdot J_{m_1+m_2}(k_0 \Delta\rho\kappa)\big]\Big\} \frac{d\kappa}{\gamma\kappa};\end{aligned} \tag{6.46}$$

(In the case of the same parity electric oscillations $E_{n_1 m_1 l_1}$, $E_{n_2 m_2 l_2}$, the function $\alpha^{HH}_{n_1 n_2}$ should be replaced by $\alpha^{EE}_{n_1 n_2}$).

Here $\Delta x = x_1 - x_2$; $\Delta y = y_1 - y_2$; $\Delta z = (z_1 - z_2)$; $\Delta\rho = \sqrt{\Delta x^2 + \Delta y^2}$; $\sin(\Delta\alpha) = \Delta y / \Delta\rho$; (x_s, y_s, z_s)—are the rectangular coordinates of the center of the s—th resonator ($s = 1, 2$); $\kappa^2 + \gamma^2 = 1$ and also for the same parity with respect to m_1, m_2 and for the different parity to φ:

$$\begin{aligned}\kappa_{12} = {} & \kappa_0 \alpha^{HH}_{n_1 n_2}(p_1, q_1; p_2, q_2) \\ & \cdot \int_0^{\infty} e^{-i\gamma k_0 |\Delta z|} \Big\{ \big(1-\gamma^2\big)^2 \frac{dP^{m_1}_{n_1}(\gamma)}{d\gamma} \cdot \frac{dP^{m_2}_{n_2}(\gamma)}{d\gamma} \\ & \cdot \big[\pm \sin(m_1 - m_2)\Delta\alpha \cdot J_{m_1-m_2}(k_0 \Delta\rho\kappa) \\ & + (-1)^{m_1 m_2} \sin(m_1 + m_2)\Delta\alpha \cdot J_{m_1+m_2}(k_0 \Delta\rho\kappa)\big] \\ & + m_1 m_2 P^{m_1}_{n_1}(\gamma) P^{m_2}_{n_2}(\gamma) \\ & \cdot \big[\pm \sin(m_1 - m_2)\Delta\alpha \cdot J_{m_1-m_2}(k_0 \Delta\rho\kappa) \\ & - (-1)^{m_1 m_2} \sin(m_1 + m_2)\Delta\alpha \cdot J_{m_1+m_2}(k_0 \Delta\rho\kappa)\big]\Big\} \frac{d\kappa}{\gamma\kappa};\end{aligned} \tag{6.47}$$

for the different parity for m_1, m_2 and the same parity with respect to φ:

$$\begin{aligned}\kappa_{12} = &-\kappa_0 \alpha_{n_1 n_2}^{HH}(p_1, q_1; p_2, q_2) \\ &\cdot \int_0^{\infty} e^{-i\gamma k_0 |\Delta z|} \Big\{ \left(1-\gamma^2\right)^2 \frac{dP_{n_1}^{m_1}(\gamma)}{d\gamma} \cdot \frac{dP_{n_2}^{m_2}(\gamma)}{d\gamma} \\ &\cdot \big[\cos(m_1 - m_2)\Delta\alpha \cdot J_{m_1-m_2}(k_0 \Delta\rho\kappa) \\ &\mp (-1)^{m_2} \cos(m_1 + m_2)\Delta\alpha \cdot J_{m_1+m_2}(k_0 \Delta\rho\kappa)\big] \\ &+ m_1 m_2 P_{n_1}^{m_1}(\gamma) P_{n_2}^{m_2}(\gamma) \\ &\cdot \big[\cos(m_1 - m_2)\Delta\alpha \cdot J_{m_1-m_2}(k_0 \Delta\rho\kappa) \\ &\pm (-1)^{m_2} \cos(m_1 + m_2)\Delta\alpha \cdot J_{m_1+m_2}(k_0 \Delta\rho\kappa)\big]\Big\} \frac{d\kappa}{\gamma\kappa}; \end{aligned} \tag{6.48}$$

for the different parity for m_1, m_2 and the different parity with respect to φ:

$$\begin{aligned}\kappa_{12} = &\kappa_0 \alpha_{n_1 n_2}^{HH}(p_1, q_1; p_2, q_2) \\ &\cdot \int_0^{\infty} e^{-i\gamma k_0 |\Delta z|} \Big\{ \left(1-\gamma^2\right)^2 \frac{dP_{n_1}^{m_1}(\gamma)}{d\gamma} \cdot \frac{dP_{n_2}^{m_2}(\gamma)}{d\gamma} \\ &\cdot \big[\mp \sin(m_1 - m_2)\Delta\alpha \cdot J_{m_1-m_2}(k_0 \Delta\rho\kappa) \\ &- (-1)^{m_2} \sin(m_1 + m_2)\Delta\alpha \cdot J_{m_1+m_2}(k_0 \Delta\rho\kappa)\big] \\ &+ m_1 m_2 P_{n_1}^{m_1}(\gamma) P_{n_2}^{m_2}(\gamma) \\ &\cdot \big[\mp \sin(m_1 - m_2)\Delta\alpha \cdot J_{m_1-m_2}(k_0 \Delta\rho\kappa) \\ &+ (-1)^{m_2} \sin(m_1 + m_2)\Delta\alpha \cdot J_{m_1+m_2}(k_0 \Delta\rho\kappa)\big]\Big\} \frac{d\kappa}{\gamma\kappa}. \end{aligned} \tag{6.49}$$

For modes $H_{n_1 m_1 l_1}$, $E_{n_2 m_2 l_2}$ and also the same parity with respect to m_1, m_2 and the same parity with φ:

$$
\begin{aligned}
\kappa_{12} = {} & \kappa_0 w_0 \alpha^{HE}_{n_1 n_2}(p_1, q_1; p_2, q_2) \\
& \cdot \int_0^\infty e^{-i\gamma k_0 |\Delta z|} \left\{ \frac{dP^{m_1}_{n_1}(\gamma)}{d\gamma} m_2 P^{m_2}_{n_2}(\gamma) \right. \\
& \cdot \big[\sin(m_1 - m_2)\Delta\alpha \cdot J_{m_1 - m_2}(k_0 \Delta\rho\kappa) \\
& \pm (-1)^{m_1 m_2} \sin(m_1 + m_2)\Delta\alpha \cdot J_{m_1 + m_2}(k_0 \Delta\rho\kappa) \big] \\
& + m_1 P^{m_1}_{n_1}(\gamma) \frac{dP^{m_2}_{n_2}(\gamma)}{d\gamma} \\
& \cdot \big[\sin(m_1 - m_2)\Delta\alpha \cdot J_{m_1 - m_2}(k_0 \Delta\rho\kappa) \\
& \left. \mp (-1)^{m_1 m_2} \sin(m_1 + m_2)\Delta\alpha \cdot J_{m_1 + m_2}(k_0 \Delta\rho\kappa) \big] \right\} \frac{\kappa d\kappa}{\gamma};
\end{aligned}
\tag{6.50}
$$

with modes $E_{n_1 m_1 l_1}$, $H_{n_2 m_2 l_2}$ and for the same parity to m_1,m_2 and the same parity with φ:

$$
\begin{aligned}
\kappa_{12} = {} & (-1)^{n_1 - n_2} \frac{\kappa_0}{w_0} \alpha^{EH}_{n_1 n_2}(p_1, q_1; p_2, q_2) \\
& \cdot \int_0^\infty e^{-i\gamma k_0 |\Delta z|} \left\{ m_1 P^{m_1}_{n_1}(\gamma) \frac{dP^{m_2}_{n_2}(\gamma)}{d\gamma} \right. \\
& \cdot \big[\sin(m_1 - m_2)\Delta\alpha \cdot J_{m_1 - m_2}(k_0 \Delta\rho\kappa) \\
& \mp (-1)^{m_1 m_2} \sin(m_1 + m_2)\Delta\alpha \cdot J_{m_1 + m_2}(k_0 \Delta\rho\kappa) \big] \\
& + \frac{dP^{m_1}_{n_1}(\gamma)}{d\gamma} m_2 P^{m_2}_{n_2}(\gamma) \\
& \cdot \big[\sin(m_1 - m_2)\Delta\alpha \cdot J_{m_1 - m_2}(k_0 \Delta\rho\kappa) \\
& \left. \pm (-1)^{m_1 m_2} \sin(m_1 + m_2)\Delta\alpha \cdot J_{m_1 + m_2}(k_0 \Delta\rho\kappa) \big] \right\} \frac{\kappa d\kappa}{\gamma};
\end{aligned}
\tag{6.51}
$$

For $H_{n_1 m_1 l_1}$, $E_{n_2 m_2 l_2}$ and also the same parity with respect to m_1, m_2 and different parity to φ:

$$\kappa_{12} = \kappa_0 w_0 \alpha^{HE}_{n_1 n_2}(p_1, q_1; p_2, q_2)$$
$$\cdot \int_0^\infty e^{-i\gamma k_0|\Delta z|}\left\{\frac{dP^{m_1}_{n_1}(\gamma)}{d\gamma} m_2 P^{m_2}_{n_2}(\gamma)\right.$$
$$\cdot \left[\mp \cos(m_1 - m_2)\Delta\alpha \cdot J_{m_1-m_2}(k_0\Delta\rho\kappa)\right.$$
$$\left.+(-1)^{m_1 m_2}\cos(m_1 + m_2)\Delta\alpha \cdot J_{m_1+m_2}(k_0\Delta\rho\kappa)\right]$$
$$- m_1 P^{m_1}_{n_1}(\gamma)\frac{dP^{m_2}_{n_2}(\gamma)}{d\gamma}$$
$$\cdot \left[\pm \cos(m_1 - m_2)\Delta\alpha \cdot J_{m_1-m_2}(k_0\Delta\rho\kappa)\right.$$
$$\left.\left.+(-1)^{m_1 m_2}\cos(m_1 + m_2)\Delta\alpha \cdot J_{m_1+m_2}(k_0\Delta\rho\kappa)\right]\right\}\frac{\kappa d\kappa}{\gamma}; \tag{6.52}$$

For modes $E_{n_1 m_1 l_1}, H_{n_2 m_2 l_2}$ and also the same parity with respect to m_1, m_2 and different parity to φ:

$$\kappa_{12} = (-1)^{n_1-n_2}\frac{\kappa_0}{w_0}\alpha^{EH}_{n_1 n_2}(p_1, q_1; p_2, q_2)$$
$$\cdot \int_0^\infty e^{-i\gamma k_0|\Delta z|}\left\{m_1 P^{m_1}_{n_1}(\gamma)\frac{dP^{m_2}_{n_2}(\gamma)}{d\gamma}\right.$$
$$\cdot \left[\mp \cos(m_1 - m_2)\Delta\alpha \cdot J_{m_1-m_2}(k_0\Delta\rho\kappa)\right.$$
$$\left.+(-1)^{m_1 m_2}\cos(m_1 + m_2)\Delta\alpha \cdot J_{m_1+m_2}(k_0\Delta\rho\kappa)\right]$$
$$- \frac{dP^{m_1}_{n_1}(\gamma)}{d\gamma} m_2 P^{m_2}_{n_2}(\gamma)$$
$$\cdot \left[\pm \cos(m_1 - m_2)\Delta\alpha \cdot J_{m_1-m_2}(k_0\Delta\rho\kappa)\right.$$
$$\left.\left.+(-1)^{m_1 m_2}\cos(m_1 + m_2)\Delta\alpha \cdot J_{m_1+m_2}(k_0\Delta\rho\kappa)\right]\right\}\frac{\kappa d\kappa}{\gamma}; \tag{6.53}$$

for $H_{n_1 m_1 l_1}, E_{n_2 m_2 l_2}$ the different parity for m_1, m_2 and the same parity with respect to φ:

$$\begin{aligned}
\kappa_{12} = \kappa_0 w_0 \alpha^{HE}_{n_1 n_2}(p_1, q_1; p_2, q_2) \\
\cdot \int_0^\infty e^{-i\gamma k_0|\Delta z|} \Bigg\{ \frac{dP^{m_1}_{n_1}(\gamma)}{d\gamma} m_2 P^{m_2}_{n_2}(\gamma) \\
\cdot \big[-\sin(m_1 - m_2)\Delta\alpha \cdot J_{m_1 - m_2}(k_0 \Delta\rho\kappa) \\
\mp (-1)^{m_2} \sin(m_1 + m_2)\Delta\alpha \cdot J_{m_1 + m_2}(k_0 \Delta\rho\kappa) \big] \\
+ m_1 P^{m_1}_{n_1}(\gamma) \frac{dP^{m_2}_{n_2}(\gamma)}{d\gamma} \\
\cdot \big[-\sin(m_1 - m_2)\Delta\alpha \cdot J_{m_1 - m_2}(k_0 \Delta\rho\kappa) \\
\pm (-1)^{m_2} \sin(m_1 + m_2)\Delta\alpha \cdot J_{m_1 + m_2}(k_0 \Delta\rho\kappa) \big] \Bigg\} \frac{\kappa d\kappa}{\gamma};
\end{aligned} \tag{6.54}$$

for $E_{n_1 m_1 l_1}, H_{n_2 m_2 l_2}$ and the different parity for m_1, m_2 and the same parity with respect to φ:

$$\begin{aligned}
\kappa_{12} = (-1)^{n_1 - n_2 - 1} \frac{\kappa_0}{w_0} \alpha^{EH}_{n_1 n_2}(p_1, q_1; p_2, q_2) \\
\cdot \int_0^\infty e^{-i\gamma k_0|\Delta z|} \Bigg\{ m_1 P^{m_1}_{n_1}(\gamma) \frac{dP^{m_2}_{n_2}(\gamma)}{d\gamma} \\
\cdot \big[-\sin(m_1 - m_2)\Delta\alpha \cdot J_{m_1 - m_2}(k_0 \Delta\rho\kappa) \\
\pm (-1)^{m_2} \sin(m_1 + m_2)\Delta\alpha \cdot J_{m_1 + m_2}(k_0 \Delta\rho\kappa) \big] \\
+ \frac{dP^{m_1}_{n_1}(\gamma)}{d\gamma} m_2 P^{m_2}_{n_2}(\gamma) \\
\cdot \big[-\sin(m_1 - m_2)\Delta\alpha \cdot J_{m_1 - m_2}(k_0 \Delta\rho\kappa) \\
\mp (-1)^{m_2} \sin(m_1 + m_2)\Delta\alpha \cdot J_{m_1 + m_2}(k_0 \Delta\rho\kappa) \big] \Bigg\} \frac{\kappa d\kappa}{\gamma};
\end{aligned} \tag{6.55}$$

for $H_{n_1 m_1 l_1}, E_{n_2 m_2 l_2}$ the different parity for m_1, m_2 and the different parity with respect to φ:

$$\kappa_{12} = \kappa_0 w_0 \alpha^{HE}_{n_1 n_2}(p_1, q_1; p_2, q_2)$$
$$\cdot \int_0^{\infty} e^{-i\gamma k_0|\Delta z|} \Big\{ m_2 P^{m_2}_{n_2}(\gamma) \frac{dP^{m_1}_{n_1}(\gamma)}{d\gamma}$$
$$\cdot \big[\pm \cos(m_1 - m_2)\Delta\alpha \cdot J_{m_1-m_2}(k_0 \Delta\rho\kappa)$$
$$-(-1)^{m_2} \cos(m_1 + m_2)\Delta\alpha \cdot J_{m_1+m_2}(k_0 \Delta\rho\kappa)\big]$$
$$+ m_1 P^{m_1}_{n_1}(\gamma) \frac{dP^{m_2}_{n_2}(\gamma)}{d\gamma}$$
$$\cdot \big[\pm \cos(m_1 - m_2)\Delta\alpha \cdot J_{m_1-m_2}(k_0 \Delta\rho\kappa)$$
$$+(-1)^{m_2} \cos(m_1 + m_2)\Delta\alpha \cdot J_{m_1+m_2}(k_0 \Delta\rho\kappa)\big]\Big\} \frac{\kappa d\kappa}{\gamma}; \tag{6.56}$$

for $E_{n_1 m_1 l_1}$, $H_{n_2 m_2 l_2}$ for different parity m_1, m_2 and the different parity with respect to φ:

$$\kappa_{12} = (-1)^{n_1 - n_2 - 1} \frac{\kappa_0}{w_0} \alpha^{EH}_{n_1 n_2}(p_1, q_1; p_2, q_2)$$
$$\cdot \int_0^{\infty} e^{-i\gamma k_0|\Delta z|} \Big\{ m_1 P^{m_1}_{n_1}(\gamma) \frac{dP^{m_1}_{n_1}(\gamma)}{d\gamma}$$
$$\cdot \big[\mp \cos(m_1 - m_2)\Delta\alpha \cdot J_{m_1-m_2}(k_0 \Delta\rho\kappa)$$
$$-(-1)^{m_2} \cos(m_1 + m_2)\Delta\alpha \cdot J_{m_1+m_2}(k_0 \Delta\rho\kappa)\big]$$
$$+ m_1 P^{m_1}_{n_1}(\gamma) \frac{dP^{m_2}_{n_2}(\gamma)}{d\gamma}$$
$$\cdot \big[\mp \cos(m_1 - m_2)\Delta\alpha \cdot J_{m_1-m_2}(k_0 \Delta\rho\kappa)$$
$$+(-1)^{m_2} \cos(m_1 + m_2)\Delta\alpha \cdot J_{m_1+m_2}(k_0 \Delta\rho\kappa)\big]\Big\} \frac{\kappa d\kappa}{\gamma}; \tag{6.57}$$

The upper sign in (6.46–6.57) corresponds to the distribution r—component of the resonator field, proportional $\sin m_1\varphi \cdot \sin m_2\varphi$, or $\sin m_1\varphi \cdot \cos m_2\varphi$; lower distribution r—component of the resonator field, proportional $\cos m_1\varphi \cdot \cos m_2\varphi$, or $\cos m_1\varphi \cdot \sin m_2\varphi$.

Here $w_0 = (\mu_0/\varepsilon_0)^{1/2}$—is the wave impedance of the open space;

$$\kappa_0 = 2 \frac{i^{1+n_1-n_2-(m_1-m_2)}}{(1+\delta_{m_2 0})} \cdot \frac{(2n_2+1)}{n_2(n_2+1)} \cdot \frac{(n_2-m_2)!}{(n_2+m_2)!};$$

$$\alpha^{HH}_{n_1 n_2}(p_1, q_1; p_2, q_2) = \frac{j_{n_1}(p_1)}{y_{n_1}(q_1)} \cdot \frac{1}{q_2} \frac{j_{n_2}(p_2)}{y_{n_2}(q_2)}$$

$$\cdot \left\{ \left[p_2^2 - n_2(n_2+1) \right] j_{n_2}^2(p_2) + \left[n_2 j_{n_2}(p_2) - p_2 j_{n_2-1}(p_2) \right]^2 \right\}^{-1};$$

$$\alpha_{n_1 n_2}^{EE}(p_1, q_1; p_2, q_2) = \varepsilon_{1r} \frac{j_{n_1}(p_1)}{y_{n_1}(q_1)} \cdot \frac{1}{q_2} \frac{j_{n_2}(p_2)}{y_{n_2}(q_2)} \cdot \left\{ \left[p_2^2 - n_2(n_2+1) \right] j_{n_2}^2(p_2) + \left[n_2 j_{n_2}(p_2) - p_2 j_{n_2-1}(p_2) \right]^2 \right\}^{-1} \tag{6.58}$$

$$\alpha_{n_1 n_2}^{HE}(p_1, q_1; p_2, q_2) = \frac{j_{n_1}(p_1)}{y_{n_1}(q_1)} \cdot \frac{1}{q_2} \frac{j_{n_2}(p_2)}{y_{n_2}(q_2)} \cdot \left\{ \left[p_2^2 - n_2(n_2+1) \right] j_{n_2}^2(p_2) + \left[n_2 j_{n_2}(p_2) - p_2 j_{n_2-1}(p_2) \right]^2 \right\}^{-1};$$

$$\alpha_{n_1 n_2}^{EH}(p_1, q_1; p_2, q_2) = \varepsilon_{1r} \frac{j_{n_1}(p_1)}{y_{n_1}(q_1)} \cdot \frac{1}{q_2} \frac{j_{n_2}(p_2)}{y_{n_2}(q_2)} \cdot \left\{ \left[p_2^2 - n_2(n_2+1) \right] j_{n_2}^2(p_2) + \left[n_2 j_{n_2}(p_2) - p_2 j_{n_2-1}(p_2) \right]^2 \right\}^{-1};$$

are the functions that determine the dependence of coupling coefficients on the parameters of oscillations and material of resonators. Where also we used $p_s = \mathrm{Re}(k_s r_s)$ and $q_s = \mathrm{Re}(k_0 r_s)$—are the real part of the characteristic parameters of the spherical resonator modes (Sect. 1.1); $k_s = \omega(\varepsilon_s \mu_0)^{1/2}$ $(s = 1, 2)$; ω—is the circular frequency; $j_n(z)$; $y_n(z)$—is the spherical Bessel and Neumann functions; $J_n(z)$—is the Bessel functions; $P_n^m(x)$—is the associated Legendre functions [2].

When deriving (6.46–6.57), we switched to the polar coordinate system, replacing the variables: $\xi = \kappa \sin \psi$; $\eta = \kappa \cos \psi$, $(\gamma = \sqrt{1 - \kappa^2})$ after which we used the integrals (D.1–D.4).

On Fig. 6.12d–f shows the dependence of the modulus of functions (6.58) on the dielectric constant of the second DR. As can be seen from these data and (6.49), the mutual coupling coefficients are asymmetric functions of the resonators. The difference between them can be significant (see, for example, Fig. 6.12f).

In the case of identical DRs the relationships (6.46, 6.47) coincides with:

$$\kappa_{12} = \frac{2 \cdot i}{1+\delta_{m0}} \cdot \frac{(2n+1)}{n(n+1)} \cdot \frac{(n-m)!}{(n+m)!} \alpha_n^H(p,q) \cdot$$
$$\cdot \left\{ \int_0^\infty \left[(1-\gamma^2) \frac{dP_n^m(\gamma)}{d\gamma} \right]^2 \right.$$
$$\cdot \left[\begin{matrix} J_0(\kappa k_0 \Delta\rho) \mp (-1)^m \cos(2m\Delta\alpha) J_{2m}(\kappa k_0 \Delta\rho) \\ (-1)^m \sin(2m\Delta\alpha) J_{2m}(\kappa k_0 \Delta\rho) \end{matrix} \right] \frac{e^{-i\gamma k_0 |\Delta z|}}{\kappa\gamma} d\kappa$$
$$+ \int_0^\infty \left[m P_n^m(\gamma) \right]^2$$
$$\left. \cdot \left[\begin{matrix} J_0(\kappa k_0 \Delta\rho) \pm (-1)^m \cos(2m\Delta\alpha) J_{2m}(\kappa k_0 \Delta\rho) \\ (-1)^{m+1} \sin(2m\Delta\alpha) J_{2m}(\kappa k_0 \Delta\rho) \end{matrix} \right] \frac{e^{-i\gamma k_0 |\Delta z|}}{\kappa\gamma} d\kappa \right\} \quad (6.59)$$

Here $m_1 = m_2 = m$, also $\kappa^2 + \gamma^2 = 1$; $\alpha_n^H(p,q)$—is the function (2.20); $p = \mathrm{Re}(k_1 r_0)$; $q = \mathrm{Re}(k_0 r_0)$; $k_1 = \sqrt{\varepsilon_{1r}} k_0$; $k_0 = \omega/c$; c—is the velocity of light; r_0—is the radius of the DR.

Top relationships of (6.59) in the square brackets are corresponds to even—even modes (6.1). The bottom relationship with the square brackets corresponds to odd—even, or even—odd modes (6.2).

Let us note here that $\Delta\rho$ is the projection of the straight line connecting the centers of the resonators onto the plane x, y, and $\Delta\alpha$ determines the angle of rotation of the $\Delta\rho$ projection in this plane relative to the axis x (Fig. 6.12b). Since the natural oscillation fields of both resonators are "tied" to the same coordinate system, the coupling coefficients explicitly depend on the parameters $\Delta\rho$, $\Delta\alpha$.

As follows from (6.59), for oscillations of one type, the parity of the coupling coefficients in the plane x, y is relatively $\Delta\alpha$ determined by the parity of the resonator fields. For oscillations of the same parity in φ'_s the mutual coupling coefficients are even in $\Delta\alpha$, and for oscillations of opposite parity the mutual coupling coefficients are odd.

In a special case, when the plane of symmetry AA' of sphere centers situated in the, parallel to the plane xz: $y_1 = y_2 = 0$ (Fig. 6.12c); $\Delta\alpha = 0$. Then, as follows from (6.47, 6.49), oscillations of the same type (magnetic—magnetic, or electric—electric) and different parities in φ, are not coupled with each other. Oscillations of different types are also not coupled: magnetic—electric, or electric—magnetic of the same parity in φ (see 6.50, 6.51, 6.54, 6.55).

The integrals (6.46–6.57) can be calculated, if used (D.16). To take advantage of this opportunity, we find the following series expansions in associated Legendre functions [2]:

$$\left[\left(1-\gamma^2\right)\frac{dP_{n_1}^{m_1}(\gamma)}{d\gamma}\cdot\frac{dP_{n_2}^{m_2}(\gamma)}{d\gamma}+m_1m_2\frac{P_{n_1}^{m_1}(\gamma)P_{n_2}^{m_2}(\gamma)}{(1-\gamma^2)}\right]=\sum_{s=|n_1-n_2|}^{n_1+n_2}a_{n_1n_2s}^{m_1m_2}P_s^{|m_1-m_2|}(\gamma);$$

$$\left[\left(1-\gamma^2\right)\frac{dP_{n_1}^{m_1}(\gamma)}{d\gamma}\cdot\frac{dP_{n_2}^{m_2}(\gamma)}{d\gamma}-m_1m_2\frac{P_{n_1}^{m_1}(\gamma)P_{n_2}^{m_2}(\gamma)}{(1-\gamma^2)}\right]=\sum_{s=|n_1-n_2|}^{n_1+n_2}b_{n_1n_2s}^{m_1m_2}P_s^{m_1+m_2}(\gamma);$$

$$\left[m_1P_{n_1}^{m_1}(\gamma)\frac{dP_{n_2}^{m_2}(\gamma)}{d\gamma}+\frac{dP_{n_1}^{m_1}(\gamma)}{d\gamma}m_2P_{n_2}^{m_2}(\gamma)\right]=\sum_{s=|n_1-n_2|}^{n_1+n_2}c_{n_1n_2s}^{m_1m_2}P_s^{|m_1-m_2|}(\gamma); \quad (6.60)$$

$$\left[m_1P_{n_1}^{m_1}(\gamma)\frac{dP_{n_2}^{m_2}(\gamma)}{d\gamma}-\frac{dP_{n_1}^{m_1}(\gamma)}{d\gamma}m_2P_{n_2}^{m_2}(\gamma)\right]=\sum_{s=|n_1-n_2|}^{n_1+n_2}d_{n_1n_2s}^{m_1m_2}P_s^{m_1+m_2}(\gamma);$$

The first two expansions are easy to determine if we use the known decompositions [5, 6]:

$$P_{n_1}^{m_1}(z)P_{n_2}^{m_2}(z)=(-1)^{m_1}\sqrt{\frac{(n_1+m_1)!(n_2+m_2)!}{(n_1-m_1)!(n_2-m_2)!}}\sum_{s=|n_1-n_2|}^{n_1+n_2}G_s^-\sqrt{\frac{(s-|m_1-m_2|)!}{(s+|m_1-m_2|)!}}P_s^{|m_1-m_2|}(z);$$

$$P_{n_1}^{m_1}(z)P_{n_2}^{m_2}(z)=\sqrt{\frac{(n_1+m_1)!(n_2+m_2)!}{(n_1-m_1)!(n_2-m_2)!}}\sum_{s=|n_1-n_2|}^{n_1+n_2}G_s^+\sqrt{\frac{(s-m_1-m_2)!}{(s+m_1+m_2)!}}P_s^{m_1+m_2}(z),$$

where

$$G_s^{\pm}=(-1)^{m_1\pm m_2}(2s+1)\begin{pmatrix} n_1 & n_2 & s \\ 0 & 0 & 0 \end{pmatrix}\begin{pmatrix} n_1 & n_2 & s \\ \pm m_1 & m_2 & -|m_1\pm m_2| \end{pmatrix}$$

and $\begin{pmatrix} a & b & c \\ \alpha & \beta & \gamma \end{pmatrix}$ is the 3-j Wigner symbol [2]. Using the recurrence relations for the Legendre polynomials [2], we obtain:

$$\begin{aligned} a_{n_1n_2s}^{m_1m_2}&=(-1)^{m_2}\sqrt{\frac{(n_1+m_1)!(n_2+m_2)!}{(n_1-m_1)!(n_2-m_2)!}}\cdot\sqrt{\frac{(s-|m_1-m_2|)!}{(s+|m_1-m_2|)!}}\cdot\Big\{m_1m_2G_s^-(m_1,m_2)\\ &-\frac{1}{2}\sqrt{(n_1-m_1)(n_1+m_1+1)(n_2-m_2)(n_1+m_1+1)}G_s^-(m_1+1,m_2+1)\\ &-\frac{1}{2}\sqrt{(n_1+m_1)(n_1-m_1+1)(n_2+m_2)(n_2-m_2+1)}G_s^-(m_1-1,m_2-1)\Big\}; \end{aligned} \quad (6.61)$$

$$\begin{aligned} b_{n_1n_2s}^{m_1m_2}&=-\sqrt{\frac{(n_1+m_1)!(n_2+m_2)!}{(n_1-m_1)!(n_2-m_2)!}}\cdot\sqrt{\frac{(s-m_1-m_2)!}{(s+m_1+m_2)!}}\cdot\Big\{m_1m_2G_s^+(m_1,m_2)\\ &+\frac{1}{2}\sqrt{(n_1-m_1)(n_1+m_1+1)(n_2+m_2)(n_2-m_2+1)}G_s^+(m_1+1,m_2-1) \end{aligned}$$

$$+\frac{1}{2}\sqrt{(n_1+m_1)(n_1-m_1+1)(n_2-m_2)(n_2+m_2+1)}G_s^+(m_1-1,m_2+1)\}.$$

The coefficients of the last two expansions (6.60) may be calculated by using

$$c_{n_1n_2s}^{m_1m_2} = \frac{(2s+1)(s-|m_1-m_2|)!}{2(s+|m_1-m_2|)!} \cdot \int_{-1}^{1}\left[m_1P_{n_1}^{m_1}(x)\frac{dP_{n_2}^{m_2}(x)}{dx}+\frac{dP_{n_1}^{m_1}(x)}{dx}m_2P_{n_2}^{m_2}(x)\right]P_s^{|m_1-m_2|}(x)dx; \tag{6.62}$$

$$d_{n_1n_2s}^{m_1m_2} = \frac{(2s+1)(s-m_1-m_2)!}{2(s+m_1+m_2)!} \cdot \int_{-1}^{1}\left[m_1P_{n_1}^{m_1}(x)\frac{dP_{n_2}^{m_2}(x)}{dx}-\frac{dP_{n_1}^{m_1}(x)}{dx}m_2P_{n_2}^{m_2}(x)\right]P_s^{m_1+m_2}(x)dx.$$

Substituting (6.60) in the (6.46–6.57) and using (D.16), obtains finally: for $H_{n_1m_1l_1}$, $H_{n_2m_2l_2}$ modes and for the same parity with respect to m_1, m_2 and for the same parity with respect to φ:

$$\begin{aligned}\kappa_{12} &= \kappa_0\cdot\alpha_{n_1n_2}^{HH}(p_1,q_1;p_2,q_2)\\ &\cdot\Big[\cos(m_1-m_2)\Delta\alpha\\ &\cdot\sum_{s=|n_1-n_2|}^{n_1+n_2} i^{|m_1-m_2|-s}a_{n_1n_2s}^{m_1m_2}P_s^{|m_1-m_2|}(\cos\Delta\theta)h_s^{(2)}(k_0\Delta r)\\ &\mp(-1)^{m_1m_2}\cos(m_1+m_2)\Delta\alpha\\ &\cdot\sum_{s=|n_1-n_2|}^{n_1+n_2} i^{m_1+m_2-s}b_{n_1n_2s}^{m_1m_2}P_s^{m_1+m_2}(\cos\Delta\theta)h_s^{(2)}(k_0\Delta r)\Big];\end{aligned} \tag{6.63}$$

for $H_{n_1m_1l_1}$, $H_{n_2m_2l_2}$ modes and for the same parity with respect to m_1, m_2 and for the different parity with respect to φ:

$$\begin{aligned}\kappa_{12} &= \kappa_0\cdot\alpha_{n_1n_2}^{HH}(p_1,q_1;p_2,q_2)\\ &\cdot\Big[\pm\sin(m_1-m_2)\Delta\alpha\\ &\cdot\sum_{s=|n_1-n_2|}^{n_1+n_2} i^{|m_1-m_2|-s}a_{n_1n_2s}^{m_1m_2}P_s^{|m_1-m_2|}(\cos\Delta\theta)h_s^{(2)}(k_0\Delta r)\\ &+(-1)^{m_1m_2}\sin(m_1+m_2)\Delta\alpha\\ &\cdot\sum_{s=|n_1-n_2|}^{n_1+n_2} i^{m_1+m_2-s}b_{n_1n_2s}^{m_1m_2}P_s^{m_1+m_2}(\cos\Delta\theta)h_s^{(2)}(k_0\Delta r)\Big];\end{aligned} \tag{6.64}$$

for $H_{n_1m_1l_1}, H_{n_2m_2l_2}$ modes and for the different parity with respect to m_1, m_2 and for the same parity with respect to φ:

$$\begin{aligned}\kappa_{12} = \kappa_0 \cdot \alpha^{HH}_{n_1n_2}(p_1, q_1; p_2, q_2) \\ \cdot \Bigg[-\cos(m_1 - m_2)\Delta\alpha \\ \cdot \sum_{s=|n_1-n_2|}^{n_1+n_2} i^{|m_1-m_2|-s} a^{m_1m_2}_{n_1n_2s} P_s^{|m_1-m_2|}(\cos\Delta\theta) h_s^{(2)}(k_0\Delta r) \\ \pm (-1)^{m_2} \cos(m_1 + m_2)\Delta\alpha \\ \cdot \sum_{s=|n_1-n_2|}^{n_1+n_2} i^{m_1+m_2-s} b^{m_1m_2}_{n_1n_2s} P_s^{m_1+m_2}(\cos\Delta\theta) h_s^{(2)}(k_0\Delta r) \Bigg];\end{aligned} \tag{6.65}$$

for $H_{n_1m_1l_1}, H_{n_2m_2l_2}$ modes and for the different parity with respect to m_1, m_2 and for the different parity with respect to φ:

$$\begin{aligned}\kappa_{12} = \kappa_0 \cdot \alpha^{HH}_{n_1n_2}(p_1, q_1; p_2, q_2) \\ \cdot \Bigg[\mp\sin(m_1 - m_2)\Delta\alpha \\ \cdot \sum_{s=|n_1-n_2|}^{n_1+n_2} i^{|m_1-m_2|-s} a^{m_1m_2}_{n_1n_2s} P_s^{|m_1-m_2|}(\cos\Delta\theta) h_s^{(2)}(k_0\Delta r) \\ - (-1)^{m_2} \sin(m_1 + m_2)\Delta\alpha \\ \cdot \sum_{s=|n_1-n_2|}^{n_1+n_2} i^{m_1+m_2-s} b^{m_1m_2}_{n_1n_2s} P_s^{m_1+m_2}(\cos\Delta\theta) h_s^{(2)}(k_0\Delta r) \Bigg];\end{aligned} \tag{6.66}$$

for $H_{n_1m_1l_1}, E_{n_2m_2l_2}$ modes and for the same parity with respect to m_1, m_2 and for the same parity with respect to φ:

$$\begin{aligned}\kappa_{12} = \kappa_0 w_0 \cdot \alpha^{HE}_{n_1n_2}(p_1, q_1; p_2, q_2) \\ \cdot \Bigg[\sin(m_1 - m_2)\Delta\alpha \\ \cdot \sum_{s=|m_1-m_2|}^{n_1+n_2} i^{|m_1-m_2|-s} c^{m_1m_2}_{n_1n_2s} P_s^{|m_1-m_2|}(\cos\Delta\theta) h_s^{(2)}(k_0\Delta r) \\ \mp (-1)^{m_2} \sin(m_1 + m_2)\Delta\alpha \\ \cdot \sum_{s=|m_1-m_2|}^{n_1+n_2} i^{m_1+m_2-s} d^{m_1m_2}_{n_1n_2s} P_s^{m_1+m_2}(\cos\Delta\theta) h_s^{(2)}(k_0\Delta r) \Bigg];\end{aligned} \tag{6.67}$$

for $E_{n_1m_1l_1}, H_{n_2m_2l_2}$ modes and for the same parity with respect to m_1, m_2 and for the same parity with respect to φ:

$$
\begin{aligned}
\kappa_{12} = {} & (-1)^{n_1-n_2}\frac{\kappa_0}{w_0}\cdot\alpha^{EH}_{n_1n_2}(p_1,q_1;p_2,q_2) \\
& \cdot\Bigg[\sin(m_1-m_2)\Delta\alpha \\
& \cdot\sum_{s=|m_1-m_2|}^{n_1+n_2} i^{|m_1-m_2|-s}c^{m_1m_2}_{n_1n_2s}P_s^{|m_1-m_2|}(\cos\Delta\theta)h_s^{(2)}(k_0\Delta r) \\
& \mp(-1)^{m_2}\sin(m_1+m_2)\Delta\alpha \\
& \cdot\sum_{s=|m_1-m_2|}^{n_1+n_2} i^{m_1+m_2-s}d^{m_1m_2}_{n_1n_2s}P_s^{m_1+m_2}(\cos\Delta\theta)h_s^{(2)}(k_0\Delta r)\Bigg];
\end{aligned}
\tag{6.68}
$$

for $H_{n_1m_1l_1}$, $E_{n_2m_2l_2}$ modes and for the same parity with respect to m_1, m_2 and for the different parity with respect to φ:

$$
\begin{aligned}
\kappa_{12} = {} & \kappa_0 w_0\cdot\alpha^{HE}_{n_1n_2}(p_1,q_1;p_2,q_2) \\
& \cdot\Bigg[\mp\cos(m_1-m_2)\Delta\alpha \\
& \cdot\sum_{s=|m_1-m_2|}^{n_1+n_2} i^{|m_1-m_2|-s}c^{m_1m_2}_{n_1n_2s}P_s^{|m_1-m_2|}(\cos\Delta\theta)h_s^{(2)}(k_0\Delta r) \\
& -(-1)^{m_1m_2}\cos(m_1+m_2)\Delta\alpha \\
& \cdot\sum_{s=m_1+m_2}^{n_1+n_2} i^{m_1+m_2-s}d^{m_1m_2}_{n_1n_2s}P_s^{m_1+m_2}(\cos\Delta\theta)h_s^{(2)}(k_0\Delta r)\Bigg];
\end{aligned}
\tag{6.69}
$$

for $E_{n_1m_1l_1}$, $H_{n_2m_2l_2}$ modes and for the same parity with respect to m_1, m_2 and for the different parity with respect to φ:

$$
\begin{aligned}
\kappa_{12} = {} & (-1)^{n_1-n_2}\frac{\kappa_0}{w_0}\cdot\alpha^{EH}_{n_1n_2}(p_1,q_1;p_2,q_2) \\
& \cdot\Bigg[\mp\cos(m_1-m_2)\Delta\alpha \\
& \cdot\sum_{s=|m_1-m_2|}^{n_1+n_2} i^{|m_1-m_2|-s}c^{m_1m_2}_{n_1n_2s}P_s^{|m_1-m_2|}(\cos\Delta\theta)h_s^{(2)}(k_0\Delta r) \\
& +(-1)^{m_1m_2}\cos(m_1+m_2)\Delta\alpha \\
& \cdot\sum_{s=m_1+m_2}^{n_1+n_2} i^{m_1+m_2-s}d^{m_1m_2}_{n_1n_2s}P_s^{m_1+m_2}(\cos\Delta\theta)h_s^{(2)}(k_0\Delta r)\Bigg];
\end{aligned}
\tag{6.70}
$$

for $H_{n_1m_1l_1}$, $E_{n_2m_2l_2}$ modes and for the different parity with respect to m_1, m_2 and for the same parity with respect to φ:

$$\kappa_{12} = \kappa_0 w_0 \cdot \alpha^{HE}_{n_1 n_2}(p_1, q_1; p_2, q_2) \\ \cdot \Bigg[-\sin(m_1 - m_2)\Delta\alpha \\ \cdot \sum_{s=|m_1-m_2|}^{n_1+n_2} i^{|m_1-m_2|-s} c^{m_1 m_2}_{n_1 n_2 s} P_s^{|m_1-m_2|}(\cos\Delta\theta) h_s^{(2)}(k_0 \Delta r) \\ \pm (-1)^{m_2} \sin(m_1 + m_2)\Delta\alpha \\ \cdot \sum_{s=m_1+m_2}^{n_1+n_2} i^{m_1+m_2-s} d^{m_1 m_2}_{n_1 n_2 s} P_s^{m_1+m_2}(\cos\Delta\theta) h_s^{(2)}(k_0 \Delta r)\Bigg]; \tag{6.71}$$

for $E_{n_1 m_1 l_1}$, $H_{n_2 m_2 l_2}$ modes and for the different parity with respect to m_1, m_2 and for the same parity with respect to φ:

$$\kappa_{12} = (-1)^{n_1-n_2-1}\frac{\kappa_0}{w_0} \cdot \alpha^{EH}_{n_1 n_2}(p_1, q_1; p_2, q_2) \\ \cdot \Bigg[-\sin(m_1 - m_2)\Delta\alpha \\ \cdot \sum_{s=|m_1-m_2|}^{n_1+n_2} i^{|m_1-m_2|-s} c^{m_1 m_2}_{n_1 n_2 s} P_s^{|m_1-m_2|}(\cos\Delta\theta) h_s^{(2)}(k_0 \Delta r) \\ \pm (-1)^{m_2} \sin(m_1 + m_2)\Delta\alpha \\ \cdot \sum_{s=m_1+m_2}^{n_1+n_2} i^{m_1+m_2-s} d^{m_1 m_2}_{n_1 n_2 s} P_s^{m_1+m_2}(\cos\Delta\theta) h_s^{(2)}(k_0 \Delta r)\Bigg]; \tag{6.72}$$

for $H_{n_1 m_1 l_1}$, $E_{n_2 m_2 l_2}$ modes and for the different parity with respect to m_1, m_2 and for the different parity with respect to φ:

$$\kappa_{12} = \kappa_0 w_0 \cdot \alpha^{HE}_{n_1 n_2}(p_1, q_1; p_2, q_2) \\ \cdot \Bigg[\pm\cos(m_1 - m_2)\Delta\alpha \\ \cdot \sum_{s=|m_1-m_2|}^{n_1+n_2} i^{|m_1-m_2|-s} c^{m_1 m_2}_{n_1 n_2 s} P_s^{|m_1-m_2|}(\cos\Delta\theta) h_s^{(2)}(k_0 \Delta r) \\ - (-1)^{m_2} \cos(m_1 + m_2)\Delta\alpha \\ \cdot \sum_{s=m_1+m_2}^{n_1+n_2} i^{m_1+m_2-s} d^{m_1 m_2}_{n_1 n_2 s} P_s^{m_1+m_2}(\cos\Delta\theta) h_s^{(2)}(k_0 \Delta r)\Bigg]; \tag{6.73}$$

for $E_{n_1 m_1 l_1}$, $H_{n_2 m_2 l_2}$ modes and for the different parity with respect to m_1, m_2 and for the different parity with respect to φ:

$$\kappa_{12} = (-1)^{n_1-n_2-1}\frac{\kappa_0}{w_0}\cdot\alpha^{EH}_{n_1n_2}(p_1, q_1; p_2, q_2)$$
$$\cdot\Big[\mp\cos(m_1-m_2)\Delta\alpha$$
$$\cdot\sum_{s=|m_1-m_2|}^{n_1+n_2} i^{|m_1-m_2|-s} c^{m_1m_2}_{n_1n_2s} P^{|m_1-m_2|}_s(\cos\Delta\theta) h^{(2)}_s(k_0\Delta r)$$
$$-(-1)^{m_2}\cos(m_1+m_2)\Delta\alpha$$
$$\cdot\sum_{s=m_1+m_2}^{n_1+n_2} i^{m_1+m_2-s} d^{m_1m_2}_{n_1n_2s} P^{m_1+m_2}_s(\cos\Delta\theta) h^{(2)}_s(k_0\Delta r)\Big]; \quad (6.74)$$

Here the $\Delta r = \sqrt{\Delta x^2 + \Delta y^2 + \Delta z^2}$ is the relative distance between the centers of the resonators; $\cos\Delta\theta = \Delta z/\Delta r$; $\cos\Delta\alpha = \Delta x/\Delta\rho$. In the spherical coordinate system of the first resonator, the $(\Delta r, \Delta\theta, \Delta\alpha)$ determines the coordinates of the center of the second resonator.

Thus, the application of expansion (6.4) in the case of spherical DRs makes it possible in general to calculate the coefficients of mutual coupling in open space for any type of DR oscillations.

In some special cases, expressions (6.46–6.57) are more convenient to represent in a different form. For example, for the eigenoscillations with $n = m$, taking into account the presentation of the Legendre polynomials [2]:

$$P^m_m(z) = \frac{(2m)!}{2^m m!}\left(1-z^2\right)^{m/2},$$

obtain:

$$\kappa_{12} = 2i\cdot\frac{m}{(m+1)}\cdot\frac{(2m+1)!}{2^{2m}(m!)^2}\alpha^H_n(p, q)$$
$$\cdot\left[\begin{array}{c}(m-1)!\sum_{s=0}^{m}\frac{m-2s}{(m-s)!s!}\cdot\frac{d^{2s}}{d(k_0\Delta r)^{2s}}h^{(2)}_0(k_0\Delta r)\mp(-1)^m\cos(2m\Delta\alpha)(sin\Delta\theta)^{2m}h^{(2)}_{2m}(k_0\Delta r)\\ (-1)^{m+1}\sin(2m\Delta\alpha)(sin\Delta\theta)^{2m}h^{(2)}_{2m}(k_0\Delta r)\end{array}\right]. \quad (6.75)$$

where it is more convenient to calculate the derivatives of the spherical Hankel functions using the representation [7].

$$\frac{d^n}{dz^n}h^{(2)}_0(z) = h^{(2)}_0(z)\cdot n!\sum_{s=0}^{n}\frac{(-i)^{n+s}}{(n-s)!}\cdot\frac{1}{z^s}$$

This expansion follows from the representation of the Hankel spherical function in terms of elementary functions [2] and Leibniz's theorem on differentiation of the product [8].

In a more general case, the coupling between magnetic and electrical types for oscillations with $n_1 = m_1$; $n_2 = m_2$, the expansion coefficients in (6.67–6.74) can be presented explicitly:

$$c_{n_1 n_2 s}^{m_1 m_2} = 2(-1)^{m_1} \cdot m_1 m_2 \cdot (2m_2 - 1)\sqrt{(2m_1 - 1)!(2m_2 - 2)!} \cdot \sqrt{\frac{(s - |m_1 - m_2|)!}{(s + |m_1 - m_2|)!}} G_s^-; \tag{6.76}$$

$$d_{m_1 m_2 s}^{m_1 m_2} = 0.$$

where the coefficients $G_s^- = G_s^-(m_1, m_2, n_1, n_2)$ for expansion (6.76) are equal:

$$G_s^- = G_s^-(m_1 - 1, m_2 - 1, m_1, m_2 - 1), \quad s \in [|m_1 - m_2 + 1|, (m_1 + m_2)]$$

Obtained relationships suitable for the calculating of the coupling coefficients for whispering gallery modes of spherical microresonators with n_s and $m_s >> 1$.

The real part of the mutual coupling coefficient is determined by the overlap of the stored field of the resonators, and the imaginary part is the magnitude of the radiations of the structure. If such overlap is mainly determined by the electrical components of the field, the real part of the coupling coefficient becomes negative, and if the overlap is mainly determined by the magnetic components of the field, it is positive. Thus, the dependence of the real part of the mutual coupling coefficient is determined by the distribution of the stored fields of DRs in the near spatial region, while the imaginary part of the coupling coefficient is determined by the radiation fields of the coupled oscillations. As can be seen from the graphs shown in Figs. 6.13 and 6.14, this statement is in good agreement with the results of the calculations.

The coupling coefficients of the whispering gallery modes obtains sufficiently large values in the near-field region. At that, the relative motion in the tangent directions leads to a complex interference of their mutual influence, determining by significant eigenmode field variation nearby their surfaces (Figs. 6.13b, d, f and 6.14b, d).

In the majority cases the imaginary part values of the coupling coefficients at least one tenth as many as it real parts.

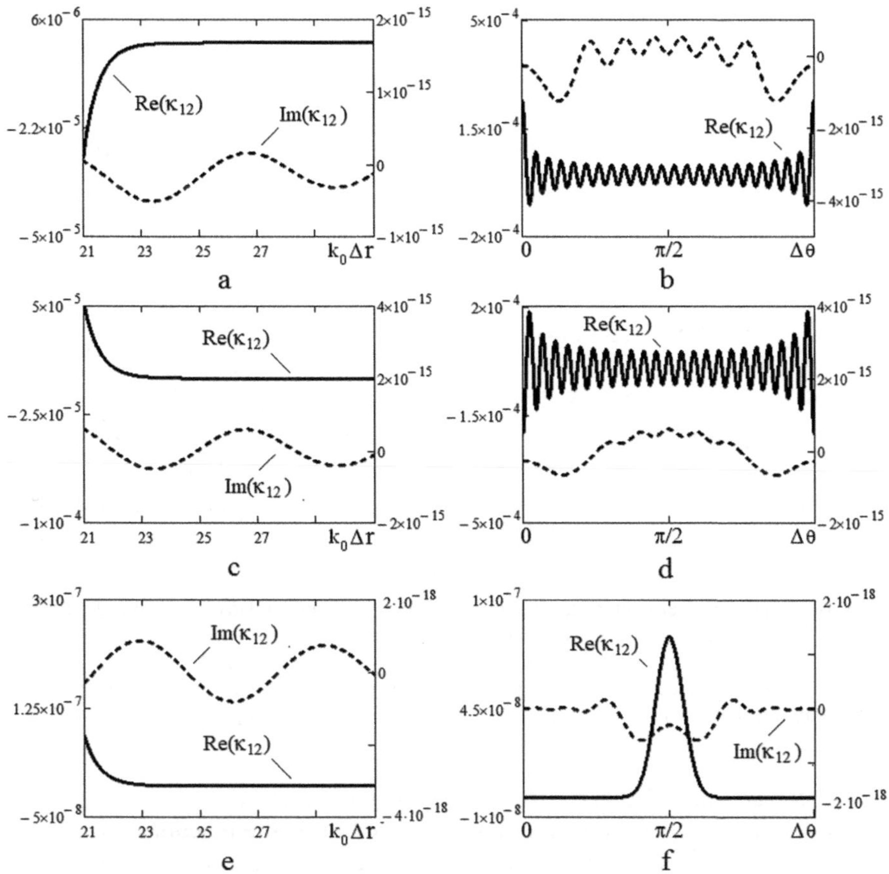

Fig. 6.13 The coupling coefficients as functions of the distance between resonator centers for the same parity whispering gallery modes: $H_{n_1m_1l_1}$, $H_{n_2m_2l_2}$ ($n_1 = 22$; $n_2 = 24$; $\varepsilon_{1r} = 4$; $\varepsilon_{2r} = 16$); **a**, **b** $m_1 = m_2 = 0$; **c**, **d** $m_1 = m_2 = 1$; **e**, **f** $m_1 = n_1$; $m_2 = n_2$. **a**, **c** $\Delta\theta = 0, 01$; **e** $\Delta\theta = \pi/2$; **b**, **d**, **f** $k_0\Delta r = 21$. The ordinate axes on the left correspond to the real part, and the y-axis to the right of the imaginary part of the coupling coefficients

6.7 Coupling Coefficients Between Basic and Higher Types Modes of Spherical Dielectric Resonators

The analytical expressions for the coupling coefficients found in Sect. 6.6 make it possible to calculate their values in the general case for any types of oscillations. Their disadvantage is the quadratic dependence on the Legendre integrands (6.46–6.57), as well as the complexity of calculating expansions (6.60), expressed through 3-j Wigner symbols. In this section we will consider in more detail the special case of coupled oscillations of spherical DR of the main magnetic types $H_{1m_1l_1}$ and DR with any types of oscillations H_{nml_2}, E_{nml_2} (Fig. 6.15a), for which integrals

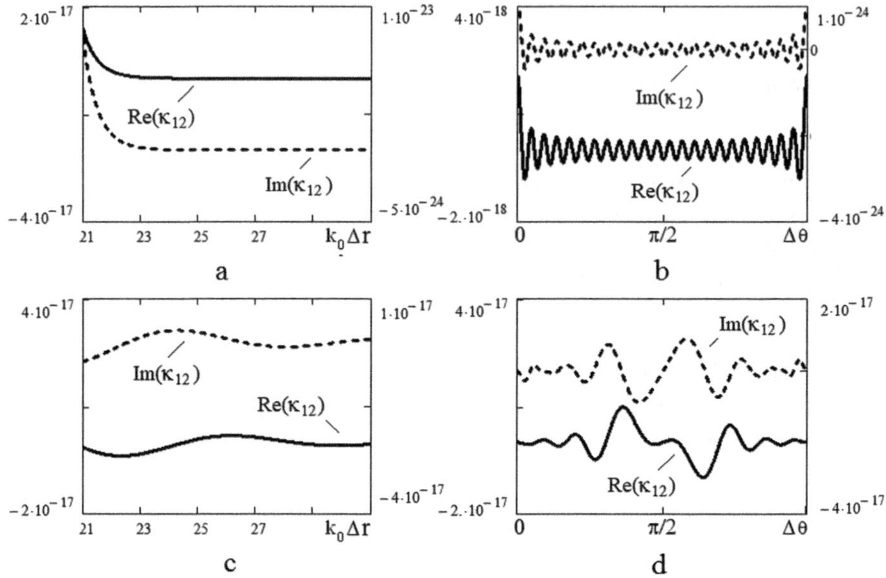

Fig. 6.14 Coupling coefficients as functions of the resonator relative coordinates for the $H_{n_1m_1l}$, $E_{n_2m_2l}$ oscillations $(n_1 = 22; n_2 = 24;\ \varepsilon_{1r} = 4;\ \varepsilon_{2r} = 16)$ $(m_1 = 1; m_2 = 1(\text{a, b}));$; $(m_1 = n_1; m_2 = n_2(\text{c, d}))$; **a** $\Delta\theta = 0, 01$; $\Delta\alpha = 0$; **c** $\Delta\theta = \Delta\alpha = \pi/4$; **b** $\Delta\alpha = 0$, **d** $\Delta\alpha = \pi/4$; $k_0\Delta r = 22$. The ordinate axes on the left and on the right correspond to the real and imaginary part of the coupling coefficients

(6.46–6.57) are expressed in relative simple analytical form. For this purpose, let us substitute particular expressions for the associated Legendre functions; $P_1^0(z) = z$; $P_1^1(z) = -(1-z^2)^{1/2}$ [9] in (6.46–6.57) and use recurrent expressions for them when calculating the corresponding integrals, according to (D.16).

We present the formulas obtained as a result of transformations for each pair of spherical DRs in the form:

with the magnetic mode H_{10l_1} in the first resonator, and magnetic modes in the second H_{nml_2} and for the same parity in φ:

$$\kappa_{12} = 2i^{m-n+1}\kappa_0\alpha_{1n}^{HH}(p_1, q_1; p_2, q_2) \cdot h_n^m(\Delta\theta, \Delta r)\cos m\Delta\alpha; \tag{6.77}$$

for the different parity in φ:

$$\kappa_{12} = 2i^{m-n+1}\kappa_0\alpha_{1n}^{HH}(p_1, q_1; p_2, q_2) \cdot h_n^m(\Delta\theta, \Delta r)\sin m\Delta\alpha; \tag{6.78}$$

with magnetic mode H_{nml_1} in the first resonator, the magnetic modes H_{10l_2} in the second and

for the same parity in φ:

$$\kappa_{12} = 2i^{-m-n+1}\kappa_0\alpha_{n1}^{HH}(p_1, q_1; p_2, q_2) \cdot h_n^m(\Delta\theta, \Delta r)\cos m\Delta\alpha; \tag{6.79}$$

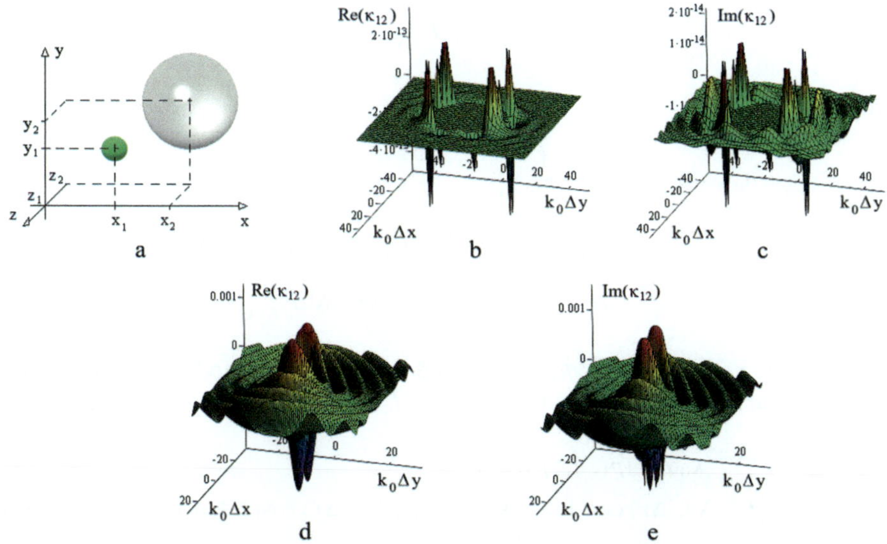

Fig. 6.15 Spherical DR with the main type of oscillations and oscillations of higher types (**a**). Distribution of the real (**b**), imaginary (**c**) part of the coupling coefficients of resonators with the main type of oscillations and with whispering gallery oscillations H_{111}, $H_{30,5,1}$ for $\varepsilon_{1r} = 25$; $\varepsilon_{2r} = 2, 25$ and $k_0|\Delta z| = 0, 1$. Coupling coefficients of resonators with the main type of oscillations and with oscillations of a bouncing ball H_{111}, $H_{1,1,10}$ for $\varepsilon_{1r} = 25$; $\varepsilon_{2r} = 25$ and $k_0|\Delta z| = 0, 1$ (**d**, **e**)

for the different parity in φ:

$$\kappa_{12} = 2i^{-m-n+1}\kappa_0\alpha_{n1}^{HH}(p_1, q_1; p_2, q_2) \cdot h_n^m(\Delta\theta, \Delta r)\sin m\Delta\alpha. \tag{6.80}$$

with magnetic mode H_{10l_1} in the first resonator, the electric modes in the second E_{nml_2} and for the same parity in φ:

$$\kappa_{12} = 2i^{m-n-2}\kappa_0 w_0\alpha_{1n}^{HE}(p_1, q_1; p_2, q_2) \cdot mP_n^m(\cos\Delta\theta)h_n^{(2)}(k_0\Delta r)\sin m\Delta\alpha; \tag{6.81}$$

for the different parity in φ:

$$\kappa_{12} = 2i^{m-n}\kappa_0 w_0\alpha_{1n}^{HE}(p_1, q_1; p_2, q_2) \cdot mP_n^m(\cos\Delta\theta)h_n^{(2)}(k_0\Delta r)\cos m\Delta\alpha; \tag{6.82}$$

with electric mode E_{nml_1} in the first resonator, the magnetic modes H_{10l_2} in the second and for the same parity in φ:

$$\kappa_{21} = 2i^{n+m-2}\frac{\kappa_0}{w_0}\alpha_{n1}^{EH}(p_1, q_1; p_2, q_2) \cdot mP_n^m(\cos\Delta\theta)h_n^{(2)}(k_0\Delta r)\sin m\Delta\alpha; \tag{6.83}$$

for the different parity in φ:

$$\kappa_{21} = 2i^{n+m-2}\frac{\kappa_0}{w_0}\alpha_{n1}^{EH}(p_1, q_1; p_2, q_2) \cdot mP_n^m(\cos\Delta\theta)h_n^{(2)}(k_0\Delta r)\cos m\Delta\alpha. \quad (6.84)$$

where

$$h_n^m(\Delta\theta, \Delta r) = \frac{1}{(2n+1)} \cdot \Big[(n+m)(n+1)P_{n-1}^m(\cos\Delta\theta)h_{n-1}^{(2)}(k_0\Delta r) + n(n-m+1)P_{n+1}^m(\cos\Delta\theta)h_{n+1}^{(2)}(k_0\Delta r)\Big]$$

For $m_1 = 1$; $n_1 = 1$; $m_2 = m$; $n_2 = n$ and with magnetic mode H_{11l_1} in the first resonator, the magnetic modes in the second H_{nml_2} and for the same parity in φ:

$$\kappa_{12} = i^{m-n+2}\kappa_0\alpha_{1n}^{HH}(p_1, q_1; p_2, q_2) \cdot \big\{f_n^m(\Delta\theta, \Delta r)\cos(1-m)\Delta\alpha \mp g_n^m(\Delta\theta, \Delta r)\cos(1+m)\Delta\alpha\big\}; \quad (6.85)$$

for the different parity in φ:

$$\kappa_{12} = i^{m-n+2}\kappa_0\alpha_{1n}^{HH}(p_1, q_1; p_2, q_2) \cdot \big\{\pm f_n^m(\Delta\theta, \Delta r)\sin(1-m)\Delta\alpha + g_n^m(\Delta\theta, \Delta r)\sin(1+m)\Delta\alpha\big\}; \quad (6.86)$$

with magnetic mode H_{nml_1} in the first resonator, the magnetic modes H_{11l_2} in the second and for the same parity in φ:

$$\kappa_{21} = i^{m-n+2}\kappa_0\alpha_{n1}^{HH}(p_1, q_1; p_2, q_2) \cdot \big\{f_n^m(\Delta\theta, \Delta r)\cos(1-m)\Delta\alpha \mp g_n^m(\Delta\theta, \Delta r)\cos(1+m)\Delta\alpha\big\}; \quad (6.87)$$

for the different parity in φ:

$$\kappa_{21} = i^{-m-n}\kappa_0\alpha_{n1}^{HH}(p_1, q_1; p_2, q_2) \cdot \big\{\pm f_n^m(\Delta\theta, \Delta r)\sin(1-m)\Delta\alpha + g_n^m(\Delta\theta, \Delta r)\sin(1+m)\Delta\alpha\big\} \quad (6.88)$$

Here

$$f_n^m(\Delta\theta, \Delta r) = \frac{1}{(2n+1)} \cdot \Big[n(n-m+1)(n+m+2)P_{n+1}^{m-1}(\cos\Delta\theta)h_{n+1}^{(2)}(k_0\Delta r) - (n+1)(n+m-1)(n+m)P_{n-1}^{m-1}(\cos\Delta\theta)h_{n-1}^{(2)}(k_0\Delta r)\Big];$$

$$g_n^m(\Delta\theta, \Delta r) = \frac{-1}{(2n+1)} \cdot \Big[nP_{n+1}^{m+1}(\cos\Delta\theta)h_{n+1}^{(2)}(k_0\Delta r) - (n+1)P_{n-1}^{m+1}(\cos\Delta\theta)h_{n-1}^{(2)}(k_0\Delta r)\Big];$$

with magnetic mode H_{11l_1} in the first resonator, the electric modes in the second E_{nml_2} and for the same parity in φ:

$$\begin{aligned}\kappa_{12} = {} & i^{m-n+1}\kappa_0 w_0 \alpha_{1n}^{HE}(p_1, q_1; p_2, q_2) \\ & \cdot \{(n+m)(n-m+1)\,P_n^{m-1}(\cos\Delta\theta)\sin(1-m)\Delta\alpha \\ & \pm (-1)^m P_n^{m+1}(\cos\Delta\theta)\sin(1+m)\Delta\alpha\} h_n^{(2)}(k_0\Delta r);\end{aligned} \quad (6.89)$$

for the different parity in φ:

$$\begin{aligned}\kappa_{12} = {} & i^{m-n+1}\kappa_0 w_0 \alpha_{1n}^{HE}(p_1, q_1; p_2, q_2) \\ & \cdot \left\{\mp(n+m)(n-m+1)P_n^{m-1}(\cos\Delta\theta)\cos(1-m)\Delta\alpha \right. \\ & \left. + P_n^{m+1}(\cos\Delta\theta)\cos(1+m)\Delta\alpha\right\} h_n^{(2)}(k_0\Delta r);\end{aligned} \quad (6.90)$$

with electric mode E_{nml_1} in the first resonator, the magnetic modes H_{11l_2} in the second and for the same parity in φ:

$$\begin{aligned}\kappa_{21} = {} & i^{n-m-1}\frac{\kappa_0}{w_0}\alpha_{n1}^{EH}(p_1, q_1; p_2, q_2) \\ & \cdot \left\{(n+m)(n-m+1)P_n^{m-1}(\cos\Delta\theta)\sin(1-m)\Delta\alpha \right. \\ & \left. \mp P_n^{m+1}(\cos\Delta\theta)\sin(1+m)\Delta\alpha\right\} h_n^{(2)}(k_0\Delta r);\end{aligned} \quad (6.91)$$

for the different parity in φ:

$$\begin{aligned}\kappa_{12} = {} & i^{n+m+1}\frac{\kappa_0}{w_0}\alpha_{n1}^{EH}(p_1, q_1; p_2, q_2) \\ & \cdot \left\{\pm(n+m)(n-m+1)P_n^{m-1}(\cos\Delta\theta)\cos(1-m)\Delta\alpha \right. \\ & \left. + P_n^{m+1}(\cos\Delta\theta)\cos(1+m)\Delta\alpha\right\} h_n^{(2)}(k_0\Delta r)\end{aligned} \quad (6.92)$$

The obtained expressions, in particular, show that azimuthally homogeneous oscillations of the magnetic and electric types ($m = 0$) are not coupled with each other at any values of the coordinates of the resonators (6.81–6.84). As follows from (6.89) and (6.91), oscillations of the same parity in φ and ($m = 1$) are also not coupled to each other in open space.

In the case of excitation of whispering gallery oscillations in the second resonator ($n \gg 1; l_2 \sim 1, 2$), the coupling coefficients are sufficiently large only near its surface (Fig. 6.15b, c). For oscillations like a bouncing ball in the second DR (Sect. 1.1), the amplitude of the coupling coefficients is much higher, and its decrease with increasing relative distance between the DR centers is much smaller (Fig. 6.15d, e).

6.8 Coupling Coefficients of Cylindrical Dielectric Resonators with Higher Types of Oscillations

The integral (6.1) can be calculated, based on early known analytical expression for the coupling coefficients, of the cylindrical DRs in the rectangular metal waveguide. In this case, required expressions can be received also by transferring the waveguide walls to the infinity. After simplifications, we obtain:

In the case of two different resonators, with the same parity of each field in φ (Fig. 6.16a) and in the indexes m_s, relative to the plane of symmetry $A - A'$: $y_s - y = 0$ $(s = 1, 2)$ (Figs. 6.16b and 6.17a), the mutual coupling coefficients can be represented in the form:

in the area: $\Delta z \geq r_1 + r_2$ (Fig. 6.17a):

$$\begin{aligned}\kappa_{1,2} = i\kappa_0^{12} \cdot \int_0^\infty & \left\{ e^{-i(m_1-m_2)\pi/2} \cos[(m_1 - m_2)\Delta\alpha] H^{(2)}_{m_1-m_2}\left(k_0\Delta\rho\sqrt{1-\xi^2}\right) \right. \\ & \left. \mp (-1)^{m_1 m_2} e^{-i(m_1+m_2)\pi/2} \cos[(m_1 + m_2)\Delta\alpha] H^{(2)}_{m_1+m_2}\left(k_0\Delta\rho\sqrt{1-\xi^2}\right) \right\} \\ & \cdot \left[F_{11}(\xi)F_{12}(\xi)^* + F_{21}(\xi)F_{22}(\xi)^*\right] \cos(k_0\Delta x\xi) d\xi; \qquad (6.93)\end{aligned}$$

where the top sign corresponds to even-mode field distributions (1.17), and the bottom one corresponds to odd-mode field distributions (1.17) relative to the plane of symmetry $\varphi = 0, \pi$; $\sin\Delta\alpha = \Delta z/\Delta\rho$; $\Delta\rho = \sqrt{\Delta y^2 + \Delta z^2}$; $\Delta x = x_2 - x_1$; $\Delta y = y_2 - y_1$; $\Delta z = z_2 - z_1$;

$$\kappa_0^{12} = \frac{32}{\varepsilon_{2r} w(2)} \cdot \frac{p_{2z}}{q_{2z}} \cdot \frac{q_{1\perp}}{q_{2\perp}}$$

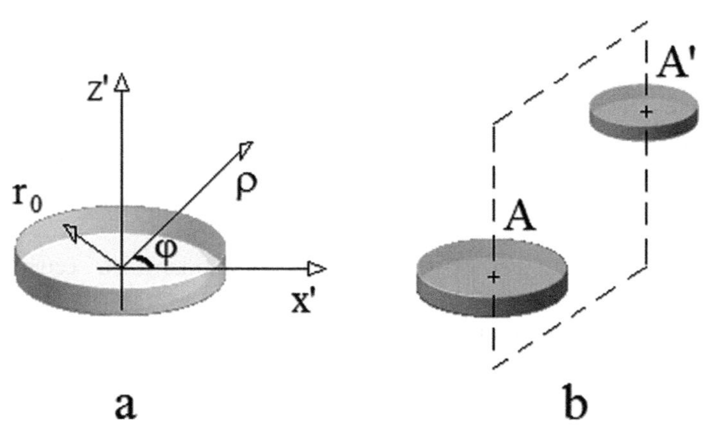

Fig. 6.16 **a**—Cylindrical dielectric resonator in the open space; **b**—Symmetry plane (AA') of two different cylindrical dielectric resonators

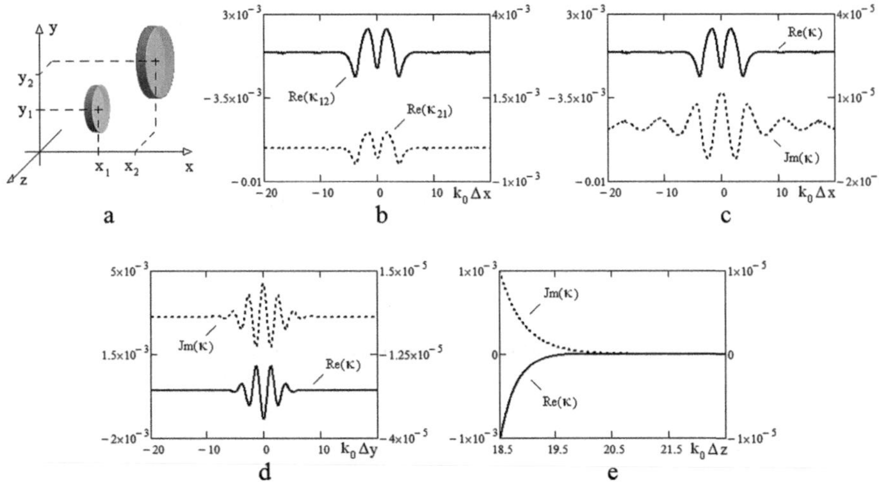

Fig. 6.17 Rectangular coordinate system and two laterally coupled different cylindrical resonators in the open space (**a**). Mutual coupling coefficients as function coordinates of the resonators with $\varepsilon_{1r} = 16$; $\varepsilon_{2r} = 9,6$; $\Delta_1 = \Delta_2 = 0,2$; for $HE^{+}_{1,m,1}$ modes; ($m_1 = 20$;$m_2 = 24$): $k_0\Delta y = 0$; $k_0\Delta z = q_{1\perp} + q_{2\perp}$ (**b**, **c**); $k_0\Delta x = 0$; $k_0\Delta z = q_{1\perp} + q_{2\perp}$ (**d**, **e**)

Here $p_{sz} = \beta_{sz}L_s/2$; $q_{s\perp} = k_0 r_s$; $q_{sz} = k_0 L_s/2$; 2r0 r_s—is the radius; L L_s—is the height; ε_{sr}—relative dielectric permittivity of the s-th resonator ($s = 1, 2$);

$$
\begin{aligned}
F_{1s}(\xi) = {} & \frac{e_{0s}}{\beta_{sz}}\Bigg\{\Big[k_s^2\sqrt{1-\xi^2}J'_{m_s}(p_{s\perp})J_{m_s}\Big(q_{s\perp}\sqrt{1-\xi^2}\Big) \\
& -k_0\beta_s J_{m_s}(p_{s\perp})J'_{m_s}\Big(q_{s\perp}\sqrt{1-\xi^2}\Big)\Big] \\
& \cdot\frac{1}{\beta_{sz}^2-(k_0\xi)^2}\begin{bmatrix}(\beta_{sz}\cos p_{sz}\sin q_{sz}\xi - k_0\xi\sin p_{sz}\cos q_{sz}\xi)\\(\beta_{sz}\sin p_{sz}\cos q_{sz}\xi - k_0\xi\cos p_{sz}\sin q_{sz}\xi)\end{bmatrix} \\
& +\Big[\beta_s J_{m_s}(p_{s\perp})J'_{m_s}\Big(q_{s\perp}\sqrt{1-\xi^2}\Big) \\
& -k_0\sqrt{1-\xi^2}J'_{m_s}(p_{s\perp})J_{m_s}\Big(q_{s\perp}\sqrt{1-\xi^2}\Big)\Big] \\
& \cdot\frac{1}{\beta_s^2-k_0^2(1-\xi^2)}\begin{bmatrix}(k_s^2\xi\sin p_{sz}\cos q_{sz}\xi - k_0\beta_{sz}\cos p_{sz}\sin q_{sz}\xi)\\(k_s^2\xi\cos p_{sz}\sin q_{sz}\xi - k_0\beta_{sz}\sin p_{sz}\cos q_{sz}\xi)\end{bmatrix}\Bigg\} \\
& - h_{0s}m_s\frac{\xi(k_s^2-k_0^2)}{p_{s\perp}k_0\sqrt{1-\xi^2}}J_{m_s}(p_{\perp})J_{m_s}\Big(q_{s\perp}\sqrt{1-\xi^2}\Big) \\
& \cdot\frac{1}{\beta_{sz}^2-(k_0\xi)^2}\begin{bmatrix}(\beta_{sz}\sin p_{sz}\cos q_{sz}\xi - k_0\xi\cos p_{sz}\sin q_{sz}\xi)\\(\beta_{sz}\cos p_{sz}\sin q_{sz}\xi - k_0\xi\sin p_{sz}\cos q_{sz}\xi)\end{bmatrix};
\end{aligned}
\tag{6.94}
$$

$$F_{2s}(\xi) = \left(k_s^2 - k_0^2\right)\left\{ e_{0s} m_s \frac{1}{p_{s\perp} k_0^2 \sqrt{1-\xi^2}} J_{m_s}(p_{s\perp}) J_{m_s}\left(q_{s\perp}\sqrt{1-\xi^2}\right) \right.$$
$$- h_{0s} \frac{1}{\beta_s^2 - k_0^2(1-\xi^2)} \cdot \left[\beta_s J_{m_s}(p_{s\perp}) J'_{m_s}\left(q_{s\perp}\sqrt{1-\xi^2}\right)\right.$$
$$\left.\left. \cdot - k_0\sqrt{1-\xi^2} J'_{m_s}(p_{s\perp}) J_{m_s}\left(q_{s\perp}\sqrt{1-\xi^2}\right)\right]\right\}$$
$$\cdot \frac{1}{\beta_{sz}^2 - (k_0\xi)^2} \begin{bmatrix} (\beta_{sz} \sin p_{sz} \cos q_{sz}\xi - k_0\xi \cos p_{sz} \sin q_{sz}\xi) \\ (\beta_{sz} \cos p_{sz} \sin q_{sz}\xi - k_0\xi \sin p_{sz} \cos q_{sz}\xi) \end{bmatrix};$$

e_{s0};h_{s0}—are the normalized amplitudes, defined from (2.31);

$$\begin{aligned} w(s) = {} & \left[(e_{s0} - h_{s0})^2 \left[J_{m_s-1}^2(p_{s\perp}) - J_{m_s}(p_{s\perp}) J_{m_s-2}(p_{s\perp})\right]\right. \\ & + (e_{s0} + h_{s0})^2 \left[J_{m_s+1}^2(p_{s\perp}) - J_{m_s}(p_{s\perp}) J_{m_s+2}(p_{s\perp})\right] \\ & \left. + 2\left(\frac{\beta_s}{k_s} h_{s0}\right)^2 \left[J_{m_s}^2(p_{s\perp}) - J_{m_s-1}(p_{s\perp}) J_{m_s+1}(p_{s\perp})\right]\right] \begin{bmatrix} 2p_{sz} + \sin 2p_{sz} \\ 2p_{sz} - \sin 2p_{sz} \end{bmatrix} \\ & + \left[\left(\frac{k_s}{\beta_{sz}} e_{s0} - \frac{\beta_{sz}}{k_s} h_{s0}\right)^2 \left[J_{m_s-1}^2(p_{s\perp}) - J_{m_s}(p_{s\perp}) J_{m_s-2}(p_{s\perp})\right]\right. \\ & + \left(\frac{k_s}{\beta_{sz}} e_{s0} + \frac{\beta_{sz}}{k_s} h_{s0}\right)^2 \left[J_{m_s+1}^2(p_{s\perp}) - J_{m_s}(p_{s\perp}) J_{m_s+2}(p_{s\perp})\right] \\ & \left. + 2\left(\frac{\beta_s}{\beta_{sz}} e_{s0}\right)^2 \left[J_{m_s}^2(p_{s\perp}) - J_{m_s-1}(p_{s\perp}) J_{m_s+1}(p_{s\perp})\right]\right] \begin{bmatrix} 2p_{sz} - \sin 2p_{sz} \\ 2p_{sz} + \sin 2p_{sz} \end{bmatrix}; \end{aligned} \tag{6.95}$$

$k_s = \sqrt{\varepsilon_{sr}} k_0$; $k_0 = \omega_0/c$; ω_0—is the circular resonance frequency; c—is the light velocity.

In the case of different cylindrical resonators with equal parity of the field distribution (see Fig. 6.18a) relative to symmetry plane $A - A'$ (Fig. 6.16b):

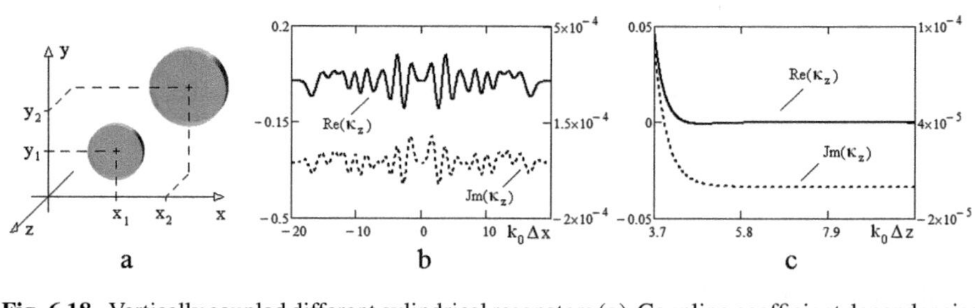

Fig. 6.18 Vertically coupled different cylindrical resonators (**a**). Coupling coefficient dependencies for the $HE^{+}_{1,m,1}$ modes;($m_1 = 20$; $m_2 = 24$):$\varepsilon_{1r} = 16$; $\varepsilon_{2r} = 9,6$ $\Delta_1 = \Delta_2 = 0,2$; $k_0\Delta y = 0$; $k_0\Delta z = 3,7$ (**b**); $k_0\Delta x = 0$; $k_0\Delta y = 1,8$ (**c**)

in the area: $\Delta z > L_1/2 + L_2/2$ obtains:

$$\kappa_z^{1,2} = i\kappa_0^{12} \cdot \int_0^\infty \frac{e^{-i\sqrt{1-\kappa^2}k_0|\Delta z|}}{\sqrt{1-\kappa^2}} \cdot \left\{(k_1^2 - k_0^2)(k_2^2 - k_0^2)C_1(\kappa)C_2(\kappa)^*\left[\cos(m_1 - m_2)\Delta\psi J_{m_1-m_2}(k_0\Delta\rho\kappa) \pm(-1)^{m_1m_2}\cos(m_1 + m_2)\Delta\psi J_{m_1+m_2}(k_0\Delta\rho\kappa)\right] + D_1(\kappa)D_2(\kappa)^*\left[\cos(m_1 - m_2)\Delta\psi J_{m_1-m_2}(k_0\Delta\rho\kappa) \mp(-1)^{m_1m_2}\cos(m_1 + m_2)\Delta\psi J_{m_1+m_2}(k_0\Delta\rho\kappa)\right]\right\}\kappa d\kappa, \tag{6.96}$$

where $\sin\Delta\psi = \Delta y/\Delta\rho$; $\Delta\rho = \sqrt{\Delta x^2 + \Delta y^2}$;

$$C_s(\kappa) = \left[e_{s0}m_s\frac{1}{p_{s\perp}k_0\kappa}J_{m_s}(p_{s\perp})J_{m_s}(q_{s\perp}\kappa) - \mathrm{h}_{s0}\frac{1}{\mathrm{r}_s\left(\beta_s^2 - k_0^2\kappa^2\right)} \cdot \left[p_{s\perp}J_{m_s}(p_{s\perp})J'_{m_s}(q_{s\perp}\kappa) - q_{s\perp}\kappa J'_{m_s}(p_{s\perp})J_{m_s}(q_{s\perp}\kappa)\right]\right\} \cdot \frac{1}{\beta_{sz}^2 - k_0^2(1-\kappa^2)}\begin{bmatrix}\beta_{sz}\sin p_{sz}\cos q_{sz}\sqrt{1-\kappa^2} - k_0\sqrt{1-\kappa^2}\cos p_{sz}\sin q_{sz}\sqrt{1-\kappa^2} \\ \beta_{sz}\cos p_{sz}\sin q_{sz}\sqrt{1-\kappa^2} - k_0\sqrt{1-\kappa^2}\sin p_{sz}\cos q_{sz}\sqrt{1-\kappa^2}\end{bmatrix}; \tag{6.97}$$

$$D_s(\kappa) = \frac{e_{s0}}{\beta_{sz}q_{s\perp}}\left\{\left[k_0^2p_{s\perp}J_{m_s}(p_{s\perp})J'_{m_s}(q_{s\perp}\kappa) - k_s^2q_{s\perp}\kappa J'_{m_s}(p_{s\perp})J_{m_s}(q_{s\perp}\kappa)\right] \cdot \frac{1}{\beta_{sz}^2 - k_0^2(1-\kappa^2)}\begin{bmatrix}k_0\sqrt{1-\kappa^2}\sin p_{sz}\cos q_{sz}\sqrt{1-\kappa^2} - \beta_{sz}\cos p_{sz}\sin q_{sz}\sqrt{1-\kappa^2} \\ \beta_{sz}\sin p_{sz}\cos q_{sz}\sqrt{1-\kappa^2} - k_0\sqrt{1-\kappa^2}\cos p_{sz}\sin q_{sz}\sqrt{1-\kappa^2}\end{bmatrix} - \frac{1}{\beta_s^2 - (k_0\kappa)^2}\left[p_{s\perp}J_{m_s}(p_{s\perp})J'_{m_s}(q_{s\perp}\kappa) - q_{s\perp}\kappa J'_{m_s}(p_{s\perp})J_{m_s}(q_{s\perp}\kappa)\right] \cdot \begin{bmatrix}k_0^2\beta_{sz}\cos p_{sz}\sin q_{sz}\sqrt{1-\kappa^2} - k_s^2k_0\sqrt{1-\kappa^2}\sin p_{sz}\cos q_{sz}\sqrt{1-\kappa^2} \\ k_s^2k_0\sqrt{1-\kappa^2}cosp_{sz}\sin q_{sz}\sqrt{1-\kappa^2} - k_0^2\beta_{sz}\sin p_{sz}\cos q_{sz}\sqrt{1-\kappa^2}\end{bmatrix}\right\} - h_{s0}m_s\frac{\left|\sqrt{1-\xi^2}\right|\left(k_s^2 - k_0^2\right)}{p_{s\perp}k_0\kappa}J_{m_s}(p_\perp)J_{m_s}(q_\perp\kappa) \cdot \frac{1}{\beta_{sz}^2 - k_0^2(1-\kappa^2)}\begin{bmatrix}\beta_{sz}\sin p_{sz}\cos q_{sz}\sqrt{1-\kappa^2} - k_0\sqrt{1-\kappa^2}\cos p_{sz}\sin q_{sz}\sqrt{1-\kappa^2} \\ k_0\sqrt{1-\kappa^2}\sin p_{sz}\cos q_{sz}\sqrt{1-\kappa^2} - \beta_{sz}\cos p_{sz}\sin q_{sz}\sqrt{1-\kappa^2}\end{bmatrix}.$$

The integral convergence provides by choice of the radical signs for $\xi > 1$: $\sqrt{\xi^2 - 1} = -i\sqrt{1-\xi^2}$ in the (6.94), as well as for $\sqrt{\kappa^2 - 1} = -i\sqrt{1-\kappa^2}$ in the (6.97).

The coupling between cylindrical resonator and the open space following from Helmholtz-Kirchhoff's integral theorem can be obtained as:

$$\kappa_s = i\tilde{k}_s = \frac{32i}{\varepsilon_{sr} w(s)} \cdot \frac{p_{sz}}{q_{sz}} \cdot \int_0^1 \left\{ \left(k_s^2 - k_0^2\right)^2 \left| C_s\left(\sqrt{1-\kappa^2}\right)\right|^2 + \left| D_s\left(\sqrt{1-\kappa^2}\right)\right|^2 \right\} d\kappa \tag{6.98}$$

Discovered relationships valid for any eigenoscillations of the cylindrical resonators in the open space, but a greatest interest presents the whispering gallery modes, as it's well known, possessed a highest possible quality.

Figures 6.17, 6.18, and 6.19 are showing coupling coefficient dependences on cylindrical resonator centers, composed of the dielectric with the relative permittivity $\varepsilon_{1r} = 16$; $\varepsilon_{2r} = 9, 6$ and the comparative dimensions $\Delta_s = L_s/2r_s = 0, 2$. Right ordinate axes on Fig. 6.17c–e; Figures 6.18 and 6.19 shows imaginary part of the coupling coefficients. It's clear, that the real part of the coupling coefficients has sufficiently visible values only in a region, where the resonator surfaces are close to each other (Figs. 6.17b–e, 6.18b–c, 6.19). Increase of the distance between resonator centers is accompanied by rapid coupling coefficient decrease. According to that, relative motion in the tangent direction leads to complex interference (Figs. 6.17b, d, 6.18b, 6.19a, c) of their mutual influence, determining by significant eigenmode field variation nearby their surface.

The greatest amount of coupling between resonators appears on its coaxial arrangement (Fig. 6.18b, c). The sign of the coupling coefficients extreme values is determined both by the azimuth numbers and mutual resonators position (Fig. 6.17c, e).

Imaginary parts of the coupling coefficients in many cases are more smooth functions on coordinates (Figs. 6.17b, d, 6.18b, 6.19a, c). For selected dielectric permittivity, its values are approximately one tenth degrees to the real parts.

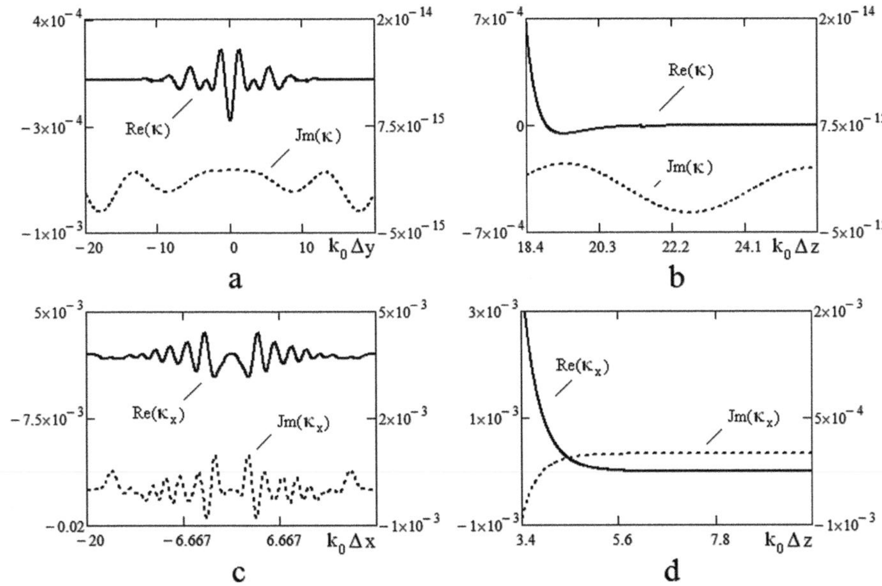

Fig. 6.19 Coupling coefficients as functions of the coordinates for different cylindrical resonators in the position showed in Fig. 6.17a (a, b) $k_0\Delta x = 0; k_0\Delta z = 1, 05(q_{1\perp} + q_{2\perp})$ (**a**); $k_0\Delta x = k_0\Delta y = 0$ (**b**); in the position showed in Fig. 6.18a (c, d); $k_0\Delta y = 0$; $k_0\Delta z = q_{1\perp} + q_{2\perp}$ (**c**); $k_0\Delta x = 0$; $k_0\Delta z = 3.7$ (**d**); $\varepsilon_{1r} = 16$; $\varepsilon_{2r} = 9, 6$; $\Delta_1 = \Delta_2 = 0, 2$, for the $EH^+_{1,m,1}$ modes

References

1. A.A. Trubin, A.S. Kupriianov, V.R. Tuz, Coupling coefficient of two rotated cylindrical dielectric resonators disposed in the free space. J. Electromagn. Waves Appl., 1731–1744 (2022)
2. M. Abramowitz, I.A. Stegun (eds.), *Handbook of Mathematical Functions with Formulas, Graphs and Mathematical Tables*, vol. 55 (National Bureau of Standarts. Applied Mathematics Ser., 1964), 830 p
3. A.A. Trubin, Mutual coupling coefficients of rotated rectangular dielectric resonators in open space. Visnyk NTUU KPI Seriia - Radiotekhnika Radioaparatobuduvannia **88**, 60–68 (2022)
4. A.S. Kupriianov, V.I. Fesenko, A.A. Trubin, V.R. Tuz, Coupling coefficients for dielectric cuboids located in free space. Appl. Opt. OSA **59**(23/10), 6918–6924 (2020)
5. H.A. Mavromatis, R.S. Alassar, A generalized formula for the integral of three associated legendre polynomials. Appl. Math. Letters **12**, 101–105 (1999)
6. S.-H. Dong, R. Lemus, The overlap integral of three associated legendre polynomials. Appl. Math. Letters **15**, 541–546 (2002)
7. Trubin A.A., *Lattices of Dielectric Resonators*, vol. 53 (Springer. Series in Advanced Microelectronics, 2016), 171 p
8. Gradshteyn I.S., Ryzhik I.M, *Table of Integrals, Series, and Products*, 7th edn (Academic Press, USA, 2007), 1171 p
9. H. Bateman, A. Erdelyi, *Higher Transcendental Functions*, vol. 1 (McGraw-Hill Book Company Inc., 1953), 302 p

Chapter 7
Coupled Oscillations of Dielectric Resonators

This chapter discusses systems of coupled DR located in a regular transmission line or in a section of a cutoff waveguide, as well as in open space.

The problem of natural oscillations of the resonator system is formulated. The field of natural oscillations is represented in general form of a superposition of the oscillation fields of partial resonators.

Using perturbation theory, a system of equations is derived for amplitudes of the field of partial DRs, obtained of a problem to determine the eigenfunctions and eigenvalues of a given coupling operator, expressed through the coupling coefficients discussed in previous sections. Normality conditions that determine the orthogonality of eigenfunctions of the coupling operator are found.

The general results obtained are illustrated by various examples of solving problems coupled oscillations of DRs located in a regular transmission line, in a section of a cutoff waveguide as well as in the open space.

In conclusion, the proposed approach to calculating parameters of coupled oscillations, based on perturbation theory, is generalized to DR systems of various shapes, possibly made from different dielectrics as well as DRs made from active material. It is noted that the found system of equations for complex amplitudes coincides with that previously derived in the case of identical DR.

Using perturbation theory, coupled oscillations of frequency-detuned DRs are modeled. The general properties of their coupled oscillations are investigated.

7.1 Perturbation Theory for Coupled Oscillations of Dielectric Resonators

Let us consider in general a system, consisting of N identical DRs, interacting with each other in some given external structure (Fig. 7.1a). Let us assume that each of the resonators, when removing the remaining resonators of the system, is characterized

A. Trubin, *Introduction to the Theory of Dielectric Resonators*, Springer Series in Advanced Microelectronics 65, https://doi.org/10.1007/978-3-031-65396-4_7

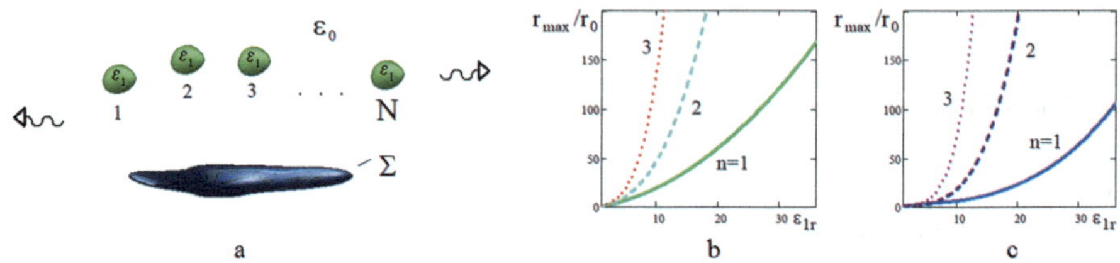

Fig. 7.1 Coupled oscillations of the N DR system in the external structure Σ. Area size dependences $r_{\max}/r_0$ of a spherical DR in open space, guaranteeing the fulfillment of inequalities (7.1) for the basis functions of the main magnetic H_{nml} (**b**) and electrical E_{nml} (**c**) types partial DR oscillations

by a given frequency ω_0 of one of its own oscillations. We denote by $(\mathbf{e}_s, \mathbf{h}_s) = \left(\mathbf{e}_s^{1,0}, \mathbf{h}_s^{1,0}\right)$ field of natural oscillations of the s-th DR, $(s = 1, 2, \ldots, N)$, where $\left(\mathbf{e}_s^1, \mathbf{h}_s^1\right)$ field in the material, and $\left(\mathbf{e}_s^0, \mathbf{h}_s^0\right)$—field of s-th DR in external space. We will assume that the indicated field components satisfy the boundary conditions on the resonator surface, as well as the radiation conditions for open resonators (Chap. 1). As we were convinced earlier by the example of solving the problem of natural oscillations of a spherical DR (Chap. 1), the functions of natural oscillations $(\mathbf{e}_s, \mathbf{h}_s)$ cannot be the basis for the expansion of the scattering field, since they are characterized by different laws of radiation in the far zone. There are several ways to solve this problem.

The total scattering field can be expanded into solutions to simpler scattering problems, for example, such as spherical waves. As we saw earlier [see (1.12), (1.13)], these solutions are close to the distribution functions of the natural oscillation field in the near zone, but satisfy the radiation conditions (1.1).

The second possible way out of this difficulty is the requirement of closeness to the exact solution of the linear combination of functions of natural oscillations $(\mathbf{e}_s, \mathbf{h}_s)$ not in the entire space as a whole, but only in a certain finite region, containing dielectric resonators. If such a solution is constructed with sufficient accuracy, the field in the remaining region can be reconstructed using Kirchhoff vector integrals [1].

Further, as basis functions, we will consider only the internal natural oscillations $(\mathbf{e}_s, \mathbf{h}_s)$ of each of the partial DRs (see Sect. 1.1), made of the same dielectric without losses. Taking into account the behavior of the field [2, 3] in the dielectric and the external region of the resonator, we use the inequalities:

for $s \neq t$

$$\begin{aligned} \left|\int_{V_t} (\mathbf{e}_t^1, \mathbf{e}_s^{0*})\mathrm{dv}\right| &<< \int_{V_t} |\mathbf{e}_t^1|^2 \mathrm{dv}; \\ \left|\int_{V_t} (\mathbf{h}_t^1, \mathbf{h}_s^{0*})\mathrm{dv}\right| &<< \int_{V_t} |\mathbf{h}_t^1|^2 \mathrm{dv}. \end{aligned} \tag{7.1}$$

Here V_t—dielectric volume of the t-th resonator, Inequalities (7.1) reflect the general property of accumulation of electromagnetic field energy in the volume of a dielectric in comparison with the energy stored in external space, characteristic of internal oscillations of the DR (Chap. 1).

Let us estimate the size of the region in which inequalities (7.1) are satisfied for each of the resonators of the system, using the radiation conditions (Sect. 1.1). As we have seen, the external field of the DR in an open spatial region, at large distances from its center, increases according to the law: $\exp(q''r/r_0)/(r/r_0)$, where $q = q' + iq''$, $(q'' > 0)$, —complex external characteristic parameter. From here we can estimate the maximum size of the region corresponding to the minimum of this function, which is achieved when $r_{\max}/r_0 = 1/q''$. On Fig. 7.1b, c, the dependence $r_{\max}/r_0$ is plotted for several lower types of oscillations of a spherical DR in open space. As can be seen from these data, in the case of basic magnetic oscillations H_{1m1}, $2r_{\max} > 100 \times (2r_0)$ already at $\varepsilon_{1r} \geq 30$.

Let us now denote by $\tilde{\omega}_s = \omega_0 + i\omega_s''$ the complex frequency of natural oscillations of the sth partial DR in the structure in the absence of other resonators, and by $\tilde{\omega} = \omega + i\omega''$—frequency of natural oscillations of the system of coupled DR as a whole. We will assume here that all partial DRs are made of lossless dielectric: $\varepsilon_1'' = 0$.

Let us consider sequentially bilinear combinations similar to (C.7), including the field of a system of coupled resonators ($\mathbf{e}$,$\mathbf{h}$), as well as fields of partial DR ($\mathbf{e}_n$, $\mathbf{h}_n$):

$$\begin{aligned}\operatorname{div}\{[\mathbf{h}_n^*, \mathbf{e}] + [\mathbf{h}, \mathbf{e}_n^*]\} &= \mathrm{i}(\omega - \omega_0)[\varepsilon_1 \mathbf{e} \cdot \mathbf{e}_n^* + \mu_0 \mathbf{h} \cdot \mathbf{h}_n^*] \\ &\quad - (\omega'' + \omega_n'')[\varepsilon_1 \mathbf{e} \cdot \mathbf{e}_n^* + \mu_0 \mathbf{h} \cdot \mathbf{h}_n^*],\end{aligned} \tag{7.2}$$

and in addition, the fields of each of the partial resonators:

$$\operatorname{div}\{[\mathbf{h}_n^*, \mathbf{e}_n] + [\mathbf{h}_n, \mathbf{e}_n^*]\} = -2\omega_n''[\varepsilon_1 |\mathbf{e}_n|^2 + \mu_0 |\mathbf{h}_n|^2]\cdot \tag{7.3}$$

We will look for a solution to the problem of natural oscillations of a system of N coupled DRs ($\mathbf{e}$, $\mathbf{h}$) in the form of an expansion in terms of the natural oscillations of the same but isolated resonators located in a given coordinates:

$$\mathbf{e} = \sum_{s=1}^{N} \mathrm{b_s} \mathbf{e}_s; \quad \mathbf{h} = \sum_{s=1}^{N} \mathrm{b_s} \mathbf{h}_s, \tag{7.4}$$

Substituting (7.4) into (7.2) and subtracting from it (7.3), multiplied by b_n, after integrating over the volume V_n of the n-th partial resonator, we find, taking into account (7.1):

$$\sum_{s=1}^{N} \frac{\mathrm{b_s}}{(1+\delta_{sn})} \cdot \oint_{S_n} \{[\mathbf{h}_n^*, \mathbf{e}_s] + [\mathbf{h}_s, \mathbf{e}_n^*]\} \mathbf{n} ds \approx 4\mathrm{i}\delta\omega \mathrm{b}_n \mathrm{w}_n - 4\omega'' \mathrm{b}_n \mathrm{w}_n \tag{7.5}$$

Here $\delta\omega = \omega - \omega_0$—shift of the real part of the natural oscillation frequency of the system relative to the natural oscillation frequency of isolated resonators; $w_n = 1/4\int_{V_n}\left(\varepsilon_1|\mathbf{e}_n|^2 + \mu_0|\mathbf{h}_n|^2\right)dv$—electromagnetic field energy stored in the n-th resonator material.

Recalling the definition of mutual coupling coefficients DR (6.1), we rewrite (7.5) in the form [4]:

$$\sum_{s=1}^{N} \kappa_{sn}\,\mathrm{b}_s - \lambda \mathrm{b}_n = 0 \quad (n = 1, 2, \ldots, N) \tag{7.6}$$

where

$$\lambda = 2(\tilde{\omega} - \omega_0)/\omega_0 = 2\left(\delta\omega/\omega_0 + \mathrm{i}\,\omega''/\omega_0\right) \tag{7.7}$$

Thus, the distribution of the amplitudes coupled oscillations of a system of identical resonators $\|\mathrm{b}_n\|$ is formulated as an eigenvalue problem for a finite-dimensional coupling operator $\mathrm{K} = \|\kappa_{sn}\|$:

$$K = \begin{pmatrix} i\tilde{k}_1 & \kappa_{21} & \kappa_{31} & \cdots & \kappa_{N1} \\ \kappa_{12} & i\tilde{k}_2 & \kappa_{32} & \cdots & \kappa_{N2} \\ \vdots & \vdots & \vdots & \vdots & \vdots \\ \kappa_{1,N-1} & \kappa_{2,N-1} & \kappa_{3,N-1} & i\tilde{k}_{N-1} & \kappa_{N,N-1} \\ \kappa_{1,N} & \kappa_{2,N} & \kappa_{3,N} & \cdots & i\tilde{k}_N \end{pmatrix} \tag{7.8}$$

Diagonal elements of the coupling operator matrix: determined only by the magnitude of the radiation of partial resonators (Sect. 2).

When deriving Eq. (7.6), the real parts $\mathrm{Re}(\kappa_{nn}) = 0$ are subtracted and do not have any effect on the coupled oscillations of the resonator system.

Equating to zero, the determinant of the system of equations (7.6),

$$\det\|\kappa_{sn} - \lambda\delta_{sn}\| = 0 \tag{7.9}$$

we find the characteristic equation, the solution of which determines the frequency splitting that arises due to the electromagnetic influence of the resonators on each other. In this case, each non-degenerate value of the frequency $\tilde{\omega}^s = \omega^s + \mathrm{i}\omega^{s''}$ $(s = 1, 2, \ldots, N)$ of the system's natural oscillations corresponds to its own column vector:

$$\left\|\mathrm{b}_t^s\right\| = \begin{pmatrix} \mathrm{b}_1^s \\ \mathrm{b}_2^s \\ \vdots \\ \mathrm{b}_N^s \end{pmatrix}, \quad (s = 1, 2, \ldots, N) \tag{7.10}$$

of the coupling operator $K(t = 1, 2, \ldots, N)$, determining the distribution of amplitudes of partial resonators. Thus, in the absence of degeneracy, a system consisting of N resonators is characterized by a $N \times N$ matrix of amplitudes of coupled oscillations:

$$B = \begin{pmatrix} b_1^1 & b_1^2 & \cdots & b_1^N \\ b_2^1 & b_2^2 & \cdots & b_2^N \\ \vdots & \vdots & \vdots & \vdots \\ b_N^1 & b_N^2 & \cdots & b_N^N \end{pmatrix} \tag{7.11}$$

We will denote the field of each of the natural oscillations of the system of coupled DRs $(\mathbf{e}^s, \mathbf{h}^s)$ by contravariant indices, and the field of partial resonators $(\mathbf{e}_t, \mathbf{h}_t)$ by covariant indices:

$$\mathbf{e}^s = \sum_{t=1}^{N} b_t^s \mathbf{e}_t; \quad \mathbf{h}^s = \sum_{t=1}^{N} b_t^s \mathbf{h}_t \tag{7.12}$$

Let us further define the scalar product, which is equal, taking into account approximation (7.1):

$$\begin{aligned} \left\langle \mathbf{e}^n, \mathbf{e}^s \right\rangle &= \int_V \left(\mathbf{e}^n, \mathbf{e}^{s*}\right) dv \approx \sum_{t=1}^{N} b_t^n b_t^{s*} \int_{Vt} |\mathbf{e}_t|^2 dv; \\ \left\langle \mathbf{h}^n, \mathbf{h}^s \right\rangle &= \int_V \left(\mathbf{h}^n, \mathbf{h}^{s*}\right) dv \approx \sum_{t=1}^{N} b_t^n b_t^{s*} \int_{Vt} |\mathbf{h}_t|^2 dv, \end{aligned} \tag{7.13}$$

where $V = \sum_{t=1}^{N} V_t$—sum of volumes occupied by partial resonators; $*$—is the complex conjugation symbol.

The resulting system of equations (7.6) has a fairly simple form, thanks to the use of approximation (7.1) and, (7.4), which includes expansion only in terms of the quality (internal) oscillations of isolated DR (Chap. 1). In reality, as is known, such a system of functions is not complete. A more rigorous approach involves the use of an extended set of basis functions, including both external natural oscillations (Sect. 1. 1) and functions of the continuous spectrum [5]. However, along this path a number of computational difficulties arise, which are a consequence of the fact that functions of the continuous spectrum, as well as fields corresponding to external oscillations, have not yet been constructed for any structure, with the exception of spherical DR in open space.

Let us now show that the eigenvalues of the coupling operator $K = \|\kappa_{sn}\|$ do not depend at all on the phase distributions of the relative amplitudes of the selected basis functions $(\mathbf{e}_s, \mathbf{h}_s)$. To do this, let's move from $(\mathbf{e}_s, \mathbf{h}_s)$ to new basis functions $(e^{i\psi_s}\mathbf{e}_s, e^{i\psi_s}\mathbf{h}_s)$ where $\psi_s$are arbitrary real constants.

Taking into account definition (6.1), we arrive at new expressions for the coefficients of mutual coupling: $\widehat{\kappa}_{sn} = e^{i(\psi_s-\psi_n)}\kappa_{sn}$. Accordingly, the matrix of the new coupling operator takes the form: $\widehat{\mathrm{K}} = \left\| e^{i(\psi_s-\psi_n)}\kappa_{\mathrm{sn}} \right\|$. The matrix $\widehat{K}$ is similar to the K matrix: $\widehat{K} = \psi^{-1}K\psi$ in the sense of [6], where $\psi = \left\| e^{i\psi_s}\delta_{sn} \right\|$—is a diagonal matrix, and therefore $\widehat{K}$ has the same characteristic equation.

Thus, we see that the new system of basis functions $(e^{i\psi_s}\mathbf{e}_s,\ e^{i\psi_s}\mathbf{h}_s)$ leads to the same frequencies of coupled oscillations of the resonators.

7.2 Coupled Oscillations of Dielectric Resonators in Regular Transmission Lines

The integral over the surface of the n-th DR (6.1) when the system is placed in a regular (Fig. 7.2a) or segment AB of an cutoff (Fig. 7.2b) waveguide is easily calculated, if we use expansion (2.3) and definition (3.1):

$$\oint_{\mathrm{Sn}} \left\{\left[\mathbf{h}_{\mathrm{n}}^*, \mathbf{e}_s\right] + \left[\mathbf{h}_s, \mathbf{e}_{\mathrm{n}}^*\right]\right\} \cdot \mathbf{n}\mathrm{ds} = 2\omega_0 w_{\mathrm{n}}(1+\delta_{\mathrm{sn}})\left(-\tilde{\mathrm{k}}_{\mathrm{sn}} + \mathrm{ik}_{\mathrm{sn}}\right) \tag{7.14}$$

where

$$\tilde{k}_{\mathrm{sn}} = \left(\tilde{k}_{\mathrm{sn}}\right)_0 \mathrm{e}^{-\mathrm{i}\Gamma|\Delta \mathrm{z}|}$$

—coefficient of mutual coupling of DR by propagating waves. Here, for simplicity, we will assume that only one type of wave can propagate in the waveguide: $\left(\tilde{k}_{\mathrm{sn}}\right)_0 = \left(\mathrm{c}_{\mathrm{s}}^{\pm}\right)_0\left(\mathrm{c}_{\mathrm{n}}^{\pm *}\right)_0/(\omega_0 w_{\mathrm{n}})$. Otherwise, it is necessary to proceed to the sum of all propagating waves, as was done in (3.40). $\left(\mathrm{c}_{\mathrm{s}}^{\pm}\right)_0$—expansion coefficients (2.3) found in a local coordinate system with the origin at the center of the s-th DR; Γ—longitudinal wave number of the line; $\Delta \mathrm{z} = \mathrm{z}_{\mathrm{s}} - \mathrm{z}_{\mathrm{n}}$.

The coefficients of mutual coupling k_{sn} were defined by us earlier in (3.1) [k_{nn} are not included in the Eq. (7.6)].

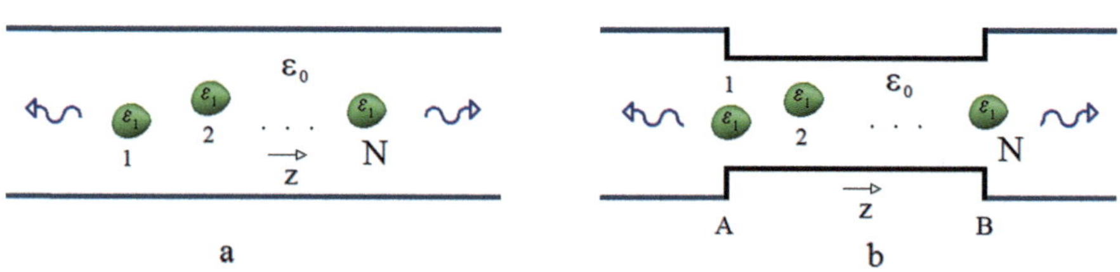

Fig. 7.2 Coupled oscillations of the DR in a regular waveguide (**a**), a section AB of a cut-off waveguide (**b**)

The resulting expression (7.14) establishes correspondence with Eq. (7.6):

$$\kappa_{sn} = k_{sn} + i\tilde{k}_{sn} \tag{7.15}$$

We discussed these coupling coefficients earlier in Chaps. 2, 3, 4, 5 and 6. Let us note here that in a regular waveguide the specified frequency splitting is determined equally by both the coefficients of mutual coupling of the resonators for damped k_{sn} and propagating $\tilde{k}_{sn}$ waves.

Let us consider, for simplicity, the case of identical partial DRs in the line. As follows from (3.1), in this case all coupling coefficients k_{sn} are real. Let us decompose the coupling matrix (7.8) into Hermitian and anti-Hermitian terms: $K = K_1 + iK_2$, where:

$$K_1 = \left\| k_{sn} + \left(\tilde{k}_{sn}\right)_0 \sin \Gamma|\Delta z| \right\|; \quad K_2 = \left\| \left(\tilde{k}_{sn}\right)_0 \cos \Gamma|\Delta z| \right\|$$

Whence it follows immediately that:

the operator matrix K is Hermitian if:

(a) All partial DRs have the same values $\tilde{k}_{nn} = \tilde{k}_0$ and are located in pairs in sections of regular waveguides at distances:

$$\Gamma \Delta z = \frac{\pi}{2}(2t+1), \quad (t = 0, 1, 2, \ldots)$$

(b) Each of the resonators is coupled only with a regular waveguide:

$$\tilde{k}_{sn} = \delta_{ns}\tilde{k}_0.$$

In these cases, the eigenvalues λ, or $\overline{\lambda} = \left(2\left(\delta\omega + i\omega''\right)/\omega_0\right) - i\tilde{k}_0$ are real [6]; the appearance of other resonators in the system changes only the frequencies of coupled oscillations.

the operator matrix K is anti-Hermitian if:

(c) Resonators are not coupled: $k_{sn} = 0$, and in a regular waveguide they are located at distances:

$$\Gamma \Delta z = t\pi \quad (t = 0, 1, 2, \ldots)$$

(d) Coupling coefficients are subject to the conditions:

$$k_{sn} = -\left(\tilde{k}_{sn}\right)_0 \sin \Gamma|\Delta z|$$

In these cases, the eigenvalues λ are imaginary; the frequencies of coupled oscillations coincide with the frequencies of partial DR, and the resonators of the system influence each other, changing only the amount of radiation.

(e) The matrix K takes the form of a cellular one: $\mathrm{K} = \mathrm{K}_1 \oplus \mathrm{K}_2 \oplus \cdots$ [6], where

$$K_s = \begin{pmatrix} i\tilde{k}_s & \kappa_s \\ \pm\kappa_s & i\tilde{k}_s \end{pmatrix}$$

—normal matrix corresponding to the natural oscillations of a pair of coupled DRs in a waveguide.

In all the listed cases, the operator is normal and the following **Theorem** is valid: If one of the conditions (a)–(e) is met, the vectors of the amplitudes of natural oscillations of the system of coupled DRs are orthogonal to each other:

$$\sum_{t=1}^{N} b_t^n b_t^{s*} = \delta_{ns} \sum_{t=1}^{N} \left|b_t^n\right|^2 \tag{7.16}$$

and the fields of various coupled oscillations (7.12) are also asymptotically orthogonal (here δ_{ns}—Kronecker symbol).

Indeed, equality (7.16) holds provided that the K operator is normal, and from (7.12), (7.13) it follows that the second statement is satisfied:

$$\left\langle \mathbf{e}^n,\ \mathbf{e}^s \right\rangle \approx \delta_{ns}\left\langle \mathbf{e}^n,\ \mathbf{e}^n \right\rangle;\ \left\langle \mathbf{h}^n,\ \mathbf{h}^s \right\rangle \approx \delta_{ns}\left\langle \mathbf{h}^n,\ \mathbf{h}^n \right\rangle \tag{7.17}$$

It is easy to see that it is almost impossible to implement a structure of more than 2 coupled resonators in a regular waveguide, described by a Hermitian coupling operator [cases (a), (b)]. While the conditions for the anti-Hermitian operator [(c), (d)] can be realized quite simply.

Let us consider particular solutions of the system of equations (7.6).

System of two coupled DRs in a regular waveguide. Let us depict, for clarity, the structure of connections between DR, indicated by circles, for non-propagating types of waves as a solid line, and for propagating waves as a wavy line. In a regular waveguide, such a diagram of coupled resonators has the form shown in Fig. 7.3a. In this case, the solution to system (7.6) has the form:

$$\begin{aligned} b_1^{1,2} &= \frac{i}{2}\left(\tilde{k}_1 - \tilde{k}_2\right) \mp \sqrt{d}; \\ b_2^{1,2} &= k_{12} + i\tilde{k}_{12} \mathrm{e}^{-i\Gamma|\Delta z|}, \end{aligned} \tag{7.18}$$

where;

$$d = \left(k_{12} + i\tilde{k}_{12}\mathrm{e}^{-i\Gamma|\Delta z|}\right) \cdot \left(k_{21} + i\tilde{k}_{21}\mathrm{e}^{-i\Gamma|\Delta z|}\right) - \frac{1}{4}\left(\tilde{k}_1 - \tilde{k}_2\right)^2$$

$$\lambda^{1,2} = \frac{i}{2}\left(\tilde{k}_1 + \tilde{k}_2\right) \mp \sqrt{d}.$$

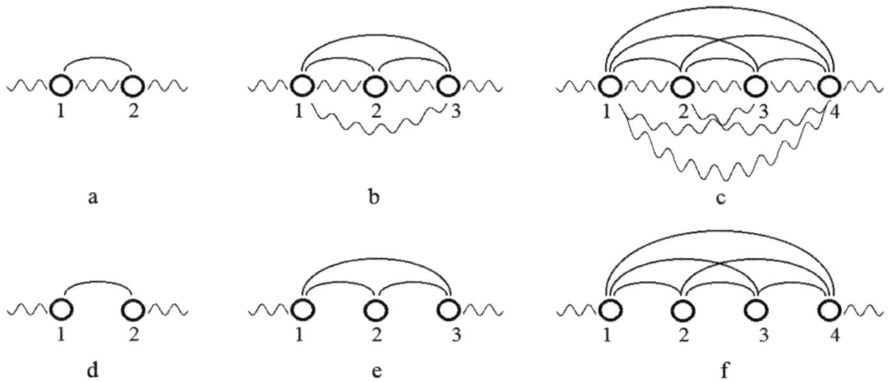

Fig. 7.3 Diagrams of coupled oscillations of two (**a**, **d**); three (**b**, **e**); four (**c**, **f**) DRs in a regular (**a**–**c**), section of the cut-off (**d**–**f**) waveguide

Here and below, for brevity, the index "0" for the coupling coefficients (7.14) in (7.18) is omitted. It is easy to verify that the amplitude orthogonality condition (7.16) is satisfied if both resonators are oriented in the same way and have the same transverse coordinates in the line.

Then

$$\tilde{k}_{12} = \tilde{k}_{21} = \tilde{k}_1 = \tilde{k}_2;\ k_{12} = k_{21}$$

and

$$\begin{aligned} \mathrm{Re}\ \lambda^{1,2} &= \mp\left(k_{12} + \tilde{k}_1 \sin \Gamma|\Delta z|\right); \\ \mathrm{Im}\ \lambda^{1,2} &= \tilde{k}_1 (1 \mp \cos \Gamma|\Delta z|); \end{aligned} \tag{7.19}$$

$$b_2^{1,2} = \mp b_1^{1,2}.$$

As follows from the solution obtained, the coupled oscillations of a system of two identical DRs are split into oscillations of two types: in-phase, characterized by the same amplitude signs, and antiphase, with amplitudes of opposite signs [8]. In this case, the field distribution of a symmetrical structure also corresponds to the condition of a magnetic or electric wall in the symmetry plane: $z = 0$. As noted above, the frequencies of such coupled oscillations in a regular waveguide depend not only on the coupling coefficient k_{12}, but also on the magnitude of the radiation.

A system of three DRs in a regular waveguide, which is symmetrical relative to the center of the second resonator $|\Delta z| = |z_2 - z_1| = |z_3 - z_2|$ (Fig. 7.3b) in a regular waveguide is characterized by the parameters:

$$\lambda^{1,2} = \frac{1}{2} k_{31} + \frac{i}{2}\left[\tilde{k}_1 \left(1 + \mathrm{e}^{-2i\Gamma|\Delta z|}\right) + \tilde{k}_2\right] \mp \sqrt{d};$$

$$\lambda^3 = -k_{31} + i\tilde{k}_1\left(1 - e^{-2i\Gamma|\Delta z|}\right), \tag{7.20}$$

where

$$\begin{aligned} d = &\left[\frac{i}{2}\left(\tilde{k}_1 - \tilde{k}_2 + \tilde{k}_1 e^{-2i\Gamma|\Delta z|}\right) + \frac{1}{2}k_{31}\right]^2 \\ &+ 2\left(k_{12} + i\tilde{k}_{12}e^{-i\Gamma|\Delta z|}\right)\left(k_{21} + i\tilde{k}_{21}e^{-i\Gamma|\Delta z|}\right); \\ b_3^{1,2} &= b_1^{1,2} = \lambda^{1,2} - i\tilde{k}_2; \quad b_2^{1,2} = 2\left(k_{12} + i\tilde{k}_{12}e^{-i\Gamma|\Delta z|}\right); \\ b_3^3 &= -b_1^3; \quad b_2^3 = 0. \end{aligned} \tag{7.21}$$

Thus, we see that in the symmetrical structure of three DRs, two in-phase $\left(b_3^{1,2} = b_1^{1,2}\right)$ and one anti-phase $\left(b_3^3 = -b_1^3\right)$ types of coupled oscillations arise. The frequencies of these oscillations are distributed equidistantly, provided:

$$k_{31} + \tilde{k}_1 \sin 2\Gamma|\Delta z| = 0 \tag{7.22}$$

In general, the frequencies of coupled oscillations are distributed in a rather complex way, even in a simple system of three resonators (Fig. 7.4). It is easy to verify that in-phase oscillations are no longer orthogonal in the sense (7.16).

System of four DRs in a regular waveguide. Let us assume that the resonators are also located in pairs symmetrically relative to the plane $z = 0$ (Fig. 7.3c). We denote the longitudinal coordinates of 2 and 3 DR by: $z_{23} = \mp\Delta z_2/2$, as well as 1 and 4—by $z_{14} = \mp(\Delta z_1 + \Delta z_2/2)$. The solution to the system of equations (7.6) shows that in this case there are two even-type oscillations:

$$\begin{aligned} b_4^+ &= b_1^+ = \lambda^+ - k_{32} - i\tilde{k}_2\left(1 + e^{-i\Gamma|\Delta z_2|}\right); \\ b_3^+ &= b_2^+ = k_{12} + k_{13} + i\tilde{k}_{12}e^{-i\Gamma|\Delta z_1|}\left(1 + e^{-i\Gamma|\Delta z_2|}\right); \end{aligned} \tag{7.23}$$

and two odd:

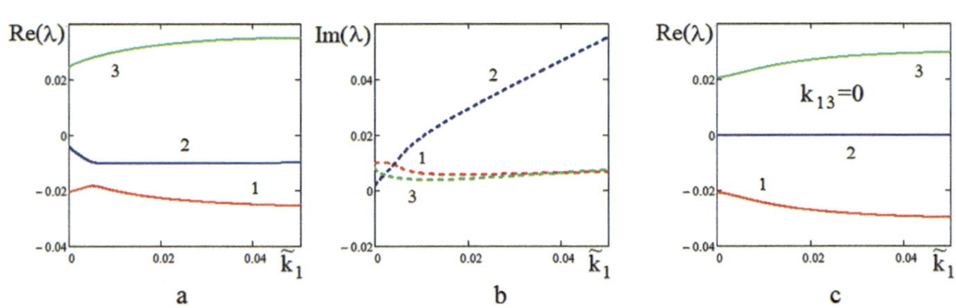

Fig. 7.4 Dependence of the relative values of the complex frequencies of coupled oscillations of 3 DR in the line on the coupling coefficient of the first resonator (**a**, **b**; $\tilde{k}_2 = \tilde{k}_3 = 0.01$; $k_{12} = 0.005$; $k_{13} = k_{23} = 0.01$; $\Gamma\Delta z_{12} = \Gamma\Delta z_{23} = 0.5\pi$; $\Gamma\Delta z_{13} = \Gamma\Delta z_{12} + \Gamma\Delta z_{23}$). In the case $k_{13} = 0$; $k_{23} = 0.01$ (**c**)

$$\begin{aligned} b_4^- &= -b_1^- = \lambda^- + k_{32} - i\tilde{k}_2\left(1 - e^{-i\Gamma|\Delta z_2|}\right); \\ b_2^- &= -b_3^- = k_{12} - k_{13} + i\tilde{k}_{12}e^{-i\Gamma|\Delta z_1|}\left(1 - e^{-i\Gamma|\Delta z_2|}\right). \end{aligned} \quad (7.24)$$

Here

$$\begin{aligned} \lambda^+ &= \frac{1}{2}(k_{41} + k_{32}) + \frac{i}{2}\left[\tilde{k}_1\left(1 + e^{-i\Gamma(2|\Delta z_1|+|\Delta z_2|)}\right) + \tilde{k}_2\left(1 + e^{-i\Gamma|\Delta z_2|}\right)\right] \mp \sqrt{d^+}; \\ \lambda^- &= -\frac{1}{2}(k_{41} + k_{32}) + \frac{i}{2}\left[\tilde{k}_1\left(1 - e^{-i\Gamma(2|\Delta z_1|+|\Delta z_2|)}\right) + \tilde{k}_2\left(1 - e^{-i\Gamma|\Delta z_2|}\right)\right] \mp \sqrt{d^-}; \\ d^\pm &= \left\{\pm\frac{1}{2}(k_{41} - k_{32}) + \frac{i}{2}\left[\tilde{k}_1\left(1 \pm e^{-i\Gamma(2|\Delta z_1|+|\Delta z_2|)}\right) - \tilde{k}_2\left(1 \pm e^{-i\Gamma|\Delta z_2|}\right)\right]\right\}^2 \\ &\quad + \left[k_{12} \pm k_{13} + i\tilde{k}_{12}e^{-i\Gamma|\Delta z_1|}\left(1 \pm e^{-i\Gamma|\Delta z_2|}\right)\right] \\ &\quad \left[k_{21} \pm k_{31} + i\tilde{k}_{21}e^{-i\Gamma|\Delta z_1|}\left(1 \pm e^{-i\Gamma|\Delta z_2|}\right)\right]. \end{aligned} \quad (7.25)$$

In a special case: $\Gamma|\Delta z_1| = \pi/2$; $\Gamma|\Delta z_2| = 0$ and $k_{41} = k_{32}$

$$\begin{aligned} &\mathrm{Im}\,\lambda^+ = \tilde{k}_2; \quad \mathrm{Im}\,\lambda^- = \tilde{k}_1; \\ &\mathrm{Re}\,\lambda^+ = k_{41} \mp \sqrt{d^+}; \quad \mathrm{Re}\,\lambda^- = -\mathrm{k}_{41} \mp \sqrt{\mathrm{d}^-}; \end{aligned} \quad (7.26)$$

where

$$d^+ = \left|k_{12} + k_{13} + 2\tilde{k}_{12}\right|^2 - \tilde{k}_2^2; \quad d^- = |k_{12} - k_{13}|^2 - \tilde{k}_1^2.$$

It follows that the frequencies of coupled oscillations of a structure consisting of 4 DR located from each other at a distance $|\Delta z| = \lambda/4$, where $\lambda = 2\pi/\Gamma$, are split in pairs symmetrically with respect to ω_0, if the coupling coefficient between the outer resonators is zero.

7.3 Coupled Oscillations of Dielectric Resonators in Cutoff Waveguides

Let us now consider the coupled oscillations of the DR located in a section AB of the evanescent waveguide (Fig. 7.2b). Let us assume that only a pair of outer resonators of such a system is coupled to regular waveguides (Fig. 7.3d–f). In this case, *K* matrix (7.8) becomes sparse.

A system of two coupled DRs in a cutoff waveguide. For two coupled DR (Fig. 7.3d) with $\tilde{k}_1 = \tilde{k}_2$, the solution to the system (7.6) takes the simple form:

$$\mathrm{Re}\,\lambda^\mp = \mp\sqrt{k_{12}k_{21}};\ \mathrm{Im}\,\lambda^\mp = \tilde{k}_1;$$

$$b_2^{1,2}/b_1^{1,2} = \mp\sqrt{k_{12}k_{21}}/k_{21}. \quad (7.27)$$

From this we see that if $k_{21} > 0$, then the frequency of the in-phase oscillation ω^+ lies above the frequency ω_0, and the frequency of the anti-phase oscillation ω^- lies below the frequency ω_0. Substituting (3.1) into (7.27), we find:

$$\left|b_2^{1,2}/b_1^{1,2}\right|^2 = w_1/w_2 \text{ or } \left|b_1^{1,2}\right|^2 w_1 = \left|b_2^{1,2}\right|^2 w_2,$$

that is, the energies stored by each of the resonators are equal to each other. In addition, this leads to an interesting conclusion that the ratio of the amplitudes of coupled oscillations in a system of two DRs in a evanescent waveguide in the considered approximation does not depend on their relative position, but is determined only by the shape and material, A similar energy equality occurs in a system of two different coupled LC circuits, pre-tuned to the same frequency.

The structure shown in Fig. 7.3d, is convenient for measuring the mutual coupling coefficient of resonators, usually defined as:

$$k = \left(\omega^+ - \omega^-\right)/\omega_0 = \sqrt{k_{12}k_{21}}. \quad (7.28)$$

System of three DRs in a cutoff waveguide. The solution to the system of equations (7.6) for the case of three DRs located in the evanescent waveguide is also symmetrical with respect to the 2nd resonator (Fig. 7.3e) for $\tilde{k}_1 = \tilde{k}_3$ has the form:

$$\lambda^{1,2} = \frac{1}{2}\left(i\tilde{k}_1 + k_{13}\right) \mp \frac{1}{2}\sqrt{d};$$
$$\lambda^3 = i\tilde{k}_1 - k_{13} \quad (7.29)$$

where the $\lambda^{1,2}$ eigenvalues correspond to even oscillations

$$b_3^{1,2} = b_1^{1,2} = \lambda^{1,2}; \quad b_2^{1,2} = 2k_{12} \quad (7.30)$$

and λ^3—odd oscillation:

$$b_3^3 = -b_1^3; \quad b_2^3 = 0.$$

Here

$$d = \left(i\tilde{k}_1 + k_{13}\right)^2 + 8k_{12}^2.$$

The frequencies of coupled oscillations are located symmetrically relative to ω_0 on the condition: $k_{13} = 0$.

A system of four coupled DRs in a cutoff waveguide. For a system of four DRs located in an evanescent waveguide 1, 4 in pairs symmetrically 2, 3 (Fig. 7.3f) with

$\tilde{k}_1 = \tilde{k}_4$ we also get two even oscillations:

$$b_4^+ = b_1^+ = \lambda^+ - k_{23}; \quad b_3^+ = b_2^+ = k_{12} + k_{13} \tag{7.31}$$

and two odd oscillations:

$$b_1^- = -b_4^- = \lambda^- + k_{23}; \quad b_2^- = -b_3^- = k_{12} - k_{13} \tag{7.32}$$

Here

$$\lambda^+ = \frac{i}{2}\tilde{k}_1 + \frac{1}{2}(k_{14} + k_{23}) \mp \sqrt{d^+};$$

$$\lambda^- = \frac{i}{2}\tilde{k}_1 - \frac{1}{2}(k_{14} + k_{23}) \mp \sqrt{d^-};$$

$$d^{\pm} = \left[\frac{i}{2}\tilde{k}_1 \pm (k_{14} - k_{23})/2\right]^2 + (k_{12} \pm k_{13})^2.$$

In general, the frequencies of these coupled oscillations are distributed in a rather complex manner (Fig. 7.5).

The solutions found allow us to determine simple conditions for pairwise equidistant frequency splitting:

if $k_{12} = 0$; $k_{34} = 0$ (Fig. 7.6a, d) and

$$k_{23} = k_{14} = \sqrt{4k_{13}^2 - \tilde{k}_1^2}; \tag{7.33}$$

if $k_{13} = 0$; $k_{24} = 0$ (Fig. 7.6b, e) and

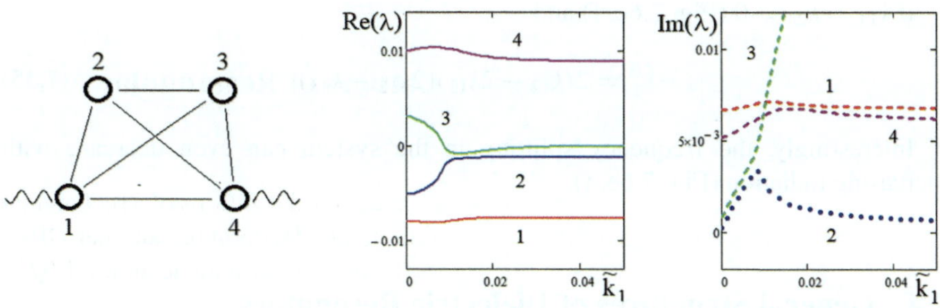

a b c

Fig. 7.5 Topology coupling of 4 DRs in a section of **a** evanescent waveguide. Relative values of the complex frequencies of coupled oscillations of 4 DR in the evanescent waveguide with variation of the coupling coefficient of the first resonator (**b**, **c**): $\widehat{k_4} = 0.013$; $k_{st} = k_{ts}$; $k_{12} = 0.005$; $k_{13} = 0$; $k_{14} = k_{24} = 0.003$; $k_{23} = 0.002$; $k_{34} = 0.01$

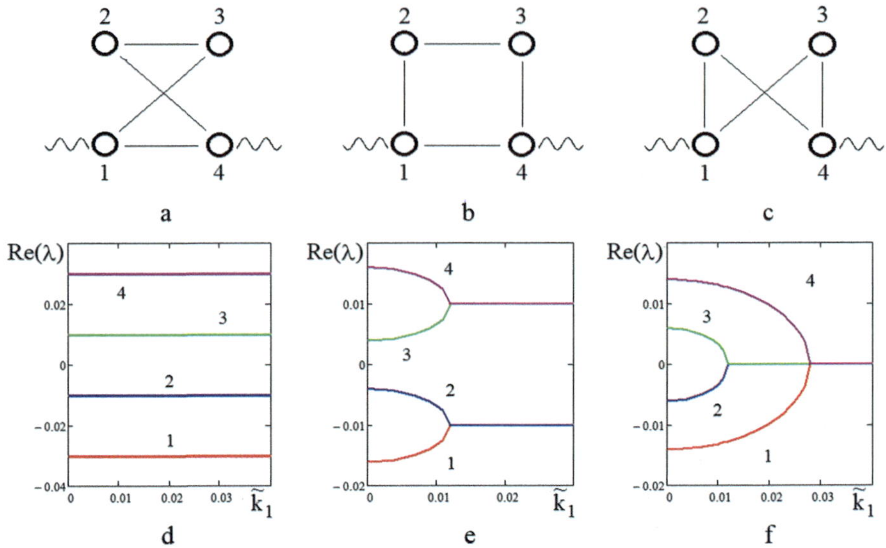

Fig. 7.6 Topologies of coupling frequency-symmetric oscillations of 4 DR in a section of a evanescent waveguide (**a–c**). Dependence of frequencies on the coupling coefficients of the outer resonators with the waveguide (**d–f**); $\overset{\frown}{k_4} = \overset{\frown}{k_1}$; $k_{st} = k_{ts}$; **d** $k_{12} = k_{34} = 0$; $k_{13} = k_{24} = 0.001$; $k_{14} = 0.03$; $k_{23} = 0.01$); **e** $k_{13} = k_{24} = 0$; $k_{12} = k_{34} = 0.006$; $k_{14} = k_{23} = 0.01$); **f** $k_{14} = k_{23} = 0$; $k_{12} = k_{34} = 0.01$; $k_{13} = k_{24} = 0.004$)

$$k_{23} = k_{14} = \sqrt{4k_{12}^2 - \tilde{k}_1^2} \tag{7.34}$$

In this case, the relative distance between two nearby frequencies becomes equal: $\delta\omega/\omega_0 = k_{14}/2$.

As well as,

if $k_{14} = k_{23} = 0$ (Fig. 7.6c, f) and

$$\tilde{k}_1^2 = 2(k_{12} - 2k_{13})(2k_{12} - k_{13}) \tag{7.35}$$

Interestingly, the frequency splitting in the system can even decrease with increasing radiation (Fig. 7.6e, f).

7.4 General Structures of Dielectric Resonators in Microwave Devices

The scattering parameters of electromagnetic waves on various systems are most often displayed using the frequency distribution of the scattering matrix [7]. In this case, the condition of symmetry of its elements in a given frequency band is often set. Let us define the condition for frequency symmetry of the scattering matrix in the form:

$$|S_{mn}(\delta\omega)| = |S_{mn}(-\delta\omega)|, \tag{7.36}$$

where $\delta\omega = \omega - \omega_0$; ω_0, ω—center and current frequency, respectively.

The requirement for frequency symmetry of the scattering matrix is most often set when designing bandpass filters on DR. The presence of uncontrolled coupling between partial filter resonators in the general case usually leads to a noticeable asymmetry in the amplitude-frequency response.

Let us consider in more detail the problem of determining DR configurations that implement the frequency-symmetric characteristics of the scattering matrix. We will show that the definition of such configurations significantly narrows the number of possible options for constructing bandpass filters, which significantly simplifies the process of their design.

Let us pose the question, under what conditions do the elements of the scattering matrix become frequency symmetric? Taking into account that the frequency characteristics of the elements of the scattering matrix in the resonant region are determined mainly by the distribution of complex frequencies of coupled oscillations of the DR system (7.6), we will require that each complex frequency of coupled oscillations, characterized by detuning $\delta\omega^1 = \omega^1 - \omega_0$ and the imaginary part ω_1'' that determines the radiation, corresponds to the frequency of coupled oscillations with detuning $\delta\omega^2 = -\delta\omega^1$ and the same imaginary part ω_1'', i.e. between these eigenvalues of the coupling operator (7.6) the condition must be satisfied:

$$\lambda^2 = -(\lambda^1)^* \tag{7.37}$$

Here "*"—complex conjugate symbol. We will call solutions λ^1, λ^2 satisfying equality (7.37) anticonjugate.

Let us now consider the characteristic equation that arises when the determinant of the system of equations is set equal to zero (7.9):

$$\lambda^N + \alpha_1\lambda^{N-1} + \alpha_2\lambda^{N-2} + \cdots + \alpha_{N-1}\lambda + \alpha_N = 0 \tag{7.38}$$

where $\alpha_s = \alpha_s(\tilde{k}, k_{12}, k_{13}, \ldots)$—known complex functions of coupling coefficients (Sect. 7.2).

Let us determine the conditions under which Eq. (7.38) will have at least one pair of anticonjugate solutions. For this purpose, we write the complex conjugate equation (7.38) for λ^1 and use equality (7.37). Let's remove the "–" sign from under the degree sign in each term. Comparing the obtained result with the original equation (7.38), we note that λ^2 it will also satisfy the same characteristic equation if the following conditions are met:

$$\alpha_{2s}^* = \alpha_{2s}; \quad \alpha_{2s+1}^* = -\alpha_{2s+1}, \tag{7.39}$$

or vice versa:

$$\alpha_{2s}^* = -\alpha_{2s}; \ \alpha_{2s+1}^* = \alpha_{2s+1}.$$

Thus, we find that equation (7.38) has at least one pair of anticonjugate solutions if it's even coefficients are real (or imaginary) and it's odd coefficients are purely imaginary (real) [8].

Now we can formulate the main

Statement: If all solutions of the characteristic equation of the coupling operator K, with the exception of real and purely imaginary ones, are pairwise anti-conjugate, then the amplitude-frequency characteristics of the scattering parameters S on a system of coupled DRs become frequency-symmetric.

Indeed, the natural oscillations of the DR system, corresponding to real λ, do not make any contribution to the scattered field. The magnitudes of the amplitudes of coupled oscillations corresponding to pairwise anti-conjugate eigenvalues are equal to each other. Moreover, pairwise symmetrical natural oscillations of the DR system corresponding to anti-conjugate solutions are characterized by the same quality factor, but differ only in the sign of the frequency difference. The superposition of such oscillations forms a symmetrical frequency response of the device as a whole.

The obtained simple relations (7.39) allow us to determine the conditions for the implementation of the frequency-symmetric characteristics of the S matrix even before solving the characteristic equation (7.38), which significantly simplifies the procedure for determining the structure of couplings of resonators and calculating parameters of their coupled oscillations.

As an example, let us consider, for simplicity, a symmetrical configuration consisting of 5 coupled DRs in a section of an evanescent waveguide, two of which can radiate to a regular transmission line.

Using conditions (7.39), we determine possible configurations of such a system that realize frequency-symmetric scattering characteristics. For this purpose, we give the characteristic equations corresponding to even oscillations of the system the form (7.38):

$$\begin{aligned}
&\lambda^3 - \left(i\tilde{k}_1 + k_{24} + k_{15}\right)\lambda^2 \\
&\quad + \left[\left(i\tilde{k}_1 + k_{15}\right)\cdot k_{24} - (k_{12} + k_{14})^2 - 2\left(k_{23}^2 + k_{13}^2\right)\right]\cdot\lambda \\
&\quad + 2\left(i\tilde{k}_1 + k_{15}\right)\cdot k_{23}^2 + 4k_{13}k_{23}(k_{12} + k_{14}) - 2k_{13}^2 k_{24} = 0.
\end{aligned}$$

From here we find that the desired symmetry will take place if the conditions are met: $k_{15} = k_{24} = 0$ and or $k_{13} = 0$ (Fig. 7.7a, d); or $k_{23} = 0$ (Fig. 7.7b, e); or $k_{12} = -k_{14}$ (Fig. 7.7c, f). In practice, the easiest to implement are filters that realizes coupling of the first type, shown in Fig. 7.7a, which we will call filters with serial connections. The corresponding dependences of the real values $\mathrm{Re}[\lambda] = 2\delta\omega/\omega_0$

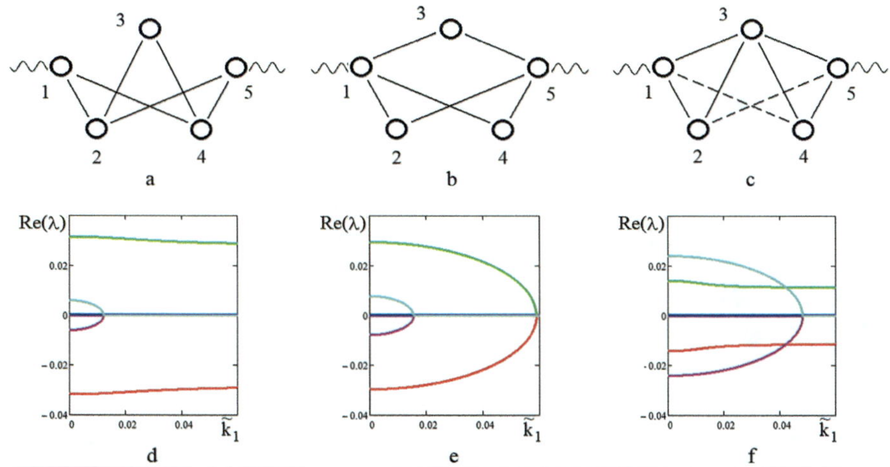

Fig. 7.7 Topologies of coupling frequency-symmetric oscillations of 5 DR in a section of an evanescent waveguide (**a–c**). Dependence of frequencies on the coupling coefficients of the outer resonators with the regular waveguide $\widehat{k_5} = \widehat{k_1}$; $k_{st} = k_{ts}$ (**d–f**); **d** $k_{13} = k_{35} = k_{24} = 0; k_{12} = k_{45} = 0.01$; $k_{14} = k_{25} = 0.004$; $k_{23} = k_{34} = 0.02$; **e** $k_{15} = k_{23} = k_{34} = k_{24} = 0$; $k_{12} = k_{45} = 0.008$; $k_{13} = k_{35} = 0.02; k_{14} = k_{25} = 0.0002$); **f** $k_{15} = k_{24} = 0$; $k_{12} = k_{45} = 0.012$; $k_{14} = k_{25} = -0.012; k_{23} = k_{34} = 0.008$

on the coupling coefficient of the resonators with a regular line (Fig. 7.7d–f) confirm the fulfillment of conditions (7.39).

7.5 Coupled Oscillations of Dielectric Resonators in Open Space

To analyze the coupled oscillations of the DR system in open space, we solve the system of equations (7.6) with coupling coefficients, discussed in Ch. 6. Let us first limit ourselves to clusters consisting of several DRs, for example, rectangular in shape (Figs. 7.8, 7.9, and 7.10).

In the general case, the solution to the system of equations (7.6) for two identical DRs has the form:

$$b_1^s = \lambda^s - i\tilde{k}_2; \quad b_2^s = \kappa_{12} \tag{7.40}$$

where

$$\lambda^{1,2} = i/2(\tilde{k}_1 + \tilde{k}_2) \mp \sqrt{\kappa_{12}\kappa_{21} - 1/4(\tilde{k}_1 - \tilde{k}_2)^2}$$

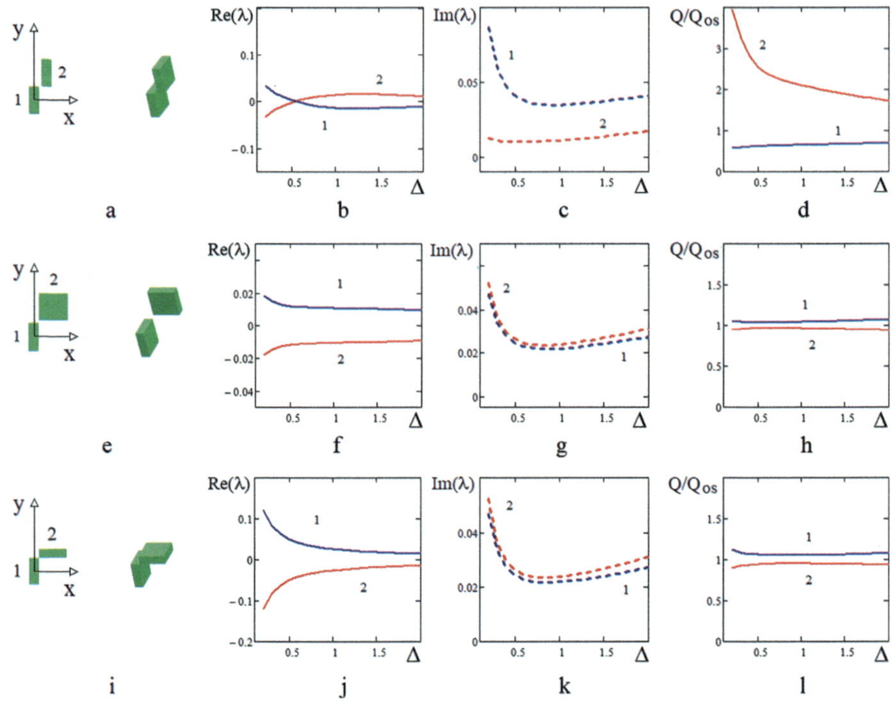

Fig. 7.8 Characteristics of natural oscillations of clusters of two rectangular DRs at different positions in open space (**a**, **e**, **i**). Dependencies of relative frequencies (**b**, **c**; **f**, **g**; **j**, **k**); dependence of the radiation quality factor (**h**, **l**) on the shape of the resonators $\Delta = L/a_0$ for H_{111} oscillations and $\varepsilon_{1r} = 36; B = 1$. Q—quality factor of the coupled oscillation of the resonator system; Q_{OS}—partial resonator quality factor

The characteristics of coupled oscillations of two rectangular DRs, calculated on the basis of (7.40) and coupling coefficients (Sect. 6.4) are shown in Fig. 7.8. Here and below, the quality factor of the s-th coupled oscillation of the DR system is defined as: $Q^s = \mathrm{Re}(\tilde{\omega}^s)/2\mathrm{Im}(\tilde{\omega}^s)$.

As can be seen from these calculations, the spatial arrangement of the resonators shown in Fig. 7.8a, e, i, is characterized by the frequency symmetry of coupled oscillations (Fig. 7.8b, f, j) mainly due to the minimization of mutual coupling. At the same time, the quality factors of coupled oscillations remain approximately the same (Fig. 7.8h, l) due to the equality of the imaginary parts of complex frequencies (Fig. 7.8g, k) in a wide range of changes in the relative sizes of the resonators. In the first configuration (Fig. 7.8a), the quality factor of the antiphase oscillation of the resonators is more than twice the quality factor of the in-phase one (Fig. 7.8d). In the second and third configurations (Fig. 7.8e, i), the quality factors of the coupled oscillations are approximately the same.

In the general case of 3 coupled DRs in open space, the analytical solution of the system of equations (7.6) is quite cumbersome. In structures with partial symmetry

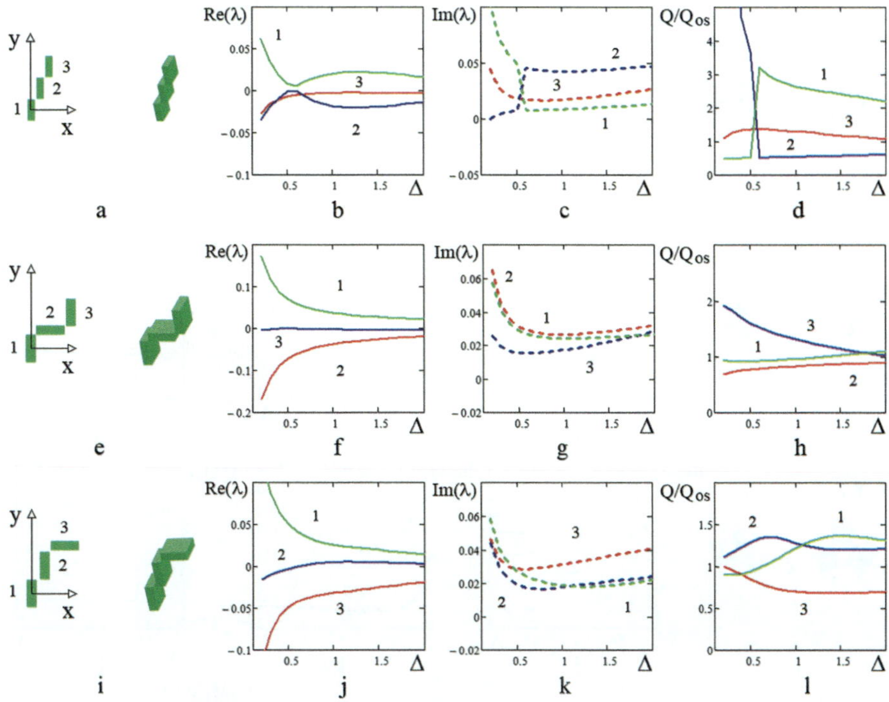

Fig. 7.9 Parameters of natural oscillations of clusters of three rectangular DRs at different locations in open space (**a**, **e**, **i**). Dependencies of relative frequencies (**b**, **c**; **f**, **g**; **j**, **k**); radiation quality factor (**d**, **h**, **l**) on the shape of the resonators $\Delta = L/a_0$ for H_{111} oscillations and $\varepsilon_{1r} = 36$; $B = 1$

between the coupling coefficients, such as those shown in Fig. 7.9a, e: $\kappa_{12} = \pm\kappa_{23}$; $\kappa_{21} = \pm\kappa_{32}$ the solutions can be simplified.

As can be seen from the results of the calculations, the frequency distribution of coupled oscillations in these cases remains close to symmetrical (Fig. 7.9b, f), however, the quality factors of coupled oscillations differ markedly from each other. (Fig. 7.9d, h). Violation of the spatial symmetry of the structure (see, for example, Fig. 7.9i) is accompanied by a less predictable distribution of complex frequencies of coupled oscillations (Fig. 7.9j–l).

Spatial structures consisting of 3 identical DRs having 2 planes of symmetry (see Fig. 7.10a, e, i) are described by the simplest solutions of the system of equations (7.6):

$$\begin{aligned}
&b_1^1 = b_2^1 = b_3^1; \quad \lambda^1 = i\tilde{k}_1 + \kappa_{21} + \kappa_{31};\\
&b_3^2 = -b_1^2 = b;\ b_2^2 = 0 \quad \lambda^2 = i\tilde{k}_1 - \kappa_{31};\\
&b_2^3 = -b_1^3 = b;\ b_3^3 = 0 \quad \lambda^3 = i\tilde{k}_1 - \kappa_{21}.\\
&b_3^4 = -b_2^4 = b;\ b_1^4 = 0 \quad \lambda^4 = i\tilde{k}_1 - \kappa_{32}.
\end{aligned} \tag{7.41}$$

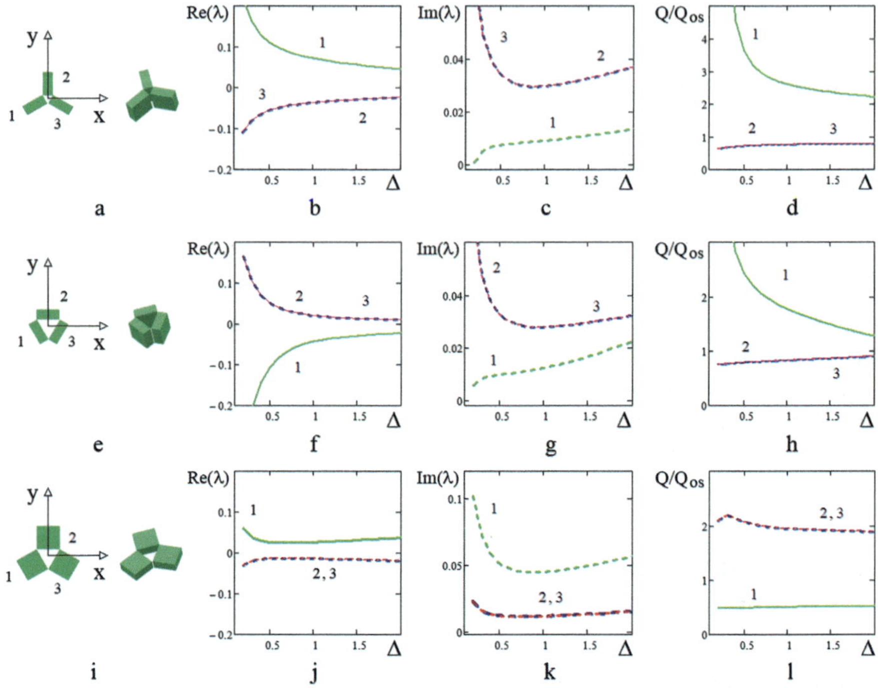

Fig. 7.10 Parameters of natural oscillations of symmetrical clusters of three rectangular DRs (**a**, **e**, **i**). Dependencies of relative frequencies (**b**, **c**; **f**, **g**; **j**, **k**); radiation quality factor (**d**, **h**, **l**) on the shape of the resonators $\Delta = L/a_0$ for the H_{111} oscillations and $\varepsilon_{1r} = 36$; $B = 1$

It is easy to see that due to the symmetry of the structure, $\lambda^2 = \lambda^3 = \lambda^4$. The paradox of the appearance of the 4-th coupled oscillation is removed if we notice that the last 3 oscillations from (7.41) are linearly dependent: $b^4 = b^2 - b^3$.

As follows from (7.41), in this case the coupled oscillations become partially degenerate (Fig. 7.10b–l) the complex frequencies of two of them coincide.

7.6 Coupled Oscillations of Lattices of Dielectric Resonators

Let us consider the general patterns of coupled oscillations arising from the solution of the system of equations (7.6) for different cases of one-, two- and three-dimensional lattices in open space, consisting of a large number of DR. For simplicity, we will assume that only no degenerate oscillations are excited in each partial resonator.

One-dimensional DR lattices. On Figs. 7.11 and 7.12 show the characteristics of coupled oscillations of linear lattices of spherical DR with the main types of magnetic oscillations H_{111} (Fig. 7.11) as well as with whispering gallery modes

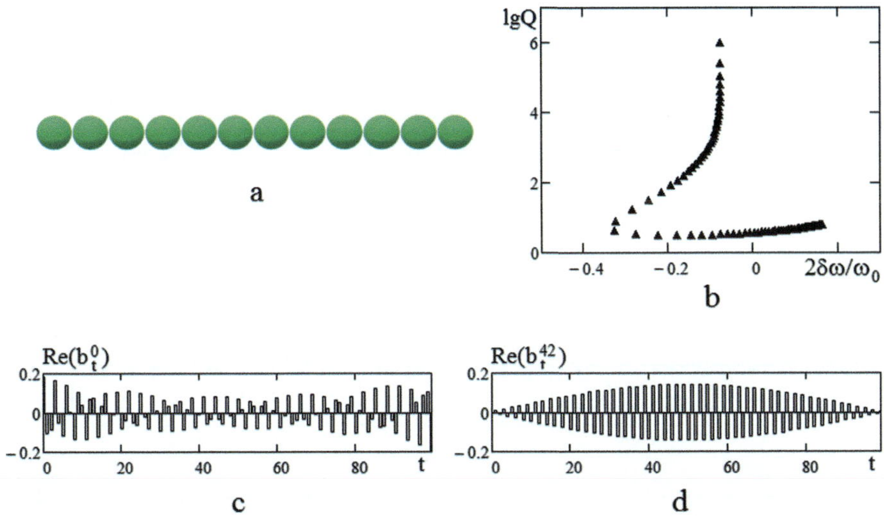

Fig. 7.11 Linear structure of spherical DR (**a**). Dependence of quality factor on frequency for coupled oscillations of a linear structure of 100 DR for oscillations H_{111} ($\varepsilon_{1r} = 9$) (**b**). Characteristic distributions of the real part of the amplitudes of partial DR for some oscillations (**c**, **d**)

$H_{30,30,1}$ (Fig. 7.12). In this case, it is convenient to depict the relative frequency detuning and the quality factor of the coupled oscillations on a plane: $2\delta\omega/\omega_0 - \lg Q$. This dependence is characteristic of each type of lattice and reflects the maximum spread of frequencies, as well as extreme values of achievable quality factors of coupled oscillations (Figs. 7.11 and 7.12b).

The distribution of the amplitudes of the partial resonators for each oscillation corresponds to the distribution of the standing wave field, in the form of a superposition of collective lattice waves, known as Bloch waves [9]. For ring lattices (Fig. 7.13), the shape of the envelope of Bloch waves is determined by the condition of periodicity of the angular distribution of amplitudes (Fig. 7.13e, f). It can be seen that the appearance of rotational symmetry in ring lattices leads to partial degeneration of coupled oscillations (see, for example, Fig. 7.13d, e), distinguished from each other by rotation by an angle equal to $\pi/2$. In this case, the symmetry of the structure again discharges the frequency spectrum of its natural oscillations. To verify this, it is enough to compare the dependences obtained for the linear and ring lattices shown in Figs. 7.11 and 7.13b.

Let us pay attention to the very high values of quality factors that can be achieved for antiphase coupled oscillations (Figs. 7.11, 7.12, and 7.13b) of large systems of resonators. At the same time, antiphase oscillations of ring lattices are characterized by the highest quality factor (Fig. 7.13b–c).

Two-dimensional DR lattices. In the cases of 2-dimensional and 3-dimensional lattices, the distribution of coupled oscillations on the frequency—quality factor plane (Figs. 7.14 , 7.15, and 7.16, b) is determined by the prevailing types of mutual coupling between partial resonators. For example, for the lattice shown in Fig. 7.14,

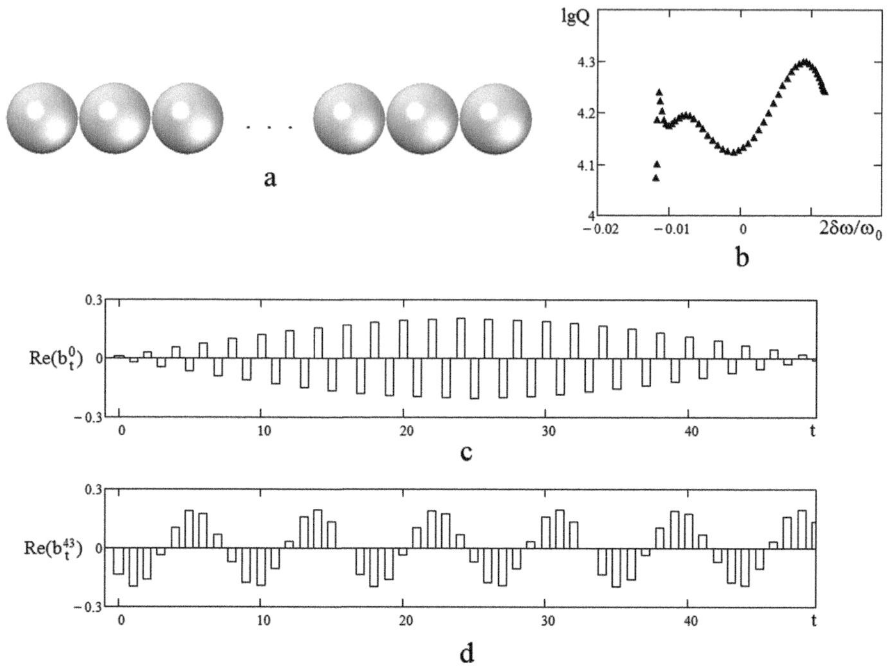

Fig. 7.12 Linear structure of spherical DRs with whispering gallery oscillations (**a**). Dependence of quality factor on frequency for coupled oscillations of a linear structure of 50 DR (**b**) for oscillations $H_{30,30,1}$ ($\varepsilon_{1r} = 2.2$) (**b**). Characteristic distributions of the real part of the amplitudes of partial DR for some oscillations (**c**, **d**)

a, all mutual coupling coefficients are determined by only one type [(6.7); Fig. 6. 2a] as well as for the 1-dimensional lattices of spherical DR considered above. The distribution of natural oscillations in these cases are similar to each other (Figs. 7.14 and 7.11b).

The mutual coupling coefficients of the lattice shown in Fig. 7.15 are determined by two types of resonator coupling [(6.7); Fig. 6.2a and (6.10); Fig. 6.2d]. In this case, the complex frequencies of coupled oscillations break up into separate branches, the number of which is determined by the number of coupled one-dimensional structures inside the two-dimensional lattice (Fig. 7.15b). The maximum values of the quality factors in this case are also realized for antiphase oscillations of neighboring partial resonators (Fig. 7.15c, d), and the amplitude distributions of partial DRs also take the form of Bloch two-dimensional standing waves. The value of the quality factor is determined mainly by the number of resonators and the dielectric constant.

Three-dimensional DR lattices. In 3-dimensional lattices, the frequencies of coupled oscillations are combined into bands (Fig. 7.16b), similar to the bands of electronic levels, for example, in metal crystals. The nature of the appearance of these bands is also due to different types of coupling between the resonators.

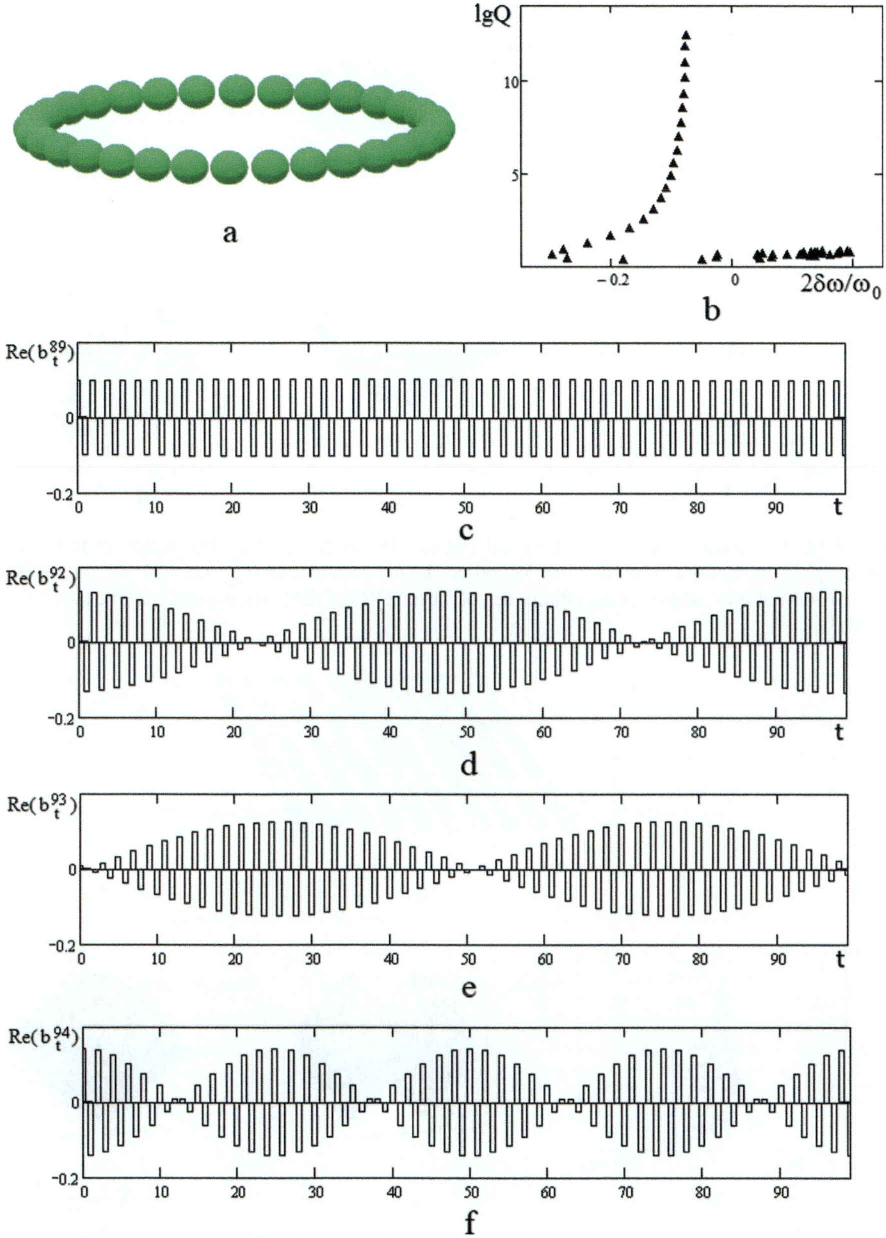

Fig. 7.13 Ring structure of spherical DR (**a**). Dependence of the quality factor on frequency for coupled oscillations of a ring structure of 100 DR (**b**) for H_{111} oscillations ($\varepsilon_{1r} = 9$). Distribution of the real part of the amplitudes of partial DRs of the highest quality oscillations (**c**–**f**)

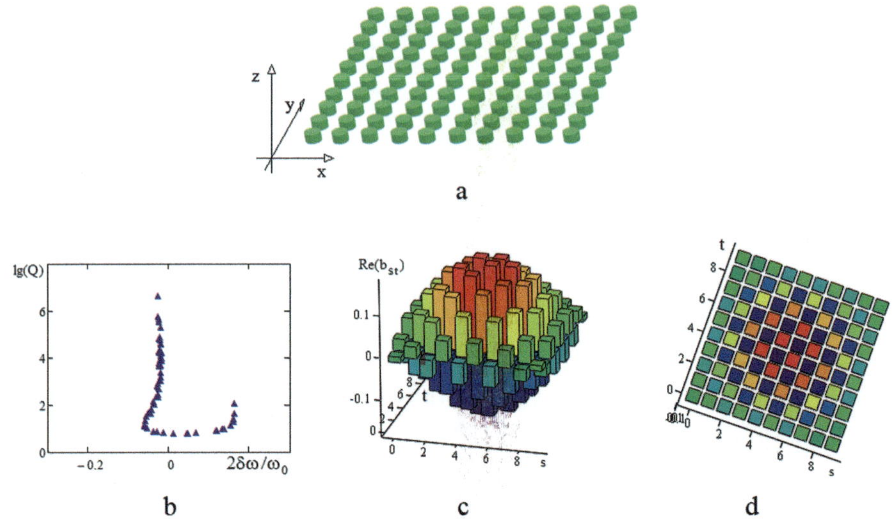

Fig. 7.14 Rectangular lattice of cylindrical DR (**a**). Dependence of quality factor on frequency for coupled oscillations of a 10 × 10 DR lattice (**b**) for oscillations H_{101}^{+} ($\Delta = 0, 4$; $\varepsilon_{1r} = 36$). Distribution of the real part of the amplitudes of partial DRs of the highest quality lattice oscillation (**c**, **d**)

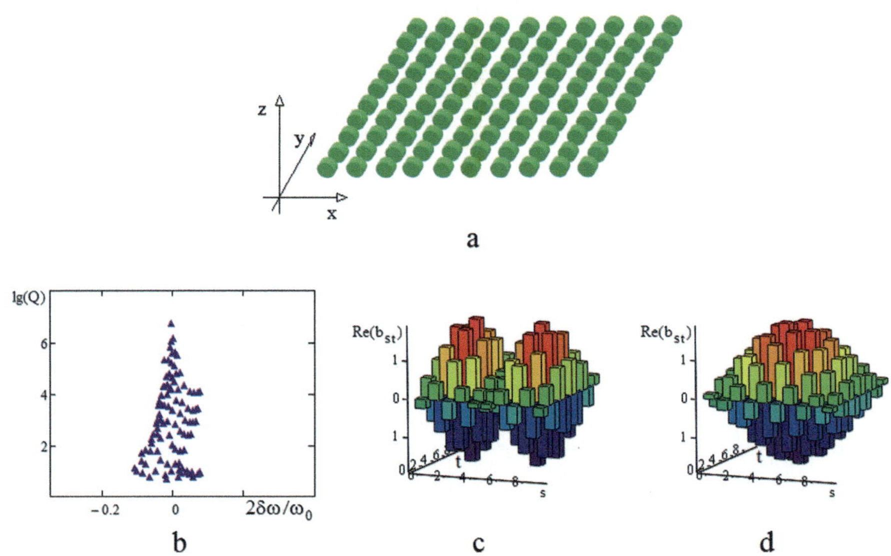

Fig. 7.15 Rectangular lattice of cylindrical DR (**a**). Dependence of quality factor on frequency for coupled oscillations of a 10 × 10 DR lattice (**b**) for H_{101}^{+} oscillations ($\Delta = 0.4$; $\varepsilon_{1r} = 36$). Distribution of the real part of the amplitudes of partial DRs of the highest quality oscillations (**c**, **d**)

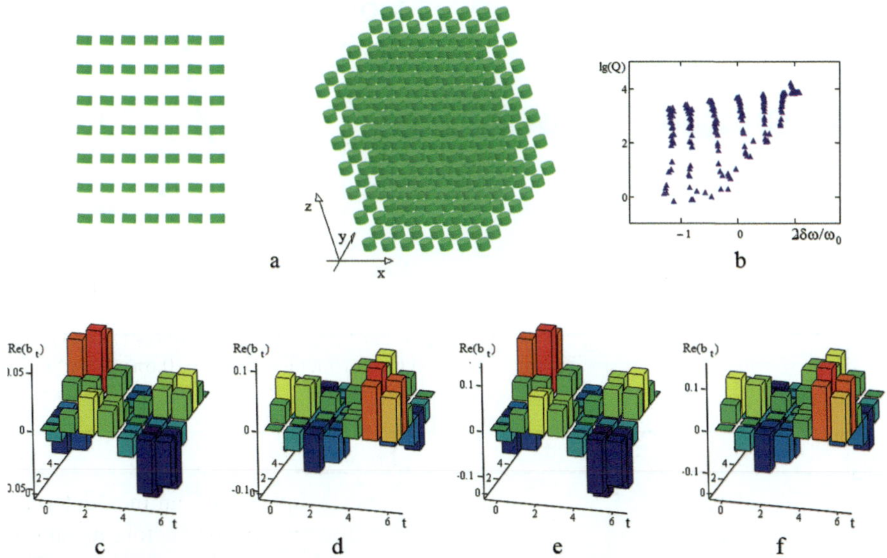

Fig. 7.16 3D rectangular lattice of 7 × 7 × 7 cylindrical DR (**a**). Parameters of coupled oscillations on the Q-factor—frequency detuning plane (b) for the H_{101}^{+} ($\Delta = 0.4$; $\varepsilon_{1r} = 36$ relative distance between adjacent resonators in the transverse plane $d = 7$ mm; in the vertical plane $d_z = 6$ mm). Distribution of the real part of the amplitudes of partial DR (**c**–**f**) in the first (**a**) 4 layers of the lattice for the highest quality oscillation (**b**)

The appearance of a vacancy of one or several adjacent resonators inside the lattice (Figs. 7.17 and 7.18a) leads to the appearance of oscillations localized near the inhomogeneity (Figs. 7.17 and 7.18b, c) [10]. The reason for the occurrence of this type of oscillation is quite clear: a system of coupled resonators in a regular line, or open space, behaves like a band-stop filter that partially reflects electromagnetic waves. The appearance of a cavity in the lattice material with a size approximately a multiple of half the wavelength allows energy to be accumulated in the adjacent volume. The higher the quality factor of such an oscillation, the greater the number of lattice resonators "screens" the field from penetration into the external region.

The disposition of a localized oscillation on the frequency-quality plane differs from the location of other coupled oscillations. The problem of optimizing such a structure with the aim of frequency separation of a localized oscillation from the remaining coupled oscillations of the resonator system is of interest. One of its solutions is the use of partially shielded grilles. Another solution is to implement a three-dimensional lattice with different relative distances between adjacent resonators in different directions. On Fig. 7.18 shows a rectangular array consisting of 7 × 7 × 7—1 DR cylindrical shape with the main magnetic types of oscillations. The distance between adjacent lattice resonators is selected in such a way that the frequency of the localized oscillation differs as much as possible from the frequencies of coupled oscillations of other types. On Fig. 7.18b. this type of oscillation is marked with a circle.

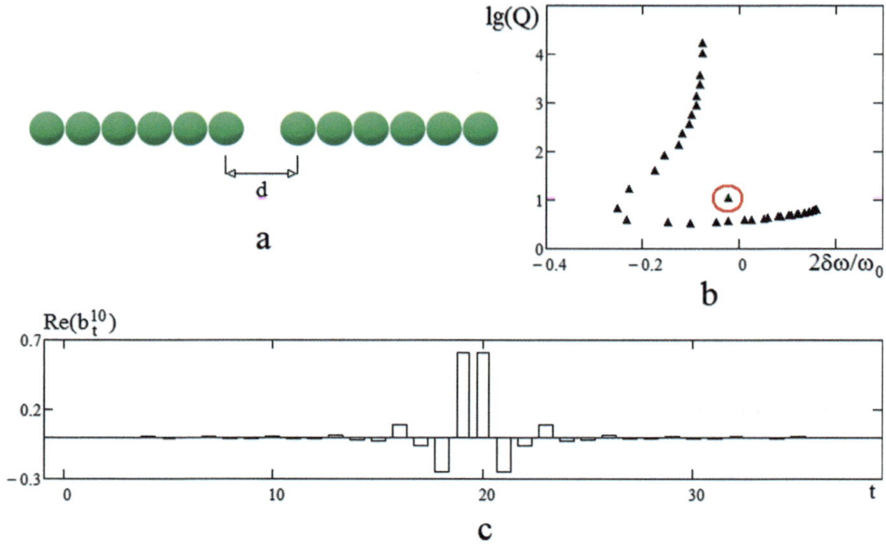

Fig. 7.17 Linear structure with a vacancy of one resonator (**a**). Dependence of the quality factor on the frequency of coupled oscillations of a structure of 39 DR (**b**) for oscillations H_{111} ($\varepsilon_{1r} = 9$; $d = 4r_0$). Distribution of the real part of the amplitudes of partial DR for an oscillation localized near a vacancy (**c**)

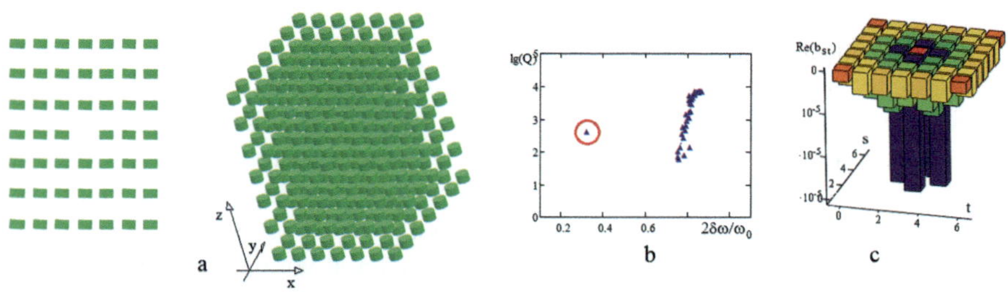

Fig. 7.18 Rectangular array of 7 × 7 × 7—1 cylindrical DRs with one resonator vacancy in the center (**a**). Dependence of the quality factor on the frequency of coupled oscillations (**b**) for H_{101}^{+} ($\Delta = 0, 4$; $\varepsilon_{1r} = 36$; relative distance between adjacent resonators in the transverse plane $d = 7$ mm; in the vertical plane $d_z = 3$ mm). Distribution of the real part of the amplitudes of partial DR located in the central plane (**c**) for oscillation localized near a vacancy

The distribution of the real part of the amplitudes of partial resonators of localized oscillation in the vacancy plane is shown in Fig. 7.18c.

7.7 Natural Oscillations of Various Dielectric Resonators in Open Space

Coupled modes of resonators made of different dielectrics, or of different sizes with different types of natural oscillations are described formally by the same system of equations (7.6), the derivation of which is carried out under the same assumptions (7.1, 7.4).

There are no losses in the dielectric of each DR. At the same time, it is also assumed that each of the isolated partial DRs has the same real part of the natural oscillation frequency as the others ω_0. Only the expressions for the mutual coupling coefficients of resonators that are included in the equation system (7.6) change (see, for example, Sects. 2.5, 2.6, 3.4, 3.6, 6.6, 6.7). In this case, the conclusions regarding the frequency symmetrization of the scattering matrix also remain valid (Sect. 7.4).

Despite such an obvious analogy in the mathematical description of coupled oscillations, DR systems of different types have interesting properties that are absent in lattices of identical resonators [11]. The most characteristic difference of such structures is the appearance of new types of partial oscillations of the sublattices (Figs. 7.19 and 7.20c, d).

As follows from the calculations performed, the frequencies of coupled oscillations fall into separate branches, reflecting oscillations of the DR sublattices made

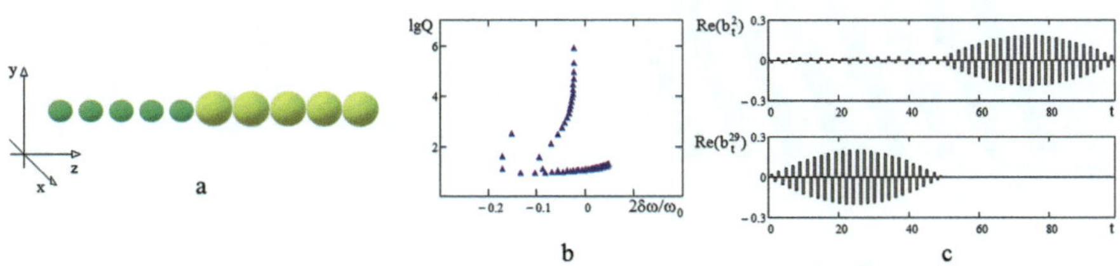

Fig. 7.19 Linear structure of spherical DRs of different dielectric constants: $\varepsilon_{1r} = 25$; $\varepsilon_{2r} = 16$ (**a**). Dependence of quality factor on frequency for coupled oscillations of a linear structure 100 DR (**b**) ($k_0\Delta z_1 = 2$; $k_0\Delta z_2 = 3$). Characteristic distributions of the real part of the amplitudes of partial DR for oscillations localized in the subsystem $\varepsilon_{2r} = 16$ and in the subsystem $\varepsilon_{1r} = 25$ (**c**)

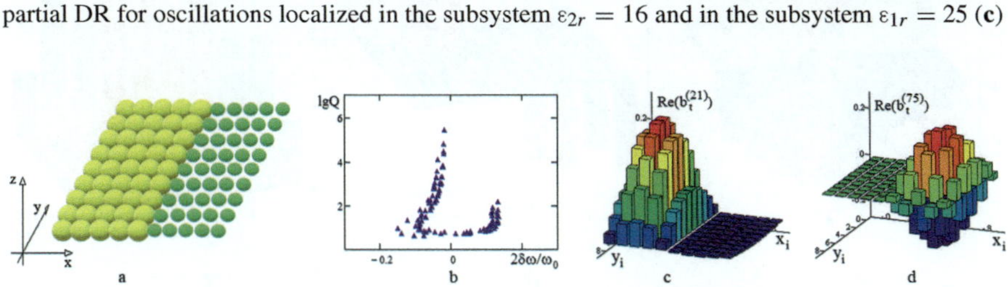

Fig. 7.20 2D structure of spherical DRs of different dielectric constants: $\varepsilon_{1r} = 16$; $\varepsilon_{2r} = 25$ (**a**). Dependence of quality factor on frequency for coupled oscillations of a 10×10 DR structure (**b**) H_{111} ($k_0\Delta x_1 = 3$; $k_0\Delta x_2 = 2$; $k_0\Delta y_1 = k_0\Delta y_2 = 2$). Characteristic distributions of the real part of the amplitudes of partial DR for oscillations localized in the subsystem $\varepsilon_{1r} = 16$ (**c**); in the subsystem $\varepsilon_{2r} = 25$ (**d**)

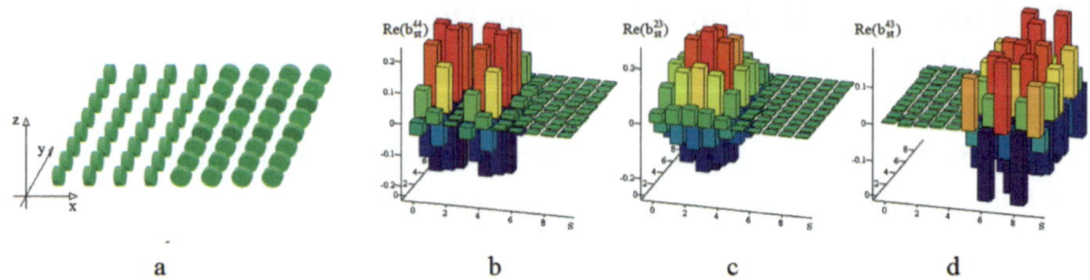

Fig. 7.21 2D structure of cylindrical DR orientation x, y (**a**) for H_{101}^{+} oscillations of a 10 × 10 DR structure ($\varepsilon_{1r} = 25$; $\Delta = 0.4$; $f_0 = 200$ GHz; $\Delta x = \Delta y = 0.5$ mm). Partial DR amplitudes for oscillations localized in different subsystems x (**b**, **c**); y (**d**)

of different materials. In this case, the condition for the existence of oscillations of DR sublattices made of different dielectrics is not mandatory.

For the appearance of additional localized oscillations, a sufficient condition is the difference in one of the material or geometric parameters in different subsystems of the resonators, for example, the difference in their spatial orientation (Figs. 7.21 and 7.22 and 7.23).

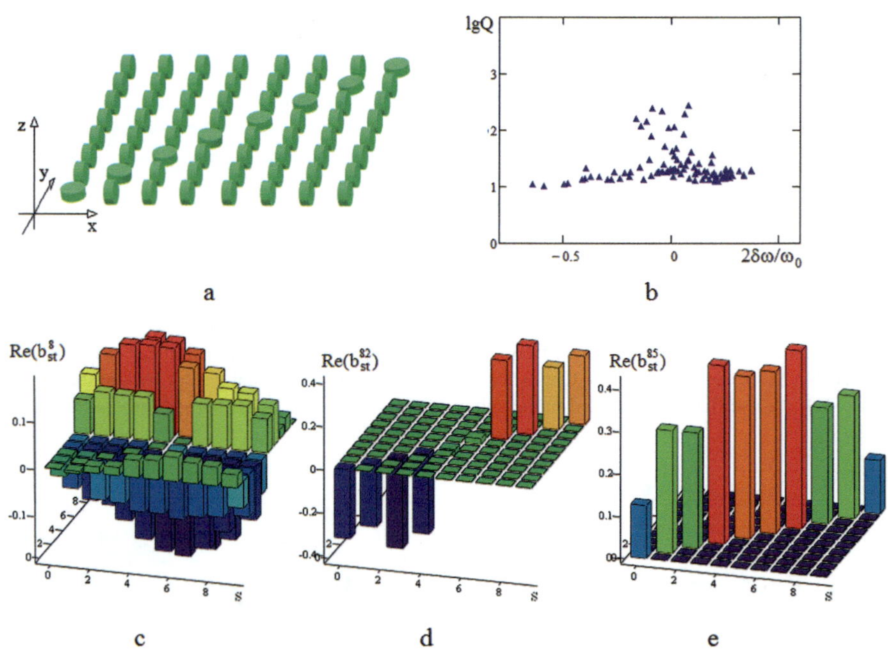

Fig. 7.22 2D structure of cylindrical DR orientation x, z (**a**). Dependence of quality factor on frequency for coupled oscillations of a 10 × 10 DR structure (**b**) ($\varepsilon_{1r} = 25$; $\Delta = 0.4$; $f_0 = 200\,\text{GHz}$; $\Delta x = \Delta y = 0.5$ mm). Characteristic distributions of partial DR amplitudes for oscillations localized in different subsystems (**c**–**e**)

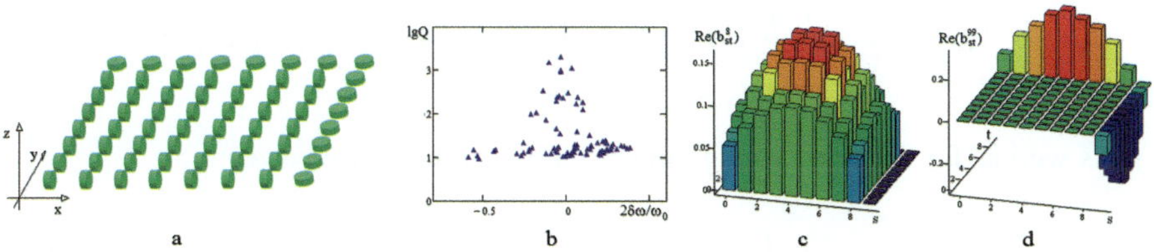

Fig. 7.23 2D structure of cylindrical DR different orientations x, z (**a**). Dependence of quality factor on frequency for coupled oscillations of a 10 × 10 DR structure (**b**) $\varepsilon_{1r} = 25$; $\Delta = 0.4$; $f_0 = 200$ GHz; $\Delta x = \Delta y = 0.5$ mm). Characteristic distributions of partial DR amplitudes for oscillations localized in different subsystems (**c**–**e**)

The distribution of field amplitudes of a subset of localized oscillations exactly reflects the distribution of resonators with different axial orientations inside the lattice (Figs. 7.21b–d, 7.22c–e and 7.23c, d). The quality factors of coupled oscillations of the sublattices decrease (Figs. 7.22 and 7.23b), which is obviously due to the smaller number of resonators taking part in the oscillations of the noted types.

7.8 Natural Oscillations of Detuned Dielectric Resonators

Let us study the patterns of coupled oscillations for the DR system, the natural frequencies of which differ from each other [12]. Let we have several various DRs, exhibit different resonance frequencies. Suppose that we know eigenoscillation field $(\mathbf{e}_s, \mathbf{h}_s)$ as well as the resonance frequencies $\omega_s (s = 1, 2, \ldots, N)$ of every DR. Let at the same time $\omega_s \neq \omega_t$. We make rough assumption that all high-Q DR fields are vary with frequency by known law $\alpha_s(\omega)$, but for simplicity we assume that their spatial distribution remain the same:

$$\begin{bmatrix} \mathbf{e}_s(\omega, \mathbf{r}) \\ \mathbf{h}_s(\omega, \mathbf{r}) \end{bmatrix} \simeq \alpha_s(\omega) \begin{bmatrix} \mathbf{e}_s(\mathbf{r}) \\ \mathbf{h}_s(\mathbf{r}) \end{bmatrix} \text{ for } \omega \simeq \omega_s. \tag{7.42}$$

At that we supposed ω_s realized maximum $\alpha_s(\omega)$ in which:

$$\alpha_s(\omega_s) = 1; \text{ and } \lim_{|\omega - \omega_s| \to \infty} \alpha_s(\omega) = 0. \tag{7.43}$$

Let us show that despite the fact that the field of natural oscillations of each of the detuned partial resonators (7.42) depends on frequency, the natural frequencies and distributions of the fields of coupled oscillations of the resonator system remain constant in the frequency domain.

Define the expansion coefficients of the DR field (2.4) on the propagation wave field of the line $\left(\mathbf{E}_t^{\pm}, \mathbf{H}_t^{\pm}\right)$ via the DR surface integrals:

$$\mathrm{c}_{\mathrm{t}}^{\mathrm{s}\pm}(\omega) = -\frac{1}{2}\oint\limits_{S_R}\left\{\left[\mathbf{e}_s^{1}(\omega),\,\mathbf{n}\right]\left(\mathbf{H}_{\mathrm{t}}^{\pm}\right)^{*} + \left[\mathbf{n},\,\mathbf{h}_{\mathrm{s}}^{1}(\omega)\right]\left(\mathbf{E}_{\mathrm{t}}^{\pm}\right)^{*}\right\}\mathrm{ds} \tag{7.44}$$

As well as for the non propagating magnetic and electric waves:

$$\begin{pmatrix}\mathrm{c}_{\mathrm{t}}^{\mathrm{s}\pm}(\omega)\\ \mathrm{d}_{\mathrm{t}}^{\mathrm{s}\pm}(\omega)\end{pmatrix} = (\mp)\frac{\mathrm{i}}{2}\oint\limits_{S_R}\left\{\left[\mathbf{e}_s^{1}(\omega),\,\mathbf{n}\right]\left(\mathbf{H}_{\mathrm{t}}^{+}\right)^{*} + \left[\mathbf{n},\,\mathbf{h}_{\mathrm{s}}^{1}(\omega)\right]\left(\mathbf{E}_{\mathrm{t}}^{+}\right)^{*}\right\}\mathrm{ds} \tag{7.45}$$

where **n**—normal to the surface of the DR. For compactness, the dependence on spatial coordinates **r** were omitted.

Then, using (7.42), we obtained:

$$c_t^{s\pm}(\omega) = \alpha_s(\omega)c_t^{s\pm};\quad d_t^{s\pm}(\omega) = \alpha_s(\omega)d_t^{s\pm}, \tag{7.46}$$

where $c_t^{s\pm}$; $d_t^{s\pm}$ are expansion coefficients (2.4) at the resonance frequency ω_s.

The energy, stored inside s-th DR at frequency ω we define as:

$$w_s(\omega) = 1/4\int\limits_{V_s}\left[\varepsilon_s|\mathbf{e}_s(\omega)|^2 + \mu_0|\mathbf{h}_s(\omega)|^2\right]dv,$$

$(s = 1, 2, \ldots, N)$. From (2.9) obtained:

$$w_s(\omega) = |\alpha_s(\omega)|^2 w_s \tag{7.47}$$

where w_s is the energy, stored s-th DR at frequency ω_s.

Define also mutual coupling coefficients of the s-th and the t-th DRs on the damped and expanding waves at frequency ω:

$$k_{st}(\omega) = \frac{-1}{\omega_t w_t(\omega)}\sum_{n=1}^{\infty}\left[c_n^{s\pm}(\omega)c_n^{t\pm}(\omega)^* - d_n^{s\pm}(\omega)d_n^{t\pm}(\omega)^*\right]$$

where n—multi-index, characterizing the non propagating waves;

$$\tilde{k}_{st}(\omega) = \frac{1}{\omega_t w_t(\omega)}c_l^{s\pm}(\omega)c_l^{t\pm}(\omega)^*$$

l—multi-index of the propagating wave.

Using (7.46) and (7.47), obtained

$$k_{st}(\omega) = \frac{\alpha_s(\omega)}{\alpha_t(\omega)}k_{st};\quad \tilde{k}_{st}(\omega) = \frac{\alpha_s(\omega)}{\alpha_t(\omega)}\tilde{k}_{st} \tag{7.48}$$

Coupling coefficients k_{st}, $\tilde{k}_{st}$ differ from (3.1) and (7.14) because functions $c_n^{s\pm}$, $c_n^{t\pm}$ are calculated at different frequencies.

If we neglected the dependence of the longitudinal wave number of the transmission line on frequency $\Gamma = \Gamma(\omega)$. For simplicity, we assume that the set of natural oscillation frequencies of partial resonators occupies a not too wide band relative to the average frequencies $\omega_0 \in [\omega_s, \omega_t]$, then:

$$k_{st}(\omega) = \frac{-1}{\omega_t w_t(\omega)} \sum_{n=1}^{\infty} \left[c_n^{s\pm}(\omega) c_n^{t\pm}(\omega)^* - d_n^{s\pm}(\omega) \mathrm{d}_n^{t\pm}(\omega)^* \right],$$

relations (7.48) will again take the form of the previously determined coupling coefficients (3.1), (7.14).

As follows from (7.48):

$$\tilde{k}_{tt}(\omega) = \tilde{k}_{tt} = \tilde{k}_t \tag{7.49}$$

The eigenoscillation problem solution of the N-DR system $(\mathbf{e}(\omega, \mathbf{r}), \mathbf{h}(\omega, \mathbf{r}))$ also can be found as a superposition of fields of the isolated resonators $(\mathbf{e}(\omega, \mathbf{r}), \mathbf{h}(\omega, \mathbf{r}))$, $(s = 1, 2, \ldots, N)$:

$$\mathbf{e}(\omega, \mathbf{r}) = \sum_{s=1}^{N} b_s(\omega) \mathbf{e}_s(\omega, \mathbf{r}); \quad \mathbf{h}(\omega, \mathbf{r}) = \sum_{s=1}^{N} b_s(\omega) \mathbf{h}_s(\omega, \mathbf{r}). \tag{7.50}$$

Here $b_s(\omega)$—is the unknown complex amplitude of the s-th resonator mode. The expansion coefficients $b_s(\omega)$ as well as the complex eigenmode DR system frequencies ω generally can be obtained from Maxwell's equations, using perturbation theory [4]. After simple transformations as in (7.1), of the eigenmodes fields $(\mathbf{e}(\omega, \mathbf{r}), \mathbf{h}(\omega, \mathbf{r}))$ and $(\mathbf{e}_s(\omega, \mathbf{r}), \mathbf{h}_s(\omega, \mathbf{r}))$ one can find, all values of $b_s(\omega)$ should satisfy the equation system:

$$\sum_{s \neq t}^{N} \left[k_{st}(\omega) + i\tilde{k}_{st}(\omega) \right] b_s(\omega) + \left[i\tilde{k}_{tt}(\omega) - \lambda_t \right] b_t(\omega) = 0, \;\; (t = 1, 2, \ldots, N)$$

Using (7.43), (7.48), (7.49), we obtained [12]:

$$\sum_{s \neq t}^{N} \alpha_s(\omega) \kappa_{st} b_s(\omega) + \alpha_t(\omega) \left(i\tilde{k}_t - \lambda_t \right) b_t(\omega) = 0 \tag{7.51}$$

where

$$\lambda_t = 2 \cdot \left(\frac{\omega - \omega_t}{\omega_t} \right), \quad (t = 1, 2, \ldots, N) \tag{7.52}$$

$\kappa_{st} = k_{st} + i\tilde{k}_{st}$; ω_t—is the real part of frequency of t-th DR mode.

When deriving Eq. (7.51), we also used the inequalities:

$$\left|\int\limits_{V_t} (\mathbf{e}_t^1, \mathbf{e}_s^0(\omega_t)^*)\,\mathrm{dv}\right| << \int\limits_{V_t} \left|\mathbf{e}_t^1\right|^2 \mathrm{dv};$$

$$\left|\int\limits_{V_t} (\mathbf{h}_t^1, \mathbf{h}_s^0(\omega_t)^*)\,\mathrm{dv}\right| << \int\limits_{V_t} \left|\mathbf{h}_t^1\right|^2 \mathrm{dv}. \tag{7.53}$$

Taking into account representation (7.42), it is easy to show that conditions (7.53) will be satisfied if the inequalities are satisfied:

$$\max_{\forall s, \forall t} |\alpha_s(\omega)| << \varepsilon_{1r} \tag{7.54}$$

The last condition will obviously be satisfied: by definition (7.43), since we previously assumed that for ω_s the function $|\alpha_s(\omega)|$ has a global maximum: $\underset{\forall \omega \neq \omega_s}{|\alpha_s(\omega)|} < 1$.

It is obvious that these arguments remain valid for the case of placing a system of frequency-detuned DR in open space.

As follows from (7.51), the frequencies of coupled oscillations in the approximation (7.42) do not depend on the $\alpha_s(\omega)$, and the functions $\alpha_s(\omega)b_s(\omega) = b_s$ must satisfy a system of equations:

$$\sum_{s \neq t}^{N} \kappa_{st} b_s + \left(i\tilde{k}_t - \lambda_t\right) b_t = 0, \tag{7.55}$$

that differs from (7.6) only in the (7.52).

Equating the determinant of (7.55) to zero, we find complex natural frequencies $\tilde{\omega} = \left(\tilde{\omega}^1, \tilde{\omega}^2, \ldots, \tilde{\omega}^N\right)$ of the coupled detuned DR system.

The quality factor of the DR system oscillation obtain also applying the known formula: $Q^s = \mathrm{Re}[\tilde{\omega}^s]/2\mathrm{Im}[\tilde{\omega}^s]$ $(s = 1, 2, \ldots, N)$,

If $|\omega_t - \omega_s| \gg \omega_s$ for every s, then, as follows from (7.43), (7.51), the t-th DR oscillations are independent of other resonators: $\lambda_t = i\tilde{k}_t$ and $\mathrm{Re}[\tilde{\omega}] = \omega_t$; $\mathrm{Im}(\tilde{\omega}) = 1/2\omega_t\tilde{k}_t$; the quality factor of the t-th mode will be: $Q^t = 1/\tilde{k}_t$.

An Eq. (7.55) allows us obtain frequencies of coupled oscillations of two detuned DRs:

$$\frac{\omega^{1,2}}{\sqrt{\omega_1\omega_2}} = \frac{1}{4}\left[\sqrt{\frac{\omega_1}{\omega_2}}\left(i\tilde{k}_1 + 2\right) + \sqrt{\frac{\omega_2}{\omega_1}}\left(i\tilde{k}_1 + 2\right) \pm \sqrt{d}\right] \tag{7.56}$$

where

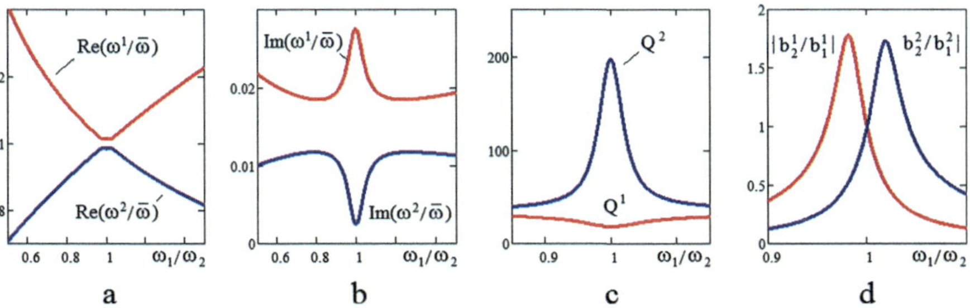

Fig. 7.24 Dependences of the relative frequencies of coupled oscillations on the resonant frequencies of 2 DR for $\tilde{k}_1 = \tilde{k}_2 = 0.03$; $\kappa_{12} = \kappa_{21} = 0.03 + i0.025$(**a, b**) $\overline{\omega} = \sqrt{\omega_1 \omega_2}$. Dependences of the quality factor (**c**) and the modulus of the relative amplitudes of the resonators (**d**) on the relative frequencies of natural oscillations

$$d = \left[\sqrt{\frac{\omega_1}{\omega_2}}(i\tilde{k}_1 + 2) + \sqrt{\frac{\omega_2}{\omega_1}}(i\tilde{k}_1 + 2)\right]^2 + 4\left[\kappa_{12}\kappa_{21} - (i\tilde{k}_1 + 2)(i\tilde{k}_1 + 2)\right]$$

$$b_1^{1,2} = \lambda_2^{1,2} - i\tilde{k}_2; \quad b_2^{1,2} = \kappa_{12}. \tag{7.57}$$

Figure 7.24 is showing the coupled oscillation parameters of two detuned DRs. One can see, that the DR frequency detuning conducted by all structure parameters variation. The real part of coupled resonance frequencies are never intersects to one another, if the real part of the mutual coupling coefficients both is no equal to zero. The quality factors achieved their extreme values for the equal DRs only (Fig. 7.24c). The relative amplitudes of the coupled oscillations are equal to one another when $\omega_1 = \omega_2$ also for the equal DRs only (d).

Using the system of equations (7.51), we have considered the lattices, consists of large number of the different Spherical DRs. Supposed, that all resonators are excited at the basic magnetic mode H_{111}. The magnetic field direction in all DR centers are oriented orthogonally to the lattice axes for one-dimensional (Figs. 7.25 and 7.27a) as well as orthogonally to the lattice plane for two-dimensional cases (Fig. 7.26a). All coupling coefficients have been calculated from (7.2).

In the cases when the DR dimensions are changed smoothed (Figs. 7.25 and 7.26a) we see that some coupled resonances had shape allocated oscillations (Figs. 7.25c, d and 7.26c–e), occupying certain segment of the lattice.

In the case of a composite lattice consisting of different types of resonators detuned in frequency (Fig. 7.27a), the coupled oscillations of the structure also break up into separate oscillations of two subsystems (Fig. 7.27c–e), the frequencies and quality factors of which form two non-intersecting branches (Fig. 7.27b).

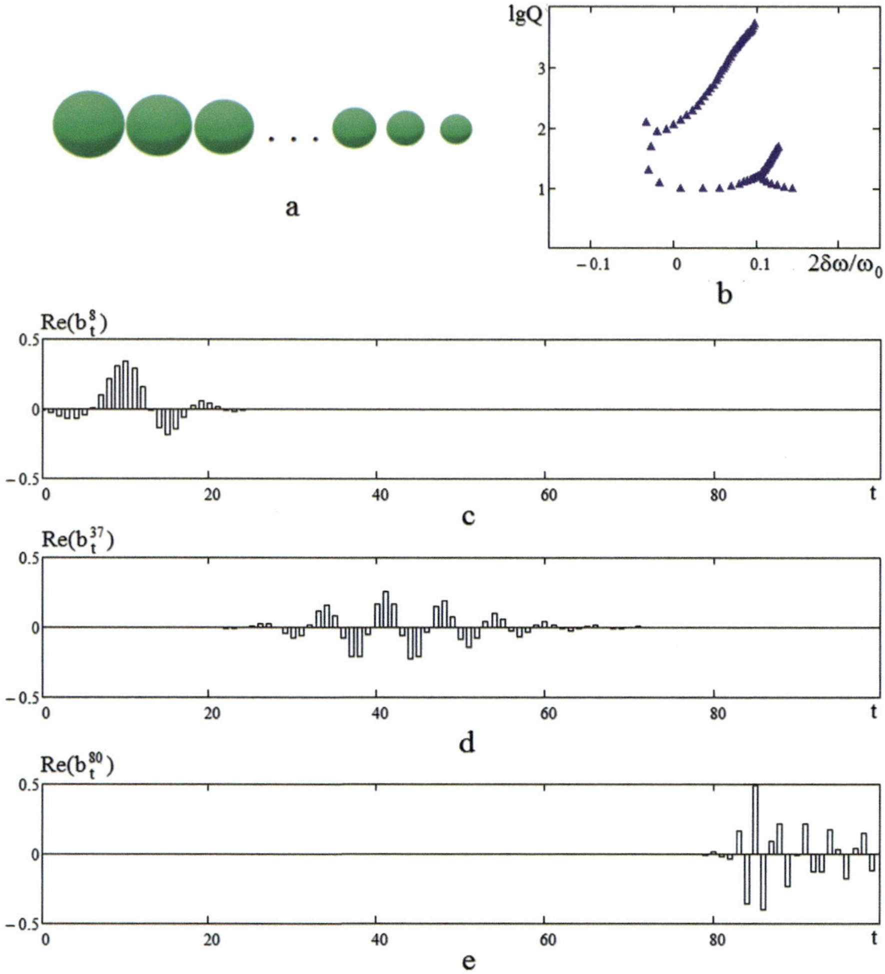

Fig. 7.25 Lattice of frequency-detuned spherical DRs of different sizes (**a**). Location of frequencies of coupled oscillations on a plane: relative frequency of coupled oscillations—logarithm of quality factor (**b**) for $\varepsilon_{1r} = 16$. for a lattice of different (blue dots) DR. Partial distributions of amplitudes of coupled oscillations of partial DR lattice (**c**–**e**)

7.9 Coupled Oscillations of Resonators with an Active Dielectric

The physical conditions in the DR material can be implemented in such a way that energy from an external source is transferred to the field excited in the resonator system. An example is a DR made an erbium-doped dielectric. Under the influence of an external electromagnetic field, in the infrared range at a selected pump frequency, erbium atoms will convert in frequency and add energy at the oscillation frequency

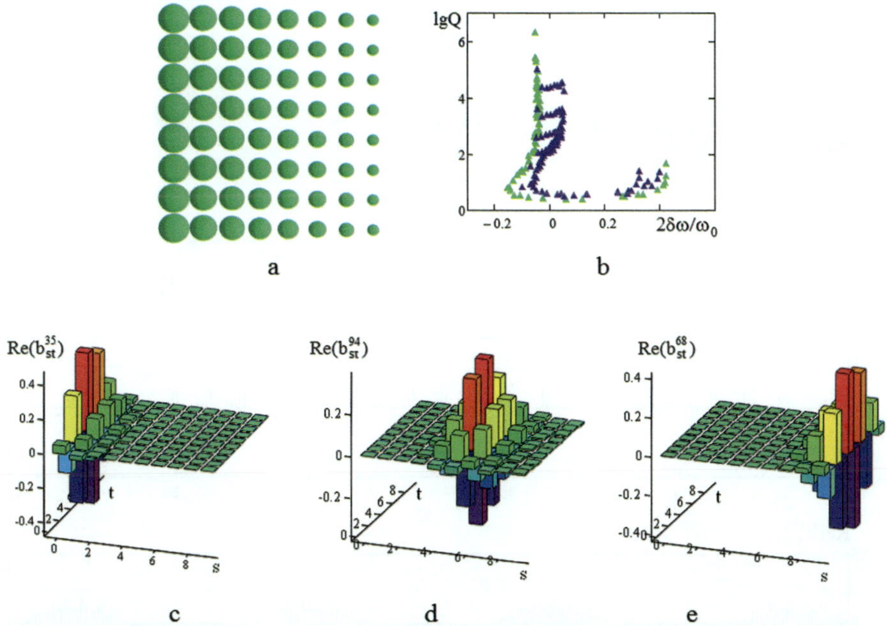

Fig. 7.26 2D lattice of frequency-detuned 10 × 10 spherical DRs of different sizes (**a**). Location of frequencies of coupled oscillations on a plane: relative frequency—logarithm of quality factor (**b**) for $\varepsilon_{1r} = 16$. for a lattice of identical ones (green dots); different DR (blue dots). Partial distributions of amplitudes of coupled oscillations of partial DR lattice (**c**–**e**)

of the resonator. From the point of view of classical electrodynamics, it is convenient to describe such resonators by introducing into consideration the concept of active dielectrics, materials in which the imaginary part of the dielectric constant can take positive values.

In Sect. 7.1, we considered resonators made of an ideal lossless dielectric whose dielectric constant is purely real. In this section, using perturbation theory, we will consider coupled oscillations of resonators, the material of which can both absorb or add energy to the field of their own oscillations, i.e. we will assume that the dielectric constant of the resonators is a complex function, the imaginary part of which can take both positive and negative values.

Using relations (C.7), (C.8) and (7.4) again, we will expand the field of coupled oscillations ($\mathbf{e}$, $\mathbf{h}$) of the system into the natural oscillations of the partial DR ($\mathbf{e}_n$, $\mathbf{h}_n$), assuming that the dielectric constant of the material of each of the resonators is a complex function that determines the absorption or release of energy in the system $\tilde{\varepsilon}_n = \varepsilon'_n - i\varepsilon''_n$. In this case we will also consider $\left|\varepsilon''_n\right| << \varepsilon'_n$.

Repeating arguments similar to (7.2)–(7.5), we obtain [14]:

$$\sum_{s\neq n}^{N} \kappa_{sn}\, b_s + \left[i\left(\tilde{k}_n + \tilde{k}_n^{\varepsilon}\right) - \lambda\right] b_n = 0; \quad (n = 1,\ 2, \ldots, N) \tag{7.58}$$

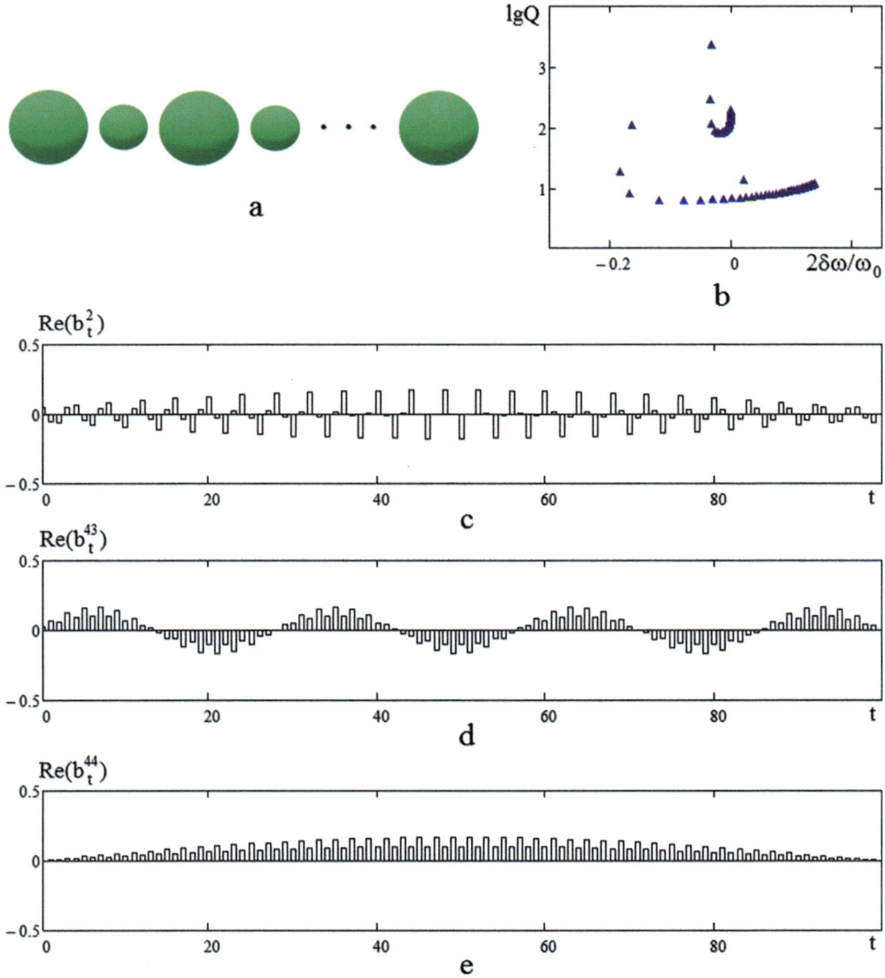

Fig. 7.27 Structure of frequency-detuned sublattices of spherical DRs of different sizes **a** $\varepsilon_{1r} = 16$. Location of frequencies of coupled oscillations on a plane: relative frequency of coupled oscillations—logarithm of quality factor (**b**). Partial distributions of amplitudes of coupled oscillations of partial DR lattice (**c–e**)

where $\tilde{k}_n^\varepsilon = P_n^\varepsilon / \omega_0 w_n$, and

$$P_n^\varepsilon = \frac{\omega_0}{2} \varepsilon_n'' \int_{V_n} |\mathbf{e}_n|^2 dv \tag{7.59}$$

determines the average power loss $\left(\varepsilon_n'' > 0\right)$ (or emit power $\left(\left(\varepsilon_n'' < 0\right)\right.$, of the n-th resonator in system. When $\varepsilon_n'' = 0$ the system of equations (7.58) coincides with (7.6).

As a result, we find that the system of equations that determines the frequencies and amplitudes of partial DR takes a form similar to (7.6) if the mutual coupling coefficients are transformed according to (7.58):

$$\kappa_{sn} \rightarrow \begin{cases} \kappa_{sn}, \ (s \neq n) \\ i(\tilde{k}_n + \tilde{k}_n^{\varepsilon}), \ (s = n) \end{cases} \tag{7.60}$$

that taking into account the complex parameters of dielectrics changes the values of the coefficients of mutual coupling between the resonators, as well as the values of the coefficients of coupling of the resonators with open space.

We see that an increase in dissipative losses $\left(\varepsilon_n'' > 0\right)$, in the perturbation theory approximation, is equivalent to an increase in radiation (7.59), (7.60), and the presence of an active dielectric $\left(\varepsilon_n'' < 0\right)$, releasing energy into the system from an external source, is equivalent to a decrease in radiation from the resonator. Increasing the relative distance between the resonators in most cases leads to $\kappa_{sn} = 0$ (Fig. 7.29)

From (7.58) and

$$Q_n = \mathrm{Re}(\tilde{\omega}_n)/2\mathrm{Im}(\tilde{\omega}_n) = 1/\left(\tilde{k}_n + \tilde{k}_n^{\varepsilon}\right)$$

we find that the quality factor of isolated resonators, in this case, is expressed in terms of quality factors determined by radiation losses and taking into account the sign in the dielectric constant by "losses" in material [13].

On Fig. 7.28a shows the solution to the system of equations (7.58) in the simplest case of two coupled DRs, one of which is made of an absorbing dielectric, and the second of an active dielectric, with varying parameters of the active material $\varepsilon_2''/\varepsilon_2'$ of the second resonator. An increase $\left|\varepsilon_2''/\varepsilon_2'\right|$ leads to a decrease in the difference in frequencies of coupled oscillations (b). From the received data Fig. 7.28c we also see that coupled oscillations are characterized by significantly different field amplitudes. The first type of oscillation corresponds to the maximum field amplitude in the active resonator (1), the second in the resonator with losses. The corresponding quality factors of coupled oscillations are shown in Fig. 7.28d.

In the case of coupled oscillations of three identical DRs (Fig. 7.28e, i), differing only in signs of the imaginary part of the dielectric constant, the relative amplitudes of the partial resonators are also distributed unevenly (Fig. 7.28g, l), Two oscillations have comparable amplitudes, but the oscillation of the third type differs from them. The frequency distribution of coupled oscillations is scaled mainly by the mutual coupling coefficients (Fig. 7.28f, k), and the distribution of quality factors (Fig. 7.28h, m) depends both on the value of the imaginary part of the dielectric constant, and the signs of the amplitudes and mutual coupling coefficients of the resonators.

In the case of specific forms and types of DR modes (Fig. 9.29), the distribution of parameters of coupled oscillations also depends on the relative location of the resonators, since the signs and amplitudes of the mutual coupling change when their coordinates change.

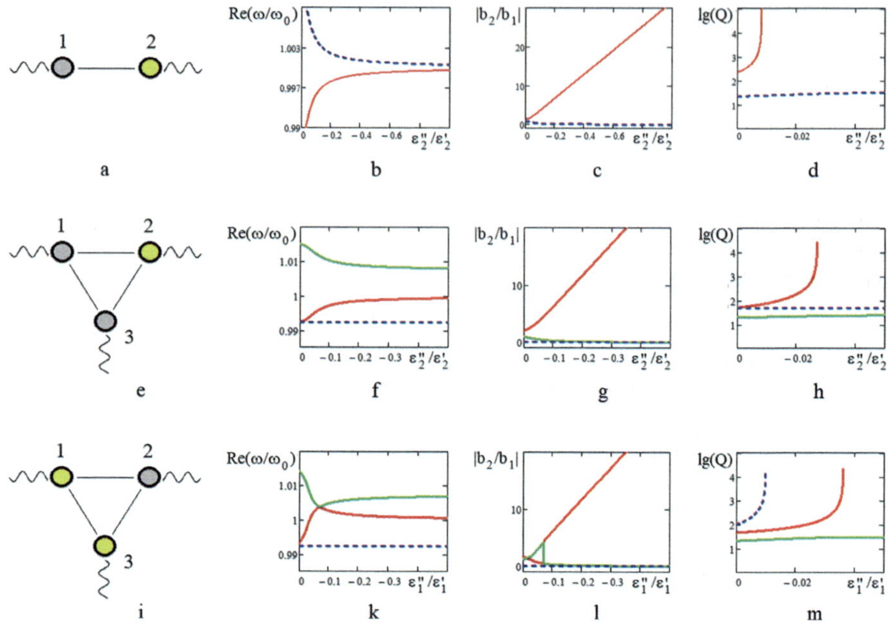

Fig. 7.28 The coupling frequencies (**b**, **f**, **k**); amplitude ratio (**c**, **g**, **l**); quality factor (**d**, **h**, **m**) as functions of the complex dielectric constant of the second DR (**a**, **e**) $(\varepsilon_1''/\varepsilon_1' = 0.01;\ \varepsilon_3''/\varepsilon_3' = \varepsilon_1''/\varepsilon_1')$; first DR (i) $(\varepsilon_2''/\varepsilon_2' = 0.01;\ \varepsilon_3''/\varepsilon_3' = \varepsilon_1''/\varepsilon_1')$. $\kappa_{12} = \kappa_{21} = 0.25 + i0.02; \tilde{k}_1 = 0.02; \tilde{k}_2 = 0.02$ (**b–d**); $\kappa_{12} = \kappa_{13} = \kappa_{23} = 0.015 + i0.01; \tilde{k}_1 = \tilde{k}_3 = 0.02; \tilde{k}_2 = 0.03$ (**f–h**; **k–m**)

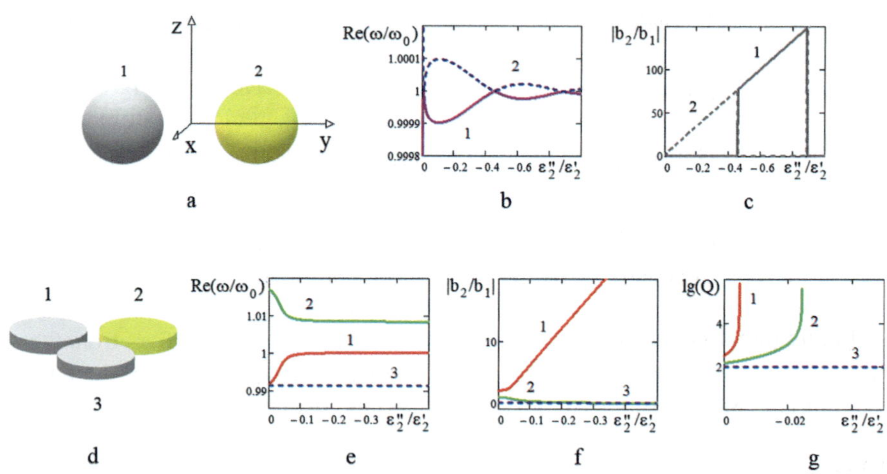

Fig. 7.29 Different spherical DR with a whispering gallery mode $H_{26,26,1}$ (**a**) $(\varepsilon'_{1r}/\varepsilon'_{2r} = 2, 4.\ \Delta x = \Delta z = 0;\ k_0\Delta y = 40)$. 3 different cylindrical DR with a whispering gallery mode $EH_{1,20,1}$ (**d**) $(\varepsilon'_{1r} = \varepsilon'_{2r} = \varepsilon'_{3r} = 9;\ \ k_0\Delta r = 22\ \ \Delta_1 = \Delta_2 = \Delta_3 = 0.2)$. The coupling frequencies (**e**); amplitude ratio (**f**); quality factor of coupled oscillations as functions of the complex dielectric constant tangent function of the second DR (**d**) $(\varepsilon_1''/\varepsilon_1' = \varepsilon_3''/\varepsilon_3' = 0.01)$

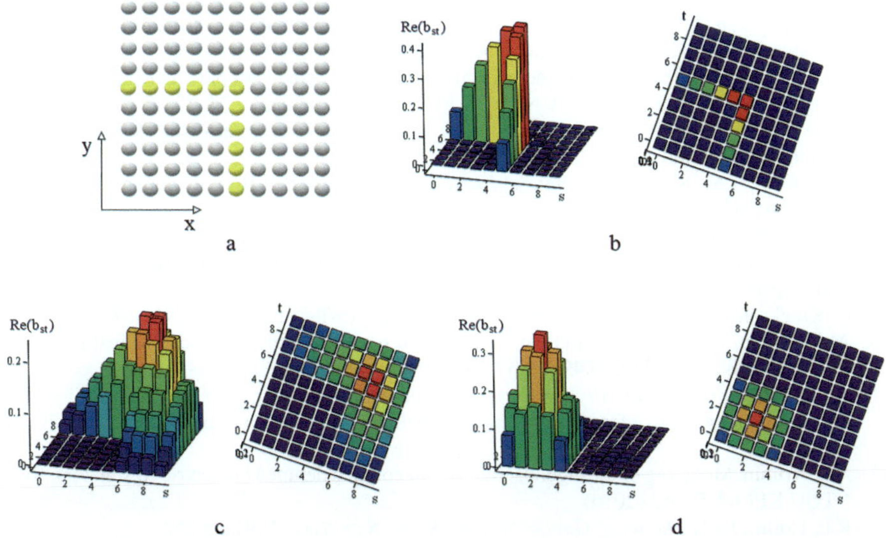

Fig. 7.30 Square lattice of 10 × 10 spherical DRs with absorbing [grey (**a**)] and active dielectric (yellow) and whispering gallery oscillations $H_{26,26,1}$ (**a**) $\left(\varepsilon'_{1r} = \varepsilon'_{2r} = 2.4;\ \varepsilon''_1/\varepsilon'_1 = 0.01;\ \varepsilon''_2/\varepsilon'_2 = -0.03.\ \Delta x = \Delta z = 0;\ k_0\Delta y = 40\right)$. Characteristic distributions of coupled oscillation amplitudes (**b**–**d**)

The above patterns of amplitude distribution confirm the previously discovered effect (7.7, 7.8), the essence of which is that coupled oscillations in complex lattices consisting of various types of DR are broken down into oscillations of individual sublattices. All resonators in each individual sublattice are characterized by the same shape and the same parameter values.

On Fig. 7.30 shows typical distributions of the amplitudes of coupled oscillations of a square lattice of spherical DR with oscillations of a whispering gallery, and consisting of resonators with an absorbing and active dielectric. A subset of active resonators forms the angular structure marked yellow. It can be seen that even with minor differences between the parameters, the distribution of amplitudes exactly repeats the pattern of distribution of coordinates of various types of DR in the lattice.

As can be seen from this example, natural oscillations of the lattice also break up into oscillations of sublattices, the distribution of the amplitudes of the DR corresponds to the distribution of their complex dielectric constant.

References

1. J.D. Jackson, *Classical Electrodynamics*, 3rd edn. (New York, John Willey and Sons, 1998), p. 810
2. J. Van Bladel, M. Verplanken, The electric-dipol resonances of ring resonator of very high permittivity. IEEE Trans. Microwave Theory Tech. **24**(2), 108–112 (1976)

3. M. Verplanken, J. Van Bladel, The magnetic-dipol resonances of ring resonators of very high permittivity. IEEE Trans. Microwave Theory Tech. **27**(4), 328–333 (1979)
4. A.A. Trubin, Study of electromagnetic oscillations in a system of coupled dielectric resonators using perturbation theory. Visnyk NTUU KPI Seriia Radiotekhnika **24**, 42–45 (1987)
5. L.A. Veinshtein, *Open resonators and open waveguides. Golem series in electromagnetics* (Golem Press, Boulder, CO, 1969), p.439
6. G.A. Korn, T.M. Korn, *Mathematical Handbook* (McGraw-Hill Book Company, New York, NY, 1968), p.943
7. J.L. Altman, *Microwave circuits* (D. Van Nostrand Company Inc., New York, NY, 1964), p.462
8. A.A. Trubin, Architecture of dielectric band pass filters with symmetric characteristics. Radioelectronics **7**, 64–69 (1998)
9. C. Kittel, *Introduction to Solid State Physics*, 8th edn. (Wiley, New York, NY, 2005), p.680
10. A.A. Trubin, Natural oscillations of vacancies in multilayer lattices of cylindrical dielectric resonators. Bull. NTUU KPI **49**, 13–17 (2012)
11. A. A. Trubin, in *Eigenoscillations lattices of different dielectric microresonators*. Proceedings of International Conference of "Modern Challenges in Telecommunications, Kiev (2019), pp. 99–102
12. A.A. Trubin, Modeling of the eigenoscillations of coupled detuned Dielectric Resonators. Bull. NTUU KPI **64**, 51–60 (2016)
13. R.E. Collin, *Field Theory of Guided Waves* (Wiley, New York, 1991), p.852
14. A. A. Trubin, in *Perturbation theory of coupled oscillations of resonators with active and absorbing dielectric*. Proceedings of Int. Conference Modern Challenges in Telecommunications, 2024. National Technical University of Ukraine "Igor Sikorsky Kyiv Polytechnic Institute", Kiev, pp. 100–102

Chapter 8
Scattering of Electromagnetic Waves on Systems of Dielectric Resonators in Transmission Lines

Problems of wave scattering on complex DR systems located in various regular transmission lines or sections of evanesced waveguides are considered.

The scattered field is represented in the form of an expansion in coupled oscillations of the DR system as a whole, discussed in detail in Chap. 7. In general form, a system of linear inhomogeneous equations is derived for the amplitudes of the field of coupled oscillations excited by an incident wave in a system of resonators. Based on the obtained solutions, general expressions for the complex coefficients of the scattering matrix of the DR system in the line are obtained in general form.

The results obtained are illustrated by examples of solving scattering problems on systems of several DRs located in various microwave transmission lines—waveguides; asymmetric strip lines, in a section of a cutoff waveguides, as well as optical DRs in Add-drop filters, side-coupled integrated spaced sequence of optical resonators (SCISSORs), Single-channel SCISSORs, Double-channel SCISSORs, Twisted double-channel SCISSORs.

In conclusion, a system of equations for the amplitudes of forced oscillations of resonators of different types, detuned in frequency, and also made of different dielectrics with loss and gain is derived. The scattering characteristics of bandpass and notch filters on various DRs are calculated.

8.1 Equations for the Scattering of Electromagnetic Waves by Systems of Coupled Dielectric Resonators

The number of dielectric structures that allow a simple analytical solution to the scattering problem is small. As a rule, these are simply connected [1, 2] dielectric bodies of regular shape located in open space. In a more general situation, one has to resort to numerical methods using a computer. Meanwhile, the capabilities of the computers themselves are also limited. This is especially noticeable when solving

A. Trubin, *Introduction to the Theory of Dielectric Resonators*, Springer Series in Advanced Microelectronics 65, https://doi.org/10.1007/978-3-031-65396-4_8

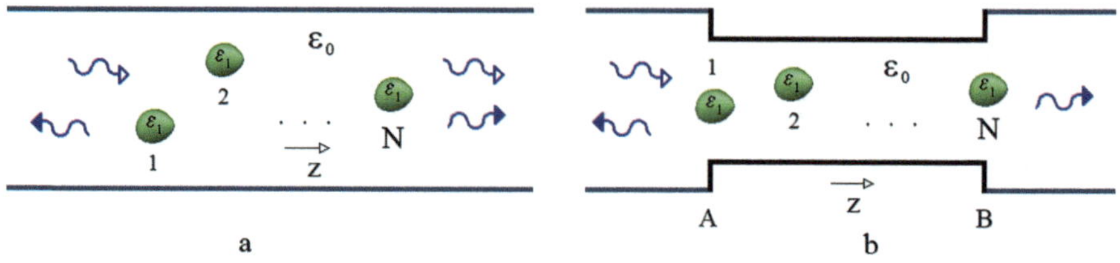

Fig. 8.1 Scattering of waves on a system of coupled DRs in a regular waveguide (**a**); section of the cutoff waveguide (**b**)

complex problems of optimizing scattering on structures consisting of a large number of resonators. Among other things, obtaining a numerical solution often tells us little about the physical processes in the structure under consideration. In this regard, as noted earlier [3], the most promising are mixed methods that combine theoretical analysis, numerical solutions and experimentation.

When solving diffraction problems on bodies consisting of high-quality dielectrics, the greatest interest is often in considering scattering parameters in a relatively narrow frequency range. In this case, the most appropriate description model seems to be the use of the classical natural frequency method [2]. Obviously, for a system of resonators, the main contribution to the solution will be made by their coupled oscillations, located in a finite frequency range.

Let us consider the general solution to the problem of scattering transmission line waves on a system of coupled DRs.

Let a line wave $\left(\mathbf{E}_l^+,\ \mathbf{H}_l^+\right)$ fall on the system of resonators (Fig. 8.1a). Further we will assume that all resonators are made of the same dielectric with known losses: $\widetilde{\varepsilon}_1 = \varepsilon_1 - i\varepsilon_1''$. We will consider such a system of resonators as one multiconnected dielectric body.

Let us denote the solution to the problem of scattering a line wave (Fig. 8.1a) on a multiply connected dielectric body, which includes the set of all resonators of the system:

$$\begin{aligned}&(\mathbf{E},\mathbf{H}) = (\mathbf{E}_1,\mathbf{H}_1)\text{ - in area }\varepsilon = \widetilde{\varepsilon}_1,\\&\text{and through }(\mathbf{E},\mathbf{H}) = (\mathbf{E}_0,\mathbf{H}_0)\text{ - outside the dielectric: }\varepsilon = \varepsilon_0.\end{aligned} \tag{8.1}$$

The scattered field (8.1) must satisfy the boundary conditions on the dielectric surface:

$$[\mathbf{E}_1 - \mathbf{E}_0,\ \mathbf{n}] = 0;\quad [\mathbf{H}_1 - \mathbf{H}_0,\ \mathbf{n}] = 0. \tag{8.2}$$

And possibly on a metal surface, for example, such as a waveguide Σ:

$$[\mathbf{E}_0,\ \mathbf{n}]_\Sigma = 0. \tag{8.3}$$

Here **n**—unit vector normal to the corresponding interface between the media.

Let us assume that the quality factor of the dielectric is sufficiently high, and the frequency range under study does not differ too much from the resonant frequency ω_0 of natural oscillations of isolated resonators:

$$\varepsilon_1''/\varepsilon_1 << 1; \quad |\omega/\omega_0 - 1| << 1. \tag{8.4}$$

We will look for a solution to the problem of $\left(\mathbf{E}_l^+, \mathbf{H}_l^+\right)$ transmission line wave scattering on the DR system in the form:

$$\begin{aligned} \mathbf{E} &\approx \mathbf{E}_l^+ + \sum_{s=1}^{N} a^s \mathbf{e}^s; \\ \mathbf{H} &\approx \mathbf{H}_l^+ + \sum_{s=1}^{N} a^s \mathbf{h}^s; \end{aligned} \tag{8.5}$$

where $(\mathbf{e}^s, \mathbf{h}^s)$ is the natural oscillation field of a system of coupled resonators (7.4).

Here we assume, for simplicity, that only one type of wave propagates in the line: $\left(\mathbf{E}_l^+, \mathbf{H}_l^+\right)$. It is obvious that expansion (8.5) satisfies the boundary conditions (8.2) and (8.3) due to the validity of (2.1) and (2.2).

Substituting (7.4) into expansions (8.5), we find the amplitudes of forced oscillations of partial resonators excited by the incident wave $\left(\mathbf{E}_l^+, \mathbf{H}_l^+\right)$.

$$A_t = \sum_{s=1}^{N} a^s b_t^s \tag{8.6}$$

Let us express bilinear forms of the (C.7) in terms of the scattering field (8.5), the field of coupled oscillations of the structure and the waveguide field. Taking into account also (7.2) and (7.3), we obtain:

$$\begin{aligned} &\sum_{s=1}^{N} a^s \left\{ \omega''^s \left(\varepsilon_1 \mathbf{e}^s \cdot \mathbf{e}^{n^*} + \mu_0 \mathbf{h}^s \cdot \mathbf{h}^{n^*} \right) + \omega \varepsilon_1'' \mathbf{e}^s \cdot \mathbf{e}^{n^*} \right. \\ &\quad \left. + i\left(\omega - \omega^s\right)\left(\varepsilon_1 \mathbf{e}^s \cdot \mathbf{e}^{n^*} + \mu_0 \mathbf{h}^s \cdot \mathbf{h}^{n^*}\right) \right\} \\ &\approx div\left\{ \left[\mathbf{h}^{n^*}, \mathbf{E}_l^+\right] + \left[\mathbf{H}_l^+, \mathbf{e}^{n^*}\right] \right\}. \end{aligned}$$

Substituting expansions (7.4) here, we integrate the resulting expressions over the volume $V = \sum_{t=1}^{N} V_t$ of the "multi-connected resonator" (Fig. 8.1a) taking into account (7.1). In the case of identical partial DRs and types of oscillations in them, we find a system of equations that relates the required coefficients a^s [4]:

$$\sum_{s=1}^{N} a^s \cdot \left(\sum_{t=1}^{N} b_t^s b_t^{n*} \right) Q_s = - \left(\sum_{t=1}^{N} b_t^n c_t^{l+} \right)^* / P_n^D. \tag{8.7}$$

Here $n = 1, 2, \ldots, N$;

$$Q_s = \omega/\omega_0 + 2iQ^D(\omega/\omega_0 - 1 - \lambda^s/2) \tag{8.8}$$

λ^s—the s-th eigenvalue of the operator K (7.6), corresponding to the s-th eigenoscillation of the multiply connected DR; $P_t^D = \omega_0\varepsilon_1''/2 \int_{V_t} \left|\mathbf{e}_t^1\right|^2 dv$—power loss in the dielectric of the t-th partial resonator, $w_t = 1/4 \int\limits_{V_t} \left(\varepsilon_1 \left|\mathbf{e}_t^1\right|^2 + \mu_0 \left|\mathbf{h}_t^1\right|^2\right) dv$—energy stored in the volume of the t-th DR; $Q^D = \varepsilon_1''/\varepsilon_1''$—dielectric quality factor.

Let us rewrite expressions (8.5), taking into account (2.3) and (2.7), in the form:

$$\lim_{z\to\infty} \left[\mathbf{E}_l^+ + \sum_{n=1}^{N} a^n \mathbf{e}^n \right] = \lim_{z\to\infty} \left[\mathbf{E}_l^+ + \sum_{n=1}^{N} a^n \sum_{s=1}^{N} b_s^n \mathbf{e}_s \right] = \mathbf{E}_l^+ \left[1 + \sum_{n,s=1}^{N} a^n b_s^n c_s^{l+} \right];$$
$$\lim_{z\to-\infty} \left[\sum_{n=1}^{N} a^n \mathbf{e}^n \right] = \lim_{z\to-\infty} \left[\sum_{n=1}^{N} a^n \sum_{s=1}^{N} b_s^n \mathbf{e}_s \right] = \mathbf{E}_l^- \left[\sum_{n,s=1}^{N} a^n b_s^n c_s^{l-} \right]. \tag{8.9}$$

Relations (8.9) determine the transmission coefficient T and reflection coefficient R of the wave $\left(\mathbf{E}_l^+, \mathbf{H}_l^+\right)$ incident on the system of resonators in the regular line:

$$T = 1 + \sum_{n,s=1}^{N} b_s^n a^n c_s^{l+};$$
$$R = \sum_{n,s=1}^{N} b_s^n a^n c_s^{l-}, \tag{8.10}$$

In this case, the coefficients $c_l^{s\pm}$ take into account the coordinates of the resonators in the line coordinate system (2.7).

We will calculate the transmission coefficient T in the transverse plane of the center of the last (N-th) line DR, and the reflection coefficient R in the plane of location of the 1-st DR (Fig. 8.1).

When one of the conditions of Theorem (7.16) is met, the solution to the system of Eqs. (8.7) takes the simplest form:

$$a^n = - \left(\sum_{t=1}^{N} b_t^n c_t^{l+} \right)^* / \left(P_n^D \sum_{t=1}^{N} \left|b_t^n\right|^2 Q_n \right). \tag{8.11}$$

Taking into account (8.10)

$$T = 1 - \sum_{n=1}^{N} \left| \sum_{s=1}^{N} b_s^n c_s^{l+} \right|^2 / \left(P_n^D \sum_{t=1}^{N} \left| b_t^n \right|^2 Q_n \right).$$

$$R = -\sum_{n=1}^{N} \left(\sum_{s=1}^{N} b_s^n c_s^{l-} \right) \cdot \left(\sum_{t=1}^{N} b_t^n c_t^{l+} \right)^* / \left(P_n^D \sum_{t=1}^{N} \left| b_t^n \right|^2 Q_n \right). \tag{8.12}$$

Equalities (8.7) are satisfied if the coefficients obey a simpler system of equations:

$$\sum_{s=1}^{N} a^s b_t^s Q_s = -Q^D \left(c_t^{l+} \right)^* / \omega_0 w_t. \tag{8.13}$$

($t = 1, 2, \ldots, N$); $Q^D = \omega_0 w_n / P_n^D$.

In the general case of different partial DRs made of the same dielectric, the solution of system (8.13) is conveniently visually expressed through the determinants of matrices containing all the amplitudes of the coupled natural oscillations of the system (7.11):

$$B = \begin{bmatrix} b_1^1 & b_1^2 & \ldots & b_1^N \\ b_2^1 & b_2^2 & \ldots & b_2^N \\ \vdots & \vdots & \ldots & \vdots \\ b_N^1 & b_N^2 & \ldots & b_N^N \end{bmatrix}; \quad C_s = \begin{bmatrix} b_1^1 & b_1^2 & \ldots & c_1^{+*}/w_1 & \ldots & b_1^N \\ b_2^1 & b_2^2 & \ldots & c_2^{+*}/w_2 & \ldots & b_2^N \\ \vdots & \vdots & \ldots & \vdots & \ldots & \vdots \\ b_N^1 & b_N^2 & \ldots & c_N^{+*}/w_N & \ldots & b_N^N \end{bmatrix};$$

(matrix C_s is a matrix B in which the s-th column is replaced by a column of "normalized" coefficients c_t^{+*}/w_t):

$$a^s = -Q^D / (\omega_0 Q_s) \frac{\det C_s}{\det B}. \tag{8.14}$$

From (8.14) it also follows that the transmission and reflection coefficients (8.10) are expressed in closed form only through the coupling coefficients between the DR (7.15) and the coupling coefficients of the resonators with the line (see Chap. 2).

8.2 Scattering by Systems of Coupled Dielectric Resonators in a Regular Waveguide: Band-Stop Filters on Dielectric Resonators

Let us now consider particular solutions of the system of Eq. (8.14). In the case of scattering of a line wave on two DRs, the transmission coefficient takes the form:

$$T = 1 - \frac{\left(b_1^1 c_1^+ + b_2 c_2^+\right)\left(b_1^2 \frac{c_2^{+*}}{w_2} - b_2 \frac{c_1^{+*}}{w_1}\right)}{b_2\left(b_1^2 - b_1^1\right)} \cdot \frac{Q^D}{\omega_0 Q_1} - \frac{\left(b_1^2 c_1^+ + b_2 c_2^+\right)\left(b_2 \frac{c_1^{+*}}{w_1} - b_1^1 \frac{c_2^{+*}}{w_2}\right)}{b_2\left(b_1^2 - b_1^1\right)} \cdot \frac{Q^D}{\omega_0 Q_2}, \tag{8.15}$$

where c_s^+—coefficient of expansion of the field of the s-th DR in the field of the incident wave of the line ($s = 1, 2$).

As follows from (8.15), if none of the normality conditions for the operator K are satisfied, then the expansion coefficients (7.4) depend on the amplitudes of all coupled oscillations. The resonator system behaves as a single connected structure.

In the particular case of identical transverse coordinates and orientations of the DR ($\tilde{k}_1 = \tilde{k}_2 = \tilde{k}$) considered above, the operator K is normal due to the fulfillment of condition e) of the theorem (Chap. 7), and the transfer coefficient takes the simple form:

$$T = 1 - \frac{\tilde{k} Q^D (1 - \cos \Gamma \Delta z)}{\tilde{k} Q^D (1 - \cos \Gamma \Delta z) + \omega/\omega^1 - 2i Q^D \left(1 - \omega/\omega^1\right)} - \frac{\tilde{k} Q^D (1 + \cos \Gamma \Delta z)}{\tilde{k} Q^D (1 + \cos \Gamma \Delta z) + \omega/\omega^2 - 2i Q^D \left(1 - \omega/\omega^2\right)} \tag{8.16}$$

where from (7.19) we determine the frequencies of coupled oscillations:

$$\omega^{1,2} = \omega_0 \left[1 \mp \left(k_{12} + \tilde{k} \sin \Gamma \Delta z\right)/2\right]. \tag{8.17}$$

On Fig. 8.2 shows the results of a comparison of the calculation of the frequency dependences of the scattering matrix coefficients: $S_{21} = 20 \log|T|$, dB, obtained from (8.15) and (8.16) with measurement data. The coupling coefficients were calculated using the formulas given in Sects. 2.2 and 3.2. The dots here and below show the measured values.

When scattering on a system of three DR, the outermost 1, 3 of which are located symmetrically relative to the second (Fig. 8.3a) we find:

$$T = 1 - \frac{\left(2 b_1^1 c_1^+ \cos \Gamma \Delta z + b_2 c_2^+\right)\left(b_2 \frac{c_1^{+*}}{w_1} \cos \Gamma \Delta z - b_1^2 \frac{c_2^{+*}}{w_2}\right)}{b_2\left(b_1^1 - b_1^2\right)} \cdot \frac{Q^D}{\omega_0 Q_1} - \frac{\left(2 b_1^2 c_1^+ \cos \Gamma \Delta z + b_2 c_2^+\right)\left(b_2 \frac{c_1^{+*}}{w_1} \cos \Gamma \Delta z - b_1^1 \frac{c_2^{+*}}{w_2}\right)}{b_2\left(b_1^2 - b_1^1\right)} \cdot \frac{Q^D}{\omega_0 Q_2} - 2\tilde{k}_1 \sin^2 \Gamma \Delta z \frac{Q^D}{Q_3}. \tag{8.18}$$

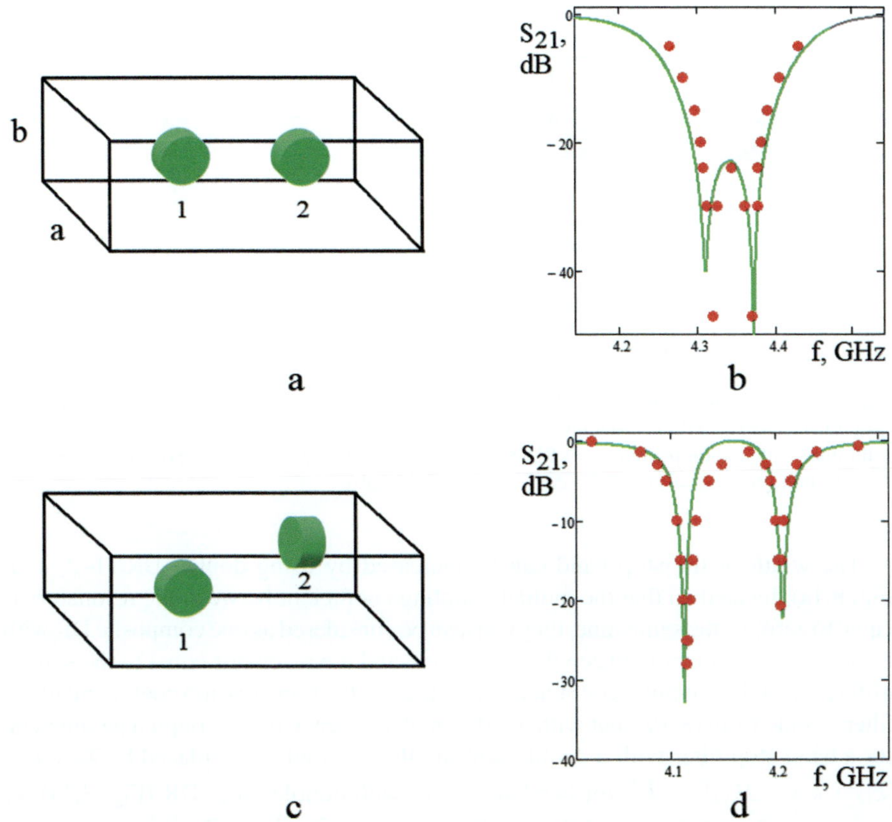

Fig. 8.2 Transmission coefficient **b**, **d** of the fundamental wave H_{10} of a rectangular waveguide ($a \times b = 58 \times 25$ mm) on two cylindrical DR (**a**, **b** $\varepsilon_{1r} = 34$; $Q^D = 3 \cdot 10^3$; $\Delta = 0, 6$; $f_1 = 4.352$ GHz; $f_2 = 4.333$ GHz, $x_1 = x_2 = a/2$; $y_1 = y_2 = 0, 3 \cdot b$; $\Delta z = |z_2 - z_1| = 21.5$ mm; **c**, **d** $\varepsilon_{1r} = 82$; $Q^D = 2400$; $\Delta = 0.25$; $f_0 = 4, 16$ GHz; $x_1 = a/2$; $x_2 = 29.6$ mm, $y_1 = y_2 = b/2$, $\Delta z = 8.5$ mm) with oscillations H^+_{101}. Solid line—calculation data; points—measured values

The expression is simplified provided that neighboring resonators are located at a distance $\Gamma \Delta z = (\pi/2)(2s+1)$ ($s = 0, 1, ...$), and the outer ones are not coupled to each other:

$$T = 1 - \frac{\sqrt{d^-} - i/2\tilde{k}_2}{\sqrt{d^-}} \cdot \frac{\tilde{k}_2 Q^D/2}{Q_1} - \frac{\sqrt{d^+} + i/2\tilde{k}_2}{\sqrt{d^+}} \cdot \frac{\tilde{k}_2 Q^D/2}{Q_2} - \frac{2\tilde{k}_1 Q^D}{Q_3}. \quad (8.19)$$

Here $d^{\pm} = 2\left|k_{12} \pm \tilde{k}_{12}\right|^2 - \left(\tilde{k}_2/2\right)$.

On Fig. 8.3 shows the results of comparing the calculation of the frequency dependences of the scattering matrix coefficients: S_{21}, dB, obtained from (8.18), with measurement data.

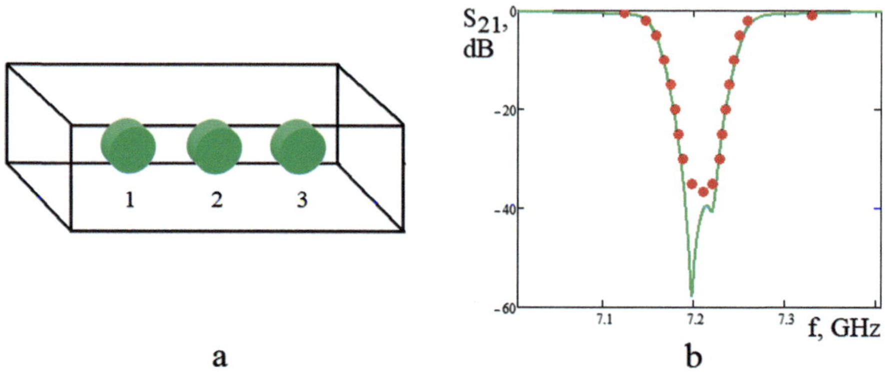

Fig. 8.3 Transmission coefficient (**b**) of a rectangular waveguide wave H_{10} on three identical cylindrical DRs (**a**) with oscillations H_{101}^{+} ($a \times b = 35 \times 15$ mm;). Dots—measured values (**b**), solid line—calculation result ($\varepsilon_{1r} = 81$; $\Delta = 0.58$; $Q^D = 10^3$; $f_0 = 7.205$ GHz; $x_1 = x_3 = a/2$; $x_2 = 13$ mm; $y_1 = y_2 = y_3 = b/2$; $\Delta z_{12} = \Delta z_{23} = 13$ mm)

The width of the stop band can be increased by using double DR (1–2; 4–5; Fig. 8.4a), located so that the mutual coupling coefficient between the resonators is equal to zero. At the same time, they can also be considered as one composite DR with a coupling coefficient between the waveguide and the resonator equal to the sum or difference of the coupling coefficients. In this case, the transmission coefficient of the filter has the form (8.19), but with a different dependence of the coupled parameters: for a band-stop filter with a double central DR: $\tilde{k}_2$ should be replaced by $2\tilde{k}_2$; $d = \left|k_{12} + k_{13} + 2\tilde{k}_{12}\right|^2 - \tilde{k}_2^2$; for band-stop filter with double outer DR (Fig. 8.4a): $\tilde{k}_1$ changes to: $\tilde{k}_2$ to $\tilde{k}_3$ and $d = 2\left(\left|k_{13} + \tilde{k}_{13}\right|^2 + \left|k_{23} + \tilde{k}_{23}\right|^2\right) - \left(\tilde{k}_3/2\right)^2$. On Fig. 8.4b shows the calculated transmission characteristic of a five-resonator band-stop filter made on cylindrical DR in a waveguide, with cross-section $a \times b = 35 \times 15$ mm. The dots show the measurement results. The transverse coordinates of the double DR were located symmetrically relative to the plane $x = a/2$: $x_{1,2} = a/2 \mp L/2$, and the coordinate of the 3-rd DR $x_3 = L/2 + 1$ mm. The maximum filter attenuation was 35 dB.

More complex structures of coupled DRs (Fig. 8.5a, b) are calculated using general solutions (8.13) and (8.14). The measurements performed show good agreement between the theoretical curves and the measurement data.

It is of interest to consider scattering on the DR system, taking into account the frequencies of their natural oscillations of higher types, for example, electric E_{1m1}. In Fig. 8.6 shows the results of calculating the scattering characteristics of the main wave H_{10} of a rectangular waveguide by 7 identical hollow spherical DRs (Sect. 1.6).

In our calculations, we took into account the contribution to the scattered field of the main oscillations of magnetic H_{111} and electrical types E_{111}. The frequency dependence of the transmission coefficient of the 7—cavity filter in the main notch

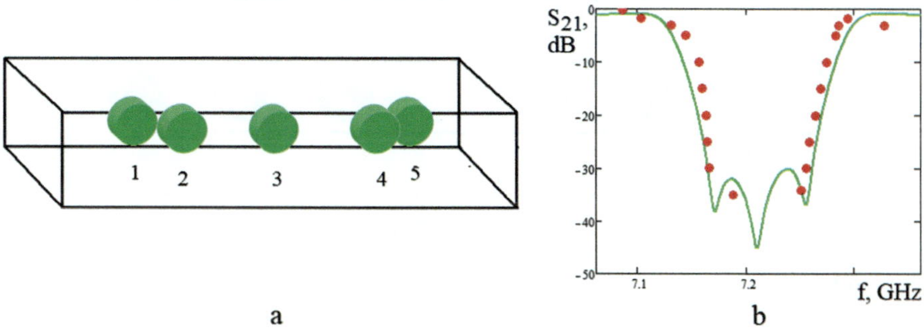

Fig. 8.4 Frequency dependence of the transmission coefficient (**b**) of the fundamental wave of a rectangular waveguide H_{10} ($a \times b = 35 \times 15$ mm; $x_1 = x_5 = 16.18$ mm; $x_2 = x_3 = x_4 = 18.82$ mm; $y_s = b/2$; $\Delta z_{13} = 25.86$ mm; $\Delta z_{23} = 12.93$ mm; DR parameters are the same as in Fig. 8.3) for five identical cylindrical DR (**a**) with oscillations H^+_{101}. Dots show measured values (**b**)

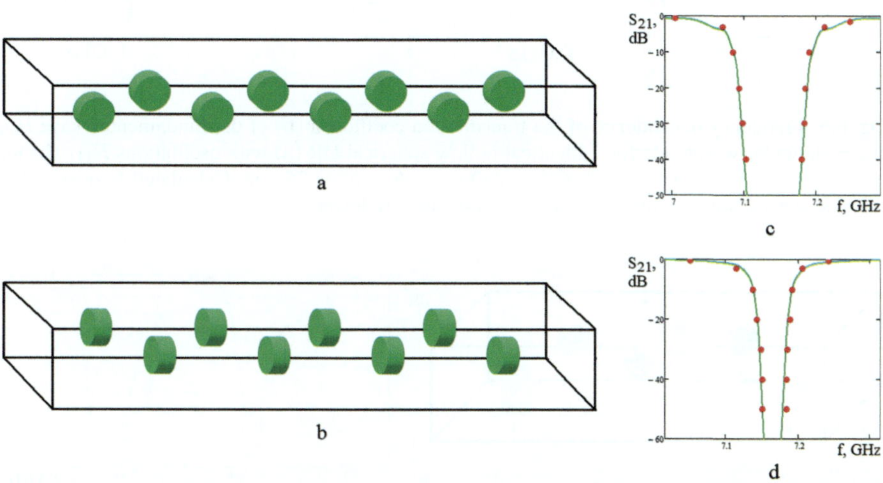

Fig. 8.5 Dependence of the transmission coefficients (**c**, **d**) of the fundamental wave of a rectangular waveguide ($a \times b = 28.5 \times 12.5$ mm) on frequency when scattered on eight identical cylindrical DRs (a: $f_0 = 7.140$ GHz; $x_{2s+1} = 12.5$ mm; $x_{2s} = 16$ mm, b: $f_0 = 7.165$ GHz; $x_{2s+1} = 8.5$ mm; $x_{2s} = 20$ mm; $y_s = b/2$; $\Delta z_{s,s+1} = \Lambda_w/4$) with oscillations H^+_{101}. (DR parameters are the same as in Fig. 8.3) Dots show measured values

band is shown in Fig. 8.6b. The calculated curves demonstrate the possibility of significant rarefaction of parasitic bands on hollow resonators (Fig. 8.6c).

On Fig. 8.7 shows the results of modeling the scattering characteristics of a composite filter implemented on 2 subsystems of detuned cylindrical DRs made of different materials. The discrepancy between the theoretical scattering curve and the measurement data is apparently due to the influence of several factors, the main one of which seems to be the neglect of the influence between the resonator subsystems on the mutual coupling coefficients.

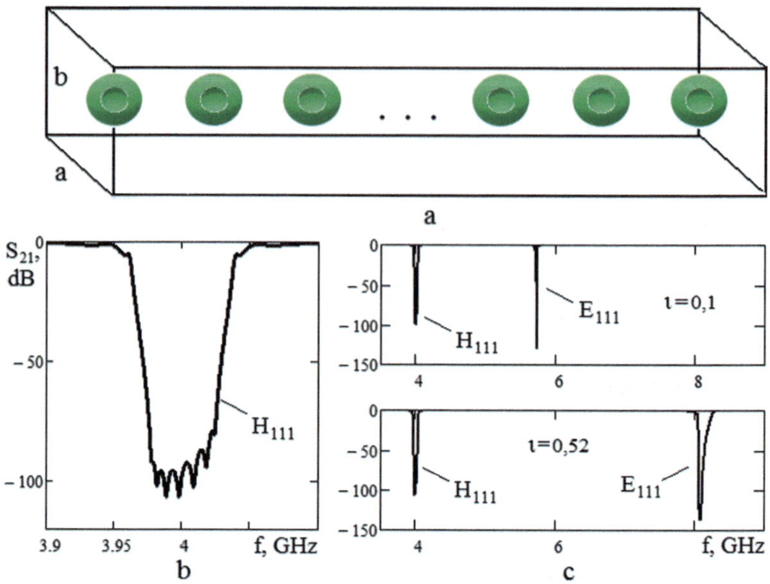

Fig. 8.6 Frequency dependence of the transmission coefficient (**b**) of the fundamental wave H_{10} of a rectangular waveguide for 7 identical hollow spherical DR (**a**) with oscillations H_{111} (**b**) for $\varepsilon_{1r} = 1$; $\varepsilon_{2r} = 81$; $r_1/r_2 = 0.52$; $Q^D = 2500$; $a \times b = 58 \times 25$ mm. Calculated frequency S_{21} dependences for solid (**c** upper) and hollow resonators (**c** lower)

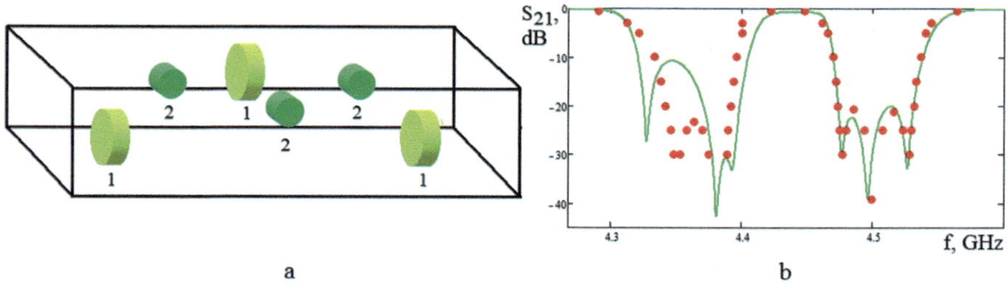

Fig. 8.7 Two-band band-stop filter on two subsystems of coupled cylindrical DR ($f_{01} = 4.368$ GHz; $f_{02} = 4.501$ GHz; $\varepsilon_{1r} = 37$; $\varepsilon_{2r} = 81$; $Q_1^D = 2150$; $Q_2^D = 1700$; $\Delta_1 = 0.588$; $\Delta_2 = 0.584$) in a rectangular waveguide ($a \times b = 58 \times 25$ mm). Frequency response of the transmission coefficient (**b**) of the fundamental wave of a rectangular waveguide H_{10}. Solid curve—calculation; points—measured values

8.3 Scattering by Systems of Coupled Dielectric Resonators in a Cutoff Waveguide. Dielectric Band Pass Filters

Scattering by coupled DR located in a section of the evanescent waveguide (A–B, Fig. 8.1b) is of the greatest practical interest, since such structures are often used in the designs of bandpass filters.

To calculate the transmission coefficients of the structures shown in Fig. 8.1b, we use Eq. (8.14).

In the case of two coupled DR (Fig. 7.2d) located in a section of the evanescent waveguide:

$$T = -\frac{\tilde{k}_{21} k_{12} Q^D}{\sqrt{d}} \left[\frac{1}{Q_1} - \frac{1}{Q_2} \right],$$
$$R = -1 + \frac{\tilde{k}_1 k_{12} Q^D}{\sqrt{d}} \left[\frac{1}{Q_1} + \frac{1}{Q_2} \right]. \tag{8.20}$$

where, formally, $\tilde{k}_{21} = \left(c_1^+\right)^* c_2^+ / \omega_0 w_1$. $d = k_{12}^2 - 1/4(\tilde{k}_1 - \tilde{k}_2)^2$.

For a three-cavity symmetrical structure (Fig. 7.2e), we find,

$$T = \frac{i\tilde{k}_1 + k_{13} - \sqrt{d}}{2\sqrt{d}} \cdot \tilde{k}_1 \frac{Q^D}{Q_1} - \frac{i\tilde{k}_1 + k_{13} + \sqrt{d}}{2\sqrt{d}} \cdot \tilde{k}_1 \frac{Q^D}{Q_2} + \tilde{k}_1 \frac{Q^D}{Q_3}.$$
$$R = -1 - \frac{i\tilde{k}_1 + k_{13} - \sqrt{d}}{2\sqrt{d}} \cdot \tilde{k}_1 \frac{Q^D}{Q_1} + \frac{i\tilde{k}_1 + k_{13} + \sqrt{d}}{2\sqrt{d}} \cdot \tilde{k}_1 \frac{Q^D}{Q_2} + \tilde{k}_1 \frac{Q^D}{Q_3} \tag{8.21}$$

Here $d = \left(i\tilde{k}_1 + k_{13}\right)^2 + 8k_{12}^2$. The eigenvalues of a system of coupled DRs are determined by expression (7.29).

Similarly, using relations (7.31)–(7.32), we find the expression for the transmission coefficient of a four-cavity bandpass filter (Fig. 7.3f):

$$T = \frac{b_1^1}{2\sqrt{d^+}} \cdot \tilde{k}_1 \frac{Q^D}{Q_1} - \frac{b_1^2}{2\sqrt{d^+}} \cdot \tilde{k}_1 \frac{Q^D}{Q_2} - \frac{b_1^3}{2\sqrt{d^-}} \cdot \tilde{k}_1 \frac{Q^D}{Q_3} + \frac{b_1^4}{2\sqrt{d^-}} \cdot \tilde{k}_1 \frac{Q^D}{Q_4},$$
$$R = -1 - \frac{b_1^1}{2\sqrt{d^+}} \cdot \tilde{k}_1 \frac{Q^D}{Q_1} + \frac{b_1^2}{2\sqrt{d^+}} \cdot \tilde{k}_1 \frac{Q^D}{Q_2} - \frac{b_1^3}{2\sqrt{d^-}} \cdot \tilde{k}_1 \frac{Q^D}{Q_3} + \frac{b_1^4}{2\sqrt{d^-}} \cdot \tilde{k}_1 \frac{Q^D}{Q_4}, \tag{8.22}$$

where indices 1, 2 indicate the parameters of even oscillations, and indices 3, 4 indicate the parameters of odd oscillations.

As follows from (7.27), in the case of two coupled DRs, the amplitude-frequency response of a symmetric filter is frequency symmetric.

On Fig. 8.8 shows the amplitude-frequency characteristics of bandpass filters made on the basis of 2 DRs of spherical and cylindrical shape, located near a circular aperture in a rectangular waveguide, cross-section $a \times b = 35 \times 15$ mm. The frequency dependence $S_{21} = 20 \log|T|$, dB, was calculated using formula (8.20), in which the DR coupling coefficients were calculated from (Chap. 2). The dots show the measurement results.

On Fig. 8.9b shows an example of calculating the transmission coefficient of a bandpass filter for 3 cylindrical DRs located in a break in an asymmetrical strip line

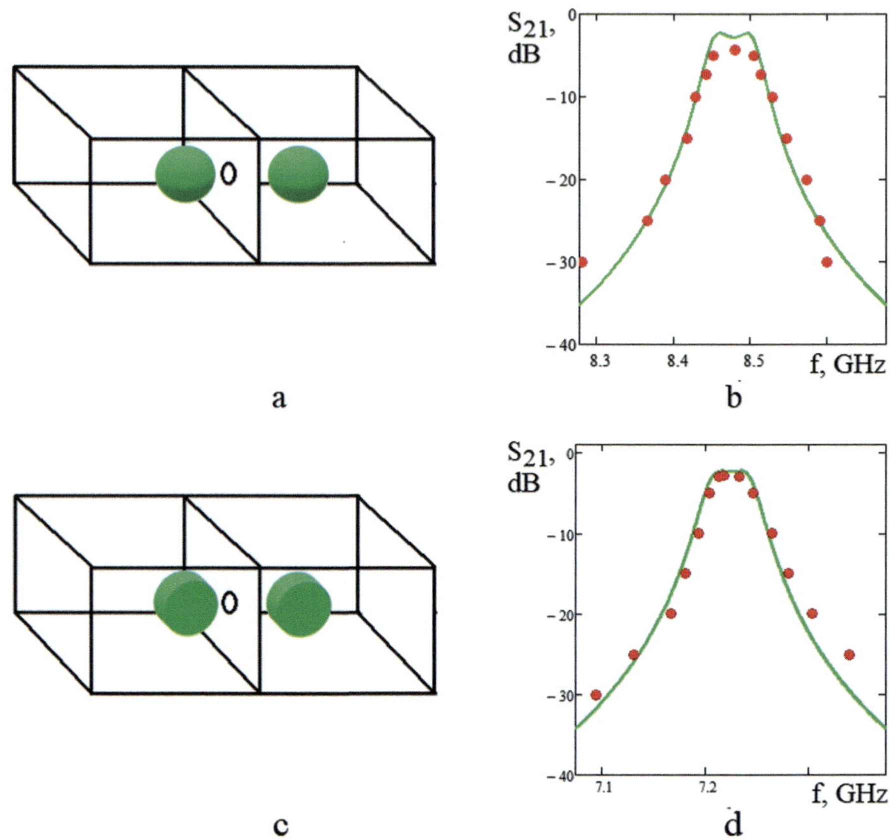

Fig. 8.8 Frequency dependences of the transmission coefficients (**b**, **d**) at scattering of the fundamental wave of a rectangular waveguide H_{10} ($a \times b = 35 \times 15$ mm; **b** $f_0 = 8.477$ GHz; **d** $f_0 = 7.225$ GHz) on two identical spherical ones (**a**); cylindrical DR (**c**) with oscillations H_{111}, H^{+}_{101}, respectively, connected through an aperture in a transverse metal screen ($\varepsilon_{1r} = 81; Q^D = 10^3$). Solid curve—calculation data; points—measured values

(Fig. 8.9a) [5]. To increase isolation, the input and output lines were oriented orthogonal to each other. Coupling coefficients were calculated using formulas (Chap. 3 and 4). The amplitude-frequency response of the filter was calculated using (8.20).

In the designs of bandpass filters, it is comfortable to use subsystems of coupled DRs that have one of the frequencies of coupled oscillations that coincides with the frequency of independent oscillations of the remaining partial DRs ω_0. In some cases, the use of such coupled DR subsystems can lead to a noticeable improvement in scattering parameters [6]. In Fig. 8.10a, b shows an example of the implementation of a 3-section bandpass filter, made using a subsystem consisting of two interconnected cylindrical DRs—2, 3.

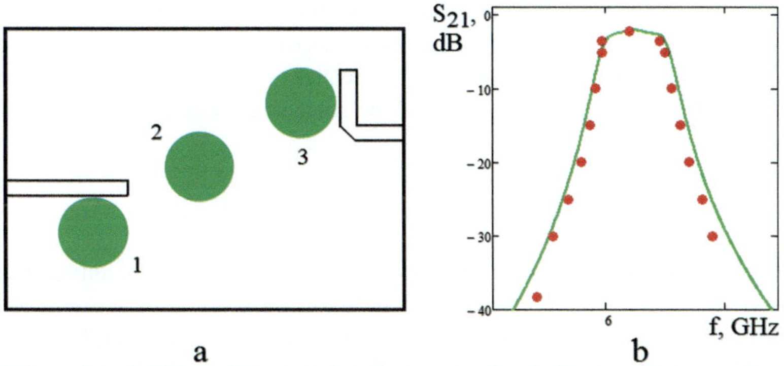

Fig. 8.9 Frequency dependence of the transmission coefficient (**b**) for three identical cylindrical DR (**a**) with H_{101}^{+} oscillations ($\varepsilon_{1r} = 81$; $Q^D = 10^3$) in a break in an asymmetrical microstrip line. Solid curve—calculation data; points—measured values

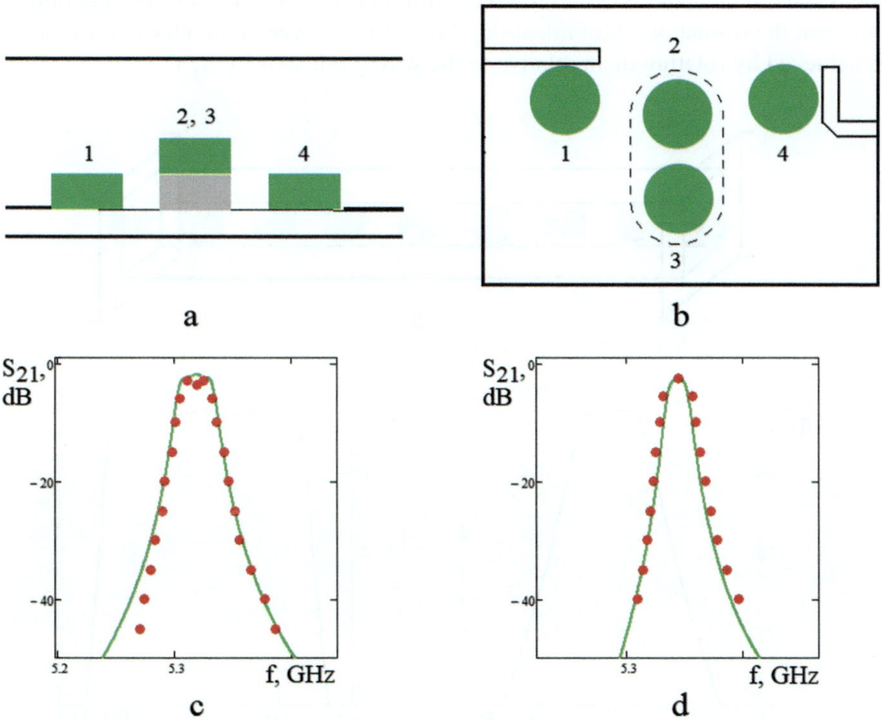

Fig. 8.10 Frequency dependence of transmission coefficients (**c**, **d**) for two cylindrical resonators 1–4 (**a**, **b**) with oscillations H_{101}^{+} ($\varepsilon_{1r} = 81$; $Q^D = 1900$) in the break of an asymmetrical microstrip line and a subsystem of two coupled DR 2–3. Solid curve—calculation data; points—measured values

When the frequency of one of the coupled oscillations of the 2nd and 3rd DRs (in-phase or antiphase) coincides with the frequencies of partial DRs 1, 4: $\omega^{\pm} = \omega_0$, such a structure implements a bandpass filter equivalent to a 3-section one. The mutual coupling coefficient between the 1-st and the subsystem of related DRs 2, 3 is equal to:

$$(k_{1p}k_{p1})^{1/2} = 1/2^{1/2}(k_{12} \pm k_{13}),$$

where the + sign corresponds to in-phase oscillations, and the sign to anti-phase oscillations. The frequency response corresponding to the excitation of common-mode oscillations in the DR subsystem 2, 3 is shown in Fig. 8.10c, and antiphase ones—in Fig. 8.10d, is characterized by increased attenuation outside the filter passband.

The calculations of the scattering characteristics on multi-link structures of coupled DRs in a section of an cut-off waveguide confirm the conclusions of the theory describing frequency-symmetric structures (Sect. 7.4). On Fig. 8.11b, c show the results of calculating 5 and 9—section filters (Fig. 8.11a) with serial couplings between the resonators. Minimization of coupling between non-adjacent resonators is achieved by rotating them relative to the waveguide axis by π/4.

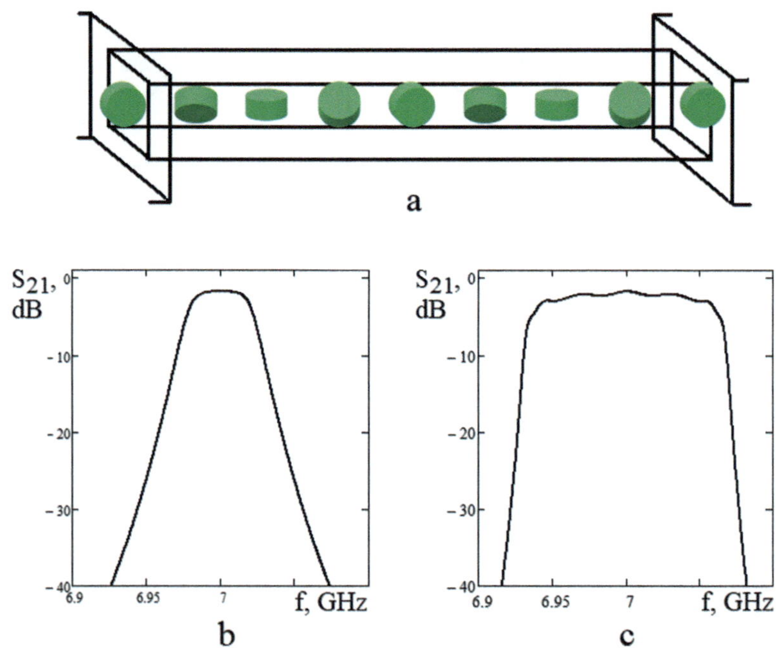

Fig. 8.11 Frequency characteristics of the transmission coefficients of five (**b**), nine (**c**) rotating cylindrical resonators with oscillations H^{+}_{101} ($\varepsilon_{1r} = 36$; $Q^D = 2 \cdot 10^3$) in a segment of a tcut-off rectangular waveguide, cross-section $a_z \times b_z = 15 \times 15$ mm (**a**)

8.4 Optical Notch Filters Based on Dielectric Microresonators

The calculation of the scattering characteristics of filters on dielectric microresonators differs from filters built on resonators with the main types of oscillations in the presence of several features. Such filters use resonators with whispering gallery oscillations (Chap. 1). Whispering gallery oscillations in cylindrical resonators are doubly degenerate; in spherical resonators, the degree of degeneracy is much higher due to the high degree of symmetry of the shape. Below, for simplicity, we will consider resonators only with doubly degenerate types of oscillations. The degeneration of natural oscillations in resonators with rotation symmetry is due to the existence in them of oscillations identical in field structure, differing from each other by rotation by an angle determined by a given azimuthally number. The presence of degenerate oscillations in resonators leads to the fact that in the transmission line their coupling coefficients of identical resonators differ from each other, and the number of oscillations taken into account in Eqs. (8.13) at least doubles. However, at the same time, the excitation of degenerate oscillations leads to a number of interesting unidirectional scattering effects, which are responsible for the emergence of a new class of so-called Add-Drop filters [8].

Following our theory (Sect. 8.1), we first consider the scattering characteristics of notch filters built according to the scheme for constructing a linear chain of microresonators located in a regular transmission line (Figs. 8.12and 8.13a). We will assume that the degenerate oscillations of each pair in each partial resonator are not coupled. We will consider one of these degenerate oscillations to be even and denote it by the index "e", and the other odd and denote it by the index "o" relative to any selected plane of symmetry. The plane of symmetry can be, for example, a plane passing through the center of the resonator and orthogonal to the axis of the transmission line.

The solution to the system of equations is formally no different from (8.13) and (8.14), except for the fact that the number of oscillations taken into account doubles, in accordance with the comments made.

The matrix of amplitudes of coupled oscillations takes the form:$(b_1^e, b_1^o, b_2^e, b_2^o, \ldots, b_N^e, b_N^o)^T$. Here T is the transposition symbol.

On Fig. 8.12b, c shows the scattering characteristics of two notch filters shown in Fig. 8.12a. The presence of coupling between adjacent resonators leads to additional ruggedness of the transmission coefficient. Such indentation can be avoided by placing the resonators on different sides of the line, reducing the coupling coefficients between them (Fig. 8.13a). However, in this case, attenuation poles appear in the frequency dependence of the reflection coefficient S_{11} (Fig. 8.13c).

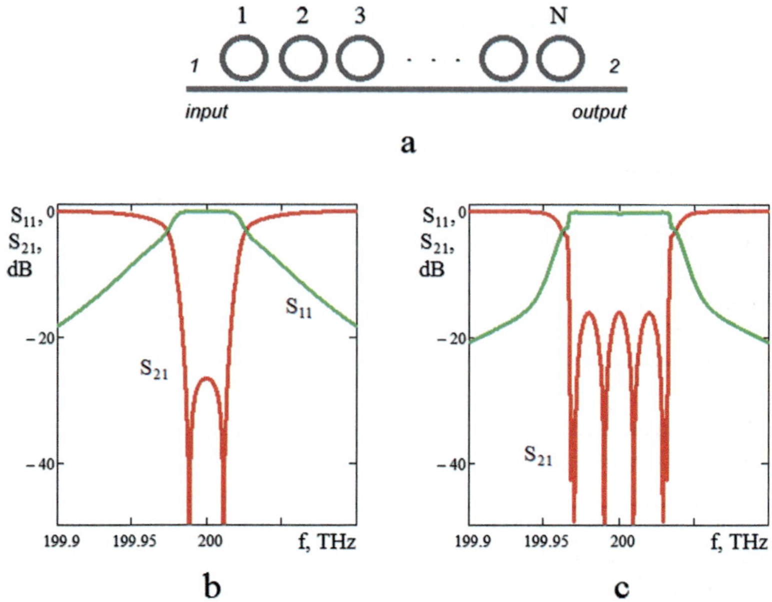

Fig. 8.12 Frequency dependence of the S-matrix module of 2 (**b**), 4 (**c**) resonator notch filter (**a**) [$f_0 = 200$ THz; $Q^D = 10^6$; $\Gamma \Delta z_{s,s+1} = 31 \cdot \pi/2$; $\tilde{k}_s^e = \tilde{k}_s^o = 2 \cdot 10^{-4}$; $k_{s,s+1}^e = k_{s+1,s}^e = 3 \cdot 10^{-4}$; $k_{s,s+1}^o = k_{s+1,s}^o = -3 \cdot 10^{-5}$ (**b**); $k_{s,s+1}^o = k_{s+1,s}^o = -2.5 \cdot 10^{-5}$ (**c**)]

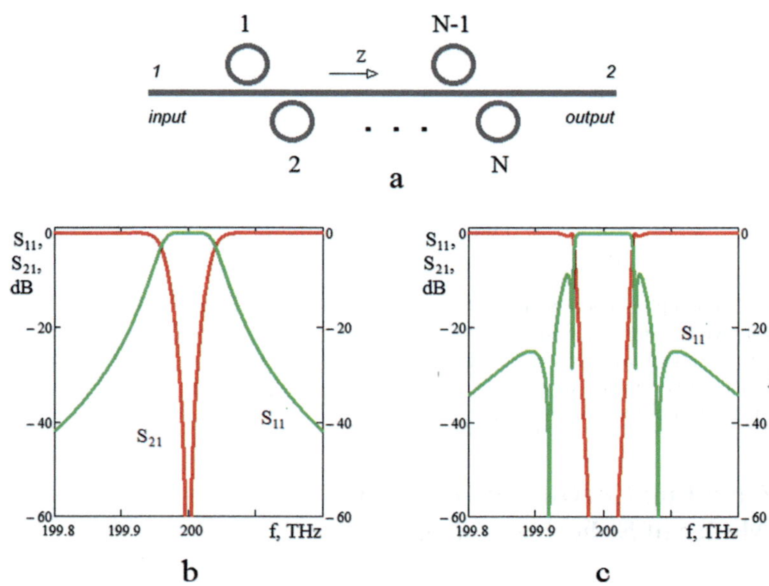

Fig. 8.13 Frequency dependences of the S-matrix module of 2 (**b**), 4 (**c**) resonator notch filters (**a**) [$f_0 = 200$ THz; $Q^D = 10^6$; $\Gamma \Delta z_{s,s+1} = 31 \cdot \pi/2$; $\tilde{k}_s^e = \tilde{k}_s^o = 2 \cdot 10^{-4}$; $k_{s,s+1}^e = k_{s+1,s}^e = 0$; $k_{s,s+1}^o = k_{s+1,s}^o = -3 \cdot 10^{-5}$ (**b**); $k_{s,s+1}^o = k_{s+1,s}^o = 0$ (**c**)]

8.5 Optical Add-Drop Filters Based on Dielectric Microresonators

Expressions (8.13) allow us to calculate more complex DR structures coupled with several different transmission lines simultaneously. In this case, the general structure of the existing transmission lines and microresonators must be taken into account both when solving the problem of their coupled oscillations, and when solving the system of equations describing the scattering of waves incident on the system along given lines.

At first, we calculated the frequencies of the coupled oscillations of the resonators, taking into account the topology of the connections between themselves and the transmission lines, as well as open space. Using a system of Eqs. (7.6), we calculated the frequencies and amplitudes of coupled oscillations of identical partial microresonators. In this case, the coupling matrix (7.8) of the microresonators for the filter shown in Fig. 8.14a, was presented in the form:

$$K = \left\| i(\tilde{k}_1^e \delta_{s1} + \tilde{k}_1^o \delta_{s2} + \tilde{k}_N^e \delta_{s(2N-1)} + \tilde{k}_N^o \delta_{s(2N)} + \tilde{k}_{OS})\delta_{sn} + \kappa_{sn}(1 - \delta_{sn}) \right\|$$

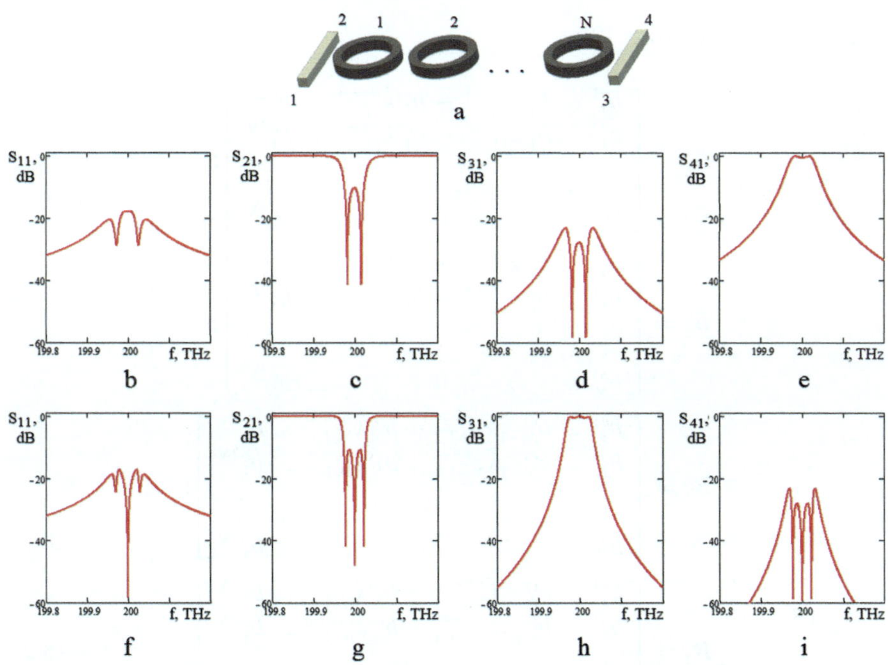

Fig. 8.14 Frequency dependences of the S-matrix of 2 (**b–e**), 3 (**f–i**) resonator Add-Drop filters (**a**) [f_0 = 200 THz; $Q^D = 10^6$; $k_{OS} = 10^{-7}$; $\tilde{k}_s^e = 2 \cdot 10^{-4}$; $\tilde{k}_s^o = 1.5 \cdot 10^{-4}$; $k_{s,s+1}^e = k_{s+1,s}^e = 2.4 \cdot 10^{-4}$; $k_{s,s+1}^o = k_{s+1,s}^o = -2.4 \cdot 10^{-4}$; (**b–e**); $k_{s,s+1}^o = k_{s+1,s}^o = -2 \cdot 10^{-4}$ (**f–i**)]

where $\tilde{k}_1^{eo}$—is the coupling coefficient of the 1-st microresonator with the transmission line 1–2 (Fig. 8.14a) on an even (or odd) oscillation; $\tilde{k}_N^{eo}$—is the coupling coefficient of the N-th microresonator with the transmission line 3–4 also on an even (odd) oscillation; $\tilde{k}_{OS}$ is the coupling coefficient of the microresonator with open space; $\kappa_{sn} = k_{12}^{e,o} = k_{21}^{e,o}$, if $|s - n| = 1$ and $\kappa_{sn} = 0$ in other cases.

The found eigenvalues λ and eigenvectors of the coupling matrix K (7.8)

$$B = \begin{bmatrix} b_1^{e1} & b_1^{e2} & \dots & b_1^{e2N} \\ b_1^{o1} & b_1^{o2} & \dots & b_1^{o2N} \\ \vdots & \vdots & & \vdots \\ b_N^{o1} & b_N^{o2} & \dots & b_N^{o2N} \end{bmatrix}, \tag{8.23}$$

were used to solve the scattering problem on a system of microresonators.

The reflection and transmission coefficients were determined from the ratios (8.13):

$$T_{1\nu}(\omega) = \delta_{2\nu} - \frac{Q^D}{\det B} \cdot \sum_{s=1}^{2N} \frac{\det B_\nu^s}{Q_s(\omega)}, \tag{8.24}$$

where for Add-drop filter [9]:

$$B_1^s = \begin{bmatrix} b_1^{e1} & \dots & b_1^{es}\tilde{k}_{11}^{ee-+} + b_1^{os}\tilde{k}_{11}^{oe-+} & \dots & b_1^{e2N} \\ b_1^{o1} & \dots & b_1^{es}\tilde{k}_{11}^{eo-+} + b_1^{os}\tilde{k}_{11}^{oo-+} & \dots & b_1^{o2N} \\ \vdots & \dots & \vdots & \dots & \vdots \\ b_N^{o1} & \dots & 0 & \dots & b_N^{o2N} \end{bmatrix};$$

$$B_2^s = \begin{bmatrix} b_1^{e1} & \dots & b_1^{es}\tilde{k}_{11}^{ee++} + b_1^{os}\tilde{k}_{11}^{oe++} & \dots & b_1^{e2N} \\ b_1^{o1} & \dots & b_1^{es}\tilde{k}_{11}^{eo++} + b_1^{os}\tilde{k}_{11}^{oo++} & \dots & b_1^{o2N} \\ \vdots & \dots & \vdots & \dots & \vdots \\ b_N^{o1} & \dots & 0 & \dots & b_N^{o2N} \end{bmatrix};$$

$$B_3^s = \begin{bmatrix} b_1^{e1} & \dots & b_{N-1}^{es}\tilde{k}_{N-1,1}^{ee++} + b_N^{os}\tilde{k}_{N,1}^{oe++} & \dots & b_1^{e2N} \\ b_1^{o1} & \dots & b_{N-1}^{es}\tilde{k}_{N-1,1}^{eo++} + b_N^{os}\tilde{k}_{N,1}^{oo++} & \dots & b_1^{o2N} \\ \vdots & \dots & \vdots & \dots & \vdots \\ b_N^{o1} & \dots & 0 & \dots & b_N^{o2N} \end{bmatrix};$$

$$B_4^s = \begin{bmatrix} b_1^{e1} & \dots & b_{N-1}^{es}\tilde{k}_{N-1,1}^{ee-+} + b_N^{os}\tilde{k}_{N,1}^{oe-+} & \dots & b_1^{e2N} \\ b_1^{o1} & \dots & b_{N-1}^{es}\tilde{k}_{N-1,1}^{eo-+} + b_N^{os}\tilde{k}_{N,1}^{oo-+} & \dots & b_1^{o2N} \\ \vdots & \dots & \vdots & \dots & \vdots \\ b_N^{o1} & \dots & 0 & \dots & b_N^{o2N} \end{bmatrix} \tag{8.25}$$

$\tilde{k}_{sn}^{ab++} = \left(c_s^{a+}c_n^{b+*}\right)/\left(\omega_0 w_n\right) = (\tilde{k}_{sn}^{ab})_0 e^{-i\Gamma(z_s-z_n)}$; $\tilde{k}_{sn}^{ab-+} = \left(c_s^{a-}c_n^{b+*}\right)/\left(\omega_0 w_n\right) = (\tilde{k}_{sn}^{ab})_0 e^{-i\Gamma(z_s+z_n)}$; $Q_s(\omega) = \omega/\omega_0 + 2iQ^D(\omega/\omega_0 - 1 - \lambda_s/2)$; Q^D—is the dielectric Q-factor of the resonators; $\omega_0 = 2\pi f_0$; f_0—is the oscillation frequency of isolated resonators; Here a, b takes values "even" or "odd" depending on the type of s-th resonator oscillations.; $c_s^{a\pm}$—is the expansion coefficient of the s-th microresonator field with a-mode on the propagating wave of the transmission line (2.3); w_n—is the energy stored in the dielectric of the n-th resonator; Γ—is the longitudinal wave number of the transmission line; z_n—is the longitudinal coordinate of the n-th resonator center. The scattering matrix coefficients: $S_{v1} = 20\lg|T_{1v}|$.

On Fig. 8.14, the frequency dependences of the S-matrix of 2 (Fig. 8.14b–e), 3 (Fig. 8.14f–i) resonator Add-Drop filters (Fig. 8.14) are shown, (a), calculated using formulas (8.24) and (8.25). The resulting curves demonstrate the known "unevenness" of transmission coefficients from input port 1 to output ports 3, 4, characteristic of Add-Drop filters with an even and odd number of resonators.

The coupling matrix (7.8) of SCISSOR (side-coupled integrated spaced sequence of resonators) shown in Fig. 8.15a, has the form:

$$K = \left\| i(\tilde{k}_{(12)s}^{a} + \tilde{k}_{(34)s}^{a} + \tilde{k}_{OS})\delta_{sn} + \kappa_{sn}(1 - \delta_{sn}) \right\|.$$

Here $\tilde{k}_{(v)s}^{a} = \tilde{k}_{(v)s}^{e,o}$—is the coupling coefficient of the s-th resonator with the v-th transmission line; $\kappa_{sn} = k_{sn} + i\tilde{k}_{sn}^{ab\pm\pm}$—is the complex coupling coefficient of the resonator.

Formally, the filter transfer coefficient is (8.24), but for the SCISSOR structure:

$$B_1^s = \begin{bmatrix} b_1^{e1} & \cdots & \sum\limits_{u=1}^{2N} b_u^s \tilde{k}_{u1}^{ae-+} & \cdots & b_1^{e2N} \\ b_1^{o1} & \cdots & \sum\limits_{u=1}^{2N} b_u^s \tilde{k}_{u1}^{ao-+} & \cdots & b_1^{o2N} \\ \vdots & \cdots & \vdots & \cdots & \vdots \\ b_N^{o1} & \cdots & \sum\limits_{u=1}^{2N} b_u^s \tilde{k}_{uN}^{ao-+} & \cdots & b_N^{o2N} \end{bmatrix}; \quad B_2^s = \begin{bmatrix} b_1^{e1} & \cdots & \sum\limits_{u=1}^{2N} b_u^s \tilde{k}_{u1}^{ae++} & \cdots & b_1^{e2N} \\ b_1^{o1} & \cdots & \sum\limits_{u=1}^{2N} b_u^s \tilde{k}_{u1}^{ao++} & \cdots & b_1^{o2N} \\ \vdots & \cdots & \vdots & \cdots & \vdots \\ b_N^{o1} & \cdots & \sum\limits_{u=1}^{2N} b_u^s \tilde{k}_{uN}^{ao++} & \cdots & b_N^{o2N} \end{bmatrix};$$

$$B_3^s = \begin{bmatrix} b_1^{e1} & \cdots & \sum\limits_{u=1}^{2N} b_u^s \tilde{k}_{u1}^{ae++} & \cdots & b_1^{e2N} \\ b_1^{o1} & \cdots & \sum\limits_{u=1}^{2N} b_u^s \tilde{k}_{u1}^{ao++} & \cdots & b_1^{o2N} \\ \vdots & \cdots & \vdots & \cdots & \vdots \\ b_N^{o1} & \cdots & \sum\limits_{u=1}^{2N} b_u^s \tilde{k}_{uN}^{ao++} & \cdots & b_N^{o2N} \end{bmatrix}; \quad B_4^s = \begin{bmatrix} b_1^{e1} & \cdots & \sum\limits_{u=1}^{2N} b_u^s \tilde{k}_{u1}^{ae-+} & \cdots & b_1^{e2N} \\ b_1^{o1} & \cdots & \sum\limits_{u=1}^{2N} b_u^s \tilde{k}_{u2}^{ao-+} & \cdots & b_1^{o2N} \\ \vdots & \cdots & \vdots & \cdots & \vdots \\ b_N^{o1} & \cdots & \sum\limits_{u=1}^{2N} b_u^s \tilde{k}_{uN}^{ao-+} & \cdots & b_N^{o2N} \end{bmatrix}. \tag{8.26}$$

Here a also takes values "even" or "odd" depending on the type of u-th microresonator oscillations.

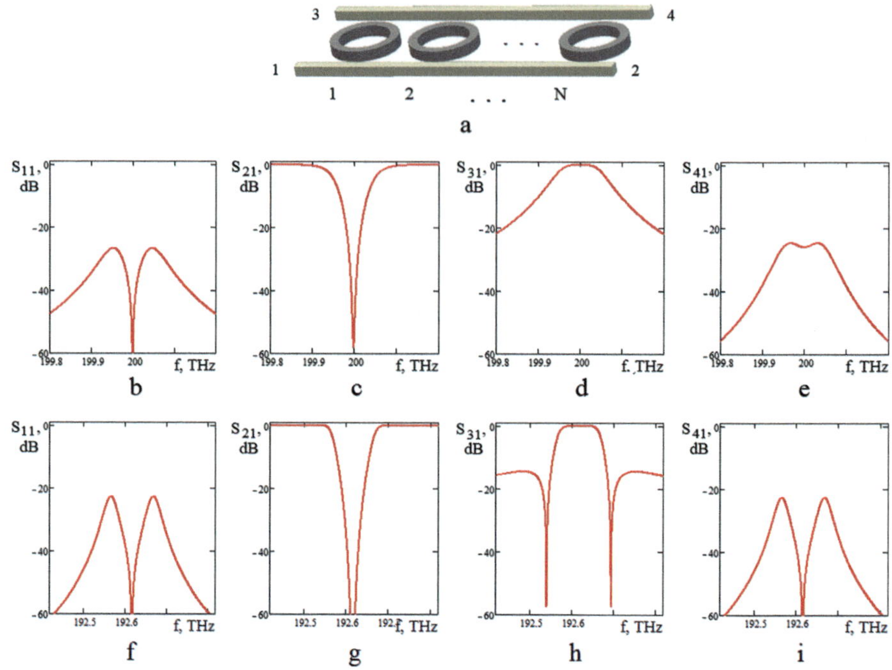

Fig. 8.15 Frequency dependences of the scattering matrix of 2 (**b–e**), 3 (**f–i**) section SCISSOR, 2 channels, coupled along the side wall of DRs (**a**) [$\Gamma\Delta z_{s,s+1} = 31 \cdot \pi/2$; $k_{OS} = 10^{-7}$; $\tilde{k}_s^e = \tilde{k}_s^o = 2 \cdot 10^{-4}$; $Q^D = 10^9$; $k_{s,s+1}^e = k_{s+1,s}^e = 2.4 \cdot 10^{-4}$; $k_{s,s+1}^o = k_{s+1,s}^o = -2 \cdot 10^{-4}$ (**b, c**); $Q^D = 10^7$; $k_{s,s+1}^e = k_{s+1,s}^e = k_{s,s+1}^o = k_{s+1,s}^o = 1 \cdot 10^{-5}$ (**f–i**)]

On Fig. 8.15, the frequency dependences of the S-matrix of 2 (Fig. 8.15b–e), 3 (Fig. 8.15f–i) section SCISSOR (Fig. 8.15a) are shown, calculated using formulas (8.24) and (8.26). The constructed calculated curves also demonstrate the difference in transmission coefficients from channel 1 to 3, 4, characteristic of this type of filter.

The coupling K and B_n^s matrices of the Double-channel SCISSOR, shown in Fig. 8.16, a, has the form:

$$K = \left\| i(\tilde{k}_s^a + \tilde{k}_{OS})\delta_{sn} + \kappa_{sn}(1 - \delta_{sn}) \right\|,$$

where $\tilde{k}_s^a = \tilde{k}_{(12)s}^a$ if $s \leq 2N$ and $\tilde{k}_s^a = \tilde{k}_{(34)s}^a$ if $s > 2N$.

Here $\kappa_{sn} = k_{sn} + i\tilde{k}_{sn}^{ab}$ also the complex mutual coupling coefficient. For microresonators with the same longitudinal coordinates $\kappa_{sn} = k_{sn} = k_V^a$, where k_V^a is the mutual coupling coefficient of "vertically coupled" microcavities (Fig. 8.16).

Also the filter transfer coefficients are (8.15), but

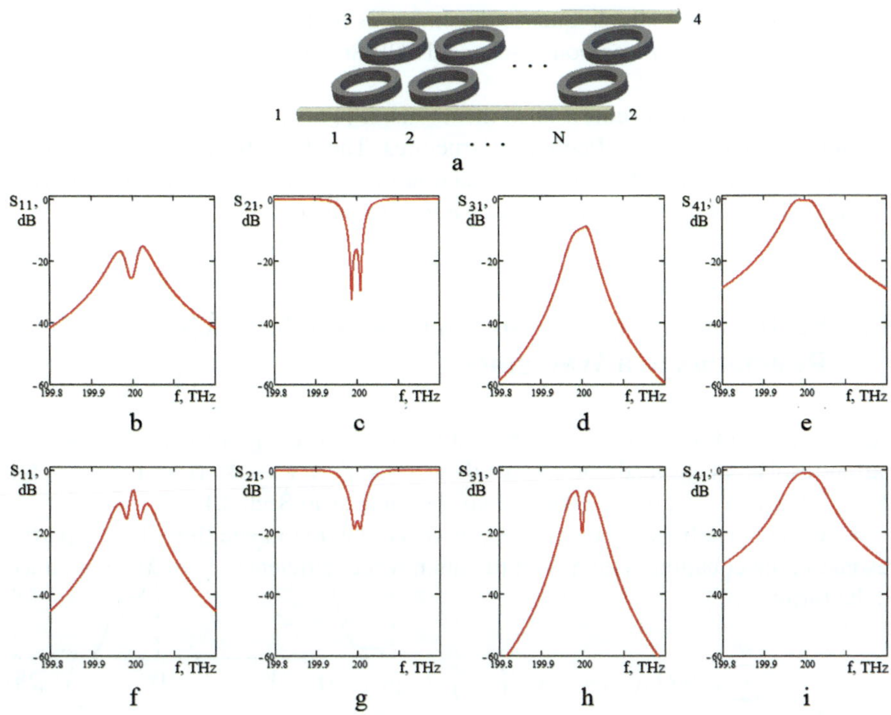

Fig. 8.16 Frequency dependences of the S-matrix of 2 (**b–e**), 3 (**f–i**) resonator filters (**a**) on a double lattice of DRs (double-channel SCISSOR) coupled along the side wall (**a**) [$Q^D = 10^9$; $\Gamma\Delta z_{s,s+1} = 31\cdot\pi/2$; $k_{OS} = 10^{-9}$; $\tilde{k}_s^e = \tilde{k}_s^o = 2.8\cdot10^{-4}$; $k^e_{s,s+1} = k^e_{s+1,s} = 3\cdot10^{-5}$; $k^o_{s,s+1} = k^o_{s+1,s} = -3\cdot10^{-5}$; $k_V^e = 1.3\cdot10^{-4}$; $k_V^o = -1.3\cdot10^{-4}$(**b–e**); $k^e_{s,s+1} = k^e_{s+1,s} = 3.5\cdot10^{-5}$; $k^o_{s,s+1} = k^o_{s+1,s} = -3.5\cdot10^{-5}$; $k_V^e = 8.2\cdot10^{-5}$; $k_V^o = -8,2\cdot10^{-5}$(**f–i**)]

$$B_1^s = \begin{bmatrix} b_1^{e1} & \dots & \sum_{u=1}^{2N} b_u^s\tilde{k}_{u1}^{ae-+} & \dots & b_1^{e4N} \\ b_1^{o1} & \dots & \sum_{u=1}^{2N} b_u^s\tilde{k}_{u1}^{ao-+} & \dots & b_1^{o4N} \\ \vdots & \dots & \vdots & \dots & \vdots \\ b_{2N}^{o1} & \dots & 0 & \dots & b_{2N}^{o4N} \end{bmatrix}; \quad B_2^s = \begin{bmatrix} b_1^{e1} & \dots & \sum_{u=1}^{2N} b_u^s\tilde{k}_{u1}^{ae++} & \dots & b_1^{e4N} \\ b_1^{o1} & \dots & \sum_{u=1}^{2N} b_u^s\tilde{k}_{u1}^{ao++} & \dots & b_1^{o4N} \\ \vdots & \dots & \vdots & \dots & \vdots \\ b_{2N}^{o1} & \dots & 0 & \dots & b_{2N}^{o4N} \end{bmatrix};$$

$$B_3^s = \begin{bmatrix} b_1^{e1} & \dots & \sum_{u=2N+1}^{4N} b_u^s\tilde{k}_{u1}^{ae++} & \dots & b_1^{e4N} \\ b_1^{o1} & \dots & \sum_{u=2N+1}^{4N} b_u^s\tilde{k}_{u1}^{ao++} & \dots & b_1^{o4N} \\ \vdots & \dots & \vdots & \dots & \vdots \\ b_{2N}^{o1} & \dots & 0 & \dots & b_{2N}^{o4N} \end{bmatrix}; \quad B_4^s = \begin{bmatrix} b_1^{e1} & \dots & \sum_{u=2N+1}^{4N} b_u^s\tilde{k}_{u1}^{ae-+} & \dots & b_1^{e4N} \\ b_1^{o1} & \dots & \sum_{u=2N+1}^{4N} b_u^s\tilde{k}_{u1}^{ao-+} & \dots & b_1^{o4N} \\ \vdots & \dots & \vdots & \dots & \vdots \\ b_{2N}^{o1} & \dots & 0 & \dots & b_{2N}^{o4N} \end{bmatrix} \quad (8.27)$$

On Fig. 8.16, shows the frequency dependences of the S-matrix of 2 (Fig. 8.16b–e), 3 (Fig. 8.16f–i) resonator Double-channel SCISSOR (Fig. 8.16a), calculated using formulas (8.24) and (8.27).

It can be seen that the constructed theory displays the main features of scattering on multi-rezonator filters of complex structures. The data obtained allow us to say that these features are due to the interaction of oscillations in degenerated types excited in each of the partial microresonator of the filter.

8.6 Scattering by Systems of Various Coupled Dielectric Resonators in a Waveguide

Consider a useful generalization of the system of scattering Eqs. (8.13) for cases when partial resonators are made of various dielectrics: $\widetilde{\varepsilon}_t = \varepsilon_t' - i\varepsilon_t''$ $(t = 1, 2, \ldots, \mathrm{N})$. We will leave the rest of the assumptions the same as in Sect. 8.1.

By using perturbation theory (C.8), after the volume integration of each partial resonator, the equation system with the unknown coefficients a^s has been obtained in the form:

$$\sum_{s=1}^{N} a^s b_t^s Q_{st}(\omega) = -Q_t^D \left(c_t^{l+}\right)^* / \omega_0 w_t; \quad (t = 1, 2, \ldots, \mathrm{N}). \tag{8.28}$$

where for different resonators, the functions are dependent on the partial DRs and the coupled oscillation numbers:

$$Q_{st}(\omega) = \omega/\omega_0 + 2iQ_t^D(\omega/\omega_0 - 1 - \lambda^s/2); \tag{8.29}$$

$Q_t^D = \omega_0 w_t / P_t^D;\ P_t^D = \omega_0 \frac{\varepsilon_t''}{2} \int_{V_t} |\mathbf{e}_t|^2 dv$—is the loss power in the dielectric of t-th DR.

The transmission T and the reflection coefficient R of different DR system in the transmission line can be obtained from (8.28) in the form:

$$T = T_0 + \sum_{u=1}^{N} \left(\sum_{s=1}^{N} b_s^u c_s^+ \right) a_u = T_0 - \frac{1}{detB(\omega)} \sum_{s=1}^{N} detB_s^+(\omega);$$
$$R = R_0 + \sum_{u=1}^{N} \left(\sum_{s=1}^{N} b_s^u c_s^- \right) a_u = R_0 - \frac{1}{detB(\omega)} \sum_{s=1}^{N} detB_s^-(\omega). \tag{8.30}$$

Here T_0, R_0 are the transmission and reflection coefficients of the transmission line without DRs;

$$B_s^{\pm}(\omega) = \begin{bmatrix} b_1^1 Q_{11}(\omega) & b_1^2 Q_{21}(\omega) & \dots & Q_1^D \sum_{u=1}^N b_u^s \tilde{k}_{u1}^{\pm+} & \dots & b_1^N Q_{N1}(\omega) \\ b_2^1 Q_{12}(\omega) & b_2^2 Q_{22}(\omega) & \dots & Q_2^D \sum_{u=1}^N b_u^s \tilde{k}_{u2}^{\pm+} & \dots & b_2^N Q_{N2}(\omega) \\ \vdots & \vdots & \dots & \vdots & \dots & \vdots \\ b_N^1 Q_{1N}(\omega) & b_N^2 Q_{2N}(\omega) & \dots & Q_N^D \sum_{u=1}^N b_u^s \tilde{k}_{uN}^{\pm+} & \dots & b_N^N Q_{NN}(\omega) \end{bmatrix};$$

$$B(\omega) = \begin{bmatrix} b_1^1 Q_{11}(\omega) & b_1^2 Q_{21}(\omega) & \dots & b_1^N Q_{N1}(\omega) \\ b_2^1 Q_{12}(\omega) & b_2^2 Q_{22}(\omega) & \dots & b_2^N Q_{N2}(\omega) \\ \vdots & \vdots & \dots & \vdots \\ b_N^1 Q_{1N}(\omega) & b_N^2 Q_{2N}(\omega) & \dots & b_N^N Q_{NN}(\omega) \end{bmatrix}; \qquad (8.31)$$

$\tilde{k}_{sn}^{++} = \left(c_s^+ c_n^{+*}\right) / (\omega_0 w_n) = (\tilde{k}_{sn})_0 e^{-i\Gamma(z_s - z_n)}$; $\tilde{k}_{sn}^{-+} = \left(c_s^- c_n^{+*}\right) / (\omega_0 w_n) = (\tilde{k}_{sn})_0 e^{-i\Gamma(z_s + z_n)}$; $w_n = 1/4 \int_{V_n} (\varepsilon_1 |\vec{e}_{1n}|^2 + \mu_0 \left|\vec{h}_{1n}\right|^2) dv$—is the energy, stored in the dielectric of the n-th resonator; $(\tilde{k}_{sn})_0$—is the coupling coefficient (4), for the propagating wave, expressed without phase difference accounting in the line; Γ—is the longitudinal wave number of the transmission line; z_s—is the longitudinal coordinate of s-th resonator center ($s = 1, 2, \ldots, N$).

It is easy to make sure that the found ratios (8.28)–(8.31) in the case of the same DRs exactly coincide with (8.13) and (8.14).

On Fig. 8.17 shows an example of the frequency characteristics of the scattering matrix of a band-stop filter built on two different cylindrical DR subsystems, calculated on the basis of relations (8.28)–(8.31). The coupling coefficients of the resonators were calculated from the relationships found earlier in Chaps. 2 and 3. $G_{21}(\omega) = \arg(T(\omega))$; $G_{11}(\omega) = \arg(R(\omega))$ (8.30).

In the case of scattering of transmission line waves on a DR system located in a section of an evanescent waveguide, the transmission and reflection coefficients retain the form (8.30), and the functions $B_s^{\pm}(\omega)$ take the form:

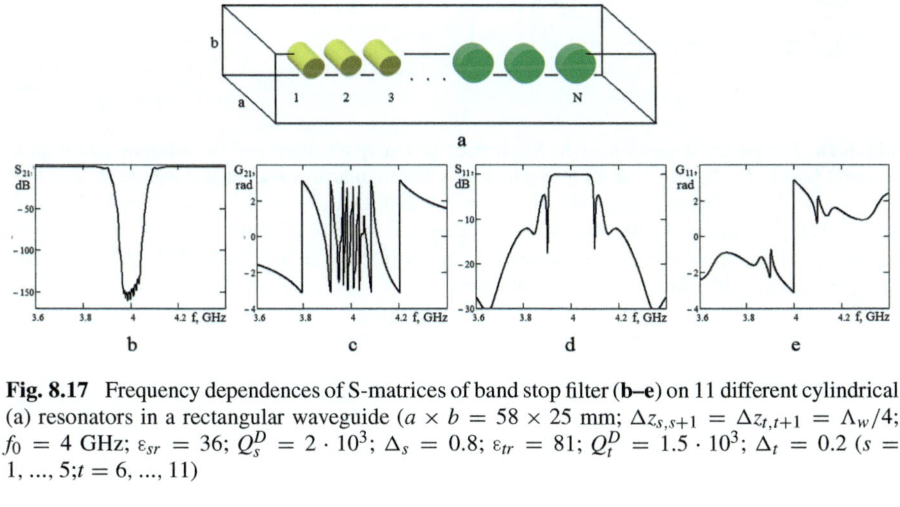

Fig. 8.17 Frequency dependences of S-matrices of band stop filter (**b–e**) on 11 different cylindrical (a) resonators in a rectangular waveguide ($a \times b = 58 \times 25$ mm; $\Delta z_{s,s+1} = \Delta z_{t,t+1} = \Lambda_w/4$; $f_0 = 4$ GHz; $\varepsilon_{sr} = 36$; $Q_s^D = 2 \cdot 10^3$; $\Delta_s = 0.8$; $\varepsilon_{tr} = 81$; $Q_t^D = 1.5 \cdot 10^3$; $\Delta_t = 0.2$ ($s = 1, \ldots, 5; t = 6, \ldots, 11$)

$$B_s^+(\omega) = \begin{bmatrix} b_1^1 Q_{11}(\omega) & b_1^2 Q_{21}(\omega) & \ldots & Q_1^D b_N^s \tilde{k}_{N1}^{++} & \ldots & b_1^N Q_{N1}(\omega) \\ b_2^1 Q_{12}(\omega) & b_2^2 Q_{22}(\omega) & \ldots & 0 & \ldots & b_2^N Q_{N2}(\omega) \\ \vdots & \vdots & \ldots & \vdots & \ldots & \vdots \\ b_N^1 Q_{1N}(\omega) & b_N^2 Q_{2N}(\omega) & \ldots & 0 & \ldots & b_N^N Q_{NN}(\omega) \end{bmatrix};$$

$$B_s^-(\omega) = \begin{bmatrix} b_1^1 Q_{11}(\omega) & b_1^2 Q_{21}(\omega) & \ldots & Q_1^D b_1^s \tilde{k}_{11}^{-+} & \ldots & b_1^N Q_{N1}(\omega) \\ b_2^1 Q_{12}(\omega) & b_2^2 Q_{22}(\omega) & \ldots & 0 & \ldots & b_2^N Q_{N2}(\omega) \\ \vdots & \vdots & \ldots & \vdots & \ldots & \vdots \\ b_N^1 Q_{1N}(\omega) & b_N^2 Q_{2N}(\omega) & \ldots & 0 & \ldots & b_N^N Q_{NN}(\omega) \end{bmatrix}, \quad (8.32)$$

If only the 1st and Nth DR are coupled to a regular waveguide.

On Fig. 8.18 shows the characteristics of the scattering of band pass filters made of resonators of different dielectric permittivity, calculated by formulas (8.30) and (8.31). The coupling factors of the resonators were calculated from the ratio of previously obtained in Sects. 2.2, 2.3, and 3.4. In addition to the possibility of varying the coupling coefficients of resonators within a wider limits, such filters also have more rarefied bands free from parasitic oscillations.

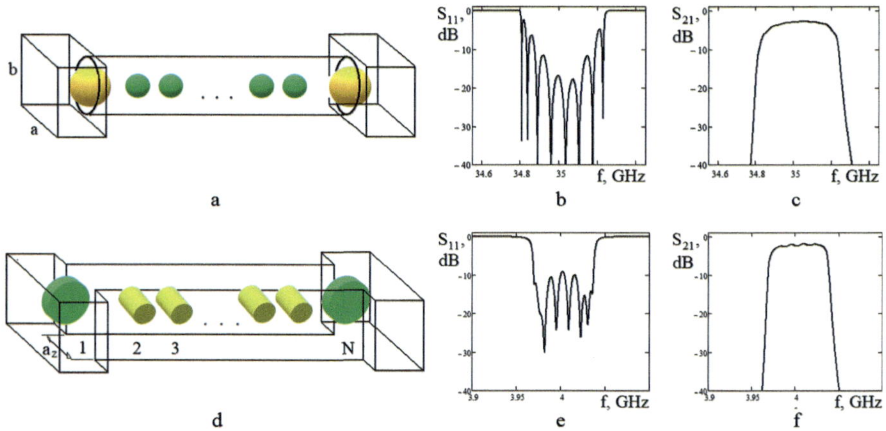

Fig. 8.18 Frequency dependences of S-matrices of bandpass filters on 10 different spherical in a round (**a–c**), 9 cylindrical (**d–f**) resonators in a rectangular evanescent waveguide (**a** $a \times b = 7 \times 3$ mm; $R = 1.5$ mm; $\Delta z_{1,2} = \Delta z_{9,10} = 2.4$ mm; $\Delta z_{s,s+1} = 2.3$ mm; $\varepsilon_{1r} = \varepsilon_{10r} = 36$; $\varepsilon_{sr} = 81$; $Q_1^D = Q_{10}^D = 3 \cdot 10^3$; $Q_s^D = 2 \cdot 10^3$; d: $a \times b = 58 \times 25$ mm; $a_z \times b_z = 20 \times 25$ mm; $\varepsilon_{1r} = \varepsilon_{9r} = 36$; $\varepsilon_{sr} = 81$; $\Delta_1 = \Delta_9 = 0,2$; $\Delta_s = 0.8$; $Q_1^D = Q_9^D = 5 \cdot 10^3$; $Q_s^D = 2 \cdot 10^3$; $\Delta z_{1,2} = \Delta z_{8,9} = 20.7$ mm; $\Delta z_{s,s+1} = 16.1$ mm)

8.7 Scattering from Systems of Detuned Dielectric Resonators

In Sect. 7.8 using perturbation theory, we built a model of coupled oscillations of frequency-detuned DR. Let us now consider the characteristics of scattering from a system of resonators detuned in frequency.

To this end, we again present the solution to the scattering problem in the form:

$$\mathbf{E}(\omega) \approx \mathbf{E}_l^+ + \sum_{s=1}^{N} a^s \mathbf{e}^s(\omega);$$
$$\mathbf{H}(\omega) \approx \mathbf{H}_l^+ + \sum_{s=1}^{N} a^s \mathbf{h}^s(\omega); \tag{8.33}$$

Equation (8.5), where $(\mathbf{e}^s(\omega),\ \mathbf{h}^s(\omega))$—field of coupled oscillations of a resonator system.

From (7.50):

$$\mathbf{e}^t(\omega) = \sum_{s=1}^{N} b_s^t(\omega)\mathbf{e}_s(\omega);$$
$$\mathbf{h}^t(\omega) = \sum_{s=1}^{N} b_s^t(\omega)\mathbf{h}_s(\omega).$$

Repeating the reasoning of the Sect. 8.1, we arrive at the equation for amplitudes:

$$\sum_{s=1}^{N} a^s \alpha_t(\omega) b_t^s(\omega) Q_{st}(\omega) = -Q_t^D \left(c_t^+\right)^* / \omega_t w_t,$$

where

$$Q_{st}(\omega) = \omega/\omega_t + 2iQ_t^D(\omega/\omega_t - 1 - \lambda^s/2);$$

$Q_t^D = \omega_t w_t / P_t^D$—loss quality factor in the dielectric of the t-th DR. Taking into account the previously found relations: $\alpha_s(\omega) b_s(\omega) = b_s$, we come to the conclusion that the equation exactly coincides with (8.29) in which ω_0 it is replaced by ω_t [10].

$$\sum_{s=1}^{N} a^s b_t^s Q_{st}(\omega) = -Q_t^D \left(c_t^+\right)^* / \omega_t w_t \tag{8.34}$$

The transmission T and the reflection coefficient R of detuned DR system in the transmission line can be also obtained from (8.28) in the form (8.30) and

(8.31) for bandstop filters and from (8.32) for band pass filters, where $\tilde{k}_{sn}^{++} = \left(c_s^+ c_n^{+*}\right) / (\omega_n w_n) = (\tilde{k}_{sn})_0 e^{-i\Gamma(z_s - z_n)}$; $\tilde{k}_{sn}^{-+} = \left(c_s^- c_n^{+*}\right) / (\omega_n w_n) = (\tilde{k}_{sn})_0 e^{-i\Gamma(z_s + z_n)}$. In this case, the amplitudes of coupled oscillations are also found from (7.55) and (7.52). Here we have also neglected the dependence $\Gamma = \Gamma(\omega)$, which in the case of large frequency detuning of partial resonators should be taken into account.

In Fig. 8.19 shows results of calculating the frequency dependences of the scattering matrix of transmission line waves on 2 (a–d), 3 (e–h) resonators detuned in frequency, performed on the basis of solving the system of Eqs. (7.55) and (8.34).

An example of calculating the frequency dependences of the scattering matrix obtained on the basis of (7.55) and (8.34), as well as (8.30) and (8.32), is shown in Fig. 8.20. It can be seen that the frequency detuning of the partial resonators affects to a greater extent the reflection characteristic of the filter (Fig. 8.20b).

The developed model can also serve as a basis for describing scattering by a single resonator with several natural oscillation modes. To do this, in the system of Eqs. (7.55), taking into account the orthogonality of the fields of natural oscillations, one should put $\kappa_{st} = 0$, and in (8.31); (8.32) represent the expansion coefficients of the resonator field into line waves, for example, in the form: $c_s^{\pm} = (\tilde{k}_s)^{1/2} e^{\mp i\pi/2s}$. For a resonator located at the origin of the line: $z_s = 0$. Then

$$\tilde{k}_{sn}^{++} = \left(c_s^+ c_n^{+*}\right) / (\omega_n w_n) = (\tilde{k}_s \tilde{k}_n)^{1/2} e^{-i\pi/2(s-n)};$$
$$\tilde{k}_{sn}^{-+} = \left(c_s^- c_n^{+*}\right) / (\omega_n w_n) = (\tilde{k}_s \tilde{k}_n)^{1/2} e^{i\pi/2(s+n)}.$$

On Fig. 8.21. a solution to the model problem of scattering on a resonator with 5 natural oscillations, with the same distance between frequencies, is presented. To describe scattering on a real resonator, it is necessary to take into account the specific values of the expansion coefficients $c_s^{\pm}$ and quality factors Q_s^D for each mode.

8.8 Scattering by Systems of Dielectric Resonators with Loss and Gain

Dielectric resonators (DR) made from an active dielectric can become new promising elements of integrated circuits in the optical and infrared ranges. The electromagnetic properties of resonators made of an active dielectric have not been fully studied at present. An important component of the DR theory is the theory of scattering, built on the basis of well-proven physical assumptions, as well as bilinear relations between the electromagnetic field vectors, directly following from Maxwell's equations. The construction of an analytical theory of scattering makes it possible to effectively design a wide class of various elements of communication systems, optical processors, quantum devices, etc.

Let us construct a solution to the problem of the wave scattering on a system of coupled active DRs, based on the perturbation theory of Maxwell's equations. using expansions (5), and the conclusions of the previous section. Let us also assume that

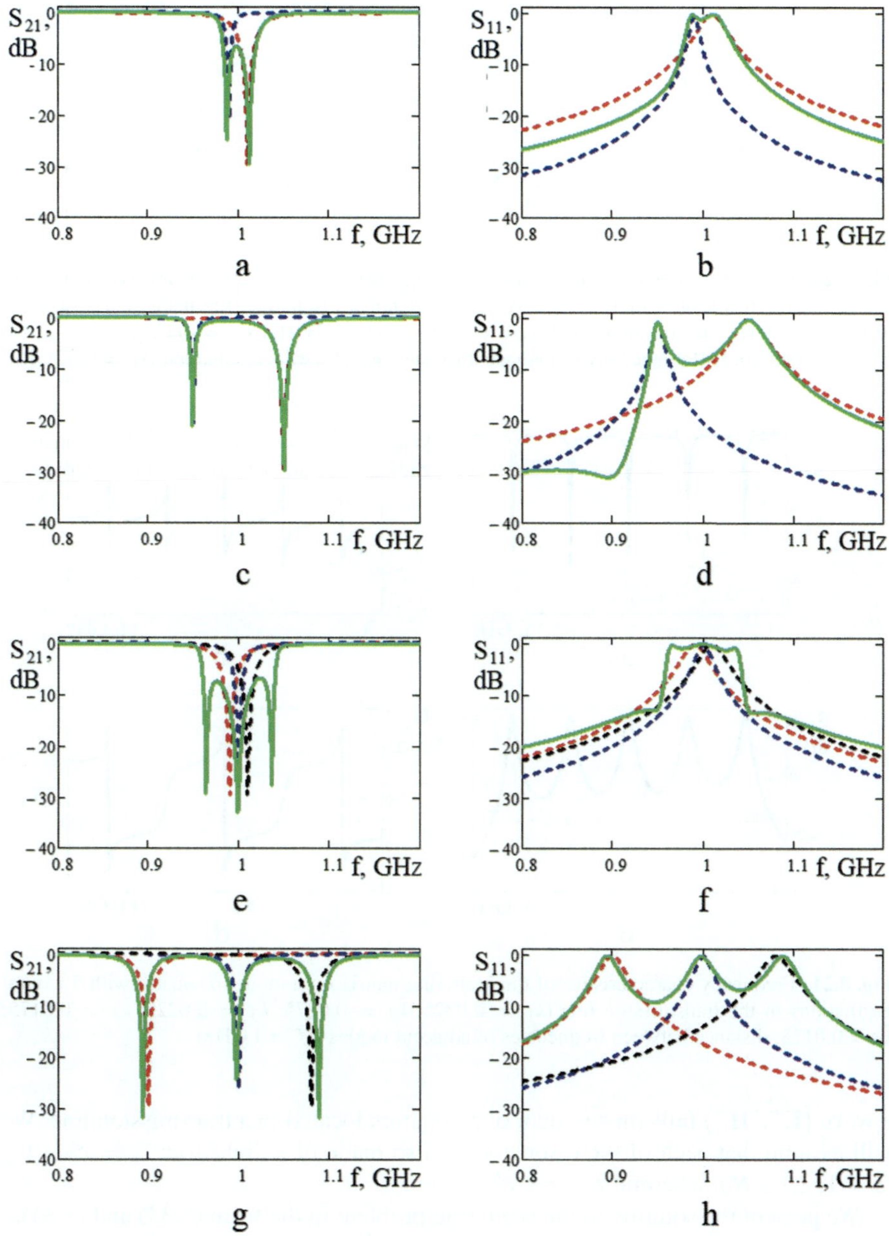

Fig. 8.19 Frequency characteristics of the scattering matrix on 2 resonators (**a–d**), 3 resonators (**e–h**) in the transmission line at different frequencies. Solid lines are the result of scattering by a system of resonators; dotted lines are the result of scattering by one resonator detuned in frequency ($\tilde{k}_1 = 0.03$; $\tilde{k}_2 = 0.01$; $k_{12} = 0.005$; $\Gamma \Delta z_{12} = \pi/2$—**a–d**); ($\tilde{k}_1 = 0.03$; $\tilde{k}_2 = 0.02$; $\tilde{k}_3 = 0.03$; $k_{12} = k_{23} = 0.003$; $k_{13} = 0.001$; $\Gamma \Delta z_{12} = \Gamma \Delta z_{23} = \pi/2$—**e–h**)

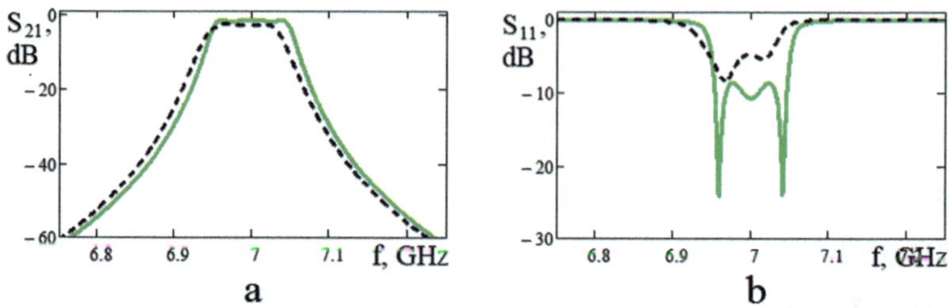

Fig. 8.20 Frequency characteristics of the scattering matrix of a 4-section bandpass filter. Solid curves are the result of scattering by a system of identical resonators; the dotted curves are the result of scattering by frequency-detuned resonators ($f_0 = 7\,\text{GHz}$; $\tilde{k}_1 = 0.012$; $\tilde{k}_2 = 0.014$; $k_{s,s+1} = 0.0092$). Difference between adjacent frequencies of partial resonators: $\Delta f = 0.02\,\text{GHz}$

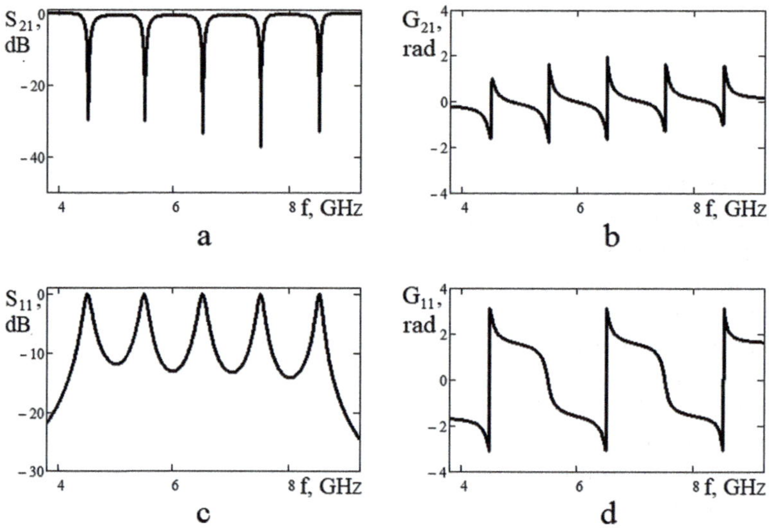

Fig. 8.21 Frequency characteristics of the scattering matrix on a model resonator with 5 natural oscillations in the transmission line ($\tilde{k}_1 = 0.0325$; $\tilde{k}_2 = 0.0275$; $\tilde{k}_3 = 0.0225$; $\tilde{k}_4 = 0.0175$; $\tilde{k}_5 = 0.0125$; distance between frequencies of adjacent modes $\Delta f = 1\,\text{GHz}$)

a wave $\left(\mathbf{E}_l^+, \mathbf{H}_l^+\right)$ falls on a system of resonators located in a transmission line. We will assume that each of the resonators is also made of a dielectric: $\tilde{\varepsilon}_s = \varepsilon_s' - i\varepsilon_s''$ $(s = 1, 2, ..., N)$. wherein $\left|\varepsilon_n''\right| << \varepsilon_n'$.

We present the solution to the scattering problem in the form (8.33) and (8.34).

Using expressions similar to (5) and (6) written for the scattered field, transmission line field, and also the fields of partial resonators, we arrive at the equation for amplitudes [11]:

$$\sum_{s=1}^{N}\left(\frac{\omega-\widetilde{\omega}^s}{\omega_0}\right)\cdot b_t^s a^s = i\frac{\omega}{\omega_0}\left(\frac{1-i\frac{\varepsilon_t''}{\varepsilon_t'-\varepsilon_0}}{2-i\frac{\varepsilon_t''}{\varepsilon_t'}}\right)\left(c_t^+\right)^*/\omega_0 w_t. \tag{8.35}$$

For a not too wide frequency band: $|\omega-\omega_0|/\omega_0 << 1$; $\omega \approx \omega_0$ and for $\varepsilon_t' >> \varepsilon_0$:

$$\sum_{s=1}^{N}(\frac{\omega-\widetilde{\omega}^s}{\omega_0})\cdot b_t^s a^s = i\left(\frac{1-i\frac{\varepsilon_t''}{\varepsilon_t'}}{2-i\frac{\varepsilon_t''}{\varepsilon_t'}}\right)\left(c_t^+\right)^*/\omega_0 w_t \tag{8.36}$$

Denoting by

$$D_s(\omega)=\frac{\omega-\widetilde{\omega}^s}{\omega_0};\quad G_t=-i\left(\frac{1-i\frac{\varepsilon_t''}{\varepsilon_t'}}{2-i\frac{\varepsilon_t''}{\varepsilon_t'}}\right), \tag{8.37}$$

We rewrite (8.36)

$$\sum_{s=1}^{N} D_s(\omega)\cdot b_t^s a^s = -G_t\cdot\left(c_t^+\right)^*/\omega_0 w_t \tag{8.38}$$

From (8.38) the transmission T and the reflection coefficient R of the DR system in the transmission line can be obtained in the form

$$T = T_0+\sum_{u=1}^{N}\left(\sum_{s=1}^{N} b_s^u c_s^+\right)a_u = T_0-\frac{1}{detA(\omega)}\sum_{s=1}^{N} detA_s^+(\omega);$$
$$R = R_0+\sum_{u=1}^{N}\left(\sum_{s=1}^{N} b_s^u c_s^-\right)a_u = R_0-\frac{1}{detA(\omega)}\sum_{s=1}^{N} detA_s^-(\omega). \tag{8.39}$$

Here T_0, R_0 are the transmission and reflection coefficients of the transmission line without DRs;

$$A_s^{\pm}(\omega)=\begin{bmatrix} b_1^1D_1(\omega) & b_1^2D_2(\omega) & \dots & G_1\sum_{u=1}^{N} b_u^s\tilde{k}_{u1}^{\pm+} & \dots & b_1^N D_N(\omega)\\ b_2^1D_1(\omega) & b_2^2D_2(\omega) & \dots & G_2\sum_{u=1}^{N} b_u^s\tilde{k}_{u2}^{\pm+} & \dots & b_2^N D_N(\omega)\\ \vdots & \vdots & \dots & \vdots & \dots & \vdots\\ b_N^1D_1(\omega) & b_N^2D_2(\omega) & \dots & G_N\sum_{u=1}^{N} b_u^s\tilde{k}_{uN}^{\pm+} & \dots & b_N^N D_N(\omega)\end{bmatrix};$$

$$A(\omega)=\begin{bmatrix} b_1^1D_1(\omega) & b_1^2D_2(\omega) & \dots & b_1^N D_N(\omega)\\ b_2^1D_1(\omega) & b_2^2D_2(\omega) & \dots & b_2^N D_N(\omega)\\ \vdots & \vdots & \dots & \vdots\\ b_N^1D_1(\omega) & b_N^2D_2(\omega) & \dots & b_N^N D_N(\omega)\end{bmatrix}; \tag{8.40}$$

for bandstop filters and

$$A_s^+(\omega)=\begin{bmatrix} b_1^1D_1(\omega) & b_1^2D_2(\omega) & \dots & G_1b_N^s\tilde{k}_{N1}^{++} & \dots & b_1^ND_N(\omega)\\ b_2^1D_1(\omega) & b_2^2D_2(\omega) & \dots & 0 & \dots & b_2^ND_N(\omega)\\ \vdots & \vdots & \dots & \vdots & \dots & \vdots\\ b_N^1D_1(\omega) & b_N^2D_2(\omega) & \dots & 0 & \dots & b_N^ND_N(\omega) \end{bmatrix};$$

$$A_s^-(\omega)=\begin{bmatrix} b_1^1D_1(\omega) & b_1^2D_2(\omega) & \dots & G_1b_1^s\tilde{k}_{11}^{-+} & \dots & b_1^ND_N(\omega)\\ b_2^1D_1(\omega) & b_2^2D_2(\omega) & \dots & 0 & \dots & b_2^ND_N(\omega)\\ \vdots & \vdots & \dots & \vdots & \dots & \vdots\\ b_N^1D_1(\omega) & b_N^2D_2(\omega) & \dots & 0 & \dots & b_N^ND_N(\omega) \end{bmatrix}; \quad (8.41)$$

Using the linearity properties of determinants of matrices, we finally find:

$$T=T_0-\frac{\omega_0}{detB}\sum_{s=1}^{N}\frac{detB_s^+}{\omega-\widetilde{\omega}^s};$$

$$R=R_0-\frac{\omega_0}{detB}\sum_{s=1}^{N}\frac{detB_s^-}{\omega-\widetilde{\omega}^s}, \quad (8.42)$$

where

$$B_s^{\pm}=\begin{bmatrix} b_1^1 & b_1^2 & \dots & G_1\sum_{u=1}^{N}b_u^s\tilde{k}_{u1}^{\pm+} & \dots & b_1^N\\ b_2^1 & b_2^2 & \dots & G_2\sum_{u=1}^{N}b_u^s\tilde{k}_{u2}^{\pm+} & \dots & b_2^N\\ \vdots & \vdots & \dots & \vdots & \dots & \vdots\\ b_N^1 & b_N^2 & \dots & G_N\sum_{u=1}^{N}b_u^s\tilde{k}_{uN}^{\pm+} & \dots & b_N^N \end{bmatrix};$$

$$B=\begin{bmatrix} b_1^1 & b_1^2 & \dots & b_1^N\\ b_2^1 & b_2^2 & \dots & b_2^N\\ \vdots & \vdots & \dots & \vdots\\ b_N^1 & b_N^2 & \dots & b_N^N \end{bmatrix}; \quad (8.43)$$

For bandstop filters and

$$B_s^+=\begin{bmatrix} b_1^1 & b_1^2 & \dots & G_1b_N^s\tilde{k}_{N1}^{++} & \dots & b_1^N\\ b_2^1 & b_2^2 & \dots & 0 & \dots & b_2^N\\ \vdots & \vdots & \dots & \vdots & \dots & \vdots\\ b_N^1 & b_N^2 & \dots & 0 & \dots & b_N^N \end{bmatrix};$$

$$B_s^- = \begin{bmatrix} b_1^1 & b_1^2 & \ldots & G_1 b_1^s \tilde{k}_{11}^{-+} & \ldots & b_1^N \\ b_2^1 & b_2^2 & \ldots & 0 & \ldots & b_2^N \\ \vdots & \vdots & \ldots & \vdots & \ldots & \vdots \\ b_N^1 & b_N^2 & \ldots & 0 & \ldots & b_N^N \end{bmatrix} \tag{8.44}$$

for the band pass filters.

Here also $\tilde{k}_{sn}^{++} = \left(c_s^+ c_n^{+*}\right) / \left(\omega_n w_n\right) = (\tilde{k}_{sn})_0 e^{-i\Gamma(z_s - z_n)}$; $\tilde{k}_{sn}^{-+} = \left(c_s^- c_n^{+*}\right) / \left(\omega_n w_n\right) = (\tilde{k}_{sn})_0 e^{-i\Gamma(z_s + z_n)}$.

In this case, frequencies and amplitudes of coupled oscillations are found from (7.58).

Let us consider bandstop filters made on loss and gain DRs (Fig. 8.22a, d, g). We assume that the natural frequency of the partial resonators of the filters is $f_1 = 200$ THz. The coupling coefficients of the resonators with the transmission line are the same: $\tilde{k}_1 = \tilde{k}_2 = 0.01$. The distance between resonators in a regular line: $\Gamma \Delta z_{12} = 31\pi/2$, where Γ is a wavenumber of line.

On Fig. 8.22 show the frequency characteristics of the scattering matrix of the bandstop filter made of two resonators with loss (b, c); bandstop filter made of two resonators with gain and loss (e, f) and bandstop filter made of two resonators with gain (h, i), obtained from (8.40, 8.42). Where $S_{21} = 20\log|T|$; $S_{11} = 20\lg|R|$; $G_{21} = \arg(T)$; $G_{11} = \arg(R)$. As can be seen from the calculations, the frequency dependences of the S-matrix in the case of resonators with losses correspond to the characteristics of a conventional notch filter made of a "bad" dielectric (b, c). For two resonators with gain and loss (e, f), the release of energy by one resonator is partially compensated by absorption in the resonator with losses (e). If both resonators consist of an active dielectric, the transmitted and reflected signal is amplified (h, i) and the filter, with appropriate settings, can be used as an amplifier.

On Fig. 8.23 shows an example of calculating from (8.42) and (8.44) the scattering characteristics of the bandpass filters placed in a break of the line (a, d, g). The obtained result shows a partially similar scattering pattern: in the case of resonators with losses in the dielectric, the frequency dependences of the S-matrix correspond to the characteristics of a bandpass filter (b, c); for active and absorbing resonators, a gain "singularity" (e, f) is observed at the resonance frequency f_0, and for active resonators, the structure also demonstrates the possibility of amplifying both reflected and transmitted signals (h, i).

The constructed model makes it possible to calculate and optimize the characteristics of scattering and more complex structures, contained big number of active resonators.

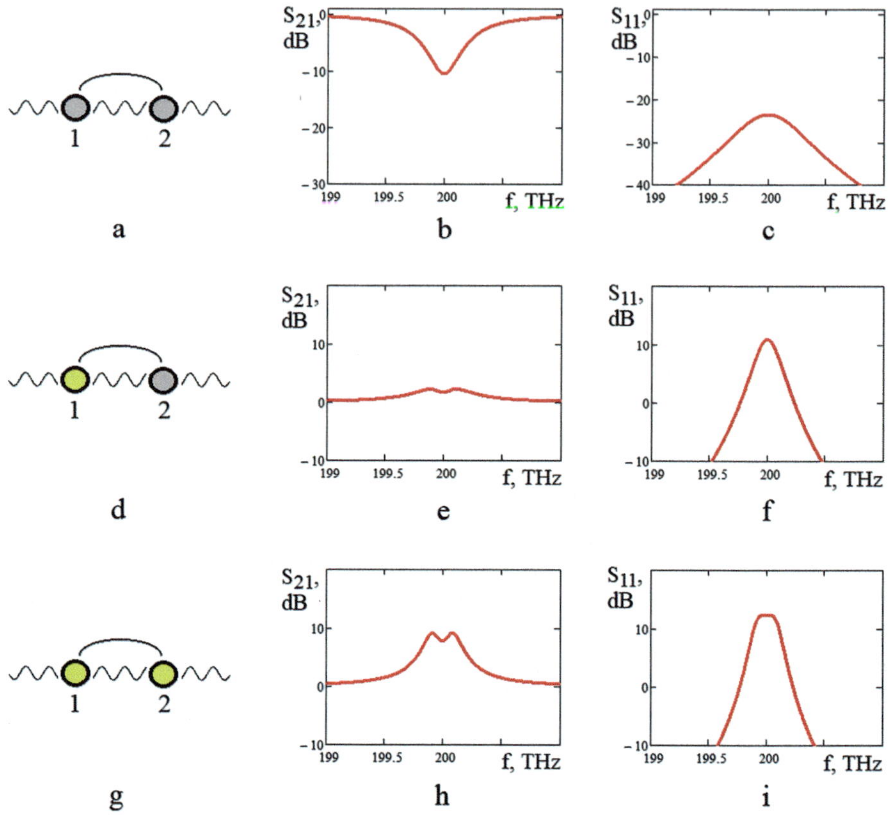

Fig. 8.22 Different loss and gain 2 DR system in a regular transmission line (**a**, **d**, **g**). Scattering characteristics of 2 DRs in the transmission line ($f_0 = 200$ THz; distance between resonators $\Gamma\Delta z_{12} = 31\pi/2$; coupling coefficients of resonators with the line $\tilde{k}_1 = \tilde{k}_2 = 0.01$; mutual coupling coefficient $k_{12} = -0.0001$). $(\varepsilon''/\varepsilon')_1 = 0.03$; $(\varepsilon''/\varepsilon')_2 = 0.01$ (**b**, **c**); $(\varepsilon''/\varepsilon')_1 = -0.03$; $(\varepsilon''/\varepsilon')_2 = 0.01$(**e**, **f**); $(\varepsilon''/\varepsilon')_1 = -0.03$; $(\varepsilon''/\varepsilon')_2 = -0.01$ (**h**, **i**)

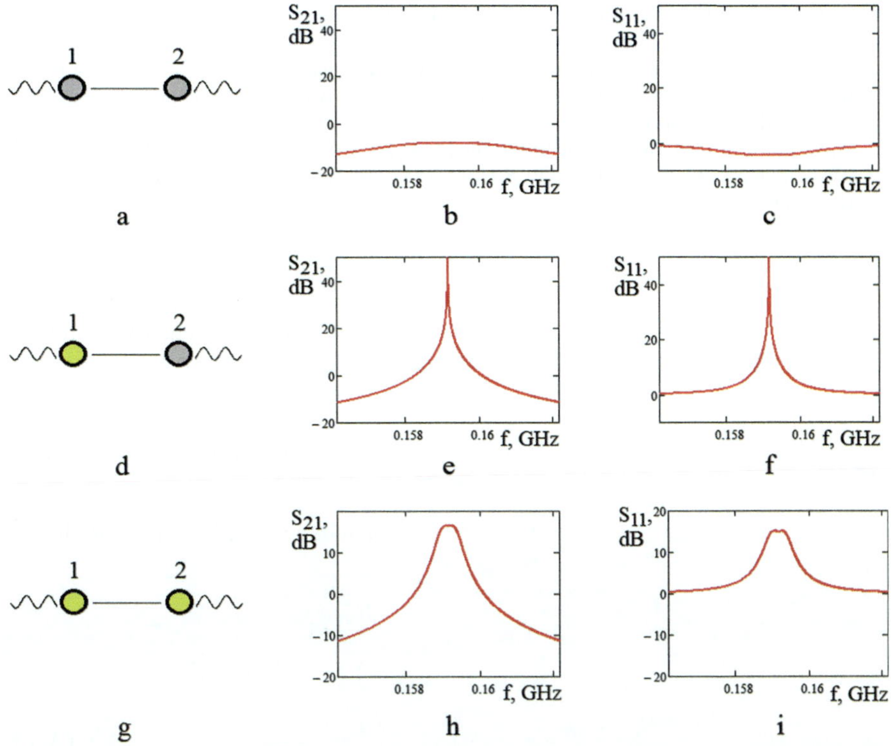

Fig. 8.23 Two loss and gain DR system in a break section of a transmission line (**a**, **d**, **g**). Scattering characteristics of 2 DRs ($f_0 = 200$ THz; coupling coefficients of resonators with the line $\tilde{k}_1 = \tilde{k}_2 = 0.01$; mutual coupling coefficient $k_{12} = 0.006$). $(\varepsilon''/\varepsilon')_1 = 0.013$;$(\varepsilon''/\varepsilon')_2 = 0.02$ (**b**, **c**); $(\varepsilon''/\varepsilon')_1 = -0.013$; $(\varepsilon''/\varepsilon')_2 = 0.02$ (**e**, **f**); $(\varepsilon''/\varepsilon')_1 = -0.013$;$(\varepsilon''/\varepsilon')_2 = -0.02$ (**h**, **i**)

References

1. G. Mie, Beltrage sur Optik truber Medien. Ann Phys **25**, 377–445 (1908)
2. M. Born, E. Wolf, *Principles of optics*, 2nd edn. (Pergamon Press, Oxford, 1984), p.859
3. E.K. Miller, A.J. Poggio, Moment-method techniques in electromagnetics from an applications viewpoint, in *Electromagnetic scattering*. ed. by P.L.E. Usienghi (Academic Press, New York, 1978), pp.315–358
4. A.A. Trubin, Scattering of electromagnetic waves on systems of coupled dielectric resonators. Radioelectronics **2**, 35–42 (1997)
5. A.A. Trubin, Calculation of characteristics of band-stop filters on dielectric resonators. Radioelectronics **9**, 69–72 (1997)
6. A.A. Trubin, Calculation of filters on cylindrical resonators in a stripline transmission line. Radioelectronics **4**, 14–19 (1998)
7. A.A. Trubin, Bandpass filters with subsystems of coupled dielectric resonators. Radioelectronics **5**, 78–80 (1999)
8. H.A. Haus, M.A. Popovic, M.R. Watts, C. Manolatou, B.E. Little, S.T. Chu, Optical resonators and filters, in *Optical microcavities*. ed. by K. Vahala (California Institute of Technology, Berkeley, CA, 2004), p.516

9. A.A. Trubin, Electrodynamics modeling of Add-drop filters on optical microresonators. Inf Telecommun Sci **1**, 30–36 (2019)
10. A.A. Trubin. Scattering of electromagnetic waves by frequency—detuned systems of dielectric resonators. Visnyk NTUU KPI Seriia—Radiotekhnika Radioaparatobuduvannia (2024) (Accepted manuscript)
11. A.A. Trubin. Scattering of electromagnetic waves by loss and gain systems of dielectric resonators. Visnyk NTUU KPI Seriia—Radiotekhnika Radioaparatobuduvannia (2024) (Preprint)

Chapter 9
Scattering of Electromagnetic Waves on Systems of Coupled Dielectric Resonators in Open Space

This chapter develops a general approach to solving scattering problems on resonant structures, which we also define through the expansion of solutions into coupled natural oscillations of a system of resonators. We first consider plane wave scattering; in the final part—solving problems of wave scattering of transmission lines "connected" to a lattice of resonators in open space. Solving these problems makes it possible to discover a number of interesting effects of wave propagation in various square DR lattices. The features of scattering on lattices of rotating resonators with the main types of magnetic natural oscillations are analyzed.

9.1 Solution of the Problem of Scattering by Coupled Dielectric Resonators in Open Space

Natural oscillations of the DR system in open space are also described by the system of equations (7.6), where the coupling coefficients are determined in the general case by expressions (6.1). Let us represent the solution to the problem of $(\mathbf{E}^+,\ \mathbf{H}^+)$ wave scattering on a DR lattice placed in open space in the form of superposition (8.5) of the incident wave and coupled oscillations (Fig. 9.1a). The system of equations for amplitudes in expansions (8.5) formally takes the form (8.13), on the right side of which the functions c_t^{l+} must be replaced by integral over the surface of the DR:

$$\mathrm{c}_t^+ = -\frac{1}{2}\oint_{S_R} \{[\mathbf{e}_t^1, \mathbf{n}](\mathbf{H}^+)^* + [\mathbf{n}, \mathbf{h}_t^1](\mathbf{E}^+)^*\}\mathrm{ds} \tag{9.1}$$

where $(\mathbf{e}_t^1, \mathbf{h}_t^1)$—field of natural oscillations in the material t-th DR ($t = 1, 2, \ldots, N$).

Let us denote by $\mathbf{P}_t = (x_t, y_t, z_t)$—the vector of rectangular coordinates of the t-th DR in the lattice (Fig. 9.1b). Typically, when solving scattering problems, one is most often interested in the field distribution in the far zone. In this case, the

A. Trubin, *Introduction to the Theory of Dielectric Resonators*, Springer Series in Advanced Microelectronics 65, https://doi.org/10.1007/978-3-031-65396-4_9

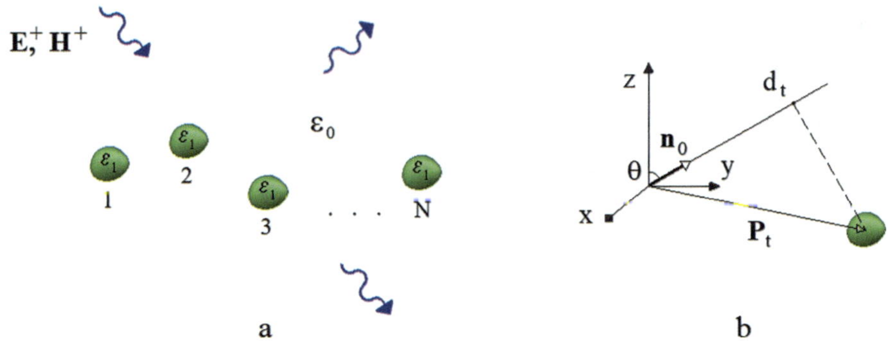

Fig. 9.1 Scattering of harmonic waves on a system of N coupled DRs in open space. Projection of coordinates of the t-th DR onto the direction of observation of the scattered wave

contribution of the field each partial resonator to the total field is determined mainly by the distribution of the field of its own oscillations in the far zone, multiplied by the amplitude found, taking into account the value of the phase factor $k_0 d_t$ in the direction of the receiver, determined by the unit vector $\mathbf{n}_0 = (\sin\theta\cos\varphi, \sin\theta\sin\varphi, \cos\theta)$ (Fig. 9.1b) in a spherical coordinate system:

$$\mathbf{e}^{\infty} = \sum_{t=1}^{N} A_t \mathbf{e}_t^{\infty} \approx \mathbf{e}_1^{\infty} \sum_{t=1}^{N} A_t e^{ik_0 d_t}$$
$$\mathbf{h}^{\infty} = \sum_{t=1}^{N} A_t \mathbf{h}_t^{\infty} \approx \mathbf{h}_1^{\infty} \sum_{t=1}^{N} A_t e^{ik_0 d_t} \tag{9.2}$$

where, A_t—amplitude of the forced oscillation of the t-th partial DR (8.6).

$$d_t = d_t(\theta, \varphi) = \mathbf{P}_t \cdot \mathbf{n}_0$$

In final form the total scattered field of the lattice, using (9.2), Eqs. (7.6) and (8.13), (8.14) can be conveniently presented as the superposition of coupled oscillations:

$$\mathbf{e}^{\infty}(\theta, \varphi) \approx \mathbf{e}_1^{\infty} \cdot \sum_{s=1}^{N} \frac{A_s(\theta, \varphi)}{Q_s(\omega)} \tag{9.3}$$

where

$$A_s(\theta, \varphi) = -\frac{Q^D \det C_s}{\omega_0 \det B}\left[\sum_{u=1}^{N} b_u^s e^{ik_0 d_u(\theta, \varphi)}\right] \tag{9.4}$$

Taking into account the determinant linearity, one obtains from (9.3), (9.4) and (8.13), (8.14) for plain waves and for identical DR:

$$A_s(\theta, \varphi) = -\frac{Q^D}{\omega_0 w_1} \cdot \frac{\left(c_1^+\right)_0^*}{\det B} \cdot \det \begin{bmatrix} b_1^1 & b_1^2 & \cdots & \sum_{u=1}^N b_u^s e^{ik_0 d_{1u}(\theta_k, \varphi_k|\theta, \varphi)} & \cdots & b_1^N \\ b_2^1 & b_2^2 & \cdots & \sum_{u=1}^N b_u^s e^{ik_0 d_{2u}(\theta_k, \varphi_k|\theta, \varphi)} & \cdots & b_2^N \\ \vdots & \vdots & \cdots & \vdots & \cdots & \vdots \\ b_N^1 & b_N^2 & \cdots & \sum_{u=1}^N b_u^s e^{ik_0 d_{Nu}(\theta_k, \varphi_k|\theta, \varphi)} & \cdots & b_N^N \end{bmatrix} \tag{9.5}$$

Whence we conclude that the phase factors $d_{tu}(\theta_k, \varphi_k|\theta, \varphi) = d_t(\theta_k, \varphi_k) + d_u(\theta, \varphi)$ in the lattice coordinate system are determined by two terms: the first is determined by the direction of incidence of the plane wave: $\mathbf{n}_k = (\sin\theta_k \cos\varphi_k, \sin\theta_k \sin\varphi_k, \cos\theta_k)$: $d_t = d_t(\theta_k, \varphi_k) = \mathbf{P}_t \cdot \mathbf{n}_k$ and the second—the direction to the observation point $\mathbf{n}_0$; w_1—is the energy, stored in the t-th DR (2.9); B—is the amplitude matrix (7.11). Here $\left(c_t^+\right)_0$—is the value of the coefficient (9.1) found in the coordinate system at the center of the t-th partial resonator.

Below we will call the $\left(c_t^+\right)_0$ dependences on the parameters of the DR and the incident wave c-functions.

Thus, from (9.5) the scattering field of the lattice in the wave zone can be written in the form [1]:

$$\mathbf{e}^\infty(\theta, \varphi) = \mathbf{e}_0 f\langle\theta_k, \varphi_k|\theta, \varphi\rangle \frac{e^{-ik_0 r}}{k_0 r} \tag{9.6}$$

where $\mathbf{e}_0$—is the unit vector, defining the polarization of the scattered electric field in the wave-zone. The function $f\langle\theta_k, \varphi_k|\theta, \varphi\rangle$ is named scattering amplitude. As follows from (9.3), the scattering amplitude is also a function of the frequency. In common case the scattering amplitude determines the coupling between incident E-field vector and the scattered E-field vector in the wave-zone, i.e. it's can be represented by the tensor of second order. In those cases, when the scattering amplitude dependence on the frequency should be taken into account, it is written in the form: $f\langle\theta_k, \varphi_k|\theta, \varphi\rangle = f\langle\omega\rangle$, where for the sake of brevity the rest variables have been omitted.

Thus, the scattered field is determined, on the one hand, by the structure of the lattice, specified by the set of vectors $\{\mathbf{P}_t\}$, and, on the other hand, by the types of oscillations of the DR, determined by the dependences of $\left(c_t^+\right)^*$ functions (9.4) on angular variables.

In cases where we are interested in the polarization characteristics of the scattered field, the electric component of the field in the wave zone was represented as the Jones vector in the spherical coordinate system (r, θ, φ), with a minimum number of independent parameters [2]:

$$\mathbf{e}^\infty = \left(\cos\beta \cdot \mathbf{n}_{0\varphi} + e^{i\delta} \sin\beta \cdot \mathbf{n}_{0\theta}\right) \cdot \frac{e^{-ik_0 r}}{k_0 r}, \tag{9.7}$$

here $\mathbf{n}_{0\varphi}$, $\mathbf{n}_{0\theta}$, the unit vectors are oriented in the directions φ, θ, respectively. It is easy to see that the partial values of the function $\beta(\theta, \varphi) = 0, \pi/2$ correspond to linear polarization; $\beta(\theta, \varphi) = \pi/4, 3\pi/4$ and $\delta(\theta, \varphi) = \pi/2, 3\pi/2$ also circular polarization of scattered waves.

9.2 Scattering of Plane Waves on Lattices of Cylindrical Dielectric Resonators in Open Space

Let us consider the general case of scattering of a plane, linearly polarized electromagnetic wave $\left(\mathbf{E}^+, \mathbf{H}^+\right)$ on a lattice consisting of cylindrical dielectric resonators (see Fig. 9.3). The field of the incident wave in a rectangular coordinate system [3] can be represented as:

$$\begin{aligned} \mathbf{E}^+ &= (\mathbf{x}_0 \cos\alpha_1 + \mathbf{y}_0 \cos\alpha_2 + \mathbf{z}_0 \cos\alpha_3) E_0 e^{-ik_0(x\cos\gamma_1 + y\cos\gamma_2 + z\cos\gamma_3)}; \\ \mathbf{H}^+ &= (\mathbf{x}_0 \cos\beta_1 + \mathbf{y}_0 \cos\beta_2 + \mathbf{z}_0 \cos\beta_3) \frac{E_0}{w_0} e^{-ik_0(x\cos\gamma_1 + y\cos\gamma_2 + z\cos\gamma_3)}. \end{aligned} \tag{9.8}$$

where $(\mathbf{x}_0, \mathbf{y}_0, \mathbf{z}_0)$—unit vectors, E_0—amplitude; $w_0 = (\mu_0/\varepsilon_0)^{1/2}$—characteristic impedance of the open space. Angles $(\alpha_1, \alpha_2, \alpha_3)$, $(\beta_1, \beta_2, \beta_3)$, determine the spatial orientation of the electric and magnetic field vector, respectively, and $(\gamma_1, \gamma_2, \gamma_3)$—direction of wave propagation, also defined by the vector $\mathbf{n}_k = (\cos\gamma_1, \cos\gamma_2, \cos\gamma_3)$.

In the general case of a plane wave incident on a cylindrical DR and for fundamental H_{101} magnetic oscillations, coefficient c_t^+ (9.1) in the resonator coordinate system shown in Fig. 9.2, a, has the form [4]:

$$\begin{aligned} \left(c_t^+\right)_0 &= -2\pi i E_0^* \cdot \frac{h_1}{\beta} \cdot \left(r_0^2 \frac{L}{2}\right) \cdot \left(k_1^2 - k_0^2\right) \frac{\cos\beta_3}{\sin\gamma_3} \\ &\cdot \frac{\left[q_\perp \sin\gamma_3 J_0(q_\perp \sin\gamma_3) J_1(p_\perp) - p_\perp J_0(p_\perp) J_1(q_\perp \sin\gamma_3)\right]}{(q_\perp \sin\gamma_3)^2 - p_\perp^2} \\ &\cdot \frac{\left[p_z \sin p_z \cos(q_z \cos\gamma_3) - q_z \cos\gamma_3 \cos p_z \sin(q_z \cos\gamma_3)\right]}{(q_z \cos\gamma_3)^2 - p_z^2}, \end{aligned} \tag{9.9}$$

where $p_\perp = \beta r_0; p_{0\perp} = \beta_0 r_0; q_\perp = k_0 r_0, q_z = k_0 L/2$—characteristic parameters of a cylindrical DR (Sect. 1.2); r_0—radius, L—dielectric resonator height, $J_n(x)$—Bessel functions of the first kind [5].

The resulting expression describes the general case of interaction of the field of a plane electromagnetic wave (9.8) with a cylindrical DR at the frequency of its main H_{101} magnetic oscillations at any angle of incidence.

Let us define the plane of incidence of the wave by the vector $\mathbf{k}_0 = k_0 \mathbf{n}_k$ and the normal vector to a given fixed grating surface. When solving scattering problems, two general cases are usually distinguished:

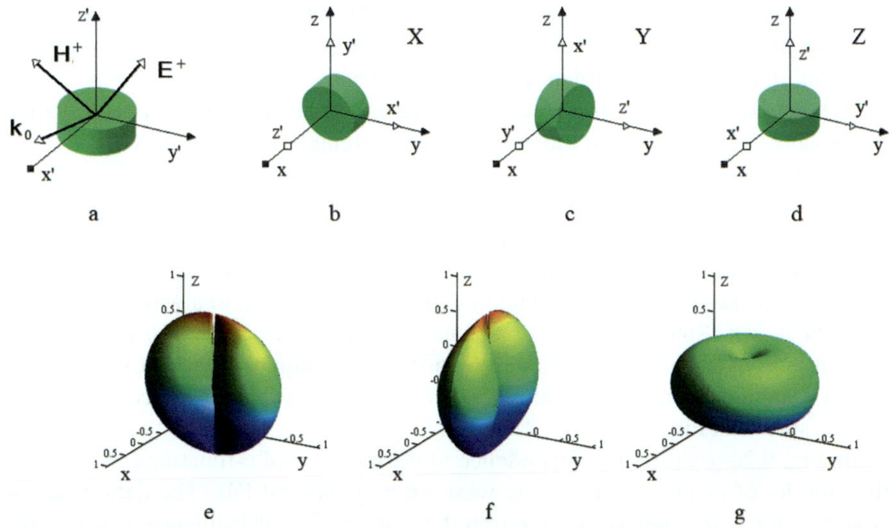

Fig. 9.2 Plane wave field vectors and cylindrical DR in local rectangular coordinate system (**a**). Standard positions of a cylindrical DR X, Y, Z in a rectangular coordinate system (**b–d**). Spatial distribution of the modulus of the c-functions for 3 standard positions (**b–e**; **c–f**; **d–g**) in the case of *p*-scattering (**e**, **f**); *s*-scattering (**g**) for H_{101} oscillation

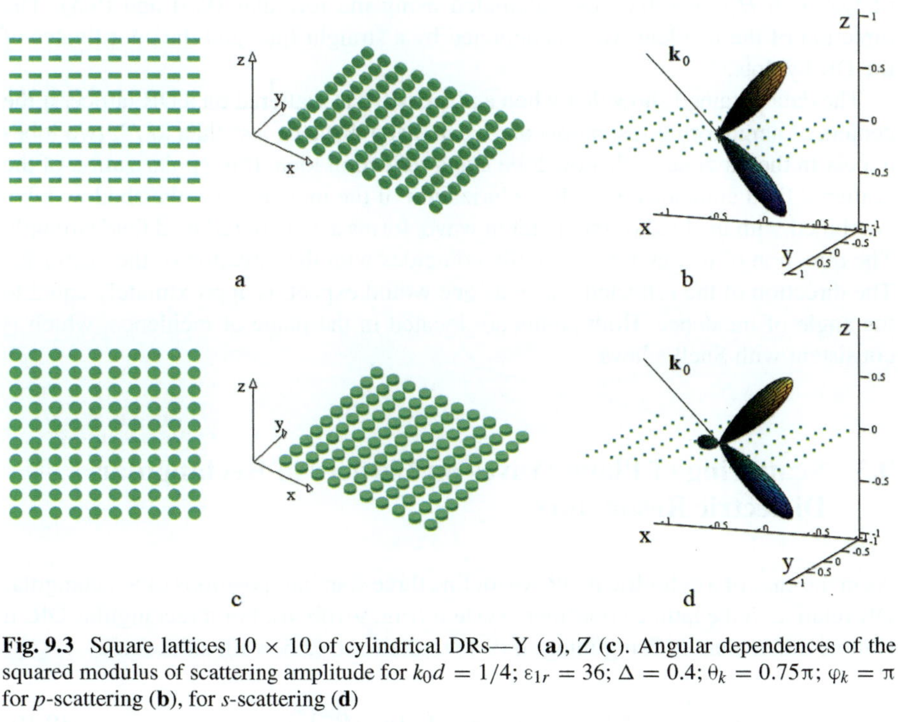

Fig. 9.3 Square lattices 10 × 10 of cylindrical DRs—Y (**a**), Z (**c**). Angular dependences of the squared modulus of scattering amplitude for $k_0d = 1/4$; $\varepsilon_{1r} = 36$; $\Delta = 0.4$; $\theta_k = 0.75\pi$; $\varphi_k = \pi$ for *p*-scattering (**b**), for *s*-scattering (**d**)

First—the vector of the magnetic field strength $\mathbf{H}^+$ of the incident wave is directed orthogonal to the plane of incidence, and the vector of the electric field strength $\mathbf{E}^+$ is located in the plane of incidence (p-scattering) and

The second case, when the electric field strength vector $\mathbf{E}^+$ is directed orthogonal to the plane of incidence, and the magnetic field strength vector $\mathbf{H}^+$ lies in the plane of incidence (s-scattering).

Below we will denote the local rectangular coordinates (x', y', z') associated with the geometry of the resonator, with the origin at its center, and the rectangular coordinates (x, y, z) that determine the coordinates of the DR centers in the lattice. The resonator axis in the local coordinate system will be assumed to always be directed along the axis z'. Since the axes of resonators in the lattice can have different directions, it is convenient to fix their orientation in the simplest cases in the form of several standard positions of the resonator axis relative to the lattice coordinate system, shown in Fig. 9.2b–d, X, Y and Z.

On Fig. 9.2e–g shows the dependence of the modulus of c-functions (9.9) on the direction $\mathbf{k}_0$ of incidence of a plane wave on a cylindrical DR. The data obtained show that for both types of polarization there are angles of incidence at which the incident wave is not coupled to the resonator. Resonant oscillations are not excited. For H_{101} oscillations, this occurs in cases where the direction of the $\mathbf{H}^+$ vector becomes orthogonal to the axis of the resonator.

On Fig. 9.3b, d show the dependences of the scattered field by square gratings of DR with H_{101} oscillations, calculated using the formulas (9.3) and (9.5). The direction of the incident wave is depicted by a straight line, and the coordinates of the DR by dots.

The data obtained show that when a plane wave is scattered on a DR lattice, if the centers of neighboring resonators are located at a distance less than $\lambda_0/4$, two lobes appear in the far zone: a shadow lobe and a reflection lobe. If the polarization of the scattered field coincides with the polarization of the incident one, the shadow lobe, combined with the field of the incident wave, forms a zone of reduced field strength. The direction of its maximum usually coincides with the direction of the vector $\mathbf{k}_0$. The direction of the reflected wave, as one would expect, is approximately equal to the angle of incidence. Both petals are located in the plane of incidence, which is consistent with Snell's laws.

9.3 Scattering of Plane Waves on Lattices of Rectangular Dielectric Resonators

As in the case of a cylindrical DR, we define three standard positions of a rectangular DR relative to the lattice coordinate system (Fig. 9.10b–d). For a rectangular DR, it is more convenient to bring integral (9.1) to the form (2.5) with an accuracy $1/Q^{\Sigma}$:

$$c_t^+ \approx \frac{i}{2}\omega(\varepsilon_1 - \varepsilon_0)\int_{V_R} (\mathbf{e}_t, (\mathbf{E}^+)^*)dv. \tag{9.10}$$

Here Q^{Σ}—is the resonator radiation Q-factor.

The results of calculating the integral (9.10) in the DR center coordinate system (x', y', z') are presented in the form [6]:

(a) For X-position (Fig. 9.4b) and

for the p-polarization of the incident wave:

$$\left(c^{+}\right)_0 = c_0 \cdot \left[\beta_y \omega_x(\cos\gamma_2)\overline{\omega}_y(\cos\gamma_3) \cdot \frac{1}{M}\cos\gamma_2\cos\gamma_3 + \beta_x \overline{\omega}_x(\cos\gamma_2)\omega_y(\cos\gamma_3)M\right]\omega_z(\cos\gamma_1) \tag{9.11}$$

s-polarization of the incident wave:

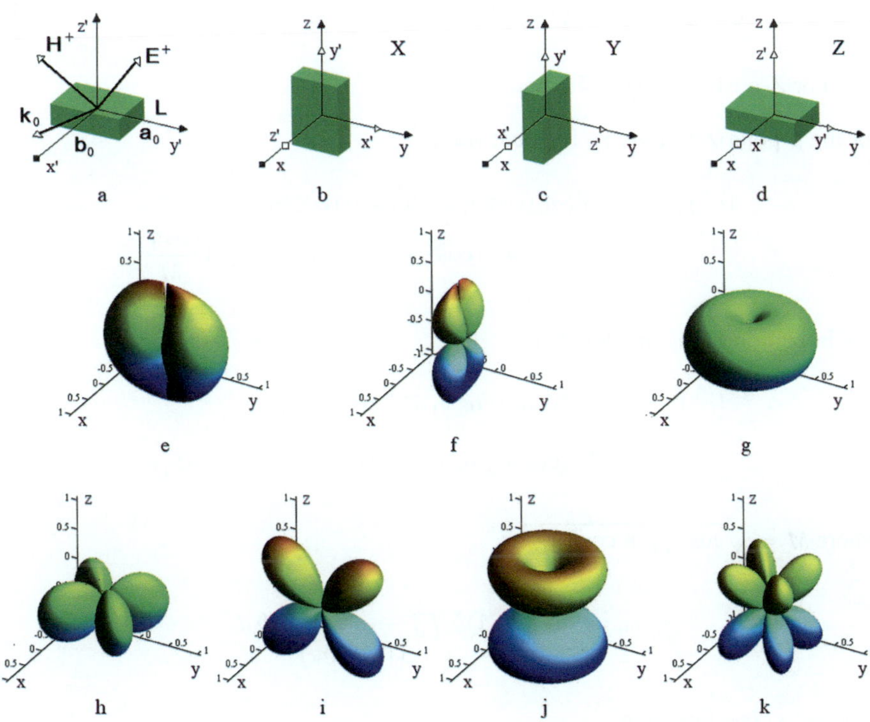

Fig. 9.4 Plane wave field vectors and rectangular DR in local coordinate system (**a**). Three orthogonal orientations of a rectangular DR: X (**b**); Y (**c**); Z (**d**) relative coordinate system (x, y, z). Module $\left|c^{+}(\theta, \varphi)\right|$ in a spherical coordinate system for three orientations of the DR with H_{ccc} oscillation relative to the lattice coordinate system X (**e**, **f**); Z (**g**) for p-scattering (**e**); for s-scattering (**f**, **g**); of the DR with H_{ccs} oscillation relative to the coordinate system X (**h**); Y (**i**); Z (**j**) for p-scattering (**h**); for s-scattering (**i**, **j**); for H_{scc} oscillation and for s-scattering (**k**)

$$\left(c^{+}\right)_0 = c_0 \cdot \beta_y \omega_x(\cos\gamma_2)\overline{\omega}_y(\cos\gamma_3)\omega_z(\cos\gamma_1) \cdot \frac{1}{M}\cos\gamma_1 \tag{9.12}$$

(b) For Y-position (Fig. 9.4c) and

for the p-polarization of the incident wave:

$$\left(c^{+}\right)_0 = c_0 \cdot \Big[\beta_y \omega_x(\cos\gamma_1)\overline{\omega}_y(\cos\gamma_3) \cdot \frac{1}{M}\cos\gamma_1\cos\gamma_3 + \beta_x\overline{\omega}_x(\cos\gamma_1)\omega_y(\cos\gamma_3)M\Big]\omega_z(\cos\gamma_2) \tag{9.13}$$

s-polarization of the incident wave:

$$\left(c^{+}\right)_0 = c_0 \cdot \beta_y \omega_x(\cos\gamma_1)\overline{\omega}_y(\cos\gamma_3)\omega_z(\cos\gamma_2) \cdot \frac{1}{M}\cos\gamma_2 \tag{9.14}$$

(c) For Z-position (Fig. 9.4d) and

for the p-polarization of the incident wave:

$$\left(c^{+}\right)_0 = c_0 \cdot \left[\beta_y \omega_x(\cos\gamma_1)\overline{\omega}_y(\cos\gamma_2)\cos\gamma_1 - \beta_x\overline{\omega}_x(\cos\gamma_1)\omega_y(\cos\gamma_2)\cos\gamma_2\right]\omega_z(\cos\gamma_3)\frac{\cos\gamma_3}{M} \tag{9.15}$$

s-polarization of the incident wave:

$$\left(c^{+}\right)_0 = -c_0 \cdot \left[\beta_y \omega_x(\cos\gamma_1)\overline{\omega}_y(\cos\gamma_2)\cos\gamma_2 + \beta_x\overline{\omega}_x(\cos\gamma_1)\omega_y(\cos\gamma_2)\cos\gamma_1\right]\omega_z(\cos\gamma_3)\frac{1}{M} \tag{9.16}$$

where $M = \sqrt{\cos^2\gamma_1 + \cos^2\gamma_2}$

$$c_0 = \frac{1}{2}(\varepsilon_{1r} - 1)h_1 E_0^* \frac{k_0^2}{\left(\beta_x^2 + \beta_y^2\right)} a_0 b_0 L$$

$\varepsilon_{1r} = \varepsilon_1/\varepsilon_0$. Functions

$$\omega_\nu(n) = \frac{1}{p_\nu^2 - (q_\nu n)^2}\begin{pmatrix} -i\left[p_\nu \cos p_\nu \sin(q_\nu n) - q_\nu n \sin p_\nu \cos(q_\nu n)\right] \\ \left[p_\nu \sin p_\nu \cos(q_\nu n) - q_\nu n \cos p_\nu \sin(q_\nu n)\right] \end{pmatrix}$$

$$\overline{\omega}_\nu(n) = \frac{1}{p_\nu^2 - (q_\nu n)^2}\begin{pmatrix} \left[p_\nu \sin p_\nu \cos(q_\nu n) - q_\nu n \cos p_\nu \sin(q_\nu n)\right] \\ i\left[p_\nu \cos p_\nu \sin(q_\nu n) - q_\nu n \sin p_\nu \cos(q_\nu n)\right] \end{pmatrix}$$

$(\nu = x, y, z)$ are the same as (2.18).

C-functions (9.11)–(9.16) determine the degree of excitation of a DR by an plane wave for any magnetic oscillation H_{nml} (1.32). Calculation of the spatial distribution of the values of the c-functions makes it possible to determine the directions of the most significant interaction of the incident waves with the resonators of the array. This is especially useful in cases of higher types of oscillations, when the distribution of the DR field has a complex spatial structure.

Figure 9.4e–k shows the angular dependence of the modulus c-functions for scattering of p-type (e, h), s-type (f–k). The relative dielectric constant of the resonator $\varepsilon_{1r} = 36$; relative sizes $b_0/a_0 = 1$; $L/a_0 = 0, 4$. Dependencies (e, f, h, k) correspond to the X orientation shown in Fig. 9.4b; dependence (i) correspond to the Y orientation shown in Fig. 9.4c; and (g, j), correspond to Z orientation (d). The lowest frequency oscillation type H_{ccc} corresponds to (1.32) with $h_z = h_1 \cos \beta_x x' \cos \beta_y y' \cos \beta_z z'$; next frequency oscillation type H_{scc} corresponds to (2) with $h_z = h_1 \sin \beta_x x' \cos \beta_y y' \cos \beta_z z'$ (1.32) and so on.

In the case of p-polarization (Fig. 9.4e, h), it is seen that there are no oscillations in the DR in the plane $\varphi = 0, \pi$, and in the case of s-polarization (Fig. 9.4f, k) at $\varphi = 1/2\pi,\ 3/2\pi$.

Figure 9.5b, d illustrates the angular dependences of the squared modulus of the scattering amplitude (9.6) for a plane p-type wave (b); s-type (d) on a square lattice (a, c) of 10×10 DR. The dots conventionally show the centers of the resonators. The straight line shows the direction of propagation of the incident wave $\mathbf{k}_0$. The relative distance between the centers of adjacent resonators is $\lambda_0/4$ (λ_0—is the wavelength in free space at the frequency of H_{ccc} resonant oscillations ω_0). As can be seen from the above data, petal 1 (Fig. 9.5b) is directed at an angle of "reflection" to the surface of the grating. Petal 2 is directed along the vector $\mathbf{k}_0$, whence, taking into account (8.5), it defines the lattice "shadow" resulting from reflection. Both petals are also located in the plane of incidence.

Let us now consider a more complex case in which the lattice resonators has degenerate types of oscillations. Degenerate oscillations arise, for example, in cubic resonators ($a_0 = b_0 = L$). The sketch of such coupled cubic DRs lattice is shown in Fig. 9.6, a. In this case, in each resonator at the lowest frequency, three eigenoscillations of the H_{ccc} magnetic type can be simultaneously excited, differing from each other by rotation by angles of $\pi/2$. All these degenerate oscillations of the DRs may be coupled to each other. This means that all $3N$ oscillations must be taken into account in the calculation (9.11–9.16); see also (6.14)–(6.17).

Figure 9.6b, c shows the result of calculating the angular dependencies of squared modulus of the scattering amplitude for a 10×10 cubic DR lattice. As can be seen from these data, an increase in the number of coupled degenerate oscillations of different "polarizations" leads to expansion and the appearance of additional lobes (Fig. 9.6b). In some cases, frequency oscillation can lead to a noticeable rearrangement of the scattered electromagnetic field.

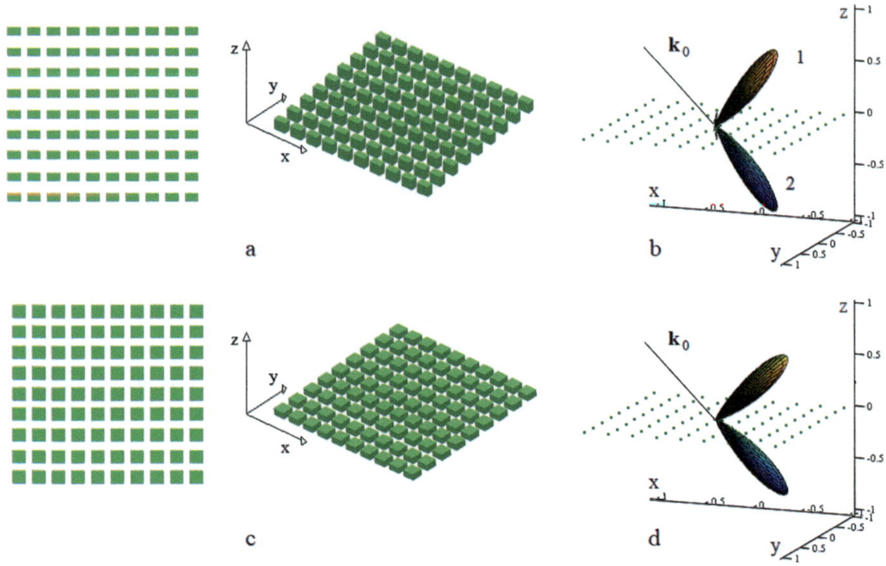

Fig. 9.5 Square lattice of rectangular DRs in position Y (**a**); in position Z (**c**). Characteristics of the scattering of a plane wave on a lattice (**a–d**) for $\theta_k = 0.75\pi$; $\varphi_k = \pi$; $k_0 d = 1/4$; $\varepsilon_{1r} = 36$; $\Delta = 0.4$; $B = 1$

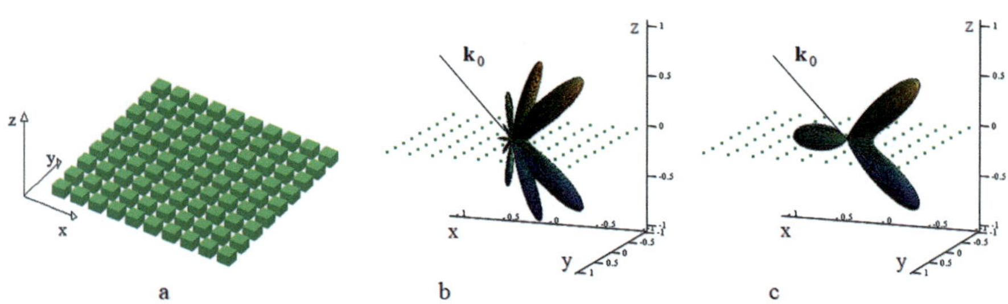

Fig. 9.6 A square lattice of a cubic shape DRs (**a**). Characteristic of the scattering of a plane wave of the p-type (**b**), s-type on the lattice (**c**) with degenerate oscillations of the DRs ($\Delta = B = 1$; $\varepsilon_{1r} = 36$; $\theta_k = 0.75\pi$; $\varphi_k = \pi$)

9.4 Scattering of Plane Waves on Lattices of Spherical Dielectric Resonators

Dielectric resonators of spherical shape, even for the main types of oscillations, have frequency degeneracy (Sect. 1.1), therefore, when decomposing the scattered field of the lattice into natural oscillations of the resonator system, coupling coefficients for all types of oscillations should be taken into account H_{1m1} ($m = 0, \pm 1$).

We have found an analytical expression for the c-function for the simple case of excitation of oscillations of magnetic types H_{1m1} in a spherical resonator in a coordinate system (x', y', z') whose z'-axis is directed along the wave vector $\mathbf{k}_0$ [7]:

$$c^{+} = 2\pi i r_0 \delta_{m1} E_0^{*} \left\{ j_1(p) \frac{d}{dq} [q j_1(q)] - j_1(p) \frac{d}{dp} [p j_1(p)] \right\} \left\{ \begin{array}{c} \sin \beta_1 \\ \cos \beta_1 \end{array} \right\} e^{ik_0 z'} \quad (9.17)$$

Here $(0, 0, z')$—is the resonator center coordinate; $m = 0, \pm 1$—azimuthal number of fundamental natural oscillations of the DR (1.2–1.3); δ_{mn}—Kronecker symbol; $j_n(z) = (\pi/2z)^{1/2} J_{n+1/2}(z)$ is the spherical Bessel function [5]. If the magnetic field of natural oscillations in the center of the resonator is directed along the y axis, this case corresponds to the upper value in the curly bracket, and if it is directed along the x axis, to the lower one.

As can be seen from (9.17), the function c^{+} is nonzero only for oscillations H_{1m1} with $m = 1$. The c^{+}-function is proportional to the projection of the magnetic field of the incident plane wave (9.8) on the direction of the magnetic field of natural oscillations in the center of the resonator.

The appearance of a coupling between degenerate oscillations with different values of azimuthally indices in the spatial structures of spherical DRs, in most cases, redistributes the lattice field, having a noticeable impact on its characteristics.

Let's consider the simplest example of scattering on a 1-dimensional linear lattice (Fig. 9.7a).

To simplify the calculation of the amplitudes of natural oscillations of the DR system, we pass to the coordinate system, the axis z of which is directed along the vector $\mathbf{k}_0$ and use (9.17). After the amplitudes of the resonators are found the scattered field is calculated by going back to the coordinate system of the lattice.

As follows from the data obtained (Fig. 9.7), in the case of s-scattering, the incident plane wave is converted by a linear lattice into one uniformly distributed in azimuth (Fig. 9.7c, e). In the case of p-polarization, the incident wave is scattered by the lattice in directions determined by its polarization relative to its spatial orientation (Fig. 9.7b, d).

Dependencies $\beta(\theta, \varphi)$; $\delta(\theta, \varphi)$ (9.7) shown in Fig. 9.7f, g, demonstrate a complex picture of the spatial distribution of the polarization of the scattered wave.

As an second example, we build a scattering model on a 10×10 square lattice shown in Fig. 9.8a.

Figure 9.8b–f shows the angular dependences of the squared scattering amplitude for typical cases for $d = \lambda_0/4$, where d—is the distance between adjacent resonators. The coupling between degenerate oscillations H_{1m1} gives rise to additional scattering directions (b, c, f) and only in special cases of angles of incidence, the grating scattering characteristic has two pronounced lobes (d, e).

It is interesting that scattering on planar structures of other shapes, for example ring ones (Fig. 9.9a), also demonstrates general properties with the patterns noted above (Fig. 9.5), inherent in 2-dimensional resonator lattices. Thus, with an oblique incidence of a plane wave on the ring structure of the DR, the appearance of shadow and reflected lobes is also observed (Fig. 9.9b–g).

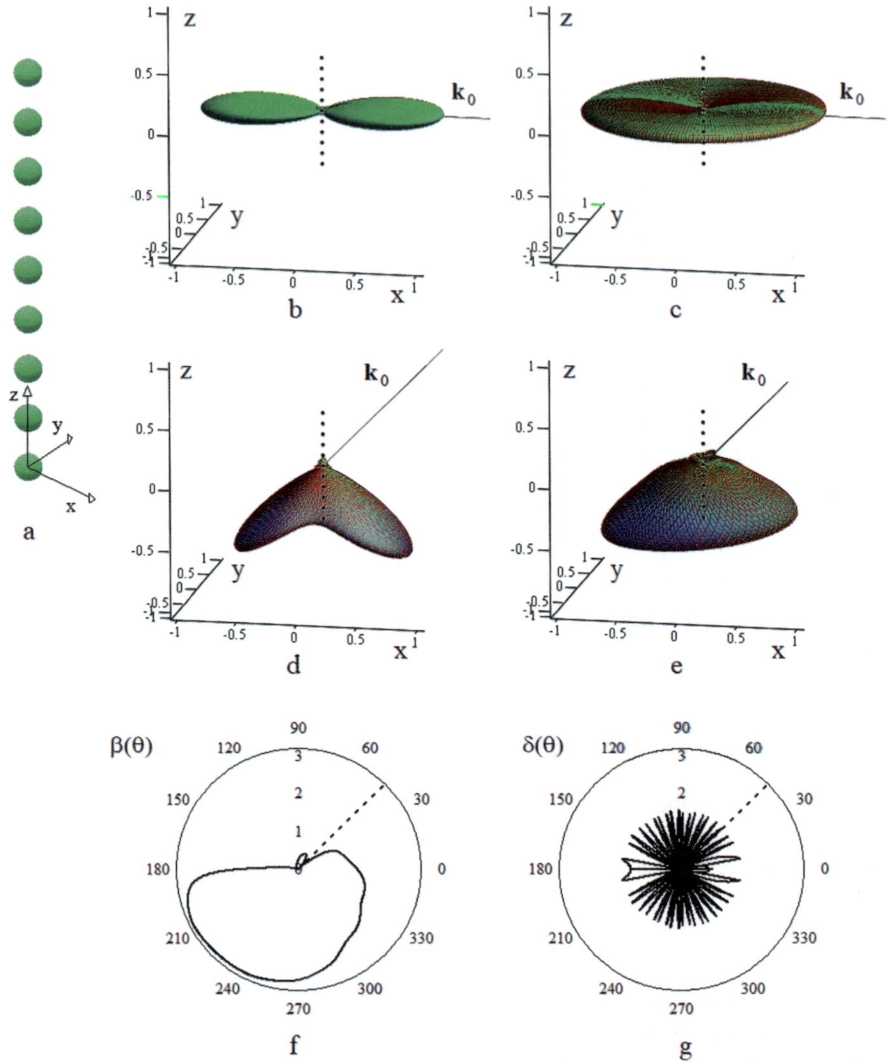

Fig. 9.7 One-dimensional lattice of spherical DR (**a**). Angular dependences of the squared modulus of scattering amplitude $|f\langle\theta_k, \varphi_k|\theta, \varphi\rangle|^2$ on the frequency f_0 of H_{1m1} DR oscillations ($d = \lambda_0/4$; $\lambda_0 = 2\pi/k_0$); $\theta_k = \pi/2$ (**b**, **c**); $\theta_k = 3\pi/4$; $\varphi_k = \pi$ (**d**, **e**); for p scattering (**b**, **d**) for s-scattering (**c**, **e**). Angular distribution of functions $\beta(\theta, \varphi)$; $\delta(\theta, \varphi)$ (9.7), that determine the polarization of the scattered wave (f, g) ($\varphi = \pi/2$) for s-scattering (Straight lines show the direction of the incident wave propagation. The dots indicate the conventional location of the lattice resonators) ($\varepsilon_{1r} = 36$)

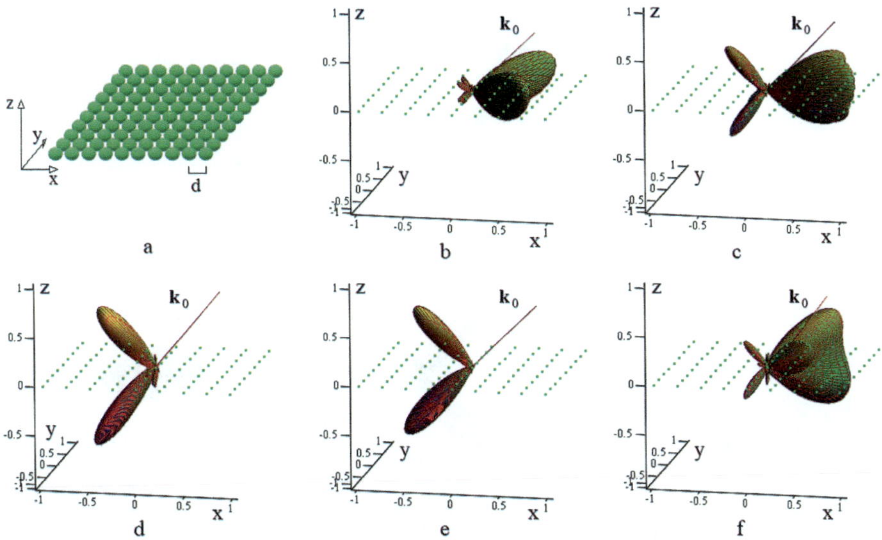

Fig. 9.8 Square lattices 10 × 10 of spherical DRs (**a**). Angular dependences of the squared modulus of scattering amplitude $|f\langle\theta_k, \varphi_k|\theta, \varphi\rangle|^2$ for $d = \lambda_0/4$ ($\lambda_0 = 2\pi/k_0$); $\varepsilon_{1r} = 36$; $\theta_k = 0.75\pi$; $\varphi_k = \pi$ (**b, e**); $\varphi_k = 1.1\pi$ (**c**); $\varphi_k = 1.2\pi$ (**d**); $\varphi_k = 1.25\pi$ (**f**); for *p*-scattering (**b–d**) for *s*-scattering (**e, f**) (Straight lines show the direction of the incident wave)

Thus, the data obtained show that the characteristics of scattering on lattices of spherical DRs have a more complex structure associated with the excitation of several types of degenerate oscillations of this type of resonators.

9.5 Features of Scattering of Plane Waves on Rotating Lattices of Dielectric Resonators

During resonant scattering on systems of coupled DRs, conditions may arise under which all lattice resonators are not coupled to the incident wave. In this case, the dielectric structure interacts with the incident field only in a non-resonant manner, which leads to the phenomenon of quasi-complete transmission of the field through the lattice. A similar effect was observed by Malyuzhinets, and was investigated for cases of scattering on dielectric lattices in [8].

Let us also consider the problem of scattering a plane electromagnetic wave $(\mathbf{E}^+, \mathbf{H}^+)$ on a lattice of dielectric resonators. In the general case, it is convenient to represent the total scattering field $(\mathbf{E}, \mathbf{H})$ in the form of a superposition:

$$\mathbf{E} = \mathbf{E}^+ + \mathbf{E}^s; \quad \mathbf{H} = \mathbf{H}^+ + \mathbf{H}^s \tag{9.18}$$

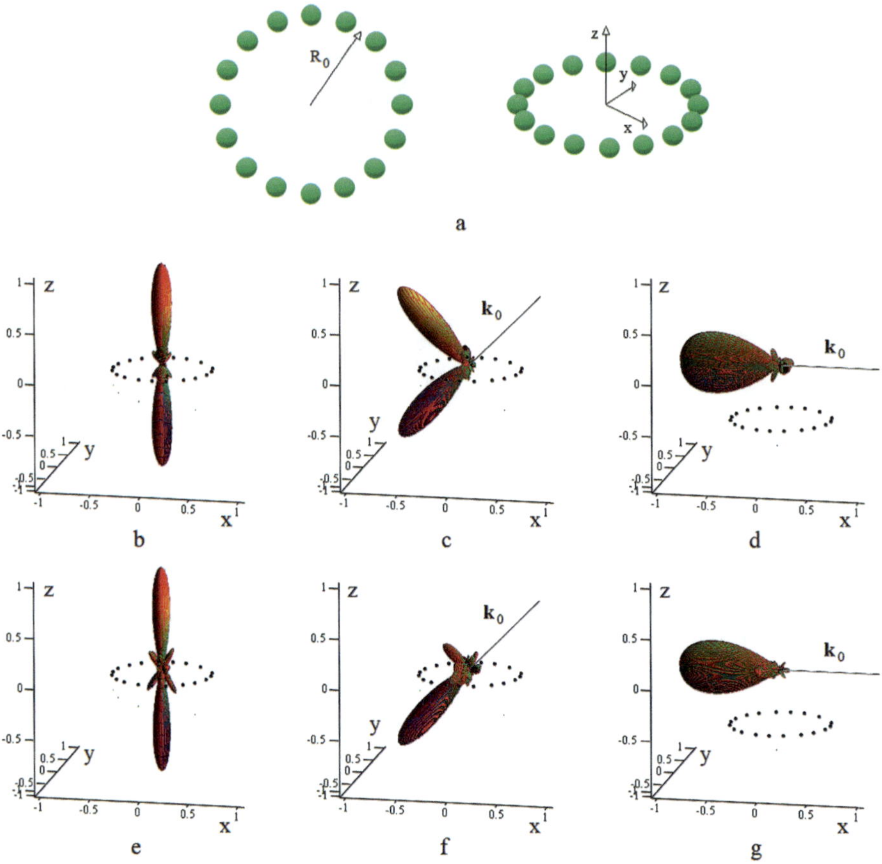

Fig. 9.9 Ring structure of 16 spherical DR (**a**). Angular dependences of the squared modulus of scattering amplitude for $k_0 R_0 = 2\pi$; $\varepsilon_{1r} = 36$; $\varphi_k = \pi$; $\theta_k = \pi$ (**b**, **e**); $\theta_k = 3\pi/4$ (**c**, **f**); $\theta_k = \pi/2$ (**d**, **g**); for *p*-scattering (**b**–**d**) for *s*-scattering (**e**–**g**)

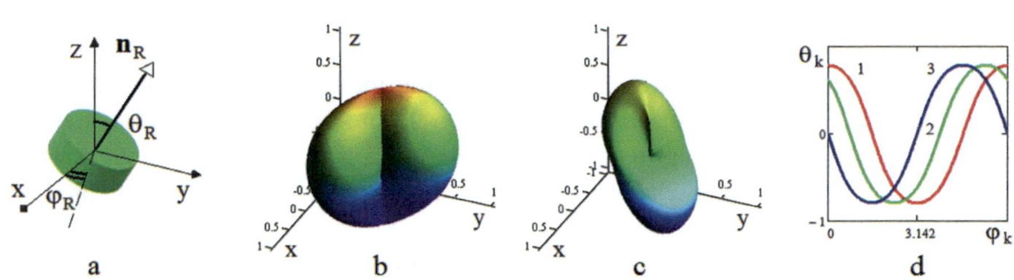

Fig. 9.10 Rotated cylindrical DR in rectangular coordinate system (**a**). Spatial distribution of the modulus of the c-function for *p*-polarization (**b**); *s*-polarization (**c**) ($\varepsilon_{1r} = 36$; $\theta_R = 0.25\pi$; $\varphi_R = 0.25\pi$). Dependences between the angles that determine the direction of incidence, at which the plane wave is not coupled to the resonator (**d**). Line 1—$\varphi_R = 0$; 2—$\varphi_R = 0.25\pi$; 3—$\varphi_R = 0.5\pi$ ($\theta_R = 0.25\pi$)

where ($\mathbf{E}^s$, $\mathbf{H}^s$)—is the field scattered by the lattice. In return, the scattered field can be decomposed into two terms, the first of which is due to the contribution of non-resonant components ($\mathbf{E}^{nr}$, $\mathbf{H}^{nr}$), and the second is the sum of the fields of coupled oscillations of the lattice resonators $\left(\mathbf{E}^L, \mathbf{H}^L\right)$:

$$\mathbf{E}^s = \mathbf{E}^{nr} + \mathbf{E}^L; \quad \mathbf{H}^s = \mathbf{H}^{nr} + \mathbf{H}^L \tag{9.19}$$

The non-resonant components of the lattice field are represented as an integral over a continuous spectrum [9], and the resonant components for the most part can be expressed as a finite sum over the coupled oscillations of the resonator system: $\left(\mathbf{e}^t, \mathbf{h}^t\right)$ $(t = 1, 2, ..., N)$ (Chap. 8) [4]:

$$\mathbf{E}^L = \sum_{t=1}^{N} a^t \mathbf{e}^t; \quad \mathbf{H}^L = \sum_{t=1}^{N} a^t \mathbf{h}^t \tag{9.20}$$

In turn, the field of coupled lattice oscillations can be represented as a superposition of the fields of partial resonators: $(\mathbf{e}_n, \mathbf{h}_n)$ $(n = 1, 2, \ldots, N)$ (Chap. 7):

$$\mathbf{e}^t = \sum_{n=1}^{N} b_n^t \mathbf{e}_n; \quad \mathbf{h}^t = \sum_{n=1}^{N} b_n^t \mathbf{h}_n \tag{9.21}$$

The field amplitudes b_n^t and a^t are determined from the system of equations (7.6, 8.13).

In most cases of resonant scattering, at frequencies near coupled oscillations, the conditions are satisfied:

$$\left|\mathbf{E}^{nr}\right| \ll \left|\mathbf{E}^L\right|; \quad \left|\mathbf{H}^{nr}\right| \ll \left|\mathbf{H}^L\right| \tag{9.22}$$

therefore, non-resonant terms in (9.19) can be neglected.

However, at certain particular values of the angles of incidence, as well as at certain spatial orientations of the DR axes, conditions may arise under which the incident wave $\left(\mathbf{E}^+, \mathbf{H}^+\right)$ is in no way coupled with the resonators. In this case, all amplitudes (9.17) $a^s = 0$ and the contribution of non-resonant scattering become decisive.

Let us consider in more detail the scattering of plane electromagnetic waves on a lattice of pseudo-rotating DRs of cylindrical and rectangular shape at the frequency of their main magnetic oscillations.

By pseudo-rotation we will understand a non-trivial rotation of the DR in which the relative spatial arrangement of the resonator axes remains constant. In the case of pseudo-rotations, the directions of the resonator axes change only relative to the lattice; in this case, the relative spatial arrangement of the axes remains unchanged, for example, parallel (Fig. 9.12a).

Below we will consider pseudo-rotations of cylindrical DRs, and also explore the conditions under which the inequalities are violated (9.20).

Let us first consider the c-functions of a cylindrical DR in the case of rotation of its axis in a selected coordinate system.

We rewrite in a more general form the analytical expression of the c-function (9.9), which defines the relationship between the plane wave field and H_{101} oscillations for arbitrary orientations of the cylindrical DR axis and arbitrary angles of incidence:

$$\left(c_t^+\right)_0 = -2\pi i E_0^* \cdot \frac{h_1}{\beta} \cdot (r_0^2 \frac{L}{2}) \cdot (k_1^2 - k_0^2) \frac{(\mathbf{n}_H, \mathbf{n}_R)}{\sqrt{1-(\mathbf{n}_k, \mathbf{n}_R)^2}} \cdot$$
$$\cdot \frac{\left[q_\perp \sqrt{1-(\mathbf{n}_k, \mathbf{n}_R)^2} J_0\left(q_\perp \sqrt{1-(\mathbf{n}_k, \mathbf{n}_R)^2}\right) J_1(p_\perp) - p_\perp J_0(p_\perp) J_1\left(q_\perp \sqrt{1-(\mathbf{n}_k, \mathbf{n}_R)^2}\right)\right]}{\left(q_\perp \sqrt{1-(\mathbf{n}_k, \mathbf{n}_R)^2}\right)^2 - p_\perp^2} \cdot$$
$$\cdot \frac{\left[p_z \sin p_z \cos(q_z(\mathbf{n}_k, \mathbf{n}_R)) - q_z(\mathbf{n}_k, \mathbf{n}_R) \cos p_z \sin(q_z(\mathbf{n}_k, \mathbf{n}_R))\right]}{(q_z(\mathbf{n}_k, \mathbf{n}_R))^2 - p_z^2} \tag{9.23}$$

Here $\mathbf{n}_H = (\cos\beta_1, \cos\beta_2, \cos\beta_3)$, unit vector defining the direction of the magnetic field of the incident wave $\left(\mathbf{E}^+, \mathbf{H}^+\right)$ in the rectangular coordinate system of the resonator (9.8); $\mathbf{n}_R = (\sin\theta_R \cos\varphi_R, \sin\theta_R \sin\varphi_R, \cos\theta_R)$—unit vector defining the direction of the DR axis; $\mathbf{n}_k$—unit vector defining the direction of propagation of the incident wave: $\mathbf{k}_0 = k_0 \mathbf{n}_k$ (Fig. 9.10a); $p_\perp = \beta r_0$; $p_z = \beta_z L/2$; $q_\perp = k_0 r_0$; $q_z = k_0 L/2$;$k_0 = \omega/c$; $k_1 = k_0\sqrt{\varepsilon_{1r}}$; ε_{1r}—relative dielectric constant of the resonator.

As follows from (9.23), the degree of interaction of the lattice resonators with the incident plane wave is in the general case determined depending on the directions of their axes $\mathbf{n}_R$; direction of fall $\mathbf{n}_k$ and in addition, the type of polarization of the incident wave $\mathbf{n}_H$.

On Fig. 9.10b, c show the angular dependences of the modules of functions (9.23) for different polarizations of a plane wave: p-scattering (the $\mathbf{E}^+$ vector lies in the plane of incidence) (c); s-scattering (the $\mathbf{E}^+$ vector is directed orthogonal to the plane of incidence) (d).

As follows from (9.23), the DR is not coupled with the incident wave if:

for p-scattering:

$$\varphi_k - \varphi_R = 0, \pi \tag{9.24}$$

for s-scattering:

$$ctg(\theta_R) tg(\theta_k) = \cos(\varphi_R + \varphi_k) \tag{9.25}$$

And also in the special case:

$$\varphi_k = -\varphi_R \quad \text{and} \quad \theta_k = \theta_R \tag{9.26}$$

The numerical solution of Eq. (9.25) is shown in Fig. 9.10d, for different directions of the DR axis. The $\left|c_t^+\right|^2$ dependences for p- and s-scattering on the (θ_k, φ_k) angles of incidence for $(\theta_R, \varphi_R) = (\pi/4, \pi/4)$ are shown in Fig. 9.10b, c, respectively.

We further conclude that the DR coupling reaches a maximum in the case if $\mathbf{n}_R = \pm\, \mathbf{n}_H$, as well as for

p-scattering, if:

$$\theta_R = \pi/2 \text{ and } \theta_k - \theta_R = \pm\pi/2 \tag{9.27}$$

s-scattering, if:

$$\varphi_k + \varphi_R = 0;\, 2\pi \text{ and } \theta_k - \theta_R = \pm\pi/2 \tag{9.28}$$

or

$$\varphi_k + \varphi_R = \pi \text{ and } \theta_R + \theta_k = \pi/2 \tag{9.29}$$

In the case of scattering of a plane wave on a rotating rectangular DR the results of calculating the integral (9.10) are presented in the form:

$$\begin{aligned}\left(c^+\right)_0 = c_0 \cdot \Big[\beta_y \omega_x\big(\big(\mathbf{n}_x^R, \mathbf{n}_k\big)\big)\overline{\omega}_y\Big(\Big(\mathbf{n}_y^R, \mathbf{n}_k\Big)\Big) \cdot \big(\mathbf{n}_E, \mathbf{n}_x^R\big) \\ -\beta_x \overline{\omega}_x\big(\big(\mathbf{n}_x^R, \mathbf{n}_k\big)\big)\omega_y\Big(\Big(\mathbf{n}_y^R, \mathbf{n}_k\Big)\Big) \cdot \big(\mathbf{n}_E, \mathbf{n}_x^R\big)\Big]\omega_z\big(\big(\mathbf{n}_z^R, \mathbf{n}_k\big)\big)\end{aligned} \tag{9.30}$$

where $\mathbf{n}_x^R, \mathbf{n}_y^R, \mathbf{n}_z^R$—orts of the local coordinate system of the DR $\left(x', y', z'\right)$ in the coordinate system (x, y, z) (Fig. 9.11a); $\mathbf{n}_E$—unit vector of the electric field of the incident wave $\left(\mathbf{E}^+,\ \mathbf{H}^+\right)$ (9.8); $\mathbf{n}_k$—unit vector defining the direction of propagation of the incident wave. The directions of the orts $\mathbf{n}_\nu^R = \mathbf{n}_\nu^R(\theta_R, \varphi_R)$ $\nu = x, y, z)$ are indicated in spherical coordinates.

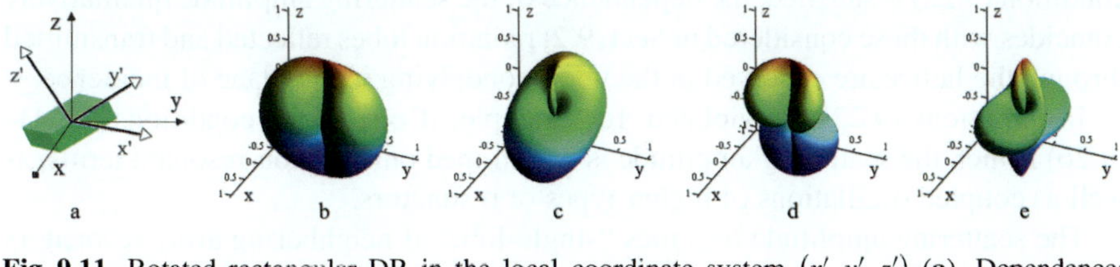

Fig. 9.11 Rotated rectangular DR in the local coordinate system $\left(x', y', z'\right)$ (**a**). Dependences $\left|c^+(\theta, \varphi)\right|$ on the directions of incidence of a plane wave $\mathbf{n}_k$ on rotated DR for $\mathbf{n}_z^R = (0.25\pi, 0)$; $\mathbf{n}_x^R = (0.75\pi, 0)$ and for H_{ccc} oscillation (**b, c**); H_{ccs} oscillation (**d, e**); for *p*-scattering (**b, d**); for *s*-scattering (**c, e**); of the DR with H_{ccs} oscillation relative to the coordinate system X (**h**); Y (**i**); Z (**j**) for p-scattering (**h**); for *s*-scattering (**i, j**); for H_{scc} oscillation and for *s*-scattering (**k**)

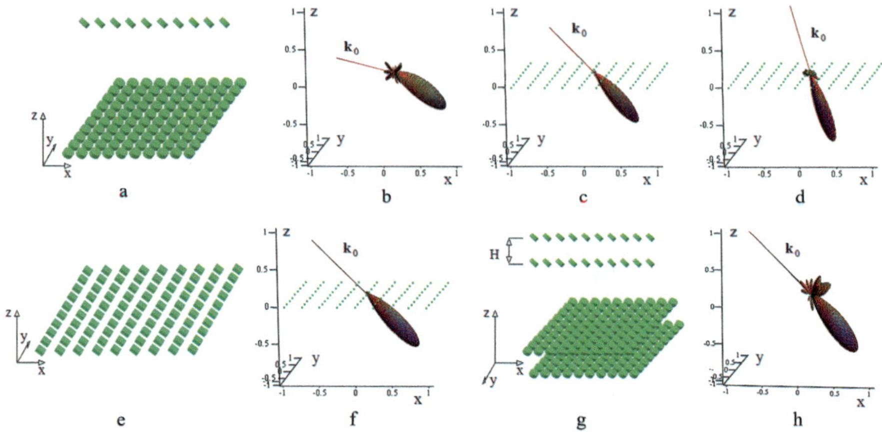

Fig. 9.12 Square lattices 10 × 10 of rotated Cylindrical (**a, g**), Rectangulare (**e**) DRs. Angular dependencies $|f\langle\theta_k, \varphi_k \mid \theta, \varphi\rangle|$ for *s*-scattering at angles of incidence: $\theta_k = 0.6\pi; 0.75\pi; 0.9\pi$; $\varphi_k = 0.25\pi$ (**b–d**); $\theta_k = 0.75\pi$; $\varphi_k = 0$ (**f, h**); when turning the DR axes: $\theta_R = \pi - \theta_k$; $\varphi_R = \varphi_k$ (The straight line shows the direction of fall $\mathbf{k}_0$)

As follows from (9.30), a rectangular DR is not coupled with an incident wave on the main magnetic type of oscillation H_{ccc} the direction of the vectors $\mathbf{n}_k$ and $\mathbf{n}_z^R$ coincides, i.e. $\left|\left(\mathbf{n}_k, \mathbf{n}_z^R\right)\right| = 1$. The second possible condition is $\left|\left(\mathbf{n}_E, \mathbf{n}_z^R\right)\right| = 1$. That is, if the z' resonator axis is directed along the $\mathbf{k}_0$ vector, or in the $\mathbf{n}_E$ direction of polarization of the electric field of the incident wave.

On Fig. 9.11b–e shows the dependences of the modulus c—the function of the rectangular DR, calculated according to (9.30) for three main types of magnetic type oscillations for $\mathbf{n}_z^R = (0.25\pi, 0)$; $\mathbf{n}_x^R = (0.75\pi, 0)$ from the directions of incidence of a plane wave. As follows from the data presented, the rotation of a rectangular DR has a particularly noticeable effect on oscillations of higher types (Fig. 9.11d, e), which have a more complex spatial field distribution.

Let us consider particular examples of scattering on a flat square lattice of pseudo-rotating cylindrical and rectangular DRs (Fig. 9.12a, e, g). In the general case, if condition (9.22) is satisfied, the dependence of the scattering amplitude qualitatively coincides with those considered in Sect. 9.2: radiation lobes reflected and transmitted through the lattice are observed in the wave zone, lying in the plane of incidence.

If conditions (9.22) are violated, for example, if one of the conditions (9.24)–(9.26) is met, the scattering amplitude is determined only by non-resonant terms, as well as coupled oscillations of higher types of resonators.

The scattering amplitude becomes "single-lobe" if neighboring array resonators are located at a distance $d \leq \lambda_0/2$, and the direction to the reflected lobe coincides with the direction of "zero radiation" of each partial resonator (9.2). The absence of a reflected wave when a plane wave falls on the interface between two media is called the Brewster effect [10]. The incident wave is re-radiated by the lattice in a resonant manner only in the direction of propagation of the incident wave, determined by the vector $\mathbf{k}_0$. An example of such scattering by a flat lattice in the direction of incidence is given in Fig. 9.12b, e. Similar patterns occur for lattices of more complex shapes of pseudo-rotating DR with basic types of oscillations H_{101} (Fig. 9.12g, h). Finally, taking into account (9.18) and (9.5), we come to the conclusion that such a lattices becomes a resonant absorber, provided that the polarizations of the incident and scattered waves are equal.

9.6 Scattering of Transmission Line Waves on Dielectric Resonator Lattices

Structures consisting of a large number of resonators have some interesting properties that are absent in homogeneous materials. In this section, applying the results obtained above, we consider scattering on DR lattices coupled with the input and output transmission lines. Below we will be mainly interested in the distribution of amplitudes of the field of forced oscillations of partial resonators installed in the passbands of the lattice.

Let's explore first a new property of DR lattices that form waveguide structures at particular frequencies of forced oscillations. Interestingly, such waveguide channels may arise in a homogeneous lattice.

The formation of a waveguide channel was first observed on a flat square lattice of cylindrical DRs with magnetic azimuthally uniform oscillations H_{101} (Fig. 9.13a). The dielectric permittivity of the DRs $\varepsilon_{1r} = 81$; $1/tg\delta = 10^5$; $\Delta = L/2r_0 = 0.44$; the coupling coefficient between the DR and transmission line $\tilde{k}_L = 0.02$. The coupling coefficients between the resonators were calculated by the formulas (6.7). In Fig. 9.13 shows the amplitude-frequency response of the lattice S_{21} coefficient of transmission between input (1) and output port (2). The red line marks the wavelength at which the formation of waveguide oscillations is detected. In Fig. 9.13c shows the distribution of the amplitude modulus (8.6) of the DRs in the lattice at the wavelength of "waveguide" resonant oscillations. It can be seen that at a given frequency the distribution of amplitudes of the resonators is established in such a way that the total field forms a "waveguide" that appears between the input and output lines. Most likely, this behavior is explained by the interference of standing Bloch waves when connecting lines to the corners of a square lattice.

Since Bloch waves arise in any periodic structures, a similar phenomenon is observed in lattices with other types of resonant oscillations, for example, whispering gallery oscillations.

In Fig. 9.14 shows the results of calculation of scattering on a square lattice of coupled spherical microresonators with magnetic oscillations H_{mml}. The coupling coefficients between the microresonators were calculated by using the results (6.75).

A similar phenomenon is observed in 3-dimensional lattices. In Fig. 9.15a shows a square lattice of cylindrical DRs with magnetic oscillations H_{101}. The distribution of the S matrix by wavelengths is shown in Fig. 9.15b. The waveguide distribution of

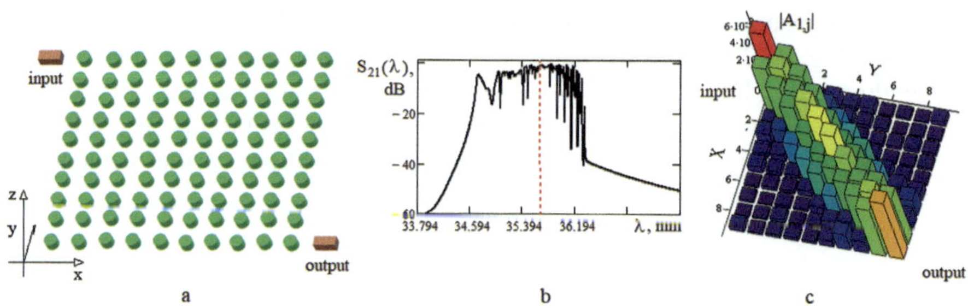

Fig. 9.13 Square lattice of cylindrical DRs with main types of oscillations H_{101}^{+} (**a**) ($\varepsilon_{1r} = 81$; $\Delta = 0.44$; $Q^D = 10^5$). Attenuation between input–output ports (**b**). Distribution modulus amplitudes of the forced oscillations at the wavelength $\lambda = 35.802$ mm (**c**)

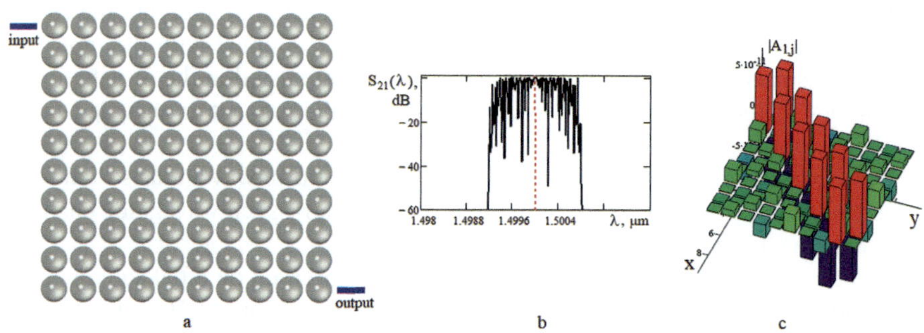

Fig. 9.14 Square lattice of Spherical microresonators with whispering gallery oscillations H_{mm1} ($m = 30$; $\varepsilon_{1r} = 4$; $Q^D = 10^{10}$) (**a**). Attenuation between input–output ports (**b**). Distribution modulus amplitudes of forced oscillations resonators at the wavelength $\lambda = 1.5\,\mu$m (**c**)

amplitudes was observed in each layer on the wavelength indicated by the dashed line (Fig. 9.15b). Thus, we can conclude that in this case a planar waveguide is formed in the diagonal plane of the lattice. Figure 9.15c shows example of distribution of the amplitude modulus in the 4-th layer of the DRs lattice.

Another exotic property of DR lattices is the possibility of electromagnetic oscillations propagating along its "surface". In solid state physics, such materials are called topological insulators. The nature of optical topological insulators is not fully understood. Accumulating experience in DR modeling suggests that possible mechanisms for the propagation of oscillations over the surface of the lattice may be its inhomogeneity or the degeneracy of oscillations of partial resonators.

To clarify the physical mechanisms of surface wave propagation, we first build an electromagnetic model of a lattice (Fig. 9.16a), consisting of ring resonators with degenerate types of oscillations.

Let us assume that the indicated gratings are excited by the input 1–2 and the transmitted wave is measured at the output 3–4 by optical transmission lines. First, let's consider the transmission coefficients between ports 12, 13, 14 (Fig. 9.16b(2)),

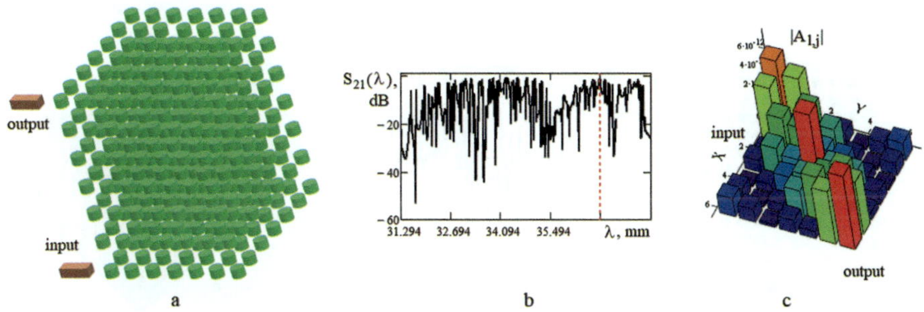

Fig. 9.15 3-dimensional square lattice of DRs (**a**). S-matrix coefficient between the input and output lines of a lattice of $7 \times 7 \times 7$ DRs with H_{101} magnetic oscillations (**b**). Waveguide distribution modes in 4 layer on $\lambda = 36.873$ mm (**c**)

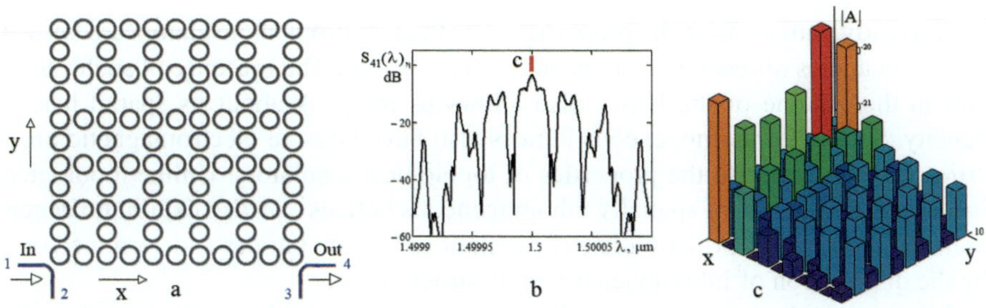

Fig. 9.16 Square lattice of optical microcavities with whispering gallery oscillations (**a**). Attenuation between ports 1–4 of the lattice (**b**). Distribution modulus amplitudes of the forced oscillations of the lattice at the central wavelength (**c**)

then in the pass band of the structure we calculate the distribution of amplitudes of forced oscillations of the resonators.

Let us assume that the coefficients of mutual coupling between neighboring resonators are known; non-adjacent resonators are not connected to each other. In each of the resonators, two types of degenerate natural oscillations are excited. Dielectric quality factor of resonators $Q^D = 10^{10}$; open space coupling coefficients $\tilde{k}_{OS} = 10^{-6}$; coupling coefficients of microresonators with lines: $\tilde{k}^e = \tilde{k}^o = 3 \cdot 10^{-4}$; mutual coupling coefficients between resonators with even types of oscillations $k^e_{s,s+1} = 5 \cdot 10^{-5}$; with odd oscillations $k^o_{s,s+1} = -2 \cdot 10^{-5}$.

The obtained simulation results showed that at the natural oscillation wavelength $\lambda_0 = 1.5\,\mu$m, shown as a red line in Fig. 9.16b, the distribution of microresonator field amplitudes approximately corresponds to the wave propagation characteristic of optical topological insulators (Fig. 9.16c). However, even when detuned from the central transmission band, the amplitude distribution of microcavities becomes chaotic. For lattices of the type considered, the bands of surface wave propagation are relatively narrow (Fig. 9.16b), however surface propagation losses are minimal.

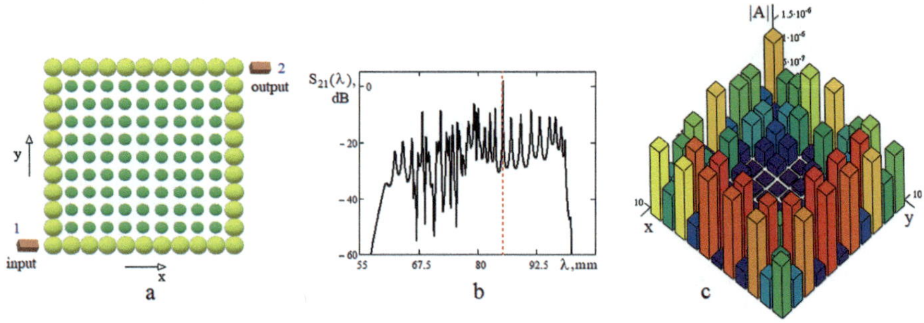

Fig. 9.17 Square lattice of various spherical DRs with main types of nondegenerate oscillations H_{101} (**a**) ($\varepsilon_{1r} = 36$; $\varepsilon_{2r} = 81$). Attenuation between ports 1–2 of the lattice (**a**). Distribution modulus amplitudes (**c**) of the forced oscillations at the wavelength 85.35 mm (**b**)

Currently, lattices with the properties of optical topological insulators are known, due to vacancies of resonators at the nodes (Fig. 9.16a). The presence of such vacancies in the volume of the lattice also allows us to speak about its spatial heterogeneity. Spatially inhomogeneous lattices also have unusual electromagnetic properties, partly similar to the properties of topological insulators. Unlike topological insulators, resonators of spatially inhomogeneous lattices can also exhibit no degenerate oscillations. The distribution of oscillations in such lattices is determined only by the distribution of inhomogeneity in its structure.

Let us consider two types of inhomogeneities that can be implemented on the surface of the DR lattice. The first type of lattices can be made from resonators consisting of different dielectrics (Fig. 9.17a) and the second type can be made from resonators of different relative sizes (Fig. 9.18a). In both examples, the natural frequencies of the isolated partial resonators are the same. In our case, the frequency of each of the lattice resonators corresponds to the main magnetic-type oscillation H_{101}. Despite the fact that the natural frequencies of the partial DRs coincide, at some frequencies (Figs. 9.17 and 9.18b) forced oscillations arise in the structure as a whole, corresponding to the propagation of waves mainly along the inhomogeneity of the lattice (Figs. 9.17and 9.18c). In this case, the main part of the power passes along the edge of the lattice.

The obtained simulation data allows us to state with a high degree of confidence that the unusual properties of optical topological insulators can be due to possible reasons:

(a) Effects of interaction of degenerate oscillations taking place in dielectric resonators with whispering gallery waves.
(b) Surface heterogeneity of the lattice structure.

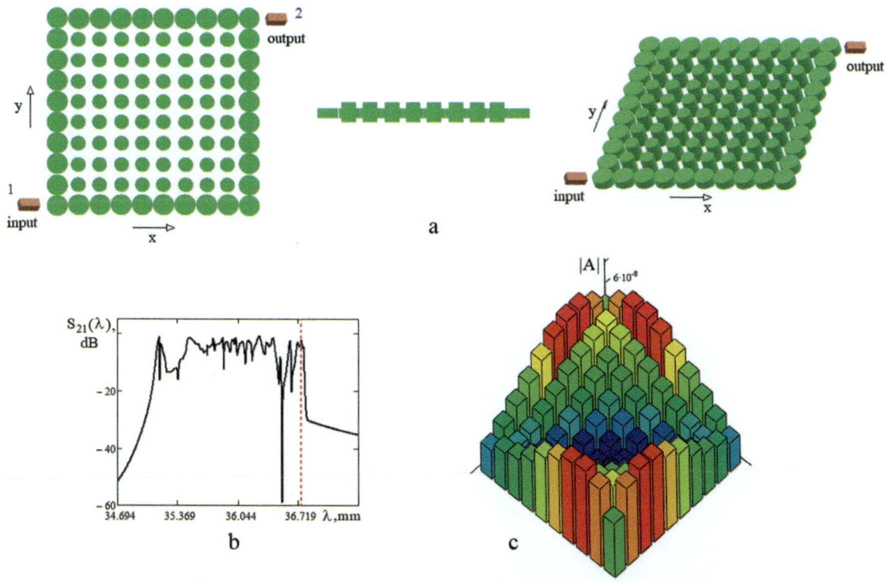

Fig. 9.18 Square lattice of various cylindrical DRs with main types of oscillations H_{101} and different compression ratios (**a**) ($\varepsilon_{1r} = 36$; $Q^D = 10^5$; $\Delta_1 = L_1/2r_1 = 0.3$; $\Delta_2 = L_2/2r_2 = 1$; L_s—height; $2r_s$—DR diameter ($s = 1,\ 2$)). Attenuation between ports 1–2 of the lattice (**a**). Distribution modulus amplitudes (**c**) of the forced oscillations of the lattice at the wavelength 36.72 mm (**b**)

References

1. J.A. Tsang, J.A. Kong, K-H. Ding, Scattering of electromagnetic waves. A Willey-Interscience Publication, vol. 1 (Wiley, New York, 2000), p. 425
2. K. Blum, *Density Matrix Theory and Applications* (Springer, Berlin, 2011), p.343
3. V.V. Nikolsky, *Electrodynamics and Radio Wave Propagation* (Nauka, Moscow, 1973), p.606
4. A.A. Trubin, *Lattices of Dielectric Resonators* (Springer, Switzerland, 2016), p.171
5. M. Abramowitz, I.A. Stegun (eds.), Handbook of Mathematical Functions With Formulas, Graphs and Mathematical Tables. Applied Mathematics, vol. 55 (National Bureau of Standards, 1964), p. 830s
6. A.A. Trubin, Scattering of electromagnetic waves on lattices of rectangular dielectric resonators. Visnyk NTUU KPI Ser. Radiotek. Radioaparatobud. **83**, 5–10 (2021)
7. A.A. Trubin, Scattering of plane electromagnetic waves by lattices of spherical dielectric resonators with degenerate lower types of natural oscillations. Visnyk NTUU KPISer. Radiotek. Radioaparatobud. **90**, 12–17 (2023)
8. V.B. Kazansky, V.R. Tuz, D.V. Yudintsev, Scattering of Gaussian wave beams by a limited sequence of alternating arrays of dielectric bars. Radio Phys. Electron. **756**(11), 82–86 (2007). ((in russian))
9. L. A. Veinshtein, Open Resonators and Open Waveguides. Golem series in electromagnetics, vol. 2 (Golem Press, 1969), p. 439
10. M. Born, E. Wolf, *Principles of Optics*, 2nd edn. (Pergamon Press, Oxford, 1964), p.855

Chapter 10
Antennas on Dielectric Resonators

The problems of scattering of electromagnetic waves of a transmission line on a system of coupled DRs also located in open space are considered. Such structures, optimized for field distribution in the wave zone, as well as matched with the supply transmission line; perform the functions of resonant-type antennas. Several classes of antenna arrays, linear, planar and 3-dimensional, are considered. Using the developed theory, the main parameters of the antennas were calculated—radiation patterns, frequency characteristics of the reflection coefficients from the side of the supply line, depending on the excitation methods, location and geometric dimensions of the resonators. The parameters of new elements, lattices of coupled DRs, placed simultaneously in transmission lines and open space, combining functions of antennas and bandpass filters are considered. It is shown that simultaneous optimization of the scattering parameters of antenna filters allows one to obtain significantly better frequency scattering characteristics.

10.1 Excitation of Electromagnetic Oscillations in Systems of Coupled Dielectric Resonators in Open Space

In the general case, an antenna built using dielectric resonators is a rather complex spatial structure, usually containing an additional metal screen, a power line and coupling elements, so an accurate analytical calculation of its parameters in most cases is a difficult practical task. However, most of the necessary parameters of such an antenna can be calculated by also applying the one described in Chaps. 7 and 8 perturbation theory.

Let's consider a multi-element antenna array on a DR. Let one (Fig. 10.1a) or several resonators (Fig. 10.1b) be connected to the transmission line L, for example, through an aperture in the metal screen of the supply line. In the future, we will also consider such a DR system as one multiply connected resonator of complex design.

A. Trubin, *Introduction to the Theory of Dielectric Resonators*, Springer Series in Advanced Microelectronics 65, https://doi.org/10.1007/978-3-031-65396-4_10

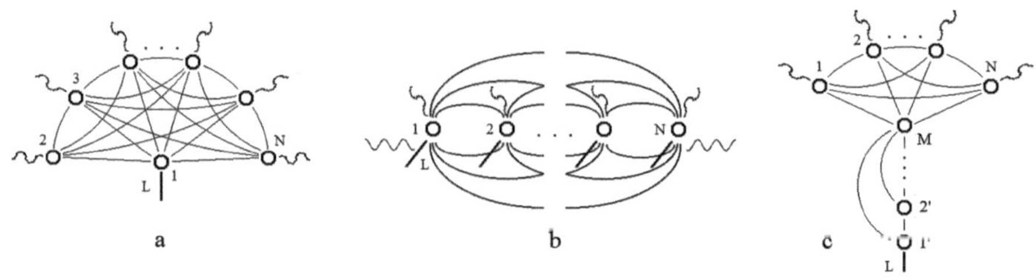

Fig. 10.1 Methods of excitation of antenna structures: through one resonator (**a**), through a system of resonators in the transmission line (**b**), through a bandpass filter (**c**)

If the geometric dimensions of the apertures are not large, and the active DRs are installed in them in the symmetry plane, the electromagnetic field of the system in the near zone is quite accurately approximated by a superposition of eigenfunctions corresponding to the coupled natural oscillations of the N-connected DR (Sect. 7). In this case, the field of natural oscillations of such N-connected resonator can also be approximately represented in the form of an expansion in natural oscillations of isolated DRs (7.4), with coordinates corresponding to their location in the antenna structure [1].

Therefore, we will also look for a solution to the problem of $\left(\mathbf{E}_l^+, \mathbf{H}_l^+\right)$ waveguide mode scattering on a N DR system in the form of an expansion in functions (8.5), in which the unknown amplitudes a^s, are found using perturbation theory for Maxwell's equations, In this case, the c_l^{t+} coefficients have the meaning of decomposing the field of the t-th active DR over the field of the line $\left(\mathbf{E}_l^+, \mathbf{H}_l^+\right)$.

Let us also assume that we know the field of each isolated resonator in the wave zone (1.1)

$$(\mathbf{e}_t, \mathbf{h}_t)_{|\mathbf{r}|\to\infty} \approx \left(\mathbf{e}_t^\infty, \mathbf{h}_t^\infty\right). \tag{10.1}$$

If the dimensions of the coupling elements are not large compared to the wavelength, then we can neglect their radiation. In this case, it is enough to represent the antenna field in the wave zone as a superposition of the fields of partial resonators:

$$\mathbf{e}^\infty \approx \sum_{t=1}^{N} A_t \mathbf{e}_t^\infty : \quad \mathbf{h}^\infty \approx \sum_{t=1}^{N} A_t \mathbf{h}_t^\infty. \tag{10.2}$$

where, similarly (8.6),

$$A_t = \sum_{s=1}^{N} a^s b_t^s \tag{10.3}$$

—amplitude of forced oscillation of the t-th DR.

The field of a system of DRs in the wave zone can be calculated approximately, taking into account the location of the resonators relative to each other, using (9.2):

$$\mathbf{e}^{\infty} \approx \mathbf{e}_1^{\infty} \sum_{t=1}^{N} A_t e^{ik_0 d_t},$$

$$\mathbf{h}^{\infty} \approx \mathbf{h}_1^{\infty} \sum_{t=1}^{N} A_t e^{ik_0 d_t}. \tag{10.4}$$

Here also $d_t = d_t(\theta, \varphi) = \mathbf{P}_t \cdot \mathbf{n}_0$; $\mathbf{n}_0 = (\sin\theta\cos\varphi, \sin\theta\sin\varphi, \cos\theta)$ unit vector determining the direction to the observation point; $\mathbf{P}_t = (x_t, y_t, z_t)$—coordinate vector of the t-th DR (Fig. 9.1b).

10.2 One-Dimensional Antenna Arrays of Dielectric Resonators

One-dimensional DR arrays are the simplest implementations of multi-element antennas. Based on them, it is possible to build unidirectional antennas, the radiation pattern of which does not depend on the azimuthally angle φ, and directional—with a pattern compressed, as a rule, in one plane, determined by the pattern of partial DR and the lattice parameters.

In a particular case, if all DRs are identical and located on the same straight line (see Fig. 10.2a, e), then the lattice field in the wave zone (10.4) in a spherical coordinate system can be represented in a simpler form:

$$\mathbf{e}^{\infty}(\theta, \varphi) \approx \mathbf{e}_1^{\infty}(\theta, \varphi) \cdot \sum_{s=1}^{N} A_s e^{ik_0 z_s \cos\theta}. \tag{10.5}$$

where $(0, 0, z_s)$ coordinates of the s-th DR on the lattice axis (Fig. 10.2a, e).

On Fig. 10.2 shows possible implementation options for antenna arrays made using cylindrical DR on the main magnetic type of oscillations H_{101} excited by a metal coupling loop. The antenna characteristics were calculated using formulas (5.21), as well as (7.6), (8.13) and Sect. 6.3. The following values were used as the initial parameters of the DR: coupling coefficient with the metal loop of the coaxial line feeding the antenna: $\tilde{k}_L = 0.025$, $f_0 = 5.5$ GHz, $\varepsilon_{1r} = 82$, $Q^D = 2 \cdot 10^3$, $\Delta = 0.44$. The relative distance between adjacent DRs for an unidirectional antenna (Fig. 10.2a) $1/2\lambda_0$, for a directional antenna (Fig. 10.2e) was equal to $3/4\lambda_0$, where λ_0—free space wavelength. The bandwidth of the reflection coefficient at the −3 dB relative to the level of minimum attenuation from the supply feeder was: for a directional antenna 1.6%, and for a non-directional antenna 3.3%. Distribution of normalized absolute values of complex amplitudes of forced oscillations of partial

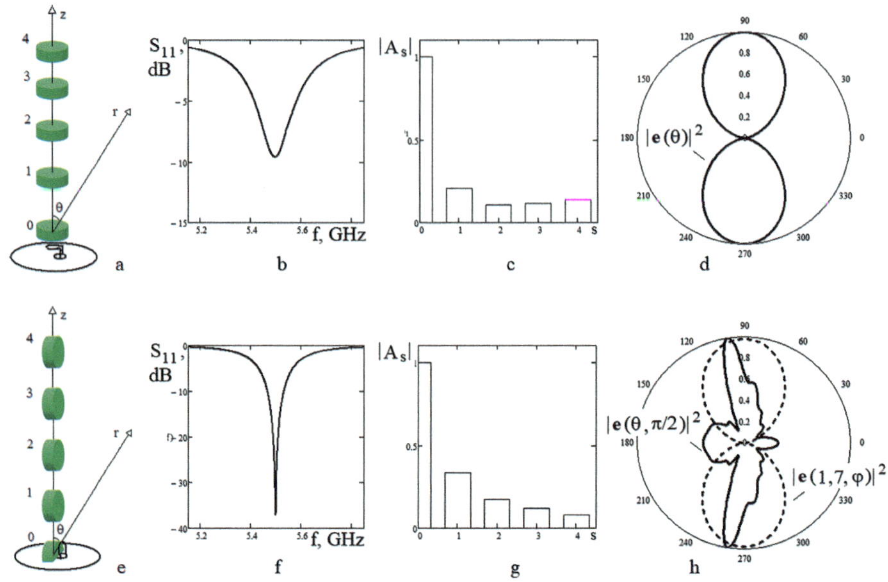

Fig. 10.2 One-dimensional lattices powered by a current loop (**a**, **e**). Frequency dependences of the reflection coefficient (**b**, **f**). Normalized amplitudes of partial resonators forced oscillations (**c**, **g**). Antenna radiation patterns (**d**, **h**). $\varepsilon_{1r} = 36$; $Q^D = 2 \cdot 10^3$; relative sizes of the resonators of the first lattice (**a**) $\Delta = 0.44$; second lattice (**e**) $\Delta = 0.2$

resonators, shown in Fig. 10.2c, g, approximately corresponds to exponential decay. The gain for one main lobe of the directional antenna (Fig. 10.2h) was approximately 11.9 dB.

A more complex version of using DRs as passive reflectors and directors of one-dimensional linear arrays is shown in Fig. 10.3a. The results of calculating the main characteristics of the antenna are shown in Fig. 10.3c, d.

Consider an array containing 9 elements and one DR as a passive reflector (Fig. 10.3a). Preliminary optimization of the antenna was carried out by varying the geometric and electrical parameters of the design, based on the requirement for maximum compression of the main lobe while simultaneously suppressing the side lobes. As a result of the numerical optimization of the design, the following parameters were obtained: $\tilde{k}_L = 0.01, f_0 = 4.2$ GHz, $\varepsilon_{1r} = 82$, $Q^D = 2 \cdot 10^3$, $\Delta = 2.5$. Here $\tilde{k}_L$-coupling coefficient between the loop and the active DR 1. The distance between DR 1 is chosen to be equal to $\lambda_0/4$ and the distance between adjacent lattice resonators is slightly larger than the diameter of the resonators $2r_0$.

The data obtained show that the use of DRs as directors and passive reflectors can lead to compression of the main lobe in both vertical and horizontal planes (Fig. 10.3d). In this case, the frequency dependence of the reflection coefficient on the line side turns out to be noticeably jagged (Fig. 10.3b), which is explained by the complex distribution of the amplitudes of forced oscillations excited in the array of coupled resonators (Fig. 10.3c).

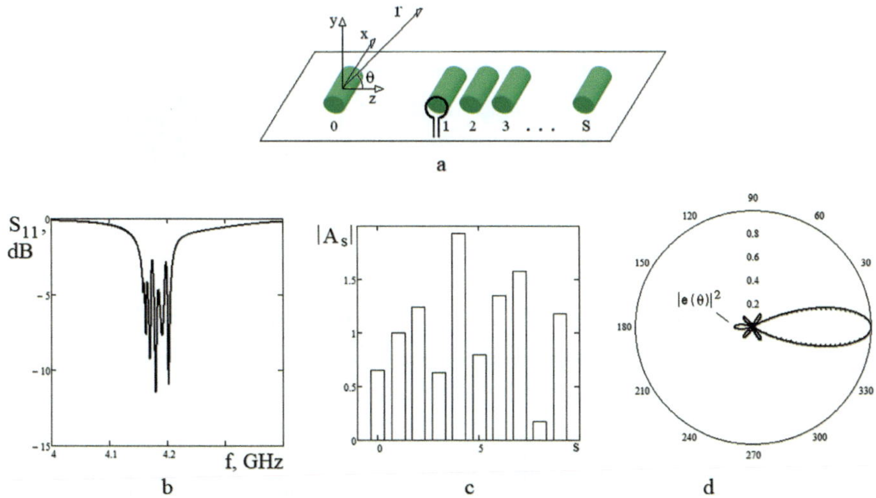

Fig. 10.3 Uda-Yagi antenna (**a**) for 10 cylindrical DRs ($\varepsilon_{1r} = 82$; $Q^D = 2 \cdot 10^3$; $\Delta = 2.5$). Frequency dependence of the reflection coefficient (**b**). Amplitude modules of partial resonators (**c**). Antenna radiation patterns (**d**)

As can be seen from the above examples, the condition of narrowing the array pattern leads to the need for spatial separation of partial DRs. The consequence of this is a decrease in the coupling coefficients between the resonators, which in turn leads to a decrease in the operating frequency band of the antenna. An effective way to expand the frequency bandwidth is to use designs that implement additional connections between partial resonators, for example, using a transmission line (Fig. 10.1b). One of the options for implementing this idea is to place a DR in one of the walls of a regular metal waveguide. In this case, the DR lattice is located parallel to the waveguide axis [2]. All resonators are coupled to each other both in open space and through propagating waves in the line. An additional increase in coupling between partial DRs can also be achieved by adding a metal wall at one end of the line. This method allows us to significantly expand the operating bandwidth of the antenna—up to 10%, even for DR made of dielectrics with increased permeability. However, in this case, the calculation of the antenna parameters becomes noticeably more complicated, since it is necessary to simultaneously take into account the interaction of the resonators in the line, as well as taking into account the influence of additional metal surfaces.

On Fig. 10.4a shows a linear array of cylindrical DR placed in the holes of the wide wall of a rectangular regular waveguide. The sizes of all DR correspond to the condition of excitation of the main type of magnetic oscillations in them H_{101}. It was also assumed that only the fundamental wave H_{10} can propagate in the waveguide. When it falls, a forced oscillation mode is established in the resonator system.

In the case of a symmetrical arrangement of cylindrical DR relative to the plane of the metal wall of the waveguide and the aperture, the partial coupling coefficients

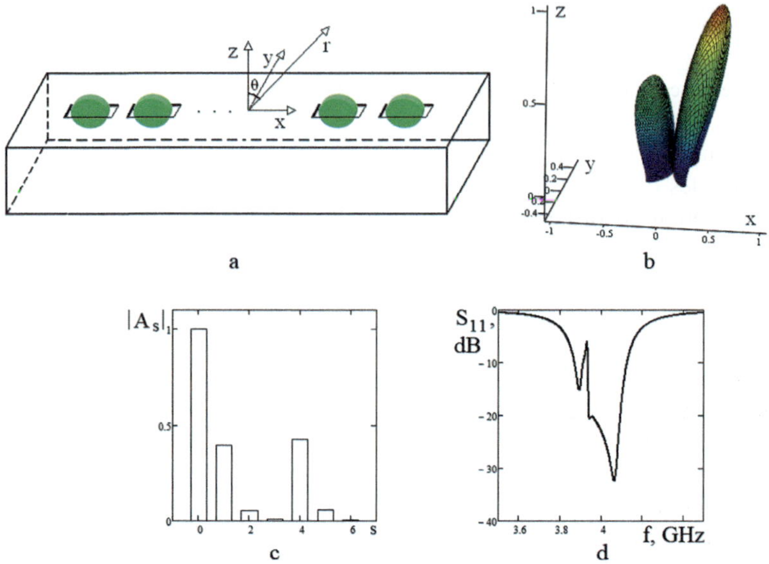

Fig. 10.4 One-dimensional lattice of 7 cylindrical DRs coupled to a rectangular waveguide (**a**) cross-section a × b = 58 × 25 mm. Main antenna radiation pattern (**b**). Normalized amplitudes of partial resonators (**c**). Parameters of lattice resonators: $\varepsilon_{1r} = 36$; $Q^D = 1.5 \cdot 10^3$; $\Delta = 0.44$. Frequency dependence of the reflection coefficient (**d**)

become equal to the sum of half values calculated on the basis of already known relationships (see Sects. 5.2 and 6.3) in open space and half values of the corresponding complex mutual coupling coefficients of the DR in the waveguide (Sects. 2.2 and 3. 2).

Calculation of the reflection coefficient on the side of the waveguide for the 7 resonator structure showed that the bandwidth of such an antenna noticeably exceeds the frequency bands of dielectric antennas of other designs.

The angular distribution of radiation has a pronounced two-lobe shape (Fig. 10.4b). The amplitudes of forced oscillations of resonators (10.3) are distributed in a more complex manner (Fig. 10.4c).

10.3 Planar Antennas Based on Coupled Dielectric Resonators

Placing an array on any flat surface makes it possible to "compress" the directivity characteristic in two mutually orthogonal planes, therefore two-dimensional arrays are of the greatest practical interest [3, 4].

In a particular case, if all DRs are identical and located in the same plane, for example (see Figs. 10.5, 10.6 and 10.7a), then the lattice field in the wave zone can be approximately represented in the form (10.4) taking into account $z_t = 0$.

On Fig. 10.5 shows the results of studying the characteristics of a simple lattice consisting of 4 DRs, one of the resonators of which is connected to a waveguide through a rectangular aperture in a thin metal screen. Three semi-cylindrical resonators are glued to the screen. The difference between the theoretical and measured radiation patterns in the E-plane is explained by the influence of the edge of the final metal shield of the short circuit, which is not taken into account in this calculation, as well as neglecting the influence of aperture radiation on the antenna characteristics.

As one would expect, wider frequency bands are available for lattices of this type, in which the active resonator coupled to the waveguide is located in the center, equidistant relative to the passive resonators (Figs. 10.6 and 10.7a, e). In this case, the distribution of passive DR amplitudes becomes more uniform. In these designs, DRs

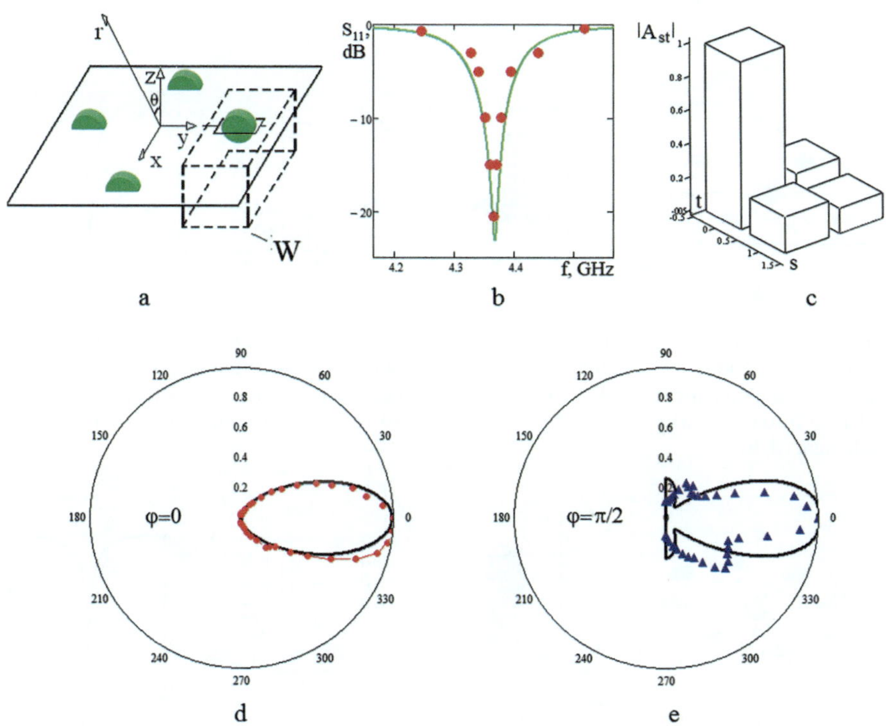

Fig. 10.5 Two-dimensional lattice of 4 semi-cylindrical DRs connected through a short circuit at the end of a rectangular waveguide (**a**) with a cross-section a × b = 58 × 25 mm. Frequency dependence of reflection coefficient (**b**). Dots show measured values. Normalized amplitude magnitudes of partial resonators (**c**). Parameters of lattice resonators (**a**) $\varepsilon_{1r} = 82$; $Q^D = 10^3$; $\Delta = 0.18$. Antenna radiation pattern (**d, e**), dots show measurement results in the H-plane (**d**); in E-plane (**e**)

with the following parameters were used: $\varepsilon_{1r} = 82$, $Q^D = 1800$, $\Delta = L/(2r_0) = 0.18$. The main magnetic type of oscillations H_{101} was excited in all resonators.

The location of the passive DR on the metal screen was chosen according to the calculation model, based on the requirement that the width of the radiation lobe in *E* and *H* plane be equal. Optimized i.e. structures are characterized by one radiation lobe orthogonal to the plane of the metal screen. In a tuned lattice of the first type (Fig. 10.6a), the coupling between the resonators is greater compared to the resonators of the lattice shown in Fig. 10.6e.

In this case, the amplitudes of the resonators are distributed more evenly (Fig. 10.6c).

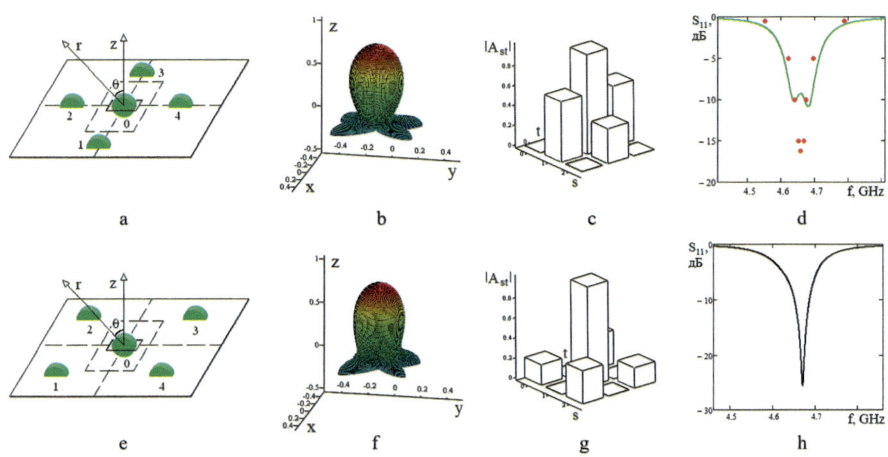

Fig. 10.6 Two-dimensional lattices of 5 semi-cylindrical DRs connected through a short-circuit aperture at the end of a rectangular waveguide (**a**, **e**) with a cross-section a × b = 58 × 25 mm. Antenna radiation patterns (**b**, **f**), Normalized amplitudes of partial resonators (**c**, **g**). Parameters of lattice resonators (**a**) $\varepsilon_{1r} = 82$; $Q^D = 10^3$; $\Delta = 0.18$. Frequency dependences of reflection coefficient (**d**, **h**)

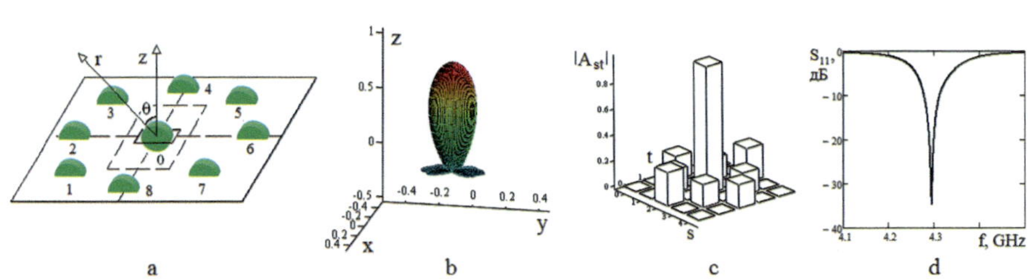

Fig. 10.7 Two-dimensional array of 9 semi-cylindrical DRs connected through an aperture at the end of a rectangular waveguide (**a**). Antenna radiation pattern (**b**), Normalized amplitudes of partial resonators (**c**). Frequency dependence of reflection coefficient (**d**). The resonator parameters are the same as in Fig. 10.6

An increase in the number of passive DR (Fig. 10.7a) leads to some narrowing of the DR (Fig. 10.7b) and an increase in the gain from approximately 10–15 dB, however, achieving better parameters in planar structures is not seems possible. Greater opportunities for improving antenna performance arise when using 3-dimensional structures.

10.4 3D Coupled Dielectric Resonator Antennas

Calculating the characteristics of 3-dimensional DR structures is fundamentally no different from the one- and two-dimensional ones discussed in the previous sections. The main difficulty arises from the significantly larger number of resonators on the basis of which antenna arrays are built in this case. To calculate the field in the wave zone, the general relation is used (10.4).

On Fig. 10.8a shows a simple 10×10 cylindrical-shaped angular reflective lattice of DR, excited by a metal wire loop with a cylindrical resonator. It is assumed that all resonators are the same, and are excited at frequencies near the main magnetic-type oscillation H_{101} ($f_0 = 10$ GHz). The dimensions of the loop are smaller than the dimensions of the resonator, so we neglected its radiation field. The distance between adjacent lattice resonators is equal to a quarter of the wavelength in open space $\lambda_0/4$. The distance between the active resonator and the lattice, as well as the "opening angle" of the grating, are optimized to obtain the best radiation characteristics. The magnitudes of forced oscillations of the DR are shown in Fig. 10.8b; frequency dependence of the reflection coefficient from the supply line—on Fig. 10.8c; directional pattern on the Fig. 10.8d. It can be seen that almost all lattice resonators are coupled to the active DR. Most of the power is reflected by the lattice towards the main lobe.

Another attractive option for constructing flat reflective lattices is the use of an irregular arrangement of resonators. An example of such a lattice is the structure shown in Fig. 10.9a. In this lattice, the distance between adjacent resonators varies only along one axis (x); along the axis y the distance between the resonators remains constant.

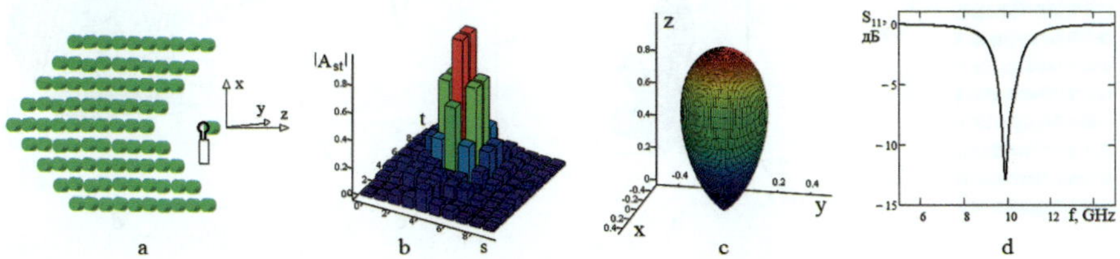

Fig. 10.8 10×10 DR (**a**) corner reflective grille fed by a metal loop resonator. Normalized amplitude magnitudes of partial resonators (**b**). Frequency dependence of reflection coefficient (**c**). Array radiation pattern (**d**), Resonator parameters: $\varepsilon_{1r} = 36$; $Q^D = 2 \cdot 10^3$; $\Delta = 0.4$

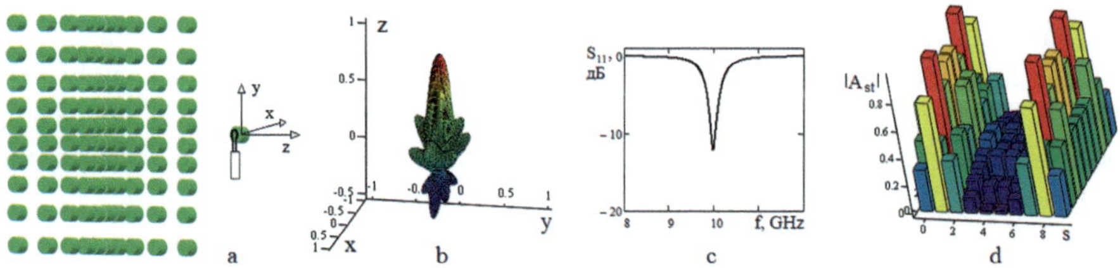

Fig. 10.9 Irregular 10 × 10 DR (**a**) lattice fed by metal loop resonator. Array radiation pattern (**b**). Frequency dependence of the reflection coefficient (**c**). Normalized amplitude magnitudes of partial resonators (**d**), Resonator parameters: $\varepsilon_{1r} = 81$; $Q^D = 10^3$; $\Delta = 0.4$

Numerical optimization of the antenna radiation parameters makes it possible to obtain the effect of unidirectional radiation, however, as can be seen from Fig. 10.9b, the lattice does not completely "reflect" the incident waves of the active DR. Part of the power is radiated by the lattice in the opposite, as well as another directions. At the same time, the level of parasitic petals also increases, which is explained by a much more complex distribution of amplitudes of partial resonators, which is established when they are aperiodicly positioned relative to each other (Fig. 10.9d).

In some cases, it is more convenient to use arrays of rotating resonators (Fig. 10.10a). Coordinated rotation of the DR axes also makes it possible to form the main lobe of the directional pattern, orthogonal to the lattice plane (Fig. 10.10b). However, as follows from Fig. 10.10b, the radiation characteristics of these structures are unsatisfactory also due to the uneven distribution of the amplitudes of forced oscillations of partial resonators (Fig. 10.10d).

In Fig. 10.11a shows a simpler version of constructing a DR antenna array, built on a ring structure [5]. The annular lattice of the DR performs the functions of a director that serves to form the main radiation lobe. Optimization of its parameters was carried out by selecting the distance between the active DR and the plane of the ring structure, as well as the radius of the ring. The frequency dependence of the reflection coefficient is shown in Fig. 10.10c, and amplitude modules of the resonators in Fig. 10.10d. We see that, as in the previous example, part of the radiation is reflected

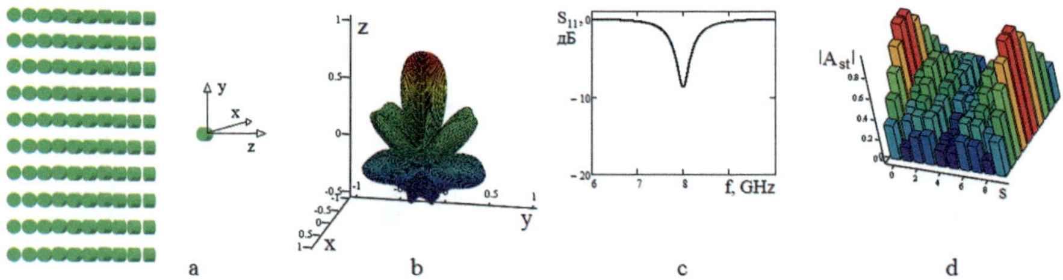

Fig. 10.10 Lattice of 10 × 10 rotatable DR (**a**). Array radiation pattern (**b**). Frequency dependence of reflection coefficient (**c**). Normalized amplitude magnitudes of partial resonators (**d**), Resonator parameters: $\varepsilon_{1r} = 36$; $Q^D = 10^3$; $\Delta = 0.44$

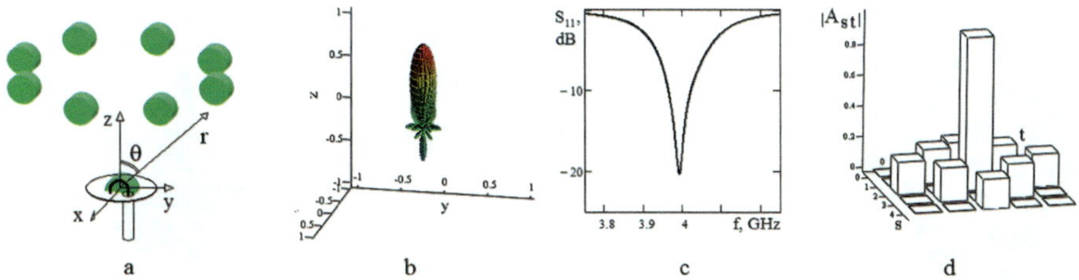

Fig. 10.11 8 DR (**a**) ring lattice fed by a metal loop resonator. Radiation pattern (**b**). Frequency dependence of reflection coefficient (**c**). Normalized amplitudes of forced oscillations of partial resonators (**d**), Resonator parameters: $\varepsilon_{1r} = 36$; $Q^D = 2 \cdot 10^3$; $\Delta = 0.4$

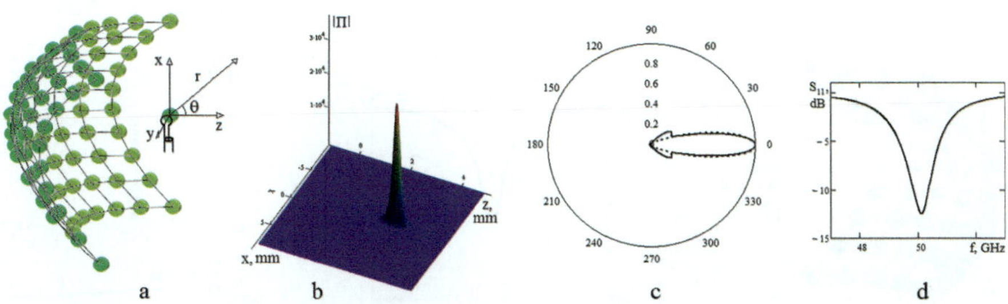

Fig. 10.12 10 × 10 DR (**a**) parabolic array fed by metal loop resonator. Spatial distribution of the Poynting vector modulus in the near zone (**b**). Radiation pattern (**c**). Frequency dependence of reflection coefficient (**d**). Resonator parameters: $\varepsilon_{1r} = 81$; $Q^D = 10^3$; $\Delta = 0.44$

by the lattice in the opposite direction; however, the level of parasitic lobes is much lower.

Parabolic DR arrays have significantly better directivity characteristics (Fig. 10.12a) [6, 7]. In this case, the field emitted by the lattice turns out to be compressed in the E and H planes not only in the far, but also in the near zone (Fig. 10.12b, c).

Parabolic DR arrays can also be built using rotating DR (Fig. 10.13a). At the same time, the directivity characteristics turn out to be somewhat worse compared to lattices of the first type, in which the coupling between partial resonators is obviously greater (Fig. 10.12a) and their radiation thanks to interference is less consistent with each other.

The directional characteristics of the arrays shown in Figs. 10.12 and 10.13a are polarization dependent. This drawback can be eliminated by using a spherical DR (Fig. 10.14a, b). On Fig. 10.14c, d show the optimized characteristics of a lattice built from 10 × 10 + 1 spherical DR with the main types of oscillations H_{1m1}. When constructing the electromagnetic model, we took into account all types of coupled oscillations of resonators H_{1m1} ($m = 0, \pm 1$). It was assumed that the active resonator is excited by a loop with a circularly polarized field. Here we also did not take into account the loop radiation field. The obtained characteristics of this type of antenna show good prospects for their possible use in various frequency ranges.

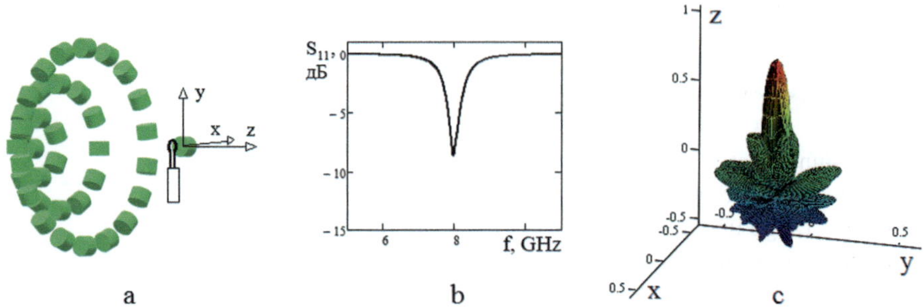

Fig. 10.13 Parabolic lattice with 38 rotatable DRs (**a**). Frequency dependence of the reflection coefficient (**b**). Array radiation pattern (**c**). Resonator parameters: $\varepsilon_{1r} = 36$; $Q^D = 10^3$; $\Delta = 0.4$

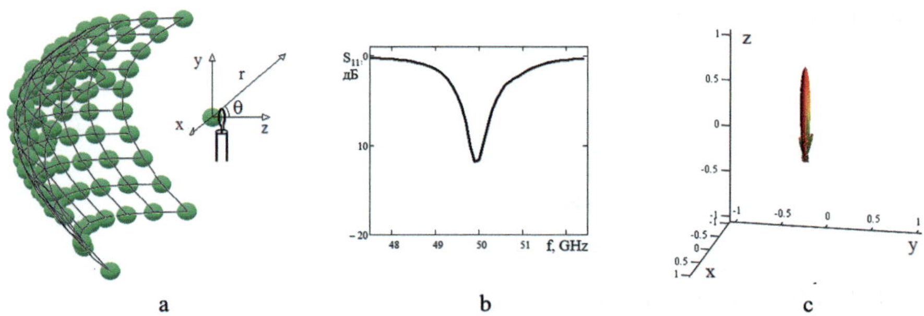

Fig. 10.14 Parabolic array of 100 + 1 Spherical DRs excited by a metal loop (**a**, **b**). Frequency dependence of reflection coefficient (**b**). Grating radiation pattern (**c**). Resonator parameters $\varepsilon_{1r} = 36$; $Q^D = 10^3$; $f_0 = 50\,GHz$; distance between the centers of neighboring resonators: $d = \lambda_0/2$

10.5 Combined Antennas-Filters Based on Coupled Dielectric Resonators

When designing antennas, several practical problems must be solved simultaneously: the necessary radiation and loss characteristics are implemented; stability of parameters in a given frequency band; the level of cross-polarization is minimized; spurious bandwidths are eliminated; electromagnetic compatibility with other systems are established. Often, these tasks cannot be solved in a single device; in this case, you have to use several consecutively connected elements, each of which performs specified functions, for example, perform frequency filtering, in the input channel. The coupling of several different elements of the input paths leads to an increase in losses, which is an undesirable phenomenon; therefore attempts are being made to combine the functions of the antenna and the input filter in one device.

The purpose of this section is to study a new type of microwave devices built on dielectric resonators, combining the functions of antennas and filters. Simultaneous utilization of several devices on the DRs in such antennas-filters (AF) allows not only reducing losses, dimensions and cost, but also eliminates the matching problems and generally simplifies tuning.

From the experience of previous studies, the structures shown in Fig. 10.1c have the best efficiency that meet the requirements of the task. Let us consider the general solution of the scattering problem on the DRs structure antenna and filter combining [8].

Suppose we have a more complex system of $N+M$ DRs, part of which is located in the segment of evanescent metal waveguide (M), and a part of it (N) is located in open space (Fig. 10.1c). The DRs $1, 2, \ldots, M$ plays the role of a bandpass filter, and the resonators $1, 2, \ldots, N$ plays the role of the antenna system. M-th DR is common to both subsystems. Let a wave $(\mathbf{E}^+, \mathbf{H}^+)$ falls on $1'$ DR of the system $1', 2', \ldots, M$ via regular waveguide.

As in the previous Sects. (8.5) we are looking for the solution of the scattering problem on the coupling DR structure:

$$\mathbf{E} \approx \mathbf{E}^+ + \sum_{s=1}^{N+M} a^s \mathbf{e}^s; \ \mathbf{H} \approx \mathbf{H}^+ + \sum_{s=1}^{N+M} a^s \mathbf{h}^s, \tag{10.6}$$

where a^s are the unknown amplitudes ($s = 1, 2, \ldots, N+M$) and $(\mathbf{e}^s, \mathbf{h}^s)$ are the sth eigenoscillation field of the coupled DR system.

The eigenoscillation field of such system we also (7.4) represented as a superposition of fields of the isolated resonators $(\mathbf{e}_u, \mathbf{h}_u)$:

$$\mathbf{e}^s = \sum_{n=1}^{N+M} b_n^s \mathbf{e}_n; \quad \mathbf{h}^s = \sum_{n=1}^{N+M} b_n^s \mathbf{h}_n. \tag{10.7}$$

The matrix $B = \left\| b_n^s \right\|$ of the amplitudes of coupled DR oscillations (7.11) should satisfy the equation system (7.6) and must be an eigenvector of $(N+M) \times (N+M)$ coupling operator K (7.8), built taking into account the coupling of all partial resonators of the filter and antenna structure.

For example, for the AFs, shown in Figs. 10.15 and 10.16a, which is a section of a evanescent waveguide connected to a regular waveguide, on the one hand, and an aperture in a metal diaphragm, on the other:

$$K = \left\| 0.5i(\tilde{k}_1 \delta_{s1} + \tilde{k}_{OS} \sum_{t=M}^{M+N} \delta_{st})\delta_{sn} + \kappa_{sn}(1 - \delta_{sn}) \right\| \tag{10.8}$$

where $\tilde{k}_1$—is the coupling coefficient of the 1-th DR with the regular waveguide; $\tilde{k}_{OS}$ is the coupling coefficient of the $M, \ldots, (M+N)$ DR with open space; $\kappa_{sn} = k_{sn}$—is the mutual coupling coefficient of the s-th and t-th resonators in the evanescent waveguide if $s < M$ and $n < M$ and κ_{sn} is the mutual coupling coefficient of split-cylindrical DRs in the open space if $s \geq M$ and $n > M$. We assumed also that the array semi-resonators (Figs. 10.15 and 10.16a) are coupled only between themselves

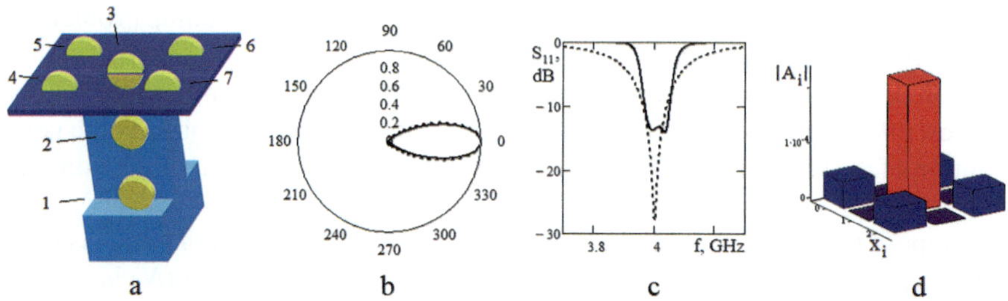

Fig. 10.15 Combined antenna-filter with 7 cylindrical DRs (**a**). Array radiation pattern (**b**). Frequency dependence of the reflection coefficient (**c**) of an antenna with a filter—solid curve; antennas without filter—dotted curve. Normalized amplitudes of partial antenna resonators (**d**). Resonator parameters: $\varepsilon_{1r} = 36$; $Q^D = 2 \cdot 10^3$; $\Delta = 0.25$. Regular waveguide cross section: $a \times b = 58 \times 25$ mm; of the cut-off waveguide: $a_z \times b_z = 33 \times 25$ mm

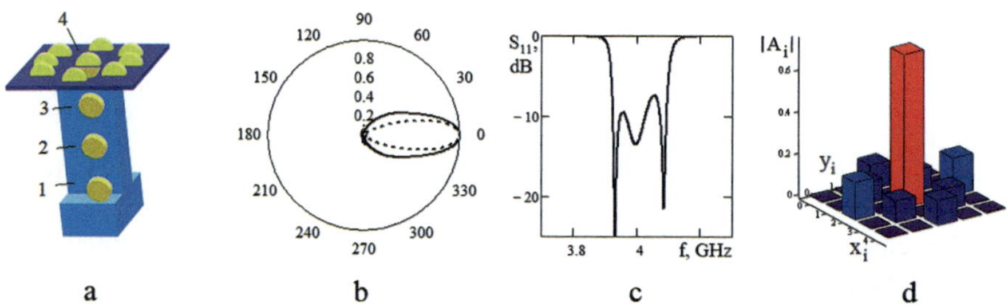

Fig. 10.16 Antenna-filter for 12 cylindrical DR (**a**). Radiation pattern (**b**) in the H-plane—solid curve; in the E-plane—dotted curve. Frequency dependence of reflection coefficient (**c**). Normalized magnitudes of amplitudes of partial antenna resonators (**d**). Resonator parameters: $\varepsilon_{1r} = 36$; $Q^D = 2 \cdot 10^3$; $\Delta = 0.2$. Regular waveguide cross section: $a \times b = 58 \times 25$ mm; cut-off waveguide cross section: $a_z \times b_z = 30 \times 25$ mm

and common resonator M. M-th DR coupled with the evanescent waveguide through the aperture in the metal screen.

The found eigenvalues λ and eigenvectors of the matrix K:

$$B = \begin{bmatrix} b_1^1 & b_1^2 & \cdots & b_1^{M+N} \\ b_2^1 & b_2^2 & \cdots & b_2^{M+N} \\ \vdots & \vdots & \cdots & \vdots \\ b_{M+N}^1 & b_{M+N}^2 & \cdots & b_{M+N}^{M+N} \end{bmatrix} \tag{10.9}$$

were used to solve the scattering problem on a system of coupled DRs.

The reflection coefficient was determined from the ratios:

$$R(\omega) = R_0 - \frac{Q^D}{\det B} \cdot \sum_{s=1}^{M+N} \frac{\det B_1^s}{Q_s(\omega)}, \tag{10.10}$$

where R_0 is the reflection coefficient of the structure without DRs;

$$B_1^s = \begin{bmatrix} b_1^1 & \dots b_1^s \tilde{k}_{11}^{-+} & \dots b_1^{M+N} \\ b_2^1 & \dots \quad 0 & \dots b_2^{M+N} \\ \vdots & \dots \quad \vdots & \dots \quad \vdots \\ b_{M+N}^1 & \dots \quad 0 & \dots b_{M+N}^{M+N} \end{bmatrix}, \tag{10.11}$$

$\tilde{k}_{11}^{-+} = (c_1^- c_1^{+*})/(\omega_0 w_1) = \tilde{k}_1$ is the coupling coefficient of the 1 DR; $Q_s(\omega) = \omega/\omega_0 + 2iQ^D(\omega/\omega_0 - 1 - \lambda_s/2)$; Q^D—is the dielectric Q-factor of the resonators; $\omega_0 = 2\pi f_0; f_0$—is the oscillation frequency of isolated DR.

The field of the antenna in the wave zone was represented in the form:

$$\begin{bmatrix} \mathbf{E} \\ \mathbf{H} \end{bmatrix} = \sum_{t=M}^{M+N} A_t \begin{bmatrix} \mathbf{e}_t^\infty \\ \mathbf{h}_t^\infty \end{bmatrix}. \tag{10.12}$$

Here $A_t = \sum\limits_{s=1}^{M+N} a^s b_t^s$—is the amplitude of the t-th DR at scattering; $(\mathbf{e}_t^\infty, \mathbf{h}_t^\infty)$—is the field of t-th isolated DR in the wave zone (10.1).

On Figs. 10.15 and 10.16, the results of AF parameter modeling are presented. The obtained data demonstrates an improvement in the squareness of the reflectance Figs. 10.15 and 10.16c, in this case, it is clear that the distribution of the radiation field in the wave zone, as well as the distribution of amplitudes of the DR antenna part of the structure does not change significantly (Figs. 10.15 and 10.16b, d).

10.6 Antennas-Filters Based on Different Dielectric Resonators

The use of different DR in one device often improves its scattering characteristics. In Figs. 10.17 and 10.18 shows the design of AF based on cylindrical DR, made of different dielectrics.

Let us examine the scattering characteristics of antenna-filters similar to those discussed in the previous Sect. (10.5), but built on the basis of the use of DR, of different shapes, or consisting of different dielectrics.

We again use expansions in coupled oscillations of the DR system, as well as the theory of scattering (Sect. 8).

Let us write a system of equations for the amplitudes of forced oscillations a^s for a more complex array of DRs of different types and topologies, shown in Fig. 10.1c:

$$\sum_{s=1}^{M+N} a^s b_t^s Q_{st}(\omega) = -(c_t^+)^* / P_t^D; \quad (s, t = 1, 2, \dots, M+N) \tag{10.13}$$

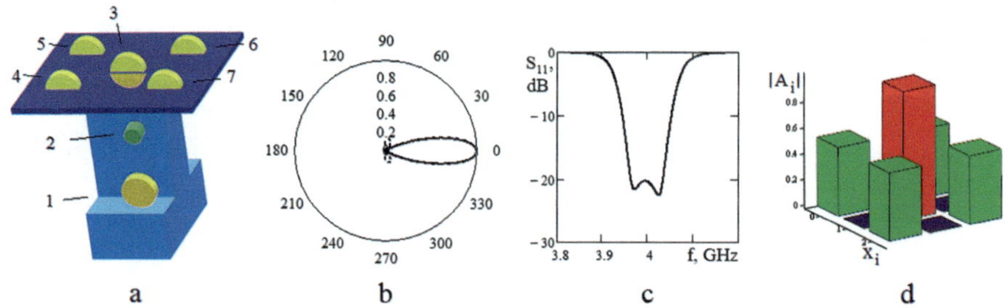

Fig. 10.17 Antenna-filter on different cylindrical DRs ($\varepsilon_{1r} = 36$; $\Delta_1 = 0, 2$; $Q_1^D = 3 \cdot 10^3$ for DRs 1, 3–7; $\varepsilon_{2r} = 82$; $\Delta_2 = 0.5$; $Q_2^D = 1 \cdot 10^3$ for DR 2). Far-field radiation pattern (**b**). Reflection coefficient as a function of the frequency (**c**). Amplitude distribution of the 3–7 DR field (**d**)

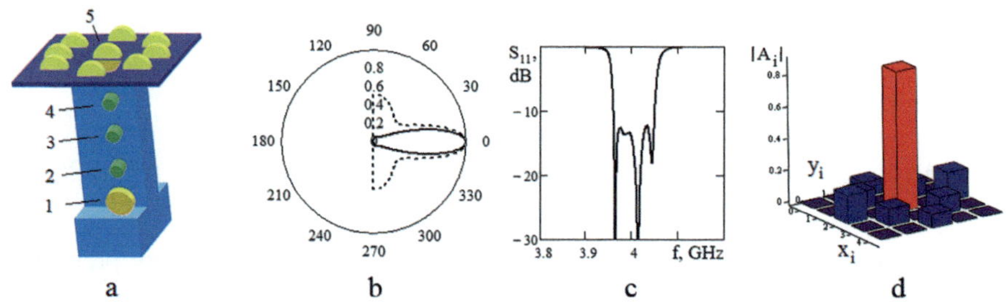

Fig. 10.18 Antenna-filter on 13 different cylindrical DRs ($\varepsilon_{1r} = 36$; $\Delta_1 = 0.44$; $Q_1^D = 3 \cdot 10^3$ for DRs 1, 5–13; $\varepsilon_{2r} = 82$; $\Delta_2 = 0.2$; $Q_2^D = 1 \cdot 10^3$ for DRs 2–4). Far-field radiation pattern (**b**). Reflection coefficient as a functions of the frequency (**c**). Amplitude distribution of the 3–7 DR field (**d**)

where b_t^s is the same amplitude of t-th resonator of s-th coupled mode of the system on the frequency $\tilde{\omega}^s$; P_t^D—is the loss power in the dielectric of the t-th DR; $\tilde{\varepsilon}_t = \varepsilon_t - i\varepsilon_t''$—is the complex dielectric permittivity of the t-th DR.

Formally, the solution of the system of Eqs. (10.13) has the form:

$$a^s(\omega) = -\det B_s(\omega)/\det B(\omega) \tag{10.14}$$

where for structures showed on Figs. 10.17 and 10.18a:

$$B_s(\omega) = \begin{bmatrix} b_1^1 Q_{11}(\omega) & \dots & (c_1^+)^*/P_1^D & \dots & b_1^{M+N} Q_{M+N,1}(\omega) \\ b_2^1 Q_{12}(\omega) & \dots & 0 & \dots & b_2^{M+N} Q_{M+N,2}(\omega) \\ \vdots & \dots & \vdots & \dots & \vdots \\ b_{M+N}^1 Q_{1,M+N}(\omega) & \dots & 0 & \dots & b_{M+N}^{M+N} Q_{M+N,M+N}(\omega) \end{bmatrix}; \tag{10.15}$$

$$B(\omega) = \begin{bmatrix} b_1^1 Q_{11}(\omega) & \dots & b_1^{M+N} Q_{M+N,1}(\omega) \\ b_2^1 Q_{12}(\omega) & \dots & b_2^{M+N} Q_{M+N,2}(\omega) \\ \vdots & \vdots & \vdots \\ b_{M+N}^1 Q_{1,M+N}(\omega) & \dots & b_{M+N}^{M+N} Q_{M+N,M+N}(\omega) \end{bmatrix};$$

and $Q_{st}(\omega)$ are the functions (8.29).

The reflection coefficient R of different DR system can be obtained from (10.14) and (10.15):

$$R(\omega) = R_0 - c_1^- \sum_{s=1}^{M+N} a^s(\omega) b_1^s, \tag{10.16}$$

where R_0 is also the reflection coefficient of the transmission line without DRs.

The field of the antenna can be represented in the form (10.12), where

$$A_t(\omega) = \sum_{s=1}^{M+N} a^s(\omega) b_t^s. \tag{10.17}$$

The radiation characteristics of the AF, calculated on basis of model (10.13–10.17), are shown in Figs. 10.17 and 10.18b–d. Comparison with the results obtained earlier (Figs. 10.15 and 10.16) showed that the radiation characteristics of the AF do not change, or change only slightly compared with antennas made only with the help of similar structures of dielectric resonators. The dependences of the reflection coefficient at the input of devices are in Figs. 10.15 and 10.16c show a increased squareness for the expansion of the frequency band. Here $S_{11}(\omega) = 20\lg|R(\omega)|$; $R(\omega)$—is the reflection coefficient (10.16). The distribution of the amplitude modules of the resonators in open space is shown in Figs. 10.17 and 10.18d. For an array of 7 DR, the amplitudes of forced oscillations are distributed more uniformly, which results in a more compressed radiation characteristic (Fig. 10.18d).

References

1. A.A. Trubin, Resonant dielectric antennas. In: 14th Int. Crimean conference "Microwave technology and telecommunication technologies (2004), pp. 366–367
2. A.A. Trubin, Microwave antenna based on a linear array of coupled dielectric resonators in a waveguide. Visnyk NTUU KPI Ser. Radiotekh. **38**, 78–83 (2009)
3. A.A. Trubin, G.S. Shmyglyuk, Modeling of antenna array parameters on cylindrical dielectric resonators. Visnyk NTUU KPI Ser. Radiotekh. **33**, 101–108 (2006)
4. A.A. Trubin, V.I. Novikov, Planar antenna for the WI-FI system on dielectric resonators. Visnyk NTUU KPI Ser. Radiotekh. **35**, 81–83 (2007)
5. A.A. Trubin, Microwave antennas based on ring arrays of dielectric resonators. Bull. NTUU KPI. Ser. Radiotekh. **44**, 77–82 (2011)

6. A.A. Trubin, Multi-element antennas based on parabolic arrays of dielectric resonators. Bull. NTUU KPI Ser. Radiotekh. **60**, 59–68 (2015)
7. V.I. Fesenko, A.S. Kupriianov, A. Sayanskiy, V.I. Sherbinin, A.A. Trubin, V.R. Tuz, Approach to analysis of all-dielectric free-form antenna systems. Opt. Exp. **27**(16), 22363–22374 (2019)
8. A.A. Trubin, Study of integrated antennas-filters on dielectric resonators. Bull. NTUU KPI. Ser. Radiotekh. **77**, 36–41 (2019)

Chapter 11
Scattering of Electromagnetic Pulses on Systems of Coupled Dielectric Resonators

Based on perturbation approach, the theory of scattering of electromagnetic pulses by DR systems is being developed. Using a number of assumptions about the pulse parameters, a system of differential equations is derived for the amplitudes of coupled oscillations in the time domain. Various special cases of solving the resulting system of equations are given, which coincide with previously obtained results. The scattering of pulses with different envelopes on DR systems located in a regular and also a section of an evanescent waveguide is considered. The patterns of changes in pulse envelopes during scattering by optical filters and DR lattices in open space, as well as the patterns of pulse generation by antenna structures made on the basis of DR are studied.

11.1 Scattering of Electromagnetic Pulses on Systems Dielectric Resonators

We consider the problem of scattering of an electromagnetic wave $\left(\mathbf{E}^{+}(\mathbf{r},t),\mathbf{H}^{+}(\mathbf{r},t)\right)$ on a system of coupled DRs. It is assume that the dependence on time is known to us. Let us represent the field of the incident pulse in the form:

$$\begin{pmatrix} \mathbf{E}^{+}(\mathbf{r},t) \\ \mathbf{H}^{+}(\mathbf{r},t) \end{pmatrix} = \begin{pmatrix} \mathcal{E}^{+}(\mathbf{r},t) \\ \mathcal{H}^{+}(\mathbf{r},t) \end{pmatrix} e^{i\Omega t}. \tag{11.1}$$

Here Ω—pulse carrier frequency. Let us assume that $(\mathcal{E}^{+}(\mathbf{r},t),\mathcal{H}^{+}(\mathbf{r},t))$ slowly varying functions of the pulse amplitude envelope in the interval $(t_0, t_0 + 2\pi/\Omega)$, those for each component of the impulse field the following inequalities are satisfied:

$$\left|d\mathcal{E}^{+}(\mathbf{r},t)/dt\right|/\left|\mathcal{E}^{+}(\mathbf{r},t)\right| \ll \Omega; \quad \left|d\mathcal{H}^{+}(\mathbf{r},t)/dt\right|/\left|\mathcal{H}^{+}(\mathbf{r},t)\right| \ll \Omega.$$

A. Trubin, *Introduction to the Theory of Dielectric Resonators*, Springer Series in Advanced Microelectronics 65, https://doi.org/10.1007/978-3-031-65396-4_11

In this case it can be shown that:

$$\frac{d\mathbf{E}^{+}}{dt} \approx i\Omega\mathbf{E}^{+}; \quad \frac{d\mathbf{H}^{+}}{dt} \approx i\Omega\mathbf{H}^{+}. \tag{11.2}$$

We are looking for the solution of the scattering problem in the form [4]:

$$\mathbf{E}(t) \approx \mathbf{E}^{+}(\mathrm{t}) + \sum_{s=1}^{N} a^{s}(t)\mathbf{e}^{s}; \quad \mathbf{H}(\mathrm{t}) \approx \mathbf{H}^{+}(\mathrm{t}) + \sum_{s=1}^{N} a^{s}(t)\mathbf{h}^{s}, \tag{11.3}$$

where $a^{s}(t)$ are the unknown time functions ($s = 1, 2, \ldots, N$) and $(\mathbf{e}^{s}, \mathbf{h}^{s})$ are the s-th eigenoscillation field of the N-DR system (Sect. 7.1). The eigenoscillation field of the N-DR system we also represented as a superposition of fields of the isolated resonators:

$$\mathbf{e}^{s} = \sum_{u=1}^{N} b_{u}^{s}\mathbf{e}_{u}; \quad \mathbf{h}^{s} = \sum_{u=1}^{N} b_{u}^{s}\mathbf{h}_{u}. \tag{11.4}$$

The DR amplitudes of coupled oscillations b_{u}^{s} should satisfy the equation system (7.6) but do not depend on time.

In this case, the amplitude of the partial resonator depends on time according to the law:

$$A_{u}(t) = \sum_{s=1}^{N} b_{u}^{s} a^{s}(t), \tag{11.5}$$

and the expansion of the scattering field takes the form:

$$\mathbf{E}(t) \approx \mathbf{E}^{+}(\mathrm{t}) + \sum_{u=1}^{N} A_{u}(t)\mathbf{e}_{u}; \quad \mathbf{H}(\mathrm{t}) \approx \mathbf{H}^{+}(\mathrm{t}) + \sum_{u=1}^{N} A_{u}(t)\mathbf{h}_{u}, \tag{11.6}$$

Consider the system of N different DRs, possibly consisting of different materials. We assume also that the fields of such different DRs also satisfy approximately the orthogonality conditions:

$$\int_{v_n} \mathbf{e}_{n}\mathbf{e}_{m}^{*}\, dv \approx \delta_{nm} \int_{v_n} |\mathbf{e}_{n}|^{2} dv; \quad \int_{v_n} \mathbf{h}_{n}\mathbf{h}_{m}^{*}\, dv \approx \delta_{nm} \int_{v_n} |\mathbf{h}_{n}|^{2} dv. \tag{11.7}$$

Here $*$—is the complex conjugation symbol; δ_{nm}—Kronecker symbol; v_{n}—is the volume of n-th DR.

Let us take into account that effective scattering of a pulse by a DR system is possible only when its frequency spectrum intersects with the set of frequencies of coupled oscillations of the resonator system. In the general case, there may be several short pulses, similar frequency sets occupied by coupled oscillations; below, for simplicity, we will limit ourselves to only one.

Thus let us assume that the frequency of the carrier pulse Ω differs slightly from the frequencies of the coupled oscillations of the resonator system $\omega^n : |\Omega - \omega^n|/\omega_0 \ll 1$; where $\omega_0 = \mathrm{Re}[\omega_s]$ $(n, s = 1, 2, \ldots, N)$ is the real part of frequencies of isolated partial resonators. Using (C.4) for the eigenoscillations of isolated and coupled DRs and the scattering field (11.6), as well as the orthogonality conditions (11.7), after integrating over the volume of each resonator, we find a system of differential equations for the amplitudes $a^s(t)$:

$$\sum_{s=1}^{N}\left[(2Q_u^D - i)\frac{d}{dt}a^s(t) - 2i\omega^s Q_u^D a^s(t)\right]b_u^s = -\frac{\omega_0}{P_u^D}\left(c_u^+(t)\right)^*. \quad (u = 1, 2, \ldots, N) \tag{11.8}$$

where $Q_u^D = \omega_0 w_u / P_u^D$—is the Q-factor of loss in dielectric of u-th DR; $w_u = 1/4 \int\limits_{v_u} (\varepsilon_u|\mathbf{e}_u|^2 + \mu_0|\mathbf{h}_u|^2)dv$—is the energy, stored in the dielectric of the u-th DR; $\varepsilon_u = \mathrm{Re}\tilde{\varepsilon}_u$; $\tilde{\varepsilon}_u = \varepsilon_u - i\varepsilon_u''$—is the complex dielectric permittivity and the μ_0 is the permeability of the u-th DR. $P_u^D = \omega_0 \frac{\varepsilon_u''}{2} \int\limits_{v_u} |\mathbf{e}_u|^2 dv$—is the loss power in the dielectric.

Here also

$$c_u^+(t) = -\frac{1}{2}\oint\limits_{s_u}\left\{[\mathbf{e}_u, \mathbf{n}]\mathbf{H}^+(\mathbf{r}, t)^* + [\mathbf{n}, \mathbf{h}_u]\mathbf{E}^+(\mathbf{r}, t)^*\right\}\mathrm{ds}; \tag{11.9}$$

is the integral over the surface s_u of u-th DR; $\mathbf{n}$—is the normal to the DR surface.

The solution of the system of Eqs. (11.8) together with (11.5) and (11.6) in the most general form determines the distribution of the field of the DR system when the time pulses of the field fall on them.

The solution of the equations system (11.8) can be found in an asymptotic form, if we take into account the presence of small parameters $\delta_v = (2Q_v^D)^{-1} \ll 1$ $(v = 1, 2, \ldots, N)$ and also matrix representation similar to (8.14) :

$$a^s(t) \approx a_0^s(t) - e^{i\omega^s t}\sum_{v=1}^{N}\delta_v \frac{\det B_v^s}{\det B}\sum_{u=1}^{N}\omega^u b_v^u \int\limits_{-\infty}^{t} e^{-i\omega^s \tau} a_0^s(\tau)\,d\tau + O(\delta_v^2), \tag{11.10}$$

where

$$a_0^s(t) = -\frac{1}{2}e^{i\omega^s t}\int\limits_{-\infty}^{t} e^{-i\omega^s \tau}\frac{\det C_s(\tau)}{\det B}\,d\tau \tag{11.11}$$

is the principal term of the expansion and

$$B = \begin{bmatrix} b_1^1 & b_1^2 & \cdots & b_1^N \\ b_2^1 & b_2^2 & \cdots & b_2^N \\ \vdots & \vdots & \cdots & \vdots \\ b_N^1 & b_N^2 & \cdots & b_N^N \end{bmatrix}; \quad B_v^s = \begin{bmatrix} b_1^1 & \cdots & 0 & \cdots & b_1^N \\ \vdots & \vdots & 0 & \vdots & \vdots \\ 0 & 0 & 1 & 0 & 0 \\ \vdots & \vdots & 0 & \vdots & \vdots \\ b_N^1 & \cdots & 0 & \cdots & b_N^N \end{bmatrix}$$

is the minor of matrix B in which s is the column and v is the row number;

$$C_s(\mathrm{t}) = \begin{bmatrix} b_1^1 & \cdots & c_1^+(t)^*/w_1 & \cdots & b_1^N \\ b_2^1 & \cdots & c_2^+(t)^*/w_2 & \cdots & b_2^N \\ \vdots & \cdots & \vdots & \cdots & \vdots \\ b_N^1 & \cdots & c_N^+(t)^*/w_N & \cdots & b_N^N \end{bmatrix}. \tag{11.12}$$

The delay of the incident pulse on DR is determined by the corresponding phase factor in the expansion coefficients $c_u^+(t)^*$. The choice of arbitrary constants in the solution of the Eq. (11.8) is determined by the physical conditions: $a^s(t) = 0$, if $\left(\mathbf{E}^+(t), \mathbf{H}^+(t)\right) = 0$.

In the particular case of identical DR and types of oscillations: $Q_u^D = Q^D$; $P_u^D = P^D$: the system (11.8) may be solved with respect to differential equations:

$$(2Q^D - i)\frac{d}{dt}a^s(t) - 2i\omega^s Q^D a^s(t) = -Q^D \frac{\det C_s(t)}{\det B}. \tag{11.13}$$

The solution of the (11.13), taking into account the linearity of the determinant $C_s(t)$, is [6, 7]:

$$a^s(t) = -\frac{Q^D}{2Q^D - i} e^{\frac{2Q^D}{2Q^D - i} i\omega^s t} \int\limits_{-\infty}^{t} e^{-\frac{2Q^D}{2Q^D - i} i\omega^s \tau} \frac{\det C_s(\tau)}{\det B} d\tau \approx a_0^s(t). \tag{11.14}$$

Thus, we can conclude that the amplitudes of coupled oscillations of identical DRs formally coincide with the amplitudes of resonators of different types found in the zero approximation under the conditions $\delta_v \ll 1$ (11.10).

Solutions (11.10); (11.14) and (11.6) allow us to directly calculate the envelopes of scattered pulses.

If the resonators are excited by a harmonic signal with a frequency $\omega : d/dt[a^s(t)] = i\omega a^s(t)$, the system of Eqs. (11.8) becomes known (8.13).

11.2 Scattering of Electromagnetic Pulses by Coupled Dielectric Resonators in a Waveguide

Let us assume in (11.1), that in the line, the incident pulse may be represented in the form:

$$\begin{pmatrix} \mathbf{E}^{+}(\mathbf{r}, t) \\ \mathbf{H}^{+}(\mathbf{r}, t) \end{pmatrix} \approx A_{in}(t) \begin{pmatrix} \boldsymbol{\mathcal{E}}^{+}(\mathbf{r}) \\ \boldsymbol{\mathcal{H}}^{+}(\mathbf{r}) \end{pmatrix} e^{i\Omega t}; \tag{11.15}$$

Then, from (11.9):

$$c_s^{+}(t) = A_{in}(t)^{*} e^{-i\Omega t} \cdot c_s^{+}, \tag{11.16}$$

where as in the (2.4)

$$c_s^{+} = -\frac{1}{2} \oint\limits_{s_s} \left\{ [\mathbf{e}_s, \mathbf{n}] \boldsymbol{\mathcal{H}}^{+}(\mathbf{r})^{*} + [\mathbf{n}, \mathbf{h}_s] \boldsymbol{\mathcal{E}}^{+}(\mathbf{r})^{*} \right\} \mathrm{ds}.$$

Scattered pulses at a sufficiently large distance from the resonators in the transmission line can also be represented in the form:

$$\begin{pmatrix} \mathbf{E}_{out}^{\pm}(\vec{r}, t) \\ \mathbf{H}_{out}^{\pm}(\vec{r}, t) \end{pmatrix} \approx A_{out}^{\pm}(t) e^{i\Omega t} \begin{pmatrix} \boldsymbol{\mathcal{E}}^{\pm}(\mathbf{r}) \\ \boldsymbol{\mathcal{H}}^{\pm}(\mathbf{r}) \end{pmatrix}. \tag{11.17}$$

And from (11.3)

$$\begin{pmatrix} \mathbf{E}(\mathbf{r}, t) \\ \mathbf{H}(\mathbf{r}, t) \end{pmatrix} \approx [A_{in}(t) \begin{pmatrix} \boldsymbol{\mathcal{E}}^{+}(\mathbf{r}) \\ \boldsymbol{\mathcal{H}}^{+}(\mathbf{r}) \end{pmatrix} + A_{out}(t) \begin{pmatrix} \boldsymbol{\mathcal{E}}^{\pm}(\mathbf{r}) \\ \boldsymbol{\mathcal{H}}^{\pm}(\mathbf{r}) \end{pmatrix}] e^{i\Omega t},$$

then envelopes of total scattered field can be represented for the bandstop filters:

$$A_{out}^{+}(t) \approx A_{in}(t) - \frac{\omega_0}{2} \sum_{s=1}^{N} \frac{\det B_s^{+}}{\det B} \{ e^{i(\omega^s - \Omega)t} \int\limits_{-\infty}^{t} e^{i(\Omega - \omega^s)\tau} A_{in}(\tau) d\tau \};$$

$$A_{out}^{-}(t) \approx -\frac{\omega_0}{2} \sum_{s=1}^{N} \frac{\det B_s^{-}}{\det B} \{ \mathrm{e}^{i(\omega^s - \Omega)t} \int\limits_{-\infty}^{t} \mathrm{e}^{i(\Omega - \omega^s)\tau} A_{in}(\tau) d\tau \}. \tag{11.18}$$

Here

$$B_s^{\pm} = \begin{bmatrix} b_1^1 & \cdots & \sum\limits_{u=1}^{N} b_u^s \tilde{k}_{u1}^{\pm+} & \cdots & b_1^N \\ b_2^1 & \cdots & \sum\limits_{u=1}^{N} b_u^s \tilde{k}_{u2}^{\pm+} & \cdots & b_2^N \\ \vdots & \cdots & \vdots & \cdots & \vdots \\ b_N^1 & \cdots & \sum\limits_{u=1}^{N} b_u^s \tilde{k}_{uN}^{\pm+} & \cdots & b_N^N \end{bmatrix}. \tag{11.19}$$

(s—is the column number of the matrix $B_s^{\pm}$ with the sum $\sum_{u=1}^{N} b_u^s \tilde{k}_{us}^{\pm+}$);

$\tilde{k}_{sn}^{++} = \left(c_s^+ c_n^{+*}\right)/(\omega_0 w_n) = (\tilde{k}_{sn})_0 e^{-i\Gamma(z_s - z_n)}$; $\tilde{k}_{sn}^{-+} = \left(c_s^- c_n^{+*}\right)(\omega_0 w_n) = (\tilde{k}_{sn})_0 e^{-i\Gamma(z_s + z_n)}$;

$(\tilde{k}_{sn})_0$—is the coupling coefficient for the propagating wave, expressed without phase difference accounting in the transmission line ($s, n = 1, 2, \ldots, N$).

for the bandpass filters:

$$A_{out}^{+}(t) \approx -\frac{\omega_0}{2} \sum_{s=1}^{N} \frac{\det B_s^{+}}{\det B} \left\{ \mathrm{e}^{i(\omega^s - \Omega)t} \int\limits_{-\infty}^{t} \mathrm{e}^{i(\Omega - \omega^s)\tau} A_{in}(\tau)\, d\tau \right\};$$

$$A_{out}^{-}(t) \approx -A_{in}(t) - \frac{\omega_0}{2} \sum_{s=1}^{N} \frac{\det B_s^{-}}{\det B} \left\{ e^{i(\omega^s - \Omega)t} \int\limits_{-\infty}^{t} e^{i(\Omega - \omega^s)\tau} A_{in}(\tau)\, d\tau \right\}. \tag{11.20}$$

$$B_s^{+} = \begin{bmatrix} b_1^1 & \cdots & b_N^s \tilde{k}_{N1}^{++} & \cdots & b_1^N \\ b_2^1 & \cdots & 0 & \cdots & b_2^N \\ \vdots & \cdots & \vdots & \cdots & \vdots \\ b_N^1 & \cdots & 0 & \cdots & b_N^N \end{bmatrix}; \quad B_s^{-} = \begin{bmatrix} b_1^1 & \cdots & b_1^s \tilde{k}_{11}^{-+} & \cdots & b_1^N \\ b_2^1 & \cdots & 0 & \cdots & b_2^N \\ \vdots & \cdots & \vdots & \cdots & \vdots \\ b_N^1 & \cdots & 0 & \cdots & b_N^N \end{bmatrix}. \tag{11.21}$$

As might be expected from (11.18), the envelopes of scattered pulses are linear functions of the incident pulse. The dependence of the envelopes is determined by the carrier frequency Ω difference from the coupled oscillations frequencies ω^s of the resonator system.

In some cases it is convenient to use the time Green's functions. Green's functions for different DR can be calculated if we use the approximate expansion of the scattered field (8.5) with a harmonic effect on the resonator system [8]:

$$A_{out}(t) e^{i\Omega t} = \int\limits_{-\infty}^{t} g(t - \tau) A_{in}(\tau) e^{i\Omega\tau}\, d\tau \tag{11.22}$$

Comparing (11.22) with (11.18), (11.20), we directly find expressions for the time Green's functions determined for the envelopes of pulses when they are scattered on the DR system in the line:

for the bandstop filter:

$$g^{+}(\tau) = \begin{cases} \delta(\tau) - \frac{\omega_0}{2}\sum\limits_{s=1}^{N}\frac{\det B_s^{+}}{\det B}e^{i\omega^s\tau}, & \tau \geq 0 \\ 0, \quad \tau < 0 \end{cases};$$

$$g^{-}(\tau) = \begin{cases} -\frac{\omega_0}{2}\sum\limits_{s=1}^{N}\frac{\det B_s^{-}}{\det B}e^{i\omega^s\tau}, & \tau \geq 0 \\ 0, \quad \tau < 0 \end{cases}, \tag{11.23}$$

As well as for the bandpass filter for resonators located in a section of the evanescent waveguide:

$$g^{+}(\tau) = \begin{cases} -\frac{\omega_0}{2}\sum\limits_{s=1}^{N}\frac{\det B_s^{+}}{\det B}e^{i\omega^s\tau}, & \tau \geq 0 \\ 0, \quad \tau < 0 \end{cases};$$

$$g^{-}(\tau) = \begin{cases} -\delta(\tau) - \frac{\omega_0}{2}\sum\limits_{s=1}^{N}\frac{\det B_s^{-}}{\det B}e^{i\omega^s\tau}, & \tau \geq 0 \\ 0, \quad \tau < 0 \end{cases}. \tag{11.24}$$

The resulting expressions exactly coincide with the expressions of the Green's functions [8], found by integrating the reflection and transmission coefficients in the complex plane of frequencies of coupled oscillations of the DR system, taking into account the causality condition.

Using relationships (11.18) and (11.20), obtains general analytical expressions for pulse shapes with rectangular envelop, duration $[t_1, t_2]$

$$A_{\text{in}}(t) = \theta(t - t_1) - \theta(t - t_2),$$

where $\theta(x)$—Heaviside step function.

(1) scattered by bandstop filter:

$$\begin{aligned} A_{out}^{+}(t) &= A_{in}(t) - [A^{+}(t, t_1) - A^{+}(t, t_2)]; \\ A_{out}^{-}(t) &= -[A^{-}(t, t_1) - A^{-}(t, t_2)], \end{aligned} \tag{11.25}$$

where

$$A^{\pm}(t, t_v) = \frac{i\omega_0}{2}\theta(t - t_v)\sum_{s=1}^{N}[\frac{\det B_s^{\pm}}{\det B}\cdot\frac{1 - e^{i(\omega^s - \Omega)\cdot(t - t_v)}}{\Omega - \omega^s}];$$

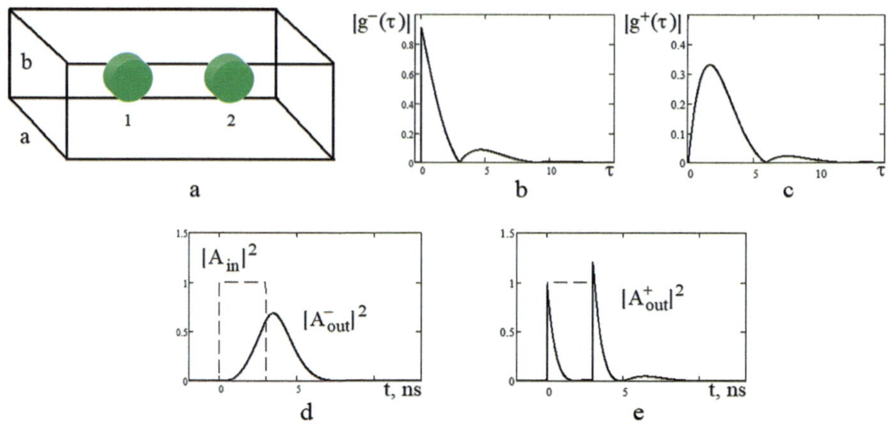

Fig. 11.1 Module of time Green's functions (**b**, **c**) of 2 coupled DRs in a rectangular waveguide (**a**) shown in Fig. 8.2a. Envelopes of the reflected (**d**), transmitted (**e**) rectangular pulse from 2 DR (**a**)

(2) scattered by bandpass filter:

$$A^{+}_{out}(t) = -[A^{+}(t, t_1) - A^{+}(t, t_2)];$$
$$A^{-}_{out}(t) = -A_{in}(t) - [A^{-}(t, t_1) - A^{-}(t, t_2)]. \qquad (11.26)$$

On Fig. 11.1 shows the results of calculating the time Green's functions (11.23) for two cylindrical DR with fundamental oscillations H_{101}, the parameters of which are shown in Fig. 8.2a. The result of calculating the envelopes of reflected and transmitted rectangular pulses is shown in Fig. 11.1d, e [1]. For comparison, the envelope of the incident pulse is shown as a dotted line; the envelope of the scattered pulse is a solid line. The shape of the scattered pulse envelopes is explained by the inertia of the rise (fall) of the resonator field under the influence of an "instantaneous" change in the field of the rectangular pulse.

An increase in the number of resonators leads to the appearance of beats (Fig. 11.2b, c, e), as well as an increase in the delay (Fig. 11.2d).

Examples of envelopes of scattered pulses on bandpass filters calculated using formulas (11.20) and (11.21) are shown in Figs. 11.3 and 11.4 for a rectangular envelope pulse by using (11.26).

For a pulse with a normal envelope distribution:

$$A_{\text{in}}(t) = \mathrm{e}^{-\frac{1+iC}{2}\cdot\left(\frac{t}{\sigma}\right)^2}; \qquad (11.27)$$

$$A^{+}_{out}(t) = -\frac{A_{\text{in}}(t)}{\sqrt{1+iC}}\sqrt{\frac{\pi}{2}\frac{\omega_0\sigma}{2}}\sum_{s=1}^{N}\frac{\det B^{+}_{s}}{\det B}\mathrm{e}^{\frac{z_s^2}{2}}\,\mathrm{erfc}\left[\frac{z_s(t)}{\sqrt{2}}\right];$$

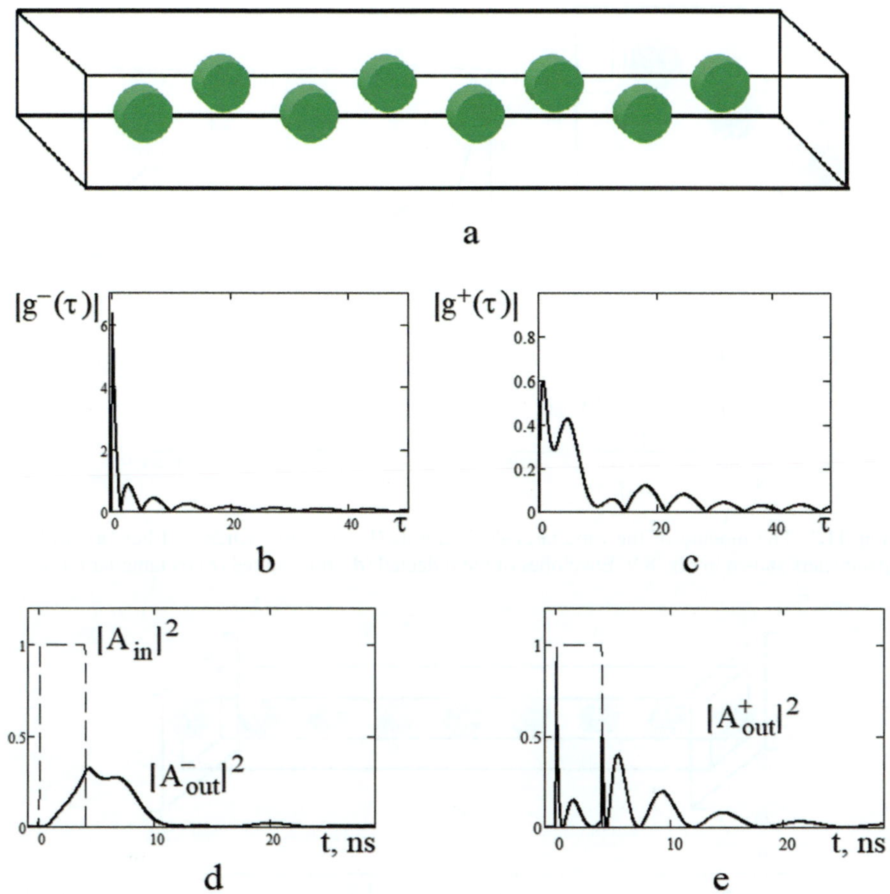

Fig. 11.2 Modulus of time Green's functions (**b**, **c**) 8 coupled DR in a rectangular waveguide (**a**) (see Fig. 8.5a). Envelopes of the reflected (**d**), transmitted (**e**) rectangular pulse from the 8 DR system (**a**)

$$A^{-}_{out}(t) = -A_{\text{in}}(t)\left\{1 - \frac{1}{\sqrt{1+iC}}\sqrt{\frac{\pi}{2}}\frac{\omega_0\sigma}{2}\sum_{s=1}^{N}\frac{\det B^{-}_{s}}{\det B}e^{\frac{z_s^2}{2}}\text{erfc}\left[\frac{z_s(t)}{\sqrt{2}}\right]\right\}$$

where $z_s(t) = -\frac{t}{\sigma_c} + i(\Omega - \omega_s)\sigma_c$, $\sigma_c = \sigma/\sqrt{1+iC}$; the constant C defines a chirp as a magnitude of frequency modulation of the pulse carrying; σ—is the half-width of the pulse envelop; *erfc*(*z*)—is the complementary error function [9].

On Fig. 11.4b, c shows the envelopes of rectangular and normal transmitted pulses scattered by a 9 resonator filter. Noteworthy is the fact that the pulse delay value weakly depends on the pulse shape, but is determined mainly by the number of resonators.

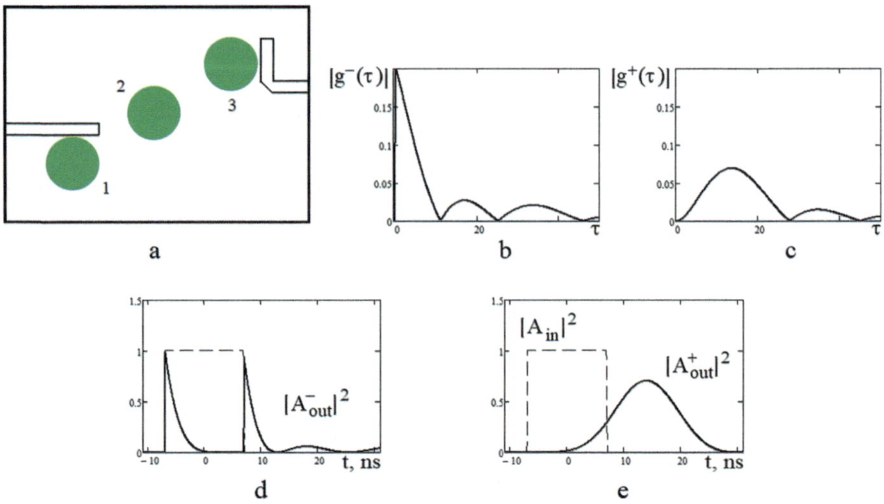

Fig. 11.3 The module of the time Green's functions (**b**, **c**) of the bandpass filter (**a**), with the parameters shown in Fig. 8.9. Envelopes of the reflected (**d**), transmitted (**e**) rectangular pulse

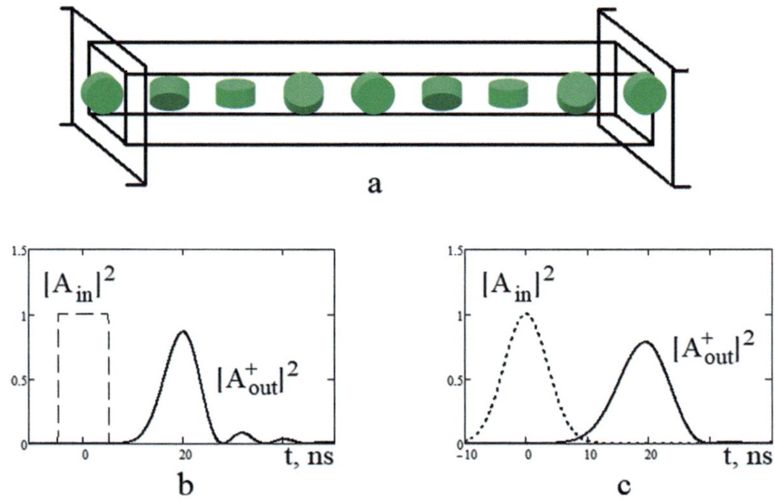

Fig. 11.4 Envelopes of transmitted pulses: rectangular (**b**), with normal amplitude distribution (**c**) when scattered on a bandpass filter by 9 rotating DR (**a**) (see Fig. 8.11)

11.3 Scattering of Electromagnetic Pulses on Add-Drop Filters

In the case of pulse scattering on a DR system with higher types of oscillations, for example, whispering gallery oscillations, the developed model can be justified only if the spatial length of the pulse envelope is significantly greater than the size of each of the resonators [see (11.6)]. At the same time, it is of interest to determine the region of parameters in which the difference between the prediction of the analytical model based on expansion (11.6) and the scattering data begins to differ most noticeably.

The simplest implementation of a filter on optical resonators with whispering gallery oscillations is band-stop filters, discussed in Sect. 8.4. On Fig. 11.5 shows the time Green's functions (b, c) [2, 3] and the scattering envelopes of a rectangular pulse (d, e) with a duration of 100 ps on a 2-cavity filter. The filter scattering characteristics are shown in Fig. 8.12. It can be seen that even for 2 resonators in the structure, beats occur, the interference of which leads to a significant distortion of the reflected pulse envelope (e).

The constructed scattering model is easily generalized to multiwire structures with DR [4, 5]. Today, similar structures are widely used to build Add-Drop filters.

Let us assume that a system of resonators is placed between two transmission lines (Figs. 11.6, 11.7 and 11.8a). We will assume that the wave scattered by the DR system propagates in a line from port 1 to port 2, and ports 3, 4 are located on another line. In this case, the general expression of the time Green's functions for Add-Drop filters is represented as:

$$g_{1\nu}(\tau) = \begin{cases} \delta(\tau)\delta_{\nu 2} - \frac{\omega_0}{2} \sum\limits_{s=M1}^{M(N)} \frac{\det B_\nu^s}{\det B} e^{i\omega_s \tau}, & \tau \geq 0 \\ 0, \quad \tau < 0 \end{cases}, \tag{11.28}$$

where $\nu = 1, 2, 3, 4$. Expressions for matrices B_ν^s, *B, as well as for M1, M(N)* are given in Sect. 8.5 for each type of filter.

Using (11.28) and (11.22) one can find the envelopes of scattered pulses at either end of line 1–4.

On Figs. 11.6, 11.7 and 11.8b–e show the results of calculating the time Green's functions for Add-Drop filters of different configurations, calculated using formulas (11.28). It was assumed that the envelopes are based on the time of arrival of the pulse at the center of the 1-st or N-th DR.

The envelopes of a rectangular pulse passing through a filter with the highest (lowest) attenuation are shown in Figs. 11.6, 11.7 and 11.8f, g. Comparing the calculation results, we see that the greatest distortion is experienced by pulses scattered Twisted double-channel SCISSOR (Fig. 11.8).

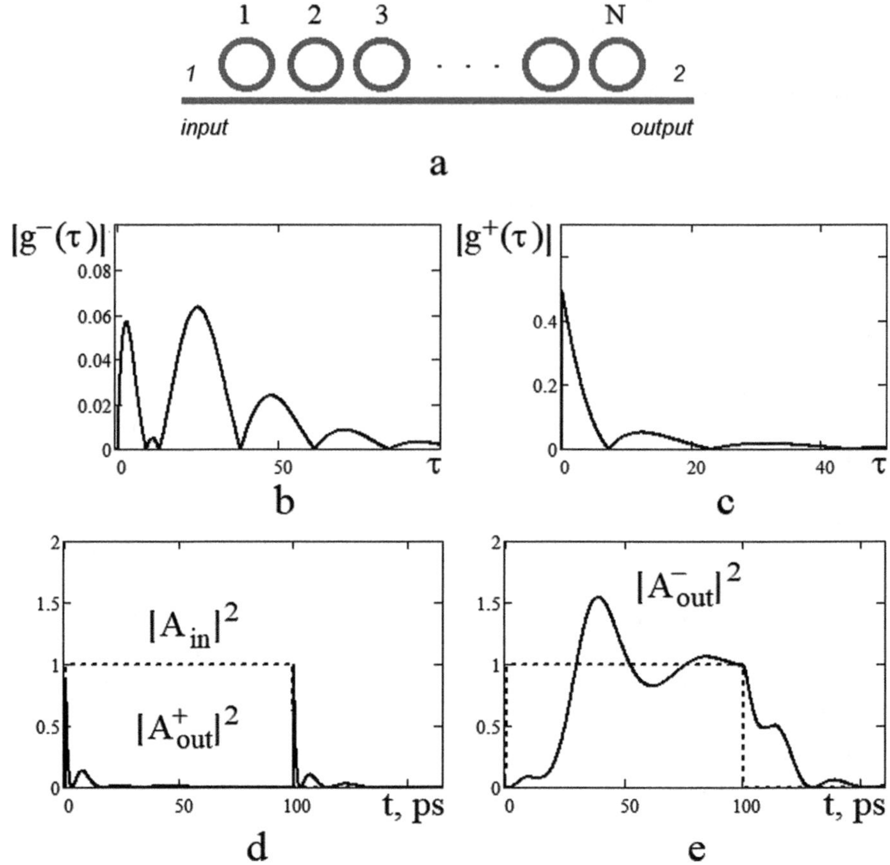

Fig. 11.5 Module of the time Green's functions (**b**, **c**) of the notch filter (**a**) on 2 ring microcavities, with the parameters shown in Fig. 8.12. Envelopes of reflected (**d**), transmitted (**e**) rectangular pulses (**d**, **e**)

11.4 Scattering of Electromagnetic Pulses on Lattices of Coupled Dielectric Resonators in Open Space

The formalization of pulse scattering on finite DR lattices in open space causes additional difficulties in the description, since the spatial distribution of the incident and scattered waves in this case belong to different classes. Incident pulses usually take the form of a superposition of plane waves, while scattered pulses are a linear combination of spherical waves. To get around this difficulty, we will consider only the pulse envelopes without taking into account changes in their actual relative amplitude and spatial distribution.

Let us considers the task of the scattering of pulses of plane waves on the DR lattices in the open space. Under fairly general assumptions, we will also seek a

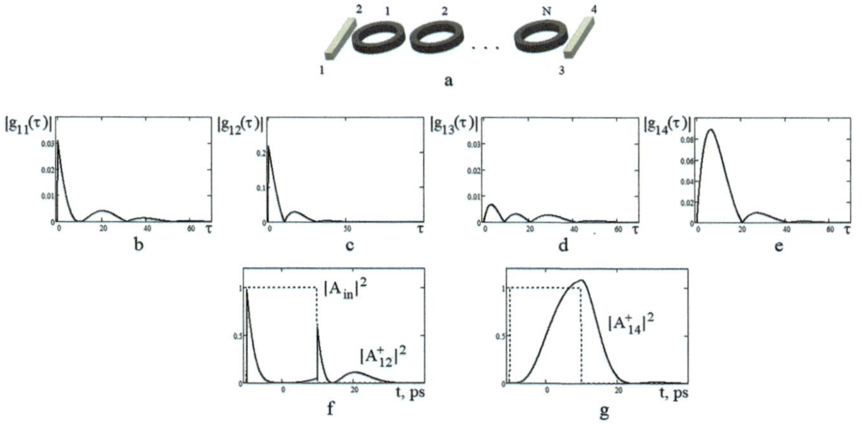

Fig. 11.6 Modules of time Green's functions (**b**, **c**) Add-Drop filter (**a**) on 2 ring microcavities, with the parameters shown in Fig. 8.14. Envelopes of a transmitted (**f**, **g**) rectangular pulse scattered by an Add-Drop filter on 2 ring microcavities

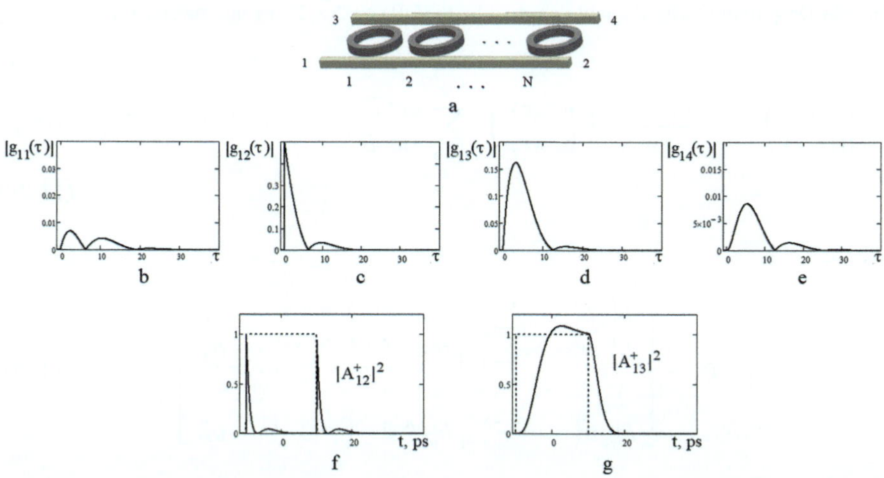

Fig. 11.7 Modules of the time Green's functions (**b**, **c**) of the filter (**a**) on 2 ring microcavities, with the parameters shown in Fig. 8.15. Envelopes of a transmitted (**f**, **g**) rectangular pulse scattered by an Add-Drop filter (Double-channel SCISSOR) on 2 ring microcavities

solution to this problem in the approximation (11.11). We find the functions c_u^+ for each resonator from (11.9). The scattered field of an array of identical DRs in the far zone is obtained from (11.6). In the case of pulse scattering with a flat spatial field distribution, we use the approximation (11.15). Then, for a lattice of identical DRs we find:

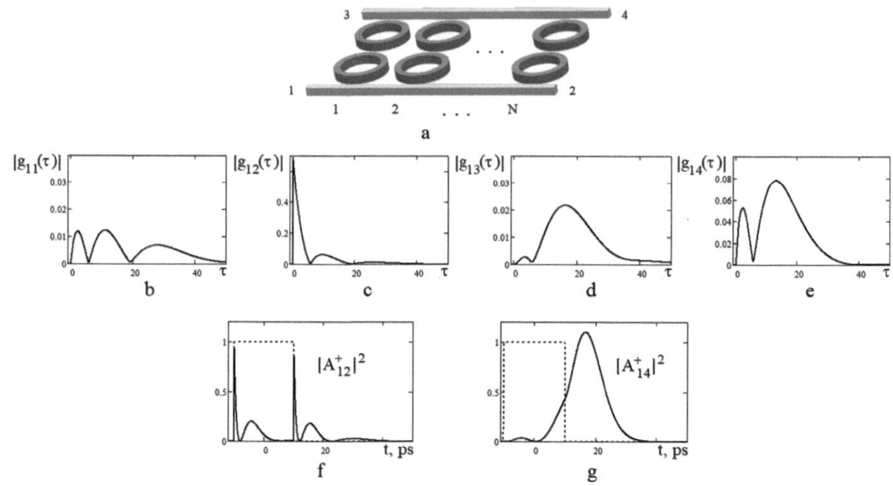

Fig. 11.8 Modules of the time Green's functions (**b–c**) of the filter (**a**) on 4 ring microcavities, with the parameters shown in Fig. 8.16. Envelopes of a transmitted (**f, g**) rectangular pulse scattered by an Add-Drop filter (Twisted double-channel SCISSOR) on 2 × 2 ring microresonators

$$\begin{pmatrix} \mathbf{E}(\mathbf{r},t) \\ \mathbf{H}(\mathbf{r},t) \end{pmatrix} = -\frac{(c_1^{+})^* e^{i\Omega t}}{2w_1} \begin{pmatrix} \mathbf{e}_1^{\infty}(\mathbf{r}) \\ \mathbf{h}_1^{\infty}(\mathbf{r}) \end{pmatrix} \cdot \sum_{s=1}^{N} \frac{\det B^s}{\det B} \{ e^{i(\omega^s - \Omega)t} \int_{-\infty}^{t} e^{i(\Omega-\omega^s)\tau} A_{in}(\tau)\, d\tau \}, \tag{11.29}$$

where

$$B^s = \begin{bmatrix} b_1^1 & b_1^2 & \cdots & \sum_{u=1}^{N} b_u^s e^{ik_0 d_{1u}(\theta_k,\varphi_k|\theta,\varphi)} & \cdots & b_1^N \\ b_2^1 & b_2^2 & \cdots & \sum_{u=1}^{N} b_u^s e^{ik_0 d_{2u}(\theta_k,\varphi_k|\theta,\varphi)} & \cdots & b_2^N \\ \vdots & \vdots & \cdots & \vdots & \cdots & . \\ b_N^1 & b_N^2 & \cdots & \sum_{u=1}^{N} b_u^s e^{ik_0 d_{Nu}(\theta_k,\varphi_k|\theta,\varphi)} & \cdots & b_N^N \end{bmatrix}. \tag{11.30}$$

Considering the appearance of the DR field in the far zone:

$$\begin{pmatrix} \mathbf{e}_1^{\infty}(\mathbf{r}) \\ \mathbf{h}_1^{\infty}(\mathbf{r}) \end{pmatrix} = \begin{pmatrix} \mathbf{e}_0(\theta,\varphi) \\ \mathbf{h}_0(\theta,\varphi) \end{pmatrix} e^{-ik_0 r}/r;$$

($r = |\mathbf{r}|$), and also (11.30), we come to the conclusion that the envelope of the scattered pulse of the lattice is determined by the directions of incidence of the plane wave (θ_k, φ_k) and the direction in which it is measured (θ, φ). The polarization of the emitted pulses is determined by the polarization of the field of partial resonators in the far zone.

Let us give the resulting expression the form (11.17),

$$\begin{pmatrix} \mathbf{E}_{out}(\mathbf{r}, t) \\ \mathbf{H}_{out}(\mathbf{r}, t) \end{pmatrix} = \begin{pmatrix} \mathbf{e}_0/r \\ \mathbf{e}_0/w_0 r \end{pmatrix} A_{out}(t) \cdot e^{i\Omega t}, \tag{11.31}$$

formally normalizing the amplitude of the scattered pulse $\mathbf{e}_0$ in such a way that the equation for the envelopes of the incident and scattered pulses (11.29) takes the form coinciding with Eqs. (11.18) for the DR system in the line:

$$A_{out}(t) \approx -\frac{\omega_0}{2} \sum_{s=1}^{N} \frac{\det B^s}{\det B} \{ e^{i(\omega^s - \Omega)t} \int_{-\infty}^{t} e^{i(\Omega - \omega^s)\tau} A_{in}(\tau)\, d\tau \}, \tag{11.32}$$

The last expression allows us to determine the time Green's function for the envelope of the scattered pulse in the form (11.24) [8]:

$$g(\tau) = \begin{cases} -\frac{\omega_0}{2} \sum\limits_{s=1}^{N} \frac{\det B^s}{\det B} e^{i\omega^s \tau}, & \tau \geq 0 \\ 0, & \tau < 0 \end{cases}; \tag{11.33}$$

On Fig. 11.9b, e shows the modules of the Green's functions for two square lattices of cylindrical DR with the main magnetic types of oscillations H_{101}. The grating parameters are shown in Fig. 9.3 (Fig. 11.9).

The envelopes of a rectangular scattered pulse were calculated in the direction of the maximum of the reflection lobe lying at the angles: $\theta = 0.25\pi$; $\varphi = \pi$.

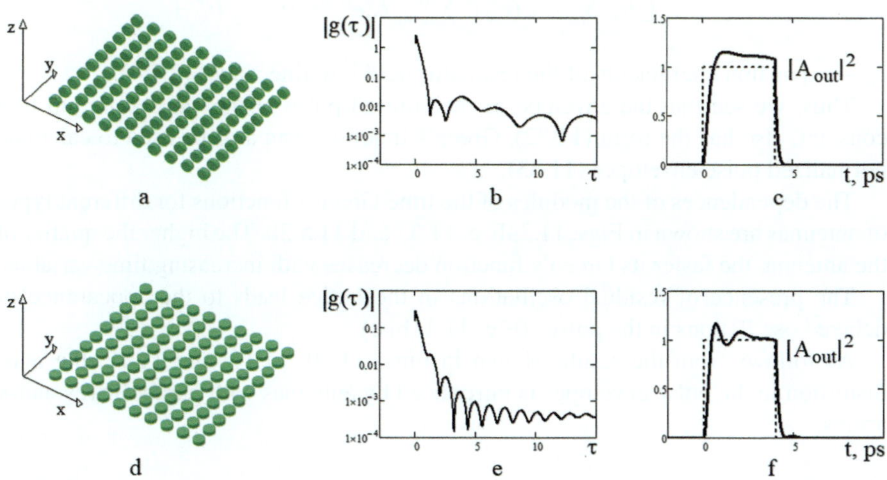

Fig. 11.9 Modules of time Green's functions (**b, d**), rectangular lattices of 10 × 10 cylindrical DR in open space (see Fig. 9.3). Envelopes of rectangular pulses reflected in the direction of the pattern maximum

11.5 Radiation of Electromagnetic Pulses by Antennas on Lattices of Coupled Dielectric Resonators

Theoretically correct engineered antennas would not inserted visibly distortions in the process of pulses passing and receiving. The antenna parameters would carry out directly, over patching different pulses shapes at their emission. For solution of this task, considers the problem of pulses propagation from side of the feeding transmission line.

Usually the antennas of DRs lattices connected on input with feeding transmission line and at that radiates it's in the open space. In the particular case when the pulse exciting each of the resonators has the same envelope [see (11.16)], the solution for the antenna far-field, consisting of identical DRs, can be given the form:

$$\begin{pmatrix} \mathbf{E}(\mathbf{r},t) \\ \mathbf{H}(\mathbf{r},t) \end{pmatrix} = -\frac{e^{i\Omega t}}{2w_1} \begin{pmatrix} \mathbf{e}_1^{\infty}(\mathbf{r}) \\ \mathbf{h}_1^{\infty}(\mathbf{r}) \end{pmatrix} \cdot \sum_{s=1}^{N} \frac{\det B^s}{\det B} \left\{ \mathrm{e}^{i(\omega^s - \Omega)t} \int_{-\infty}^{t} \mathrm{e}^{i(\Omega - \omega^s)\tau} A_{in}(\tau)\, d\tau \right\}, \tag{11.34}$$

where

$$B^s = \begin{bmatrix} b_1^1 & b_1^2 & \cdots & (c_1^+)^* \sum_{u=1}^{N} b_u^s e^{ik_0 \mathrm{d}_u(\theta,\varphi)} & \cdots & b_1^N \\ b_2^1 & b_2^2 & \cdots & (c_2^+)^* \sum_{u=1}^{N} b_u^s e^{ik_0 \mathrm{d}_u(\theta,\varphi)} & \cdots & b_2^N \\ \vdots & \vdots & \cdots & \vdots & \cdots & \vdots \\ b_N^1 & b_N^2 & \cdots & (c_N^+)^* \sum_{u=1}^{N} b_u^s e^{ik_0 \mathrm{d}_u(\theta,\varphi)} & \cdots & b_N^N \end{bmatrix}. \tag{11.35}$$

c_u^+—expansion coefficients of the resonator field into line waves (2.1.4).

Thus, we see that the envelope of the emitted pulse in the wave zone, up to a constant, also has the form (11.32). Green's functions can also be used to calculate normalized pulse envelopes (11.33).

The dependences of the modules of the time Green's functions for different types of antennas are shown in Figs. 11.34b, e, 11.35 and 11.5.3b. The higher the quality of the antenna, the faster its Green's function decreases with increasing time variable.

The presence of residual oscillations in the lattice leads to the appearance of delayed oscillations in the pulses (Fig. 11.11b, c).

As follows from the results shown in Figs. 11.10, 11.11, and 11.12, the least distortion in the pulse envelopes is introduced by antennas with parabolic resonator arrays.

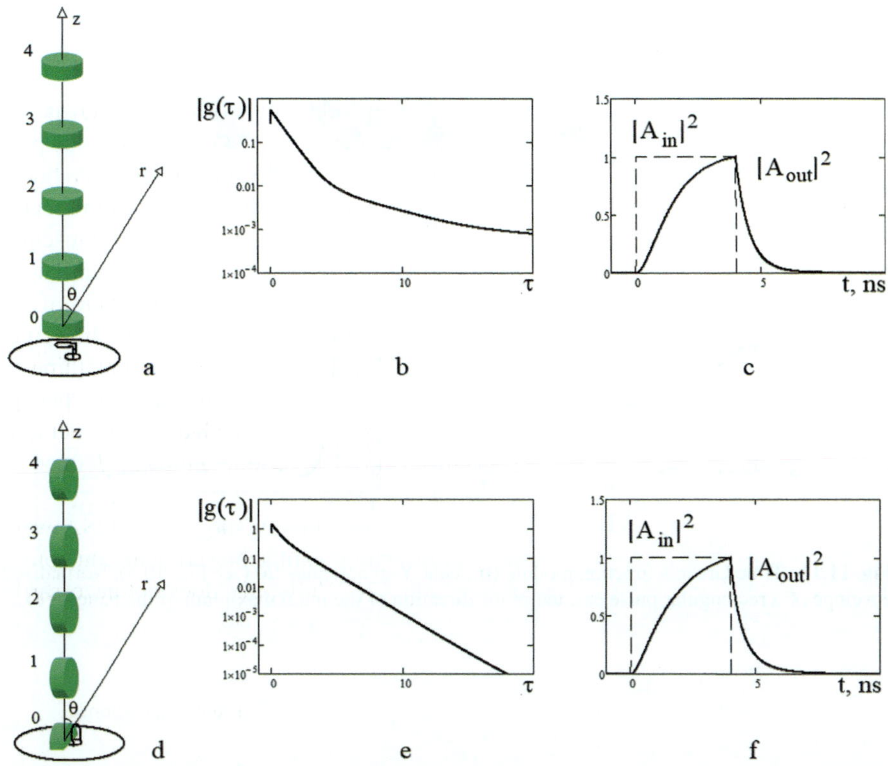

Fig. 11.10 Module of time Green's functions (**b**, **e**), linear antenna arrays of cylindrical DR (**a**, **d**) (see Fig. 10.2). Normalized envelopes of rectangular pulses emitted in the direction of the maximum radiation pattern (**c**, **f**)

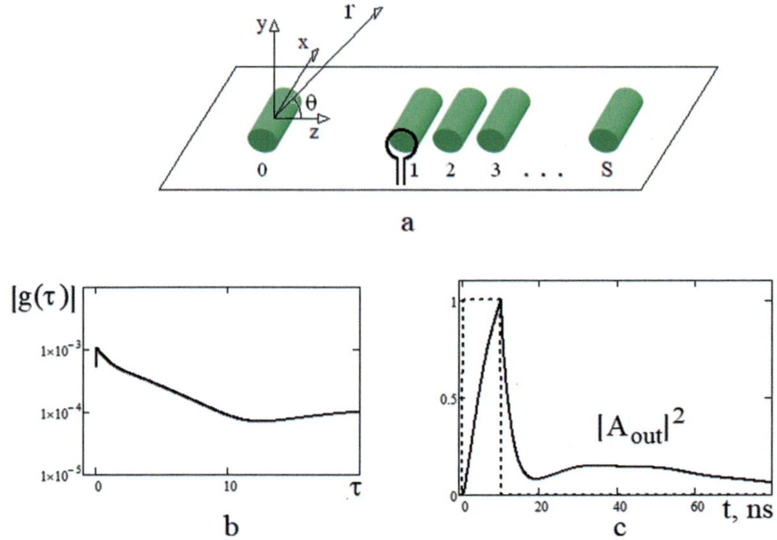

Fig. 11.11 Time Green's function module (**b**), Uda-Yagi antennas (**a**) (see Fig. 10.3). Normalized envelope of a rectangular pulse emitted in the direction of the main maximum of the pattern (**c**)

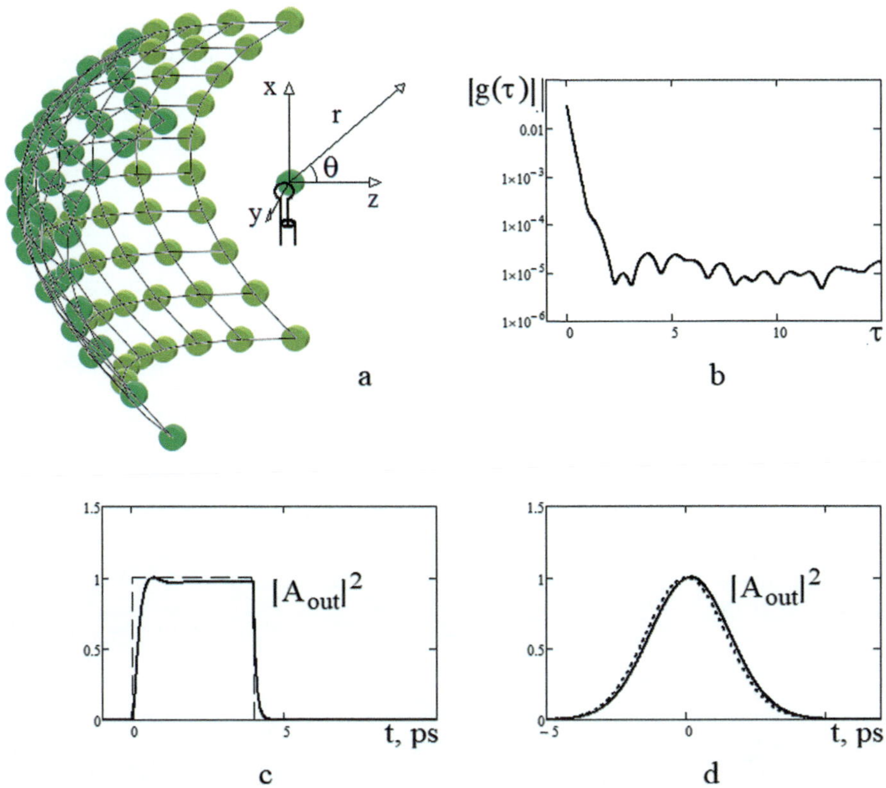

Fig. 11.12 Module of time Green's function (**b**), parabolic antenna (**a**) (see Fig. 10.10). Normalized envelopes of pulses with rectangular and normal shapes emitted in the direction of the main maximum of the pattern (**c**, **d**)

References

1. A.A. Vlasenko, A.A. Trubin, Analysis of changes in the shape of pulses during their transmission through bandpass filters of optical communication systems. Electron. Commun. **2**(67), 51–56 (2012)
2. A.A. Trubin, Scattering of infrared optical pulses on the band-stop filters on ring dielectric microresonators. Electron. Commun. **1**(72), 26–31 (2013)
3. A.A. Trubin, Scattering of the optical pulses on multiple micro-resonator band-stop filters. Electron. Commun. **19**(4), 21–25 (2014)
4. A.A. Trubin, Scattering of electromagnetic pulses on different dielectric resonator systems. Bull. NTUU KPI Ser. Radiotechn. **73**, 25–29 (2018)
5. A.A. Trubin, Scattering of optical pulses by add-drop filters on dielectric microresonators. Visnyk NTUU KPI Radiotechn. **82**, 29–35 (2020)
6. G.A. Korn, T.M. Korn, *Mathematical Handbook* (McGraw-Hill Book Company, New York, NY, 1968), p.943
7. E. Kamke, *Handbook of Ordinary Differential Equations* (Akad Verl-Ges Geest und Por-tig, Leipzig, 1959), p.576
8. A.A. Trubin, *Lattices of Dielectric Resonators* (Springer, Cham, 2016), p.171

9. M. Abramowitz, I.A. Stegun (eds.), Handbook of mathematical functions with formulas, graphs and mathematical tables. In: Applied Mathematics Series, vol. 55 (National Bureau of Standards, 1964), p. 830

Appendix A

Normalized Waves of Hollow Metal Waveguides

When deriving formulas for coupling coefficients, we use the normalization of the transverse field components ($\mathbf{E}_k$, $\mathbf{H}_k$) of the complete system functions of a hollow metal waveguide [1]

$$\mathbf{E}_{\mathrm{k}}^{\pm} = \mathbf{E}_{\mathrm{k}} \mathrm{e}^{\mp i\Gamma \mathrm{z}}; \quad \mathbf{H}_{\mathrm{k}}^{\pm} = \mathbf{H}_{\mathrm{k}} \mathrm{e}^{\mp i\Gamma \mathrm{z}};$$

where

$$\int\limits_{\mathrm{s}_\perp} \left[\mathbf{E}_{\mathrm{k}}, \mathbf{H}_{\mathrm{n}}^{*}\right] \mathbf{z}_0 \mathrm{ds} = \delta_{\mathrm{kn}} \mathrm{w}_{\mathrm{n}} / |\mathrm{w}_{\mathrm{n}}|; \tag{A.1}$$

$s_\perp$—waveguide cross section; $\Gamma = \Gamma_n$—longitudinal wave number; $\mathbf{z}_0$—unit vector directed along the waveguide axis; w_n—characteristic impedance; $w_n = w_n^H = \omega\mu_0/\Gamma_n$; $w_n = w_n^E = \Gamma_n/\omega\varepsilon_0$—for magnetic, electric waves, respectively. If $\omega > \omega_n$, Γ_n, w_n—are real and positive; when $\omega < \omega_n$, $\Gamma_n = -i|\Gamma_n|$; $w_n/|w_n| = (\pm)i$, where the " + " sign corresponds to magnetic waves, and the "−" sign to electric waves.

Rectangular waveguide cross section $a \times b$. Magnetic waves H_{su} normalized according to (A.1) have the form:

$$\begin{gathered}
\mathrm{H}_{\mathrm{x}}^{\pm} = \pm \mathrm{i} \tfrac{\Gamma}{|\Gamma|} \mathrm{h}_{\mathrm{su}}^{0} \chi_{\mathrm{sx}} \sin \chi_{\mathrm{sx}} \mathrm{x} \cos \chi_{\mathrm{uy}} \mathrm{y} \mathrm{e}^{\mp \mathrm{i}\Gamma \mathrm{z}}; \\
\mathrm{H}_{\mathrm{y}}^{\pm} = \pm \mathrm{i} \tfrac{\Gamma}{|\Gamma|} \mathrm{h}_{\mathrm{su}}^{0} \chi_{\mathrm{uy}} \cos \chi_{\mathrm{sx}} \mathrm{x} \sin \chi_{\mathrm{uy}} \mathrm{y} \mathrm{e}^{\mp \mathrm{i}\Gamma \mathrm{z}}; \\
\mathrm{H}_{\mathrm{z}}^{\pm} = \mathrm{h}_{\mathrm{su}}^{0} \tfrac{\chi^2}{|\Gamma|} \cos \chi_{\mathrm{sx}} \mathrm{x} \cos \chi_{\mathrm{uy}} \mathrm{y} \mathrm{e}^{\mp \mathrm{i}\Gamma \mathrm{z}}; \\
\mathrm{E}_{\mathrm{x}}^{\pm} = \mathrm{i} \mathrm{h}_{\mathrm{su}}^{0} \left|\mathrm{w}_{\mathrm{su}}^{\mathrm{H}}\right| \chi_{\mathrm{uy}} \cos \chi_{\mathrm{sx}} \mathrm{x} \sin \chi_{\mathrm{uy}} \mathrm{y} \mathrm{e}^{\mp \mathrm{i}\Gamma \mathrm{z}}; \\
\mathrm{E}_{\mathrm{y}}^{\pm} = -\mathrm{i} \mathrm{h}_{\mathrm{su}}^{0} \left|\mathrm{w}_{\mathrm{su}}^{\mathrm{H}}\right| \chi_{\mathrm{sx}} \sin \chi_{\mathrm{sx}} \mathrm{x} \cos \chi_{\mathrm{uy}} \mathrm{y} \mathrm{r}^{\mp \mathrm{i}\Gamma \mathrm{z}}; \\
\mathrm{E}_{\mathrm{z}}^{\pm} = 0;
\end{gathered} \tag{A.2}$$

A. Trubin, *Introduction to the Theory of Dielectric Resonators*, Springer Series in Advanced Microelectronics 65, https://doi.org/10.1007/978-3-031-65396-4

electric waves E_{su}:

$$\begin{aligned}
H_x^{\pm} &= \mp i e_{su}^0 \frac{\chi_{uy}}{\left|w_{su}^E\right|} \sin \chi_{sx} x \cos \chi_{uy} y e^{\mp i\Gamma z};\\
H_y^{\pm} &= \pm i e_{su}^0 \frac{\chi_{sx}}{\left|w_{su}^E\right|} \cos \chi_{sx} x \sin \chi_{uy} y e^{\mp i\Gamma z};\\
H_z^{\pm} &= 0;\\
E_x^{\pm} &= \mp \frac{\Gamma}{|\Gamma|} e_{su}^0 \chi_{sx} \cos \chi_{sx} x \sin \chi_{uy} y e^{\mp i\Gamma z};\\
E_y^{\pm} &= \mp \frac{\Gamma}{|\Gamma|} e_{su}^0 \chi_{uy} \sin \chi_{sx} x \cos \chi_{uy} y e^{\mp i\Gamma z};\\
E_z^{\pm} &= \mp i e_{su}^0 \frac{\chi^2}{|\Gamma|} \sin \chi_{sx} x \sin \chi_{uy} y e^{\mp i\Gamma z}.
\end{aligned} \tag{A.3}$$

Here $\chi_{sx} = s\pi/a$; $\chi_{uy} = u\pi/b$; $\chi = \left(\chi_{sx}^2 + \chi_{uy}^2\right)^{1/2}$;

$$\begin{aligned}
h_{su}^0 &= \frac{2}{\chi}\left[\frac{|\Gamma|}{\omega\mu_0 ab}\right]^{1/2} \frac{1}{(1+\delta_{s0}+\delta_{u0})^{1/2}};\\
e_{su}^0 &= \frac{2}{\chi}\left[\frac{|\Gamma|}{\omega\varepsilon_0 ab}\right]^{1/2} (1-\delta_{s0}-\delta_{u0});
\end{aligned}$$

$s, u = (0), 1, 2, \ldots$. The expression s,u = (0) means that if s = 0, u ≠ 0, or u = 0, s ≠ 0.

A coaxial waveguide formed by two ideally conducting cylinders with a central radius of R_1; external conductor R_2 [1, 2]. Let us denote by

$$\begin{aligned}
C_v(\chi\rho) &= J_v(\chi\rho) - \frac{J_u'(\chi R_1)}{Y_u'(\chi R_1)} Y_v(\chi\rho);\\
D_v(\chi\rho) &= J_v(\chi\rho) - \frac{J_u(\chi R_1)}{Y_u(\chi R_1)} Y_v(\chi\rho);
\end{aligned} \tag{A.4}$$

where $J_u'(x), Y_u'(x)$—derivatives of Bessel functions of the first and second kind, respectively [3]. The field components of a coaxial waveguide in a cylindrical coordinate system (ρ, α, z) have the form:

for magnetic waves H_{su}

$$
\begin{aligned}
H_{\rho}^{\pm} &= \mp N^{su}\tfrac{i\Gamma\chi}{\omega\mu_0}C_u'(\chi\rho)\begin{Bmatrix}\sin u\alpha\\ \cos u\alpha\end{Bmatrix}e^{\mp i\Gamma z};\\
H_{\alpha}^{\pm} &= \mp N^{su}\tfrac{iu\Gamma}{\omega\mu_0\rho}C_u(\chi\rho)\begin{Bmatrix}\cos u\alpha\\ -\sin u\alpha\end{Bmatrix}e^{\mp i\Gamma z};\\
H_{z}^{\pm} &= N^{su}\tfrac{\chi^2}{\omega\mu_0}C_u(\chi\rho)\begin{Bmatrix}\sin u\alpha\\ \cos u\alpha\end{Bmatrix}e^{\mp i\Gamma z};\\
E_{\rho}^{\pm} &= -iN^{su}\tfrac{u}{\rho}C_u(\chi\rho)\begin{Bmatrix}\cos u\alpha\\ -\sin u\alpha\end{Bmatrix}e^{\mp i\Gamma z};\\
E_{\alpha}^{\pm} &= iN^{su}\chi C_u'(\chi\rho)\begin{Bmatrix}\sin u\alpha\\ cosu\alpha\end{Bmatrix}e^{\mp i\Gamma z};\\
E_{z}^{\pm} &= 0;
\end{aligned}
\tag{A.5}
$$

for electric waves E_{su}

$$
\begin{aligned}
H_{\rho}^{\pm} &= \pm iK^{su}\tfrac{u}{\rho}D_u(\chi\rho)\begin{Bmatrix}\sin u\alpha\\ \cos u\alpha\end{Bmatrix}e^{\mp i\Gamma z};\\
H_{\alpha}^{\pm} &= \pm iK^{su}\chi D_u'(\chi\rho)\begin{Bmatrix}\cos u\alpha\\ -\sin u\alpha\end{Bmatrix}e^{\mp i\Gamma z};\\
H_{z}^{\pm} &= 0;\\
E_{\rho}^{\pm} &= iK^{su}\tfrac{\Gamma\chi}{\omega\varepsilon_0}D_u'(\chi\rho)\begin{Bmatrix}\cos u\alpha\\ -\sin u\alpha\end{Bmatrix}e^{\mp i\Gamma z};\\
E_{\alpha}^{\pm} &= -iK^{su}\tfrac{u\Gamma}{\omega\varepsilon_0\rho}D_u(\chi\rho)\begin{Bmatrix}\sin u\alpha\\ \cos u\alpha\end{Bmatrix}e^{\mp i\Gamma z};\\
E_{z}^{\pm} &= \mp K^{su}\tfrac{\chi^2}{\omega\varepsilon_0}D_u(\chi\rho)\begin{Bmatrix}\cos u\alpha\\ -\sin u\alpha\end{Bmatrix}e^{\mp i\Gamma z};
\end{aligned}
\tag{A.6}
$$

Here

$$
\begin{aligned}
N^{su} &= \left[\frac{2\omega\mu_0}{\pi(1+\delta_{u0})|\Gamma|}\right]^{1/2}\Big/\left\{C_u^2(\chi R_2)\left[(\chi R_2)^2-u^2\right]-C_u^2(\chi R_1)\left[(\chi R_1)^2-u^2\right]\right\}^{1/2};\\
K^{su} &= \left[\frac{2\omega\varepsilon_0}{\pi(1+\delta_{u0})|\Gamma|}\right]^{1/2}\Big/\left\{\left[\chi R_2 D_{u-1}(\chi R_2)\right]^2-\left[\chi R_1 D_{u-1}(\chi R_1)\right]\right\}^{1/2};
\end{aligned}
\tag{A.7}
$$

The χ parameter is defined as the s-th root of the equation:

$$J_n'(\chi R_2)Y_n'(\chi R_1)-J_n'(\chi R_1)Y_n'(\chi R_2)=0; \tag{A.8}$$

for magnetic waves;

$$J_n(\chi R_2)Y_n(\chi R_1)-J_n(\chi R_1)Y_n(\chi R_2)=0 \tag{A.9}$$

for electric waves.

A TEM wave can propagate in a coaxial waveguide:

$$E_\rho^{\pm} = J\frac{w_0}{2\pi\rho}e^{\mp ik_0 z};$$

$$H_\alpha = \pm J\frac{1}{2\pi\rho}e^{\mp ik_0 z}, \tag{A.10}$$

where $w_0 = (\mu_0/\varepsilon_0)^{1/2}$; amplitude

$$J = \left[\frac{2\pi}{w_0 \ln(R_2/R_1)}\right]^{1/2}. \tag{A.11}$$

A circulare waveguide with a metal wedge. Expressions for the field of a circular waveguide containing a metal ideally conducting wedge with an opening $\alpha = \pm\alpha_w$ have the form (A.5, A.6). In this case, the order of the cylindrical functions $C_u(\chi\rho)$ and $D_u(\chi\rho) = J_u(\chi\rho)$ are determined from the boundary conditions $E_\rho^{\pm}(\alpha = \pm\alpha_w) = 0$:

$$u = \left\{\begin{matrix}(\pi/2\alpha_w)(2t+1)\\ (\pi/\alpha_w)t\end{matrix}\right\}, \quad (t = 0, 1, 2, \ldots). \tag{A.12}$$

where in $R_1 = 0$, $R_2 = R$ and $\chi R = j'_{u,s}$ for magnetic; $\chi R = j_{u,s}$ for electric waves, and

$$N^{su} = \left[\frac{2\omega\mu_0}{\alpha_w(1+\delta_{u0})|\Gamma|\left(j'^2_{u,s} - u^2\right)}\right]^{1/2} \cdot \frac{1}{\left|J_u\left(j'_{u,s}\right)\right|};$$

$$K^{su} = \left[\frac{2\omega\varepsilon_0}{\alpha_w(1+\delta_{u0})|\Gamma|}\right]^{1/2} \cdot \frac{1}{\left|j_{u,s}J'_{u,s}\left(j_{u,s}\right)\right|}. \tag{A.13}$$

Here $j_{u,s}\left(j'_{u,s}\right)$—$s$—th root of the (derivative) Bessel function [3].

In the case of a circular homogeneous waveguide, the field expressions have the same form; at that, the order of the Bessel functions takes integer values; in (A.13) we should put: $\alpha_w = \pi$.

Appendix B

Normalized Waves of Microstrip Lines

Let's consider the fields of some types of microstrip lines most widely used in microwave integrated circuits [4–6]. The cross sections of such air and dielectric filling lines are shown in Fig. B.1.

A slot line. The line (Fig. B.1a) with a slot of width $k_0 w << 1$ in a metal screen deposited on a dielectric plate $\varepsilon_\Lambda \approx 10 \div 20$ [4]. In the quasi-static approximation, the field of the main type can be represented in the form of an expansion in plane waves H—type:

$$H_x = \mp\frac{\gamma}{\eta}\int_{-\infty}^{\infty} a(\eta,\xi)e^{-i\eta k_0\left(\xi x+\sqrt{1-\xi^2 y}\right)}\xi d\xi\cdot e^{\mp i\gamma k_0 z};$$

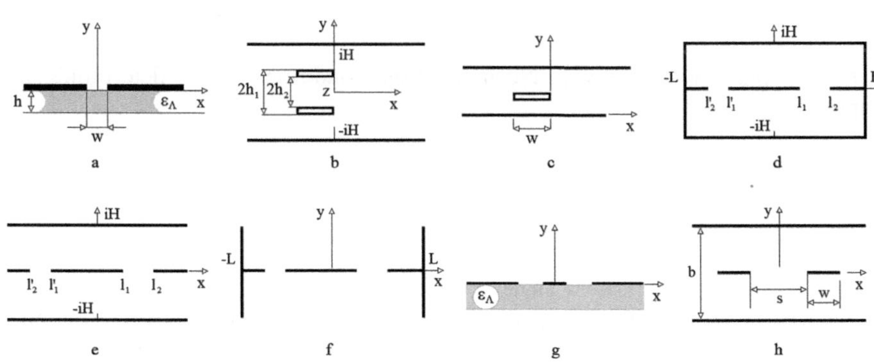

Fig. B.1 Cross sections of various types microstrip lines

A. Trubin, *Introduction to the Theory of Dielectric Resonators*, Springer Series in Advanced Microelectronics 65, https://doi.org/10.1007/978-3-031-65396-4

$$H_y = \mp\frac{\gamma}{\eta}\int_{-\infty}^{\infty} a(\eta,\xi)e^{-i\eta k_0\left(\xi x+\sqrt{1-\xi^2}y\right)}\sqrt{1-\xi^2}d\xi\cdot e^{\mp i\gamma k_0 z};$$

$$H_z = \int_{-\infty}^{\infty} a(\eta,\xi)e^{-i\eta k_0\left(\xi x+\sqrt{1-\xi^2}y\right)}d\xi\cdot e^{\mp i\gamma k_0 z}; \tag{B.1}$$

$$E_x = -\frac{w_0}{\eta}\int_{-\infty}^{\infty} a(\eta,\xi)e^{-i\eta k_0\left(\xi x+\sqrt{1-\xi^2}y\right)}\sqrt{1-\xi^2}d\xi\cdot e^{\mp i\gamma k_0 z};$$

$$E_y = \frac{w_0}{\eta}\int_{-\infty}^{\infty} a(\eta,\xi)e^{-i\eta k_0\left(\xi x+\sqrt{1-\xi^2}y\right)}\xi d\xi\cdot e^{\mp i\gamma k_0 z};$$

$$E_z = 0,$$

where $\eta^2+\gamma^2=1$. The spectral function in the case of an infinitely thin screen has the form [4]:

$$a(\eta,\xi) = -H_0/\pi J_0\left(k_0\frac{w}{2}\eta\xi\right)\cdot\left\{\begin{array}{ll} 1/\sqrt{1-\xi^2}; & (y>0) \\ \frac{K\eta}{\sqrt{\varepsilon_\Lambda-\gamma^2-(\eta\xi)^2}}\cdot\frac{(1+M)e^{ik_0h\sqrt{\varepsilon_\Lambda-\gamma^2-(\eta\xi)^2}}}{1-Me^{2ik_0h\sqrt{\varepsilon_\Lambda-\gamma^2-(\eta\xi)^2}}}; & (y<-h)\end{array}\right\}; \tag{B.2}$$

where

$$M = M(\xi) = \frac{\sqrt{\varepsilon_\Lambda-\gamma^2-(\eta\xi)^2}-K\eta\sqrt{1-\xi^2}}{\sqrt{\varepsilon_\Lambda-\gamma^2-(\eta\xi)^2}+K\eta\sqrt{1-\xi^2}}; \tag{B.3}$$

$K = (\varepsilon_\Lambda-\gamma^2)/(1-\gamma^2)$. Normalization conditions (A.1) determines the amplitude

$$|H_0| = Z^{1/2}\eta^2 k_0/(2w_0); \tag{B.4}$$

Here Z—characteristic impedance; $w_0=(\mu_0/\varepsilon_0)^{1/2}$; $k_0=\omega\cdot(\varepsilon_0\mu_0)^{1/2}$.

A strip lines. Let us consider strip lines formed by a system of conductors with a connectivity greater than or equal to two (Fig. B.1b–h). Such lines may be non-uniform in cross section—contain layers of dielectric. If the frequencies are not too high, then the main wave of the strip line can be considered as a quasi-TEM wave and the theory of functions of a complex variable can be used for calculating the parameters [5]. In this approximation, the presence of a dielectric in the line is taken into account by introducing into consideration a homogeneous medium with an effective dielectric constant ε_{eff}. The assumption of the existence of a TEM wave allows us to express the field components through a harmonic function $\varphi(\tilde{z})$ of a complex variable $\tilde{z} = x + iy$:

$$\mathbf{E} = -grad\,\varphi \cdot e^{\mp i\sqrt{\varepsilon_{eff}}k_0 z}; \quad \mathbf{H} = \mp\frac{1}{w}[\mathbf{z}_0, \mathbf{E}]. \tag{B.5}$$

Here $w = (\mu_0/\varepsilon_0\varepsilon_{\text{eff}})^{1/2}$; $\mathbf{z}_0$—unit vector directed along the line z axis. The normalization condition (A.1) leads to a simple relation: $V = \sqrt{Z_0}$, where V—potential difference between the current-carrying strip and the screen; Z_0—is the line impedance.

In the area of harmony Ω: $r \le R$, where $r = |\tilde{z}|$, the function $\varphi(\tilde{z})$ can be expressed in Poisson integral form [6]. In this case, without taking into account the longitudinal dependence, the field components can be represented in the form of an expansion:

$$\mathrm{E}_x = -\frac{1}{\pi}\int_0^{2\pi} \varphi\left(\mathrm{R}e^{i\tau}\right)\sum_{s=1}^{\infty}\frac{s}{r}\left(\frac{r}{R}\right)^s \cos[s\tau - (s-1)\alpha]d\tau;$$

$$\mathrm{E}_y = -\frac{1}{\pi}\int_0^{2\pi} \varphi\left(\mathrm{R}e^{i\tau}\right)\sum_{s=1}^{\infty}\frac{s}{r}\left(\frac{r}{R}\right)^s \sin[s\tau - (s-1)\alpha]d\tau. \tag{B.6}$$

Calculation of the parameters of a shielded coupled strip line with wide plates of arbitrary thickness (Fig. B.1b) allows for a more general consideration and, as a special case, obtaining expressions for the transverse field components of an asymmetric shielded strip line [4]:

$$\mathbf{E} = \begin{pmatrix} E_x \\ E_y \end{pmatrix} = \begin{pmatrix} \mathrm{Re} \\ \mathrm{Im} \end{pmatrix} \{-i/d\tilde{z}^*/d\omega\}.$$

The conformal mapping of a plane $\tilde{z}$ to a plane ω has the form [5]:

$$\begin{aligned}\frac{\pi}{H}\tilde{z}(\omega) = {} & arch\frac{\omega(a_1 + a_2 + 2) + 2a_1a_2 + a_1 + a_2}{(1-\omega)(a_1 - a_2)} \\ & - \frac{h_2}{H}arch\frac{\omega(a_1 + a_2) + 2a_1a_2}{\omega(a_1 - a_2)} \\ & - \left(1 - \frac{h_1}{H}\right)arch\frac{2\omega + a_1 + a_2}{a_1 - a_2}.\end{aligned} \tag{B.7}$$

where for an odd (opposite sign of the potentials V of the current-carrying strips) type of waves:

$$\omega = \exp(\pi/Vw); \tag{B.8}$$

and waves of even type (equal in sign of strip potentials):

$$\omega = ch^2[\pi w/(2V)]. \tag{B.9}$$

Here $a_1 = u_0/(H-h_1)^2$; $a_2 = h_2^2/u_0$; $2u_0 = 2Hh_1 - h_1^2 - h_2^2 + \sqrt{(h_1^2 - h_2^2)[(2H-h_1)^2 - h_2^2]}$; h_2, h_1—height; $h_1 - h_2$—thickness; w—strip width.

The field distribution of an odd-type waves satisfies the condition of an electric wall in the plane: $y = 0$, therefore the given relations can be used to approximate the calculation of the field components of an asymmetric strip line (Fig. B.1c).

For infinitely thin strips: $h_1 = h_2 = h$, the field of such a line takes a simpler form:

$$\mathbf{E} = \frac{\mathrm{Re}}{\mathrm{Im}}\left\{\frac{iV}{H-h}\left(\frac{\omega-1}{\omega+a}\right)^*\right\}; \quad \mathbf{E} = \frac{\mathrm{Re}}{\mathrm{Im}}\left\{\frac{iV}{H-h}\left(\frac{\omega-1}{\omega+a}\sqrt{\frac{\omega}{\omega-1}}\right)^*\right\}. \tag{B.10}$$

for odd and even waves, respectively. In this case, the conformal mapping has the form:

$$\begin{aligned} \tilde{z}(\omega) = \frac{1}{\pi}(H-h)[a\{\ln(-\omega) - \ln|a|\} \\ -(1+a)\{\ln(1-\omega) - \ln|1+a|\}] + ih; \end{aligned} \tag{B.11}$$

where $a = h/(H-h)$. The given approximate relations are suitable for calculating the parameters of lines with wide stripes: $w \geq 0,7h$ when the interaction of the opposite edges of the plates can be neglected [5].

Shown in Fig. B.1, varieties of coplanar (d–g) and coupled symmetric (h) strip lines make it possible to obtain simple accurate expressions for the transverse dependences of the TEM wave field. In the case of a shielded ribbed strip line (Fig. B.1d) with uniform filling, the electric field components have the form [5]:

$$\begin{aligned} \mathbf{E} = \frac{\mathrm{Re}}{\mathrm{Im}}\Bigg\{\frac{\mathrm{V}}{2\mathrm{L}} \cdot \frac{K(k)}{\sqrt{\kappa}K'(k)}\sqrt{[sn(l_1) - sn(l_1')][sn(l_2') - sn(l_2)]} \\ \cdot \sqrt{\frac{(1-k^2\omega_1^2)(1-\omega_1^2)}{[\omega_1 - sn(l_1)][\omega_1 - sn(l_2)][\omega_1 - sn(l_1')][\omega_1 - sn(l_2')]}}\Bigg\}. \end{aligned} \tag{B.12}$$

where

$$\kappa = \left(\sqrt{\lambda} - \sqrt{\lambda - 1}\right)^2; \quad \lambda = \frac{[sn(l_1) - sn(l_2')] \cdot [sn(l_2) - sn(l_1')]}{[sn(l_1) - sn(l_1')] \cdot [sn(l_2) - sn(l_2')]}; \tag{B.13}$$

$$Z_0 = 30\pi/\sqrt{\varepsilon_\Lambda} \cdot K'(\kappa)/K(\kappa)$$

characteristic impedance; conformal mapping:

$$\omega_1 = sn(\tilde{z}) = sn(K(k)\tilde{z}/L, k);$$

where $K'(k)/H = K(k)/L$; $K(k)$, $K'(k) = K(k')$—complete elliptic integrals $\left(k' = \sqrt{1-k^2}\right)$; $sn(u, k)$—elliptic Jacobi function [3].

In exceptional cases, the field of a strip line can be expressed in terms of elementary functions:

(a) If $L = \infty$ (Fig. B.1e) $k = 1$; $K(k) = \infty$; $K'(k) = \pi/2$; $sn(K(k)\tilde{z}/L, 1) = th(\pi\tilde{z}/2H)$; $(\kappa+1)^2 = 4\lambda\kappa$ and

$$\mathbf{E} = \begin{matrix}\mathrm{Re}\\ \mathrm{Im}\end{matrix}\left\{\frac{\pi V}{4H}\cdot\frac{i}{\sqrt{\kappa}K'(\kappa)}\sqrt{\left[t(l_1)-t(l_1')\right]\left[t(l_2)-t(l_2')\right]}\right.$$
$$\left.\cdot\frac{1-t^2(\tilde{z})}{\sqrt{[t(\tilde{z})-t(l_1)][t(\tilde{z})-t(l_2)]\left[t(\tilde{z})-t(l_1')\right]\left[t(\tilde{z})-t(l_2')\right]}}\right\}. \quad \text{(B.14)}$$

Here for brevity: $th(\pi\tilde{z}/2H) \equiv t(\tilde{z})$; $th(\pi l_i/2H) \equiv t(l_i)$;

$$\lambda = \frac{\left[t(l_1)-t(l_2')\right]\left[t(l_2)-t(l_1')\right]}{\left[t(l_1)-t(l_1')\right]\left[t(l_2)-t(l_2')\right]}.$$

If $l_2 = -l_2^1 = \infty$, relations (B.14) turn into expressions for the transverse dependences of the field of a symmetric strip line:

$$\mathbf{E} = \begin{matrix}\mathrm{Re}\\ \mathrm{Im}\end{matrix}\left\{\frac{\pi V}{4H}\cdot\frac{1}{K'(\kappa)}\frac{1}{\sqrt{sh^2\frac{\pi\tilde{z}}{2H}-\kappa^2 ch^2\frac{\pi z}{2H}}}\right\}, \quad \text{(B.15)}$$

where $w = l_1 - l_1'$; $\kappa = th(\pi w/4H)$.

In the limit $H = \infty$, taking into account $\lim\limits_{H\to\infty} H\, th(\pi\tilde{z}/2H) = \pi\tilde{z}/2$, we obtain from (B.14) the field of an open coplanar line (Fig. B.1g).

(b) If in (B.12) we put $H = \infty$ (Fig. B.1f), then $k = 0$; $K = \pi/2$; $sn(K(k)\tilde{z}_i/L, 0) = \sin\pi\tilde{z}_i/2L \equiv s(\tilde{z}_i)$ and

$$\mathbf{E} = \begin{matrix}\mathrm{Re}\\ \mathrm{Im}\end{matrix}\left\{\frac{\pi V}{4L}\cdot\frac{i}{\sqrt{\kappa}K'(\kappa)}\sqrt{\left[s(l_1)-s(l_1')\right]\left[s(l_2)-s(l_2')\right]}\cdot\right.$$
$$\left.\cdot\sqrt{\frac{1-s^2(\tilde{z})}{[s(\tilde{z})-s(l_1)]\cdot[s(\tilde{z})-s(l_2)]\cdot\left[s(\tilde{z})-s(l_1')\right]\cdot\left[s(\tilde{z})-s(l_2')\right]}}\right\}; \quad \text{(B.16)}$$

$$\lambda = \frac{\left[s(l_1)-s(l_2')\right]\left[s(l_2)-s(l_1')\right]}{\left[s(l_1)-s(l_1')\right]\left[s(l_2)-s(l_2')\right]}.$$

Consider the fields of coupled symmetric shielded strip lines [5] (Fig. B.1h). The characteristic impedance of such a line for waves of even (e), odd (o) types also has the form (B.13), where:

$$\kappa = \begin{bmatrix} \kappa_e \\ \kappa_o \end{bmatrix} = th\frac{\pi w}{2b} \cdot \begin{bmatrix} th \\ cth \end{bmatrix} \left(\frac{\pi}{2} \cdot \frac{w+s}{b} \right); \tag{B.17}$$

$b = 2H$—plane waveguide height; w—strip width; s—distance between stripes. The field of such a line, according to [5], is represented in the form:

$$\mathbf{E}_e = \frac{\text{Re}}{\text{Im}} \left\{ 2\pi \frac{V}{K'(b)} \cdot \frac{i}{\kappa' \sqrt{p_e}} \cdot \frac{sh(\pi \tilde{z}/b)}{\sqrt{\left[ch^2 \frac{\pi \tilde{z}}{b} - ch^2 \frac{\pi s}{2b}\right] \cdot \left[ch^2 \frac{\pi \tilde{z}}{b} - ch^2 \frac{\pi}{2b}(2w+s)\right]}} \right\}. \tag{B.18}$$

for even waves and

$$\mathbf{E}_o = -2\pi \frac{\text{Re}}{\text{Im}} \left\{ \frac{V}{K'b} \cdot \frac{1}{\kappa' \sqrt{p_0}} \cdot \frac{ch(\pi \tilde{z}/b)}{\sqrt{\left[sh^2 \frac{\pi \tilde{z}}{b} - sh^2 \frac{\pi s}{2b}\right] \cdot \left[sh^2 \frac{\pi \tilde{z}}{b} - sh^2 \frac{\pi}{2b}(2w+s)\right]}} \right\}, \tag{B.19}$$

for odd waves.

Here

$$p_e = \frac{1}{ch\frac{\pi s}{2b} ch\frac{\pi}{2b}(2w+s)}; \quad p_o = \frac{1}{sh\frac{\pi s}{2b} sh\frac{\pi}{2b}(2w+s)}. \tag{B.20}$$

Appendix C

Perturbation Theory for Maxwell's Equations

Let the first electromagnetic field are satisfying equations:

$$\begin{aligned} rot\mathbf{E}_1 &= -\mu_0 \partial \mathbf{H}_1/\partial t. \\ rot\mathbf{H}_1 &= \tilde{\varepsilon}_1 \partial \mathbf{E}_1/\partial t; \end{aligned} \tag{C.1}$$

the second field are vary in the time as: $\mathbf{E}_2 = \mathbf{E}_2 e^{i\tilde{\omega}_2 t}$; $\mathbf{H}_2 = \mathbf{H}_2 e^{i\tilde{\omega}_2 t}$ and can be presented as:

$$rot\mathbf{E}_2 = -i\tilde{\omega}_2\mu_0\mathbf{H}_2; \quad rot\mathbf{H}_2 = i\tilde{\omega}_2\tilde{\varepsilon}_2\mathbf{E}_2. \tag{C.2}$$

In the general case we suppose, that a circular frequency $\tilde{\omega}_s = \omega_s' + i\omega_s''$ and the dielectric permittivity $\tilde{\varepsilon}_s = \varepsilon_s' - i\varepsilon_s''$ are the complex, the magnetic permittivity is a real $\mu = \mu_0$ parameter, and $\omega_s'' \ll \omega_s'$; $\varepsilon_s'' \ll \varepsilon_s'$.

Combining (C.1, C.2) and using known vector relation:

$$div[\mathbf{F}, \mathbf{G}] = (\mathbf{G}, rot\mathbf{F}) - (\mathbf{F}, rot\mathbf{G}), \tag{C.3}$$

obtain:

$$\begin{aligned} div\{[\mathbf{E}_1, \mathbf{H}_2^*] + [\mathbf{E}_2^*, \mathbf{H}_1]\} &= i\tilde{\omega}_2^*\tilde{\varepsilon}_2^*(\mathbf{E}_1, \mathbf{E}_2^*) \\ &\quad - \tilde{\varepsilon}_1(\partial \mathbf{E}_1/\partial t, \mathbf{E}_2^*) \\ &\quad + i\tilde{\omega}_2^*\mu_0(\mathbf{H}_1, \mathbf{H}_2^*) - \mu_0(\partial \mathbf{H}_1/\partial t, \mathbf{H}_2^*). \end{aligned} \tag{C.4}$$

Let us supposes also, that
$\partial \mathbf{E}_1/\partial t \approx i\omega \mathbf{E}_1$, as well as $\partial \mathbf{H}_1/\partial t \approx i\omega \mathbf{H}_1$ and

$$\omega \approx \tilde{\omega}_2^* \quad (\mathrm{Im}(\tilde{\omega}_2^*) \ll 1) \tag{C.5}$$

A. Trubin, *Introduction to the Theory of Dielectric Resonators*, Springer Series in Advanced Microelectronics 65, https://doi.org/10.1007/978-3-031-65396-4

Under these assumptions from (C.4) obtain:

$$div\{[\mathbf{E}_1, \mathbf{H}_2^*] + [\mathbf{E}_2^*, \mathbf{H}_1]\} = i\omega(\tilde{\varepsilon}_2^* - \tilde{\varepsilon}_1)(\mathbf{E}_1, \mathbf{E}_2^*) \quad \text{(C.6)}$$

If both fields change in time according to the harmonic law we obtain from (C.4)

$$\begin{aligned} &div\{[\mathbf{E}_1, \mathbf{H}_2^*] + [\mathbf{E}_2^*, \mathbf{H}_1]\} \\ &= i(\tilde{\omega}_2^* - \tilde{\omega}_1)[\varepsilon'(\mathbf{E}_1, \mathbf{E}_2^*) + \mu_0(\mathbf{H}_1, \mathbf{H}_2^*)] \\ &\quad - \varepsilon''(\tilde{\omega}_1 + \tilde{\omega}_2^*)(\mathbf{E}_1, \mathbf{E}_2^*) \end{aligned} \quad \text{(C.7)}$$

To denoted through $\omega_0 = (\omega_1' + \omega_2')/2$, rewrites (C.7) in the form:

$$\begin{aligned} &\frac{1}{\omega_0} div\{[\mathbf{E}_1, \mathbf{H}_2^*] + [\mathbf{E}_2^*, \mathbf{H}_1]\} \\ &= i\frac{(\tilde{\omega}_2^* - \tilde{\omega}_1)}{\omega_0}[\varepsilon'(\mathbf{E}_1, \mathbf{E}_2^*) + \mu_0(\mathbf{H}_1, \mathbf{H}_2^*)] \\ &\quad - \varepsilon''\frac{\tilde{\omega}_1 + \tilde{\omega}_2^*}{\omega_0}(\mathbf{E}_1, \mathbf{E}_2^*) \end{aligned} \quad \text{(C.8)}$$

If the field $(\mathbf{E}_1, \mathbf{H}_1)$ is not strong different from the $(\mathbf{E}_2, \mathbf{H}_2)$, then for $\omega_s'' \ll \omega_s'$ $(s = 1, 2)$, the parameter $\xi = (\tilde{\omega}_2^* - \tilde{\omega}_1)/\omega_0$ will be a small at modulus: $|\xi| \ll 1$ and the expression can be applying as base for perturbation theory.

In the some cases at the outset it's more comfortable to resolve the task, in which the dielectric loses are absent: $\varepsilon'' = 0$, then the equation (C.8) is a more simplifying:

$$\begin{aligned} \frac{1}{\omega_0} div\big\{[\mathbf{E}_1, \mathbf{H}_2^*] + [\mathbf{E}_2^*, \mathbf{H}_1]\big\} = i\frac{(\tilde{\omega}_2^* - \tilde{\omega}_1)}{\omega_0}\big[&\varepsilon'(\mathbf{E}_1, \mathbf{E}_2^*) \\ &+\mu_0(\mathbf{H}_1, \mathbf{H}_2^*)\big]. \end{aligned} \quad \text{(C.9)}$$

Appendix D

Some Integral Representations of Special Functions

In the case of expansion of solutions to Maxwell's equation in cylindrical and spherical coordinate systems over the waves of circular and rectangular waveguides, and strip transmission lines, as well as in open space, we used a number of definite integrals. Some of them are well known, others are generalizations of known ones, or are a consequence of their transformations.

Definite integrals of trigonometric functions are a consequence of the well-known Bessel integral [7, 8]. Let a, b, c, d be real constants, and $\sin\psi_0 = b/\sqrt{a^2+b^2}$. Then

$$\int_{-\pi}^{\pi} \begin{Bmatrix} \sin m\varphi \\ \cos m\varphi \end{Bmatrix} \sin(c + a\sin\varphi)\cos(d + b\cos\varphi)d\varphi$$

$$= 2\pi J_m\left(\sqrt{a^2+b^2}\right)\cdot \begin{Bmatrix} \cos c\left[\cos d\cos m\psi_0 \sin^2 m\frac{\pi}{2} - \sin d \sin m\psi_0 \cos^2 m\frac{\pi}{2}\right] \\ \sin c\left[\cos d\cos m\psi_0 \cos^2 m\frac{\pi}{2} - \sin d \sin m\psi_0 \sin^2 m\frac{\pi}{2}\right] \end{Bmatrix}; \quad \text{(D.1)}$$

$$\int_{-\pi}^{\pi} \begin{Bmatrix} \sin m\varphi \\ \cos m\varphi \end{Bmatrix} \sin(c + a\sin\varphi)\sin(d + b\cos\varphi)d\varphi$$

$$= 2\pi J_m\left(\sqrt{a^2+b^2}\right)\cdot \begin{Bmatrix} \cos c\left[\sin d\cos m\psi_0 \sin^2 m\frac{\pi}{2} + \cos d \sin m\psi_0 \cos^2 m\frac{\pi}{2}\right] \\ \sin c\left[\sin d\cos m\psi_0 \cos^2 m\frac{\pi}{2} + \cos d \sin m\psi_0 \sin^2 m\frac{\pi}{2}\right] \end{Bmatrix}; \quad \text{(D.2)}$$

$$\int_{-\pi}^{\pi} \begin{Bmatrix} \sin m\varphi \\ \cos m\varphi \end{Bmatrix} \cos(c + a\sin\varphi)\cos(d + b\cos\varphi)d\varphi$$

A. Trubin, *Introduction to the Theory of Dielectric Resonators*, Springer Series in Advanced Microelectronics 65, https://doi.org/10.1007/978-3-031-65396-4

$$= 2\pi J_m\left(\sqrt{a^2+b^2}\right)\cdot\left\{\begin{matrix}\sin c\left[\sin d\sin m\psi_0\cos^2 m\frac{\pi}{2}-\cos d\cos m\psi_0\sin^2 m\frac{\pi}{2}\right]\\ \cos c\left[\cos d\cos m\psi_0\cos^2 m\frac{\pi}{2}-\sin d\sin m\psi_0\sin^2 m\frac{\pi}{2}\right]\end{matrix}\right\}; \quad \text{(D.3)}$$

$$\int_{-\pi}^{\pi}\left\{\begin{matrix}\sin m\varphi\\ \cos m\varphi\end{matrix}\right\}\cos(c+a\sin\varphi)\sin(d+b\cos\varphi)d\varphi$$

$$= 2\pi J_m\left(\sqrt{a^2+b^2}\right)\cdot\left\{\begin{matrix}-\sin c\left[\sin d\cos m\psi_0\sin^2 m\frac{\pi}{2}+\cos d\sin m\psi_0\cos^2 m\frac{\pi}{2}\right]\\ \cos c\left[\sin d\cos m\psi_0\cos^2 m\frac{\pi}{2}+\cos d\sin m\psi_0\sin^2 m\frac{\pi}{2}\right]\end{matrix}\right\}; \quad \text{(D.4)}$$

$$\int_{-\pi}^{\pi}\left\{\begin{matrix}\sin m\varphi\\ \cos m\varphi\end{matrix}\right\}\sin(c+a\sin\varphi)e^{\pm ib\cos\varphi}d\varphi$$

$$= 2\pi J_m\left(\sqrt{a^2+b^2}\right)\cdot\left\{\begin{matrix}\cos c\left[\cos m\psi_0\sin^2 m\frac{\pi}{2}\pm i\sin m\psi_0\cos^2 m\frac{\pi}{2}\right]\\ \sin c\left[\cos m\psi_0\cos^2 m\frac{\pi}{2}\pm i\sin m\psi_0\sin^2 m\frac{\pi}{2}\right]\end{matrix}\right\}; \quad \text{(D.5)}$$

$$\int_{-\pi}^{\pi}\left\{\begin{matrix}\sin m\varphi\\ \cos m\varphi\end{matrix}\right\}\cos(c+a\sin\varphi)e^{\pm ib\cos\varphi}d\varphi$$

$$= 2\pi J_m\left(\sqrt{a^2+b^2}\right)\cdot\left\{\begin{matrix}-\sin c\left[\cos m\psi_0\sin^2 m\frac{\pi}{2}\pm i\sin m\psi_0\cos^2 m\frac{\pi}{2}\right]\\ \cos c\left[\cos m\psi_0\cos^2 m\frac{\pi}{2}\pm i\sin m\psi_0\sin^2 m\frac{\pi}{2}\right]\end{matrix}\right\}; \quad \text{(D.6)}$$

$$\int_{-\pi}^{\pi}\left\{\begin{matrix}\sin m\varphi\\ \cos m\varphi\end{matrix}\right\}\sin(d+b\cos\varphi)e^{\pm ia\sin\varphi}d\varphi$$

$$= 2\pi J_m\left(\sqrt{a^2+b^2}\right)\cdot\left\{\begin{matrix}\pm i\left[\sin d\cos m\psi_0\sin^2 m\frac{\pi}{2}+\cos d\sin m\psi_0\cos^2 m\frac{\pi}{2}\right]\\ \sin d\cos m\psi_0\cos^2 m\frac{\pi}{2}+\cos d\sin m\psi_0\sin^2 m\frac{\pi}{2}\end{matrix}\right\}; \quad \text{(D.7)}$$

$$\int_{-\pi}^{\pi}\left\{\begin{matrix}\sin m\varphi\\ \cos m\varphi\end{matrix}\right\}\cos(d+b\cos\varphi)e^{\pm ia\sin\varphi}d\varphi$$

$$= 2\pi J_m\left(\sqrt{a^2+b^2}\right)\cdot\left\{\begin{matrix}\pm i\left[\cos d\cos m\psi_0\sin^2 m\frac{\pi}{2}-\sin d\sin m\psi_0\cos^2 m\frac{\pi}{2}\right]\\ \cos d\cos m\psi_0\cos^2 m\frac{\pi}{2}-\sin d\sin m\psi_0\sin^2 m\frac{\pi}{2}\end{matrix}\right\}; \quad \text{(D.8)}$$

Adding integral (D.5) multiplied by i with (D.6) we obtain:

$$\int_{-\pi}^{\pi} \begin{Bmatrix} \sin m\varphi \\ \cos m\varphi \end{Bmatrix} e^{\pm i(a\sin\varphi + b\cos\varphi)} d\varphi$$
$$= 2\pi J_m(\sqrt{a^2+b^2}) \cdot \begin{Bmatrix} -\sin m\psi_0 \cos^2 m\frac{\pi}{2} \pm i\cos m\psi_0 \sin^2 m\frac{\pi}{2} \\ \cos m\psi_0 \cos^2 m\frac{\pi}{2} \pm i\sin m\psi_0 \sin^2 m\frac{\pi}{2} \end{Bmatrix} \quad \text{(D.9)}$$

Definite integrals containing Bessel functions are a consequence of Gegenbauer's addition theorem [8]:

$$\int_{0}^{\pi} \frac{\sin}{\cos}[z\cos\varphi\cos\theta]\, J_m(z\sin\varphi\sin\theta) P_n^m(\cos\theta)\sin\theta\, d\theta$$
$$= 2\frac{\sin}{\cos}\left[(n-m)\frac{\pi}{2}\right] P_n^m(\cos\varphi) j_n(z). \quad \text{(D.10)}$$

In a complex form:

$$\int_{0}^{\pi} \mathrm{e}^{\pm iz\cos\varphi\cos\theta} J_m(z\sin\varphi\sin\theta) P_n^m(\cos\theta)\sin\theta\, d\theta$$
$$= 2(\pm i)^{n-m} P_n^m(\cos\varphi) j_n(z). \quad \text{(D.11)}$$

where $P_n^m(x)$—Legendre function [3]; $j_n(z)$—spherical function [3].

Passing to the limit $\varphi = 0$ in (D.11) and taking into account the asymptotic behavior of the Bessel and Legendre functions [7, 8], we find:

$$\int_{0}^{\pi} \mathrm{e}^{\pm iz\cos\theta} \sin^m\theta\, P_n^m(\cos\theta)\sin\theta\, d\theta$$
$$= 2\mathrm{e}^{\pm i\frac{\pi}{2}(n+m)} \frac{(n+m)!}{(n-m)!} \cdot \frac{j_n(z)}{z^m}. \quad \text{(D.12)}$$

The following integrals are a generalization of the known ones obtained in special cases. Let z_1, z_2 be complex constants and $\sin\tilde{\varphi}_0 = z_1/\sqrt{z_1^2+z_2^2}$; $\cos\tilde{\varphi}_0 = z_2/\sqrt{z_1^2+z_2^2}$. Then the following integral holds:

$$i_m \equiv \int_{-\pi}^{\pi} \left\{ \begin{array}{c} \sin m\varphi \\ \cos m\varphi \end{array} \right\} e^{iz_1 \sin\varphi + iz_2 \cos\varphi} d\varphi$$

$$= 2\pi e^{im\frac{\pi}{2}} J_m\left(\sqrt{z_1^2 + z_2^2}\right) \cdot \left\{ \begin{array}{c} \sin m\tilde{\varphi}_0 \\ \cos m\tilde{\varphi}_0 \end{array} \right\}. \tag{D.13}$$

We carry out the proof by induction: for $m = 0$ equality (D.13) follows from the Jacobi-Anger expansion [3];

$$i_1 = -i \left\{ \begin{array}{c} \partial/\partial z_1 \\ \partial/\partial z_2 \end{array} \right\} i_0; \quad \text{and} \quad i_{m+1} = -i\frac{\partial}{\partial z_2} i_m - i\frac{\partial}{\partial z_1} i_m^+$$

follows from differentiation formulas and recurrent formulas for the trigonometric function and Bessel functions [7]. Here the " + " sign means swapping trigonometric functions and $\varphi \to -\varphi$.

From the well-known representation of the two-dimensional Green's function:

$$\int_{-\infty}^{\infty} e^{-i\left(\xi a + \sqrt{1-\xi^2} b\right)} \frac{d\xi}{\sqrt{1-\xi^2}} = \pi H_0^{(2)}\left(\sqrt{a^2 + b^2}\right),$$

Obtain:

$$\int_{-\infty}^{\infty} \left\{ \begin{array}{c} \sin m\tilde{\varphi} \\ \cos m\tilde{\varphi} \end{array} \right\} e^{-i\left(\xi a + \sqrt{1-\xi^2} b\right)} \frac{d\xi}{\sqrt{1-\xi^2}}$$

$$= \pi e^{-im\frac{\pi}{2}} \left\{ \begin{array}{c} \sin m\psi_0 \\ \cos m\psi_0 \end{array} \right\} H_m^{(2)}\left(\sqrt{a^2 + b^2}\right). \tag{D.14}$$

where $\cos\tilde{\varphi} \equiv \xi$; $\sin\tilde{\varphi} = \sqrt{1-\xi^2}$; a, b—real constants: $\sin\psi_0 = b/\sqrt{a^2+b^2}$. The proof of (D.14) is similar.

The following integral can be considered as a generalization of the Sommerfeld integral [3]:

$$\int_0^{\infty} e^{-ib\sqrt{1-t^2}} J_0(ct) \frac{t\,dt}{\sqrt{1-t^2}} = i\frac{e^{-i\sqrt{b^2+c^2}}}{\sqrt{b^2+c^2}}; \tag{D.15}$$

$$
\begin{aligned}
i_n^m &\equiv \int\limits_0^\infty e^{-ib\sqrt{1-t^2}} J_m(ct) P_n^m\left(\sqrt{1-t^2}\right) t\,dt/\sqrt{1-t^2} \\
&= e^{-i\frac{\pi}{2}(n-m)} P_n^m\left(b/\sqrt{b^2+c^2}\right) h_n^{(2)}\left(\sqrt{b^2+c^2}\right).
\end{aligned} \tag{D.16}
$$

Here also b, c—real constants; $h_n^{(2)}(z)$—spherical Hankel function of the second kind [7]; $P_n^m(z)$—Legendre functions [9]. The proof of (D.16) is also carried out by induction and follows from the recurrence relations for the Legendre functions, cylindrical functions [3, 9] as well as

$$
\begin{aligned}
i_{n+1}^m &= \frac{1}{n-m+1}\left[i(2n+1)\frac{d}{db} i_n^m - (n+m) i_{n-1}^m\right]; \\
i_n^{m+1} &= \sum_{s=m+1}^{n} {}'' (2s-1)\left[\frac{d}{dc} i_{s-1}^m - \frac{m}{c} i_{s-1}^m\right].
\end{aligned}
$$

The symbol $\sum''$ means that the summation is carried out in steps 2.

We also used partial values of known integrals:

Poisson [3]:

$$
\int\limits_0^1 \cos bx/\sqrt{1-x^2}dx = \frac{\pi}{2}J_0(b); \quad \int\limits_0^1 \cos bx\sqrt{1-x^2}dx = \frac{\pi}{2b}J_1(b); \tag{D.17}
$$

Sonin [3, 7, 8]

$$
\int\limits_0^1 J_0\left(a\sqrt{1-x^2}\right)\cos bx dx = j_0\left(\sqrt{a^2+b^2}\right);
$$

$$
\int\limits_0^1 \sqrt{1-x^2} J_1\left(a\sqrt{1-x^2}\right)\cos bx dx = \frac{a}{\sqrt{a^2+b^2}} j_1\left(\sqrt{a^2+b^2}\right);
$$

$$
\begin{aligned}
\int\limits_0^1 (1-x^2) J_0\left(a\sqrt{1-x^2}\right)\cos bx dx &= \frac{a^2}{a^2+b^2} j_0\left(\sqrt{a^2+b^2}\right) \\
&\quad + \frac{2b^2-a^2}{\left(a^2+b^2\right)^{3/2}} j_1\left(\sqrt{a^2+b^2}\right);
\end{aligned} \tag{D.18}
$$

Sonin-Gegenbauer [3]

$$\int_0^\infty e^{-a\sqrt{t^2+z^2}}/\sqrt{t^2+z^2}\cos bt dt = K_0\left(z\sqrt{a^2+b^2}\right);$$

$$\int_0^\infty e^{-a\sqrt{t^2+z^2}}\cos bt dt = \frac{a\,z}{\sqrt{a^2+b^2}}K_1\left(z\sqrt{a^2+b^2}\right); \tag{D.19}$$

$$\int_0^\infty \sqrt{t^2+z^2}e^{-a\sqrt{t^2+z^2}}\cos bt dt = \frac{(a\,z)^2}{a^2+b^2}K_0\left(z\sqrt{a^2+b^2}\right) + z\frac{a^2-b^2}{\left(a^2+b^2\right)^{3/2}}K_1\left(z\sqrt{a^2+b^2}\right);$$

As well as

$$\int_0^\infty e^{-ia\sqrt{1-t^2}}/\sqrt{1-t^2}\cos(bt)dt = \frac{\pi}{2}H_0^{(2)}\left(\sqrt{a^2+b^2}\right);$$

$$\int_0^\infty e^{-ia\sqrt{1-t^2}}\cos(bt)dt = -i\frac{\pi}{2}\frac{a}{\sqrt{a^2+b^2}}H_1^{(2)}\left(\sqrt{a^2+b^2}\right); \tag{D.20}$$

$$\int_0^\infty e^{-ia\sqrt{1-t^2}}\sqrt{1-t^2}\cos(bt)dt = \frac{\pi}{2}\left[\frac{a^2}{(a^2+b^2)}H_0^{(2)}\left(\sqrt{a^2+b^2}\right) + \frac{b^2-a^2}{\left(a^2+b^2\right)^{3/2}}H_1^{(2)}\left(\sqrt{a^2+b^2}\right)\right].$$

Using [8], we derive the following relations between the integrals:

$$\int_1^\infty \cos bx K_0\left(c\sqrt{x^2-1}\right)dx - \frac{\pi}{2}\int_0^1 \cos bx Y_0\left(c\sqrt{1-x^2}\right)dx = -\frac{\pi}{2}y_0\sqrt{b^2+c^2}; \tag{D.21}$$

$$\int_1^\infty \cos bxK_1\left(c\sqrt{x^2-1}\right)\sqrt{x^2-1}dx$$
$$-\frac{\pi}{2}\int_0^1 \cos bxY_1\left(c\sqrt{1-x^2}\right)\sqrt{1-x^2}dx$$
$$= -\frac{\pi}{2}\frac{c}{\sqrt{b^2+c^2}}y_1\left(\sqrt{b^2+c^2}\right); \quad \text{(D.22)}$$

$$\int_1^\infty \cos bxK_0\left(c\sqrt{x^2-1}\right)\cdot x^2dx - \frac{\pi}{2}\int_0^1 \cos bxY_0\left(c\sqrt{1-x^2}\right)\cdot x^2dx$$
$$= -\frac{\pi}{2}\left[\frac{c^2-2b^2}{\left(b^2+c^2\right)^{3/2}}y_1\left(\sqrt{b^2+c^2}\right) + \frac{b^2}{b^2+c^2}y_0\left(\sqrt{b^2+c^2}\right)\right], \quad \text{(D.23)}$$

where $K_n(x)$—modified Bessel function [7]; $Y_n(x)$—Bessel function of the second kind [7]; $y_n(x)$—spherical Bessel function of the second kind [7].

From [8] we also find:

$$\int_0^\infty \cos bxH_0^{(2)}\left(c\sqrt{a^2-x^2}\right)dx = h_0^{(2)}\left(a\sqrt{b^2+c^2}\right); \quad \text{(D.24)}$$

$$\int_0^\infty \sin bxH_0^{(2)}\left(c\sqrt{a^2-x^2}\right)xdx = \frac{ab}{\sqrt{b^2+c^2}}h_1^{(2)}\left(a\sqrt{b^2+c^2}\right); \quad \text{(D.25)}$$

$$\int_0^\infty \cos bxH_1^{(2)}\left(c\sqrt{a^2-x^2}\right)\sqrt{a^2-x^2}dx = \frac{ac}{\sqrt{b^2+c^2}}h_1^{(2)}\left(a\sqrt{b^2+c^2}\right). \quad \text{(D.26)}$$

Here $H_n^{(2)}(z)$—Hankel function [7]; $h_n^{(2)}(z)$—spherical Hankel function of the second kind [7].

The generalized integral sine function is defined as:

$$S(a,x) = \int_0^x \sin\sqrt{a^2+t^2}/\sqrt{a^2+t^2}dt, \quad \text{(D.27)}$$

and the incomplete Struve function is defined in Poisson form:

$$H_\nu(w,z) = 2\frac{z^\nu}{A_\nu}\int_0^w \sin(z\cos\theta)\sin^{2\nu}\theta\, d\theta, \quad \text{(D.28)}$$

where $A_\nu = 2^\nu \Gamma(\nu + 1/2)\Gamma(1/2)$; $\Gamma(z)$—gamma function [3].

From the definition (D.28), using a change of variables, it is easy to show:

$$\int_0^y \frac{\sin a\sqrt{y^2 - t^2}}{\sqrt{y^2 - t^2}} \cos bt dt = \int_0^{\pi/2} \sin(ay\cos\varphi)\cos(by\sin\varphi)d\varphi$$
$$= \frac{\pi}{2} H_0\left(w_0, y\sqrt{a^2 + b^2}\right). \qquad \text{(D.29)}$$

where $\sin w_0 = a/\sqrt{a^2 + b^2}$.

Using the expansion of the Bessel function of half index in Lommel form [7], we obtain:

$$S(a, x) = \sqrt{\frac{\pi}{2}} x \sum_{m=0}^{\infty} \frac{\left(-x^2/2\right)^m}{m!(2m+1)} \cdot \frac{J_{m+1/2}(a)}{a^{m+1/2}}. \qquad \text{(D.30)}$$

Functions (D.27, D.28) arise when summing series containing products of the Bessel function using integral representations:

$$\sum_{k=0}^{\infty} \sin(2k+1)a \cdot J_{k+1/2}(w) J_{k+1/2}(z) = \frac{1}{2} H_0(w_0, t_1); \qquad \text{(D.31)}$$

$$\sum_{k=0}^{\infty} \cos(2k+1)a \cdot J_{k+1/2}(w) J_{k+1/2}(z) = \frac{1}{\pi} S\left(t_1, 2\sqrt{wz}\cos a\right), \qquad \text{(D.32)}$$

where $t_1 = \left(w^2 + z^2 - 2wz\cos 2a\right)^{1/2}$; $\sin w_0 = 2\sqrt{wz}\sin a/t_1$. In the first case, the product of Bessel functions under the sum sign is represented by the Hardy integral [9] and using (D.29); in the second—by the Neumann integral [9] and (D.27).

The following relations are obtained based on the definition of the generalized integral sine function and changing the order of integration:

$$\int_0^1 \left(1 - x^2\right)^{\beta - 1} S(ax, bx) x dx$$
$$= 2^{\beta - 3/2} \sqrt{\pi} \Gamma(\beta) \int_0^b \frac{J_{\beta+1/2}\sqrt{a^2 + t^2}}{\left(\sqrt{a^2 + t^2}\right)^{\beta + 1/2}} dt; \qquad \text{(D.33)}$$

$$\int_0^1 \left(1 - x^2\right)^{\beta - 1} S(ax, bx) \cdot dx/x$$

$$= \frac{1}{2} B(1/2, \beta) \int_0^b 1F_2\left(1/2;\ 3/2,\ \beta + 1/2;\ -1/4\left(a^2 + t^2\right)\right) dt, \tag{D.34}$$

где $B(\alpha, \beta) = \Gamma(\alpha)\Gamma(\beta)/\Gamma(\alpha + \beta)$; $\Gamma(\alpha)$—gamma function [3]; $1F_2(\alpha;\ \beta, \gamma;\ x)$—hypergeometric function [9].

For β integers, we find the integral (D.34) explicitly, using the properties of derivative functions [3], as well as the expansion (D.30):

$$\int_0^1 \left(1 - x^2\right)^{n-1} S(ax, bx) \cdot x dx = 2^{n-1}(n-1)! \left(-\frac{1}{a} \cdot \frac{d}{da}\right)^n S(a, b). \tag{D.35}$$

Similarly for $\beta = n + 1/2$ semi-integers, taking into account (D.33)

$$\int_0^1 \left(1 - x^2\right)^{n-1/2} S(ax, bx) \cdot x dx$$

$$= \sqrt{\pi} 2^{n-1} \Gamma(n + 1/2) \left(-\frac{1}{a} \cdot \frac{d}{da}\right)^{n+1} \int_0^b J_0\left(\sqrt{a^2 + t^2}\right) dt. \tag{D.36}$$

Improper integrals of the generalized integral sine functions, using representation (D.14), are expressed in terms of integrals of Hankel functions of the second kind:

$$\int_0^\infty S(ax, bx) \cdot e^{-ic\sqrt{1-x^2}} \frac{x dx}{\sqrt{1 - x^2}} = -\frac{\pi}{2} \int_0^b \frac{H_1^{(2)}\left(\sqrt{a^2 + c^2 + t^2}\right)}{\sqrt{a^2 + c^2 + t^2}} dt;$$

$$\frac{d}{da} \int_0^\infty S(ax, bx) \cdot e^{-ic\sqrt{1-x^2}} \frac{x dx}{\sqrt{1 - x^2}} = a \frac{\pi}{2} \int_0^b \frac{H_2^{(2)}\left(\sqrt{a^2 + c^2 + t^2}\right)}{a^2 + c^2 + t^2} dt; \tag{D.37}$$

$$\frac{d}{db} \int_0^\infty S(ax, bx) \cdot e^{-ic\sqrt{1-x^2}} \frac{x dx}{\sqrt{1 - x^2}} = -\frac{\pi}{2} \cdot \frac{H_1^{(2)}\left(\sqrt{a^2 + c^2 + b^2}\right)}{\sqrt{a^2 + c^2 + b^2}}.$$

The last expressions are convenient to use for studying asymptotic expansions.

References

1. V.V. Nikolsky, *Electrodynamics and Radio Wave Propagation* (Nauka, 1973), 606 p (in Russian)
2. R.E. Collin, *Field Theory of Guided Waves* (Wiley Interscience, 1991), 852 p
3. M. Abramowitz, I.A. Stegun, *Handbook of Mathematical Functions with Formulas, Graphs and Mathematical Tables*, vol. 55 (National Bureau of Standarts, Applied Mathematics Ser., 1964), 830 p
4. E.I Nefedov, A.T. Fialcovsky, *Strip Transmission Lines* (Nauka, 1980), 312 p (in Russian)
5. V.M. Sedyh (ed.), *Striplines and Microwave Devices* (Visha shkola, Harkov, 1974), 275 p (in Russian)
6. S. Stoilow, Teoria Functiilor de o Variabila Complexa, ed. by Academiei Repubicii Romane, vol. 1, 364 p., vol. 2, 415 p (1954)
7. G.N. Watson, *A Treatise on the Theory of Bessel Functions* (University Press, Cambridge, 1945), p.804
8. H. Bateman, A. Erdelyi, *Higher Transcendental Functions*, vol. 1 (McGraw-Hill Book Company, New York, 1953), p.302
9. H. Bateman, A. Erdelyi, *Higher Transcendental Functions*, vol. 2 (McGraw-Hill Book Company, New York, 1953), p.365

A. Trubin, *Introduction to the Theory of Dielectric Resonators*, Springer Series in Advanced Microelectronics 65, https://doi.org/10.1007/978-3-031-65396-4

GPSR Compliance

The European Union's (EU) General Product Safety Regulation (GPSR) is a set of rules that requires consumer products to be safe and our obligations to ensure this.

If you have any concerns about our products, you can contact us on ProductSafety@springernature.com

In case Publisher is established outside the EU, the EU authorized representative is:

Springer Nature Customer Service Center GmbH
Europaplatz 3
69115 Heidelberg, Germany

Batch number: 10732031

Printed by Printforce, the Netherlands